Genomics-Assisted Crop Improvement

Genomics-Assisted Crop Improvement

Vol. 1: Genomics Approaches and Platforms

Edited by

Rajeev K. Varshney
ICRISAT, Patancheru, India

and

Roberto Tuberosa
University of Bologna, Italy

Springer

A C.I.P. Catalogue record for this book is available from the Library of Congress.

ISBN 978-90-481-7599-4
ISBN 978-1-4020-6295-7 (e-book)

Published by Springer,
P.O. Box 17, 3300 AA Dordrecht, The Netherlands.

www.springer.com

Printed on acid-free paper

CONTENTS

FOREWORD TO THE SERIES: GENOMICS-ASSISTED CROP IMPROVEMENT

Genetic markers and their application in plant breeding played a large part in my research career, so I am delighted to have the opportunity to write these notes to precede the two volumes on 'Genomics-Assisted Crop Improvement'. Although I am not so old, I go right back to the beginning in 1923 when Karl Sax described how 'factors for qualitative traits' (today's genetic markers) could be used to select for 'size factors' (today's QTLs and genes for adaptation). But it was clear to me 40 years ago that even then plant breeders clearly understood how genetic markers could help them - if only they actually had the markers and understood the genetics underlying their key traits. It was not clear to me that it was going to take until the next century before marker-aided selection would become routine for crop improvement.

In the 1960s only 'morphological' markers were available to breeders. As a research student at Aberystwyth, I worked with Des Hayes at the Welsh Plant Breeding Station when he was trying to develop an F_1 hybrid barley crop based on a male sterility gene linked to a DDT resistance gene. The idea was to link the male fertile allele with susceptibility and then kill the fertile plants off in segregation populations by dousing the field with DDT. Rachel Carson's 'Silent Spring' ensured that idea never flew.

Then I moved to the Plant Breeding Institute in Cambridge where anyone working alongside the breeders in those early days could not help but be motivated by breeding. Protein electrophoresis raised the first possibility of multiple neutral markers and we were quick to become involved in the search for new isozyme markers in the late 1970s and early 1980s. Probably only the linkage between wheat endopeptidase and eyespot resistance was ever used by practical breeders, but we had an immense amount of fun uncovering the genetics of a series of expensive markers with hardly any polymorphism, all of which needed a different visualisation technology!

During this same period, of course, selection for wheat bread-making quality using glutenin subunits was being pioneered at the PBI, and is still in use around the world. These were the protein equivalent of today's 'perfect' or 'functional' markers for specific beneficial alleles. Such markers - although of course DNA-based, easy and economical to use, amenable to massively high throughput and available for all key genes in all crops - are exactly where we want to end up.

Proteins were superseded by RFLPs and in 1986 we set out to make a wheat map, only with the idea of providing breeders with the effectively infinite number of mapped neutral markers that they had always needed. We revelled in this massively expensive job, funded by a long-suffering European wheat breeding industry, of creating the first map with a marker technology so unwieldy that students today would not touch it with a bargepole, let alone plant breeders. This was, of course, before the advent of PCR, which changed everything.

The science has moved quickly and the past 20 years have seen staggering advances as genetics segued into genomics. We have seen a proliferation of maps, first in the major staples and later in other crops, including 'orphan' species grown only in developing countries. The early maps, populated with isozyme markers and RFLPs, were soon enhanced with more amenable PCR-based microsatellites, which are now beginning to give way to single nucleotide polymorphisms. These maps and markers have been used, in turn, to massively extend our knowledge of the genetic control underlying yield and quality traits. The relatively dense maps have allowed whole genome scans which have uncovered all regions of the genome involved in the control of key adaptive traits in almost all agricultural crops of any significance.

More amazing is the fact that we now have the whole genome DNA sequences of not one but four different plant genomes - *Arabidopsis*, rice, popular and sorghum. Moreover, cassava, cotton, and even maize could be added to the list before these volumes are published. Other model genomes where sequencing has been started include Aquilegia (evolutionary equidistant between rice and arabidopsis), Mimulus (for its range of variation) and Brachypodium (a small-genome relative of wheat and barley).

Two other components deserve mention. The first is synteny, the tendency for gene content and gene order to be conserved over quite distantly related genomes. Ironically, synteny emerged from comparisons between early RFLP maps and probably would not have been observed until we had long genomic sequences to compare had we started with PCR-based markers that require perfect DNA primer sequence match. The possibility of being able to predict using genetic information and DNA sequence gained in quite distantly related species has had a remarkable unifying effect on the research community. Ten years ago you could work away at your own favourite crop without ever talking to researchers and breeders elsewhere. Not so today. Synteny dictates that genome researchers are part of one single global community.

The second component is the crop species and comparative databases that we all use on a daily basis. The selfless curators, that we have all taken for granted, deserve mention and ovation here because, while the rest of us have been having fun in the lab, they have been quietly collecting and collating all relevant information for us to access at the press of a button. This is a welcome opportunity to acknowledge these unsung heroes, and of course, their sponsors.

The practical application of markers and genomics to crop improvement has been much slower to emerge. While endopeptidase and the glutenin gels continue to see use in wheat breeding, marker-aided selection (MAS) using DNA markers has, in both public breeding and the multinationals, emerged only in the last few years and examples of new varieties that have been bred using MAS are still few and far between. This will change, however, as the cost of marker data points continues to plummet and the application of high-throughput methods moves the technology from breeding laboratories to more competitive outsourced service providers.

The post-RFLP period and the new opportunities for deployment of economical high-throughput markers are the subjects of these volumes. The first volume deals with platforms and approaches while the second covers selected applications in a range of crop plants. The editors, Rajeev Varshney and Roberto Tuberosa, are to be congratulated on bringing together an authorship of today's international leaders in crop plant genomics.

The end game, where plant breeders can assemble whole genomes by manipulating recombination and selecting for specific alleles at all key genes for adaptation is still a very long way off. But these two volumes are a unique opportunity to take stock of exactly where we are in this exciting arena, which is poised to revolutionise plant breeding.

Mike Gale, FRS
John Innes Foundation Emeritus Fellow
John Innes Centre
Norwich
United Kingdom

FOREWORD

Who would have believed only two decades ago that plant scientists would have access to nearly the complete genetic code of numerous plant species, including major crop species. The idea of having ready access to whole genome sequences encompassing 140 million bases of the model plant Arabidopsis seemed like science fiction, let alone having available even larger genomes such as rice at 430 Mb or maize at 2500 Mb (the same size as the human genome). And then proceeding to identify variation beyond what was anticipated, such as the 2.6 SNPs (Single Nucleotide Polymorphisms) per kb in rice. The number of strains of various species with literally hundreds of thousands of inserts, allowing the association of sequence and trait, increased at an unanticipated rate. Who would have believed only a decade ago that we would be capable of analyzing the expression of genes across the whole genome and matching that profile with traits of interest. And now the area of metabolomics is allowing even more meaningful explanations of the genetic control of important traits.

This book brings all of these advances in genomics to the forefront and prepares the plant scientists for the next decade. Important technologies are discussed such as association mapping, simulation modeling, and development of appropriate populations including the advanced backcross and introgression-lines for incorporating traits into useful genetic materials. Such approaches are facilitating the identification of traits that are not obvious simply from observing the plant phenotype, and they provide ways to extract new and useful traits from wild related species. Comparing the genomic information across broadly-related species has generated important evolutionary information. In addition, the common occurrence of duplicated segments and large gene families with partially redundant or tissue- and developmentally-specific expression will lead to information fundamental to plant performance.

Methods for the identification of genes underlying traits are improving every day. The association of allelic variation in a candidate gene and a trait is leading to much greater understanding of the genetic control of traits. Numerous transcription factors and even non-coding sequences are being implicated as the basis of considerable genetic variation. Forward and reverse genetics are both found to be very useful in revealing gene-trait associations.

The tremendous expansion of genomic analytical approaches along with efforts to reduce the cost, together with appropriate statistical designs and analyses, are making it easier and more expeditious to use the ever-increasing sequence information to identify useful genes. This body of knowledge in plant genomics and its myriad of applications are nicely reflected in this book.

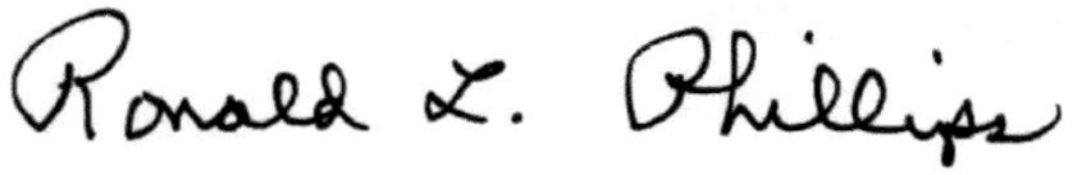

Ronald L. Phillips
Regents Professor
and
McKnight Presidential Chair in Genomics
University of Minnesota
St. Paul, MN
USA

PREFACE

Genomics, dealing with the collection and characterization of genes and analysis of the relationships between gene activity and cell function, is a rapidly evolving, interdisciplinary field of study aimed at understanding and exploiting the biological information encoded in DNA. The genomics toolbox includes automated genetic and physical mapping, DNA sequencing, bioinformatics software and databases, transcriptomics, proteomics, metabolomics, and high-throughput profiling approaches. Indeed, the past two decades have witnessed spectacular advances in genomics. For example, at the dawn of the genomics era, Arabidopsis was chosen as the first model genome for sequencing, which was then quickly followed by the sequencing of other model genomes (rice for monocots, Medicago and Lotus for legume crops and poplar for tree species) and crop species (soybean, cassava, sorghum, etc.). While new crops (e.g. maize, wheat, finger millet, etc.) are being added to the list for sequencing the genome or gene space, the generated sequence data are being analyzed in parallel for characterizing the genes and validating their functions through comparative and functional genomics approaches including bioinformatics, transcriptomics, and genetical genomics. Candidate genes are becoming increasingly useful for the development of markers for assaying and understanding functional diversity, association studies, allele mining, and most importantly, marker-assisted selection. Therefore, genomics research has great potential to revolutionize the discipline of plant breeding in order to face the challenges posed by feeding an ever-growing human population expected to top 10 billion by 2050, while decreasing the environmental footprint of agriculture and preserving the remaining biodiversity.

Several high-throughput approaches, genomics platforms, and strategies are currently available for applying genomics to crop breeding. However, the high costs invested in, and associated with, genomics research currently limit the implementation of genomics-assisted crop improvement, especially for autogamous and/or minor and orphan crops. This book presents a number of articles illustrating different contributions which genomics can offer to unravel the path from genes to phenotypes and vice versa, and how this knowledge can help to improve crops' performance and reduce the impact of agriculture on the environment. Each article shows how structural and/or functional genomics can improve our capacity to unveil and deploy natural and artificial allelic variation for the benefit of plant breeders. Volume 1, entitled "Genomics Approaches and Platforms", presents state-of-the-art genomic resources and platforms and also describes the strategies and approaches for effectively exploiting genomics research for crop improvement. Volume 2, entitled "Genomics Applications in Crops", presents a number of case studies of important crop and plant species that summarize both the achievements and limitations of genomics research for crop improvement.

More than 90 authors, representing 16 countries from five continents have contributed 16 chapters for Volume I and 18 chapters for Volume II (see Appendix I). The editors are grateful to all the authors, who not only provided a timely review of the published research work in their area of expertise but also shared their unpublished results to offer an updated view. We also appreciate their cooperation in meeting the deadlines, revising the manuscripts, and in checking the galley proofs. While editing this book, we received strong support from many reviewers (see Appendix II) who provided useful suggestions for improving the manuscripts. We would like to thank our colleagues and research scholars, especially Yogendra, Rachit, Mahender, Priti, and Spurthi working at ICRISAT for their help in various ways. Nevertheless, we take responsibility for any errors that might have crept in inadvertently during the editorial work.

The cooperation and encouragement received from Jacco Flipsen and Noeline Gibson of Springer during various stages of the development and completion of this project, together with the help of Rajeshwari Pal of Integra Software Services for typesetting and correcting the galley proofs, have been instrumental for the completion of this book and are gratefully acknowledged. We also recognize that our editorial work took away precious time that we should have spent with our respective families. RKV acknowledges the help and support of his wife, Monika and son, Prakhar (Kutkut) who allowed their time to be taken away to fulfill RKV's editorial responsibilities in addition to research and other administrative duties at ICRISAT. Similarly, RT is grateful to his wife Kay for her precious help in editing and proof-reading a number of manuscripts.

We hope that our efforts will help those working in crop genomics as well as conventional plant breeding to better focus their research plans for crop improvement programs. The book will also help graduate students and teachers to develop a better understanding of this fundamental aspect of modern plant science research. Finally, we would appreciate receiving readers' feedback on the errors and omissions, if any, as well as their suggestions, so that a future revised and updated edition, if planned, may prove more useful.

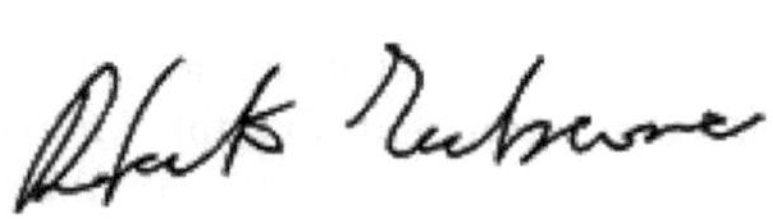

Rajeev K. Varshney

Roberto Tuberosa

COLOR PLATES

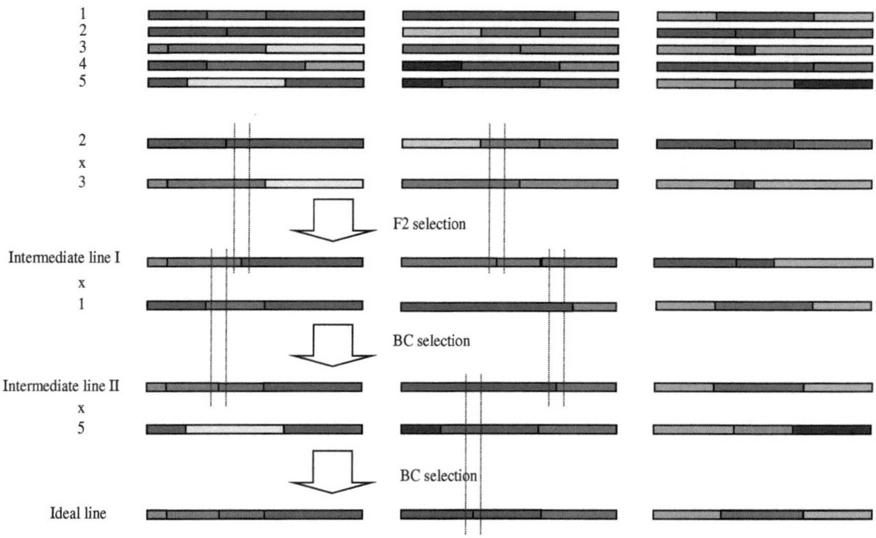

Plate 1. **The principle of Breeding by Design.** Subsequent crosses and selections using markers lead to the desired superior elite line genotype starting from a collection of 5 parental lines. Dotted lines indicate marker positions used to select for the desired recombinants *(see Fig. 5 on page 51)*
(Note: Reprinted from: Trends Plant Sci. 8, Peleman J-D, Rouppe van der Voort J, Breeding by Design, 330-334 © (2003), with permission from Elsevier)

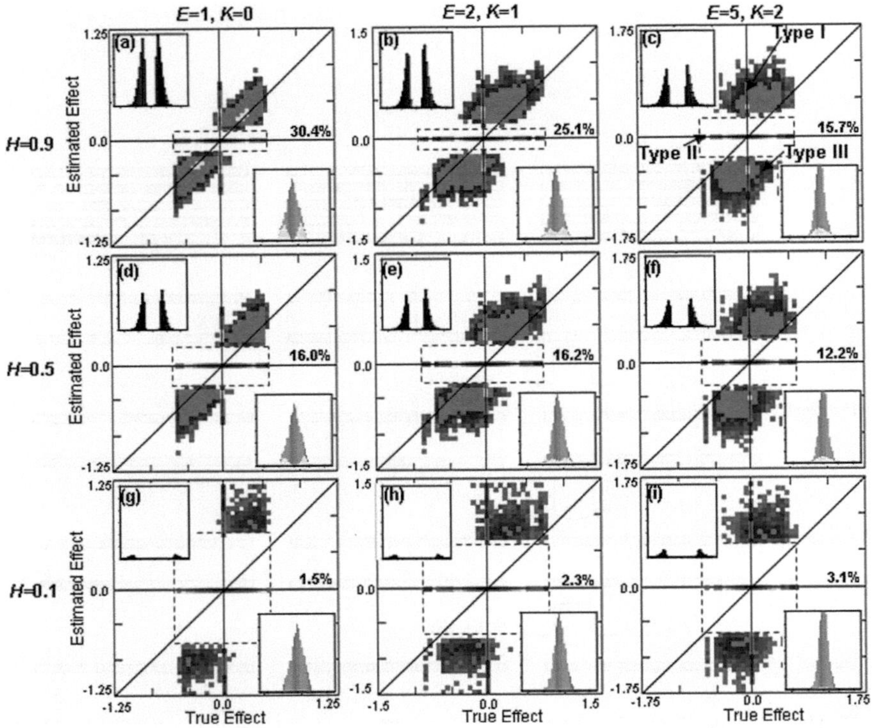

Plate 2. True and estimated additive QTL effects for three genetic models ($E(NK)$=1(48:0); $E(NK)$=2(48:1); $E(NK)$=5(48:2)) and three levels of heritability (H = 0.9, 0.5, 0.1). Results are shown as a heat plot, using true and estimated QTL detected by Composite Interval Mapping (CIM) in the bi-parental mapping populations. For each sub-panel, the results are displayed for the 50 genetic parameterizations and 20 bi-parental replications (i.e. 1000 data sets). Colors range from cyan through dark red. Type I, II and III errors are highlighted by arrows. Type I errors represent cases where QTL were falsely detected in a given map region (i.e. false positives), Type II errors represent cases where the true QTL were not detected by CIM, and Type III errors represent cases where the QTL were corrected detected but the estimated favorable allele was incorrectly defined. The percentage of true QTL detected is listed in each sub-panel *(see Fig. 8 on page 80)*

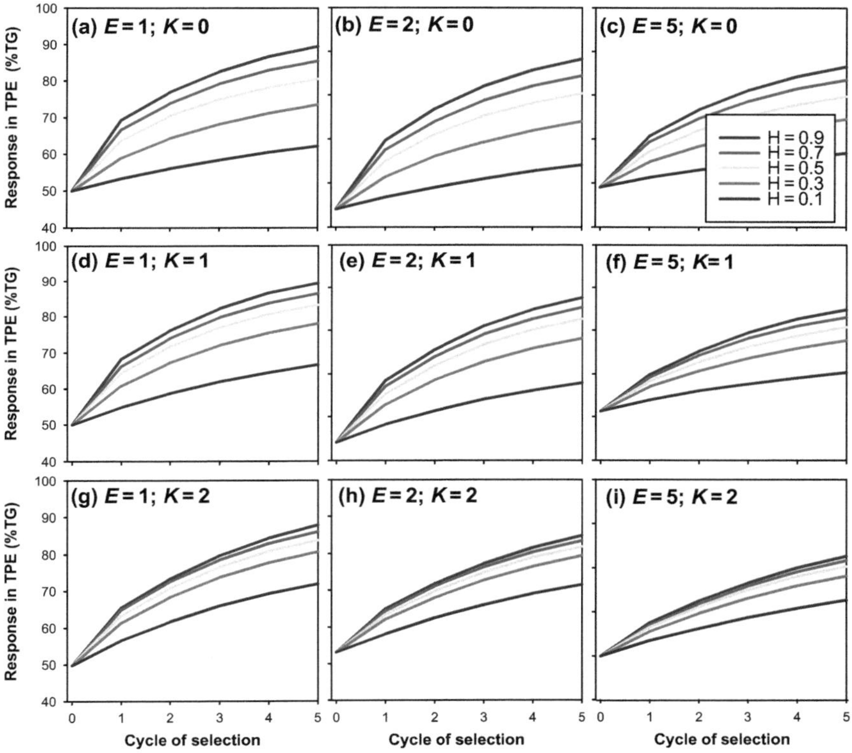

Plate 3. Average response in the TPE for the nine genetic models (factorial combinations of E and K) and five levels of heritability over five cycles of selection. For each combination, the results were computed from the 50 genetic parameterizations and 20 breeding replications (i.e. 1,000 simulations) *(see Fig. 10 on page 83)*

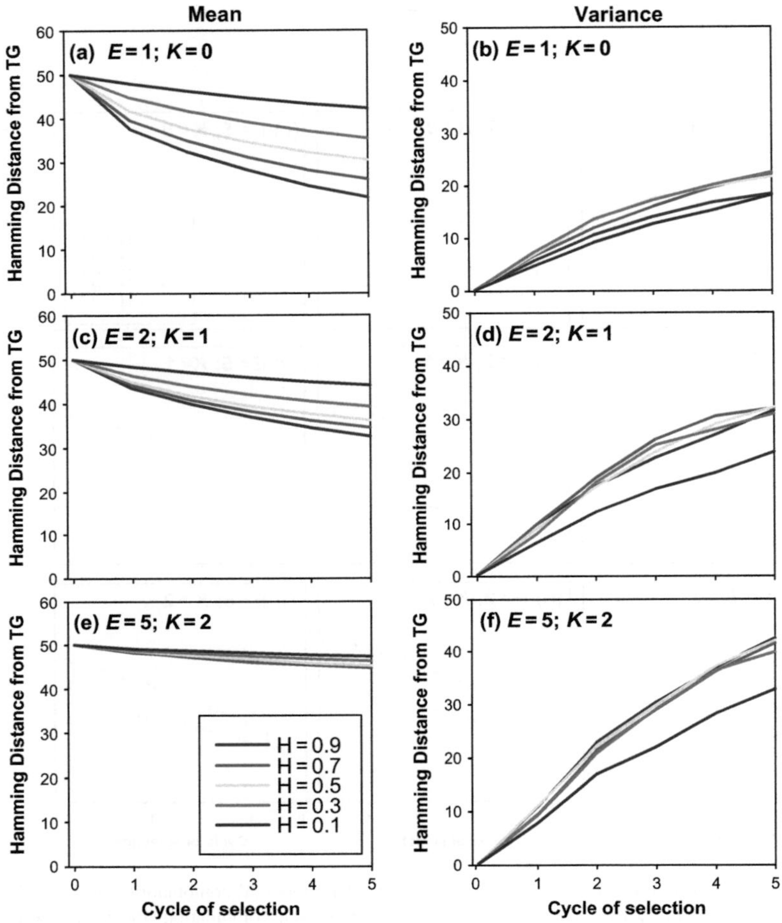

Plate 4. Hamming distances (HD) of hybrid combinations from the target genotype for three genetic models (*E(NK)* = 1(48:0); *E(NK)* = 2(48:1); *E(NK)* = 5(48:2)) and five levels of heritability (*H* = 0.9, 0.7, 0.5, 0.3, 0.1), over five cycles of selection. For each combination, the results were computed from the 50 genetic parameterizations and 20 breeding replications (i.e. 1000 simulations) *(see Fig. 11 on page 84)*

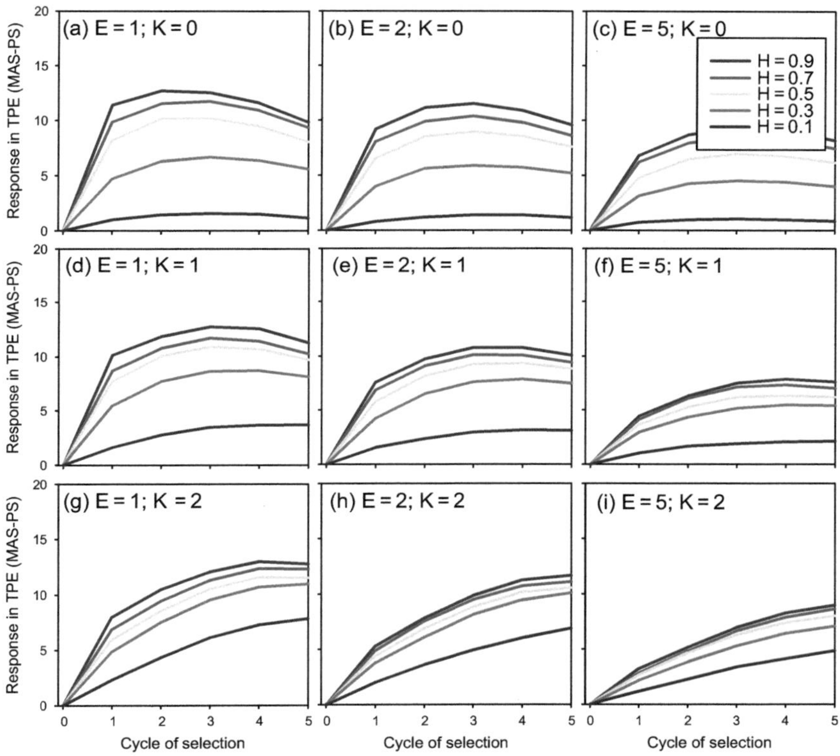

Plate 5. Difference in the average response (Marker-assisted selection – Phenotypic selection) for the nine genetic models (factorial combinations of *E* and *K*) and five levels of heritability over five cycles of selection. For each combination, the results were computed from the 50 genetic parameterizations and 20 breeding replications (i.e. 1000 simulations each breeding strategy) *(see Fig. 12 on page 85)*

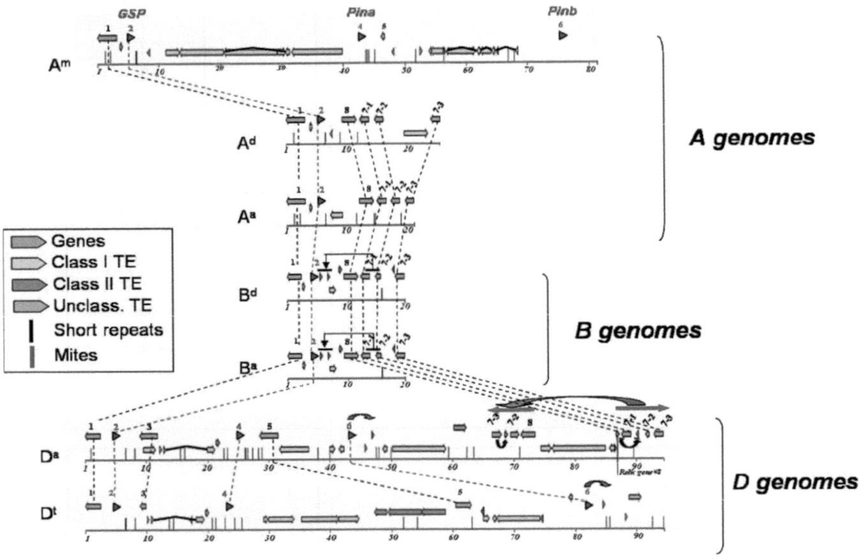

Plate 6. Microcolinearity studies at the Hardness locus in wheat (adapted from Chantret et al. 2005) Schematic representation of BAC sequence comparisons at the wheat *Ha* locus from the A (A^m : *T. monococcum;* A^a : *T. aestivum*; A^d : *T. durum*) , B (B^a : *T. aestivum* ; B^d : *T. durum*) and D (D^a : *T. aestivum*; D^t : *Ae. tauschii*) genomes in different polyploidy context. Genes (CDS) (light blue), class I TEs (yellow), class II TEs (green), unclassified elements (gray), MITEs (red), and short repeats (black) are indicated. Orthologous CDS between the different genomes are linked by dashed bars whereas CDS duplications and deletion events are indicated by arrows. The *GSP*, *Pina* and *Pinb* genes that were lost in tetraploid wheat following polyploidization are highlighted in red and are numbered respectively as gene 2, 4, 6 *(see Fig. 8 on page 187)*

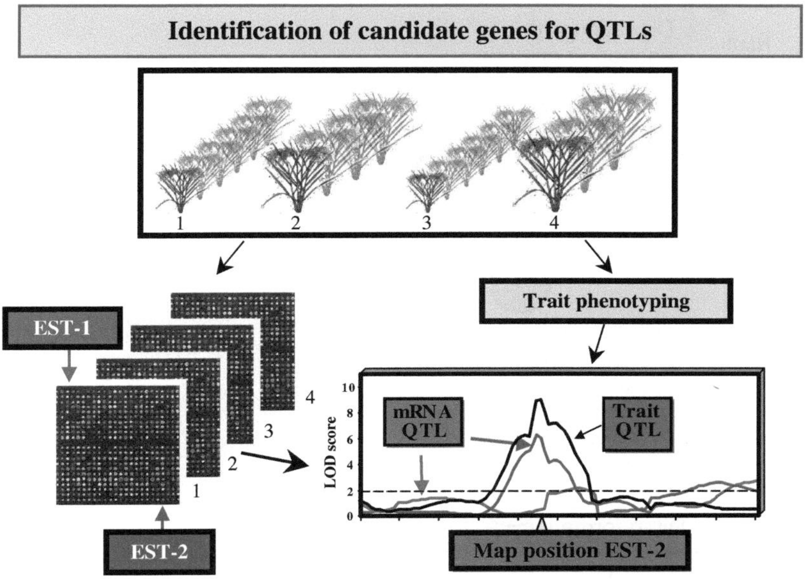

Plate 7. Expression profiling of a mapping population at the mRNA level via microarray analysis to identify expression QTLs (eQTLs) for specific cDNA and therefore genes. Correspondence between an eQTL peak for a specific cDNA (e.g. cDNA-2) and a QTL peak for a trait causally linked to the function of the protein encoded by the cDNA provides circumstantial evidence supporting the role of the cDNA as a candidate gene for the target trait *(see Fig. 1 on page 217)*

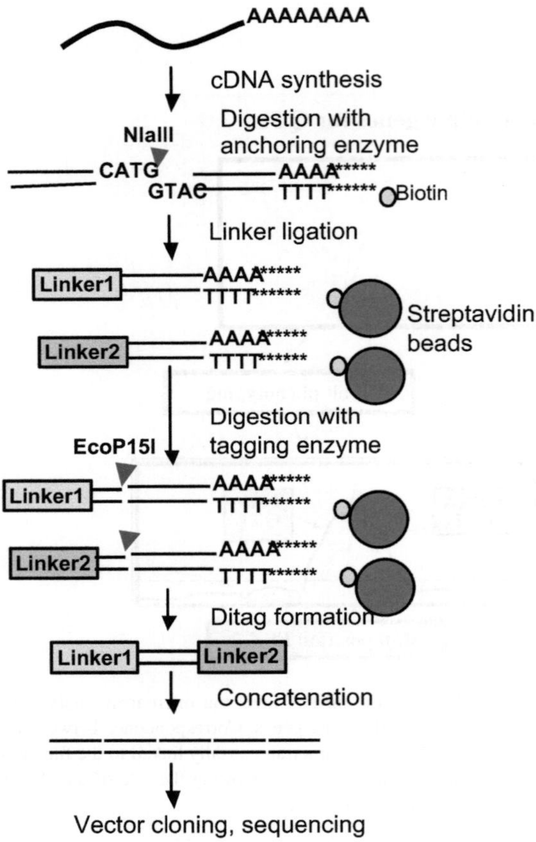

Plate 8. Schematic diagram of SAGE procedure (see text for details) *(see Fig. 10 on page 230)*

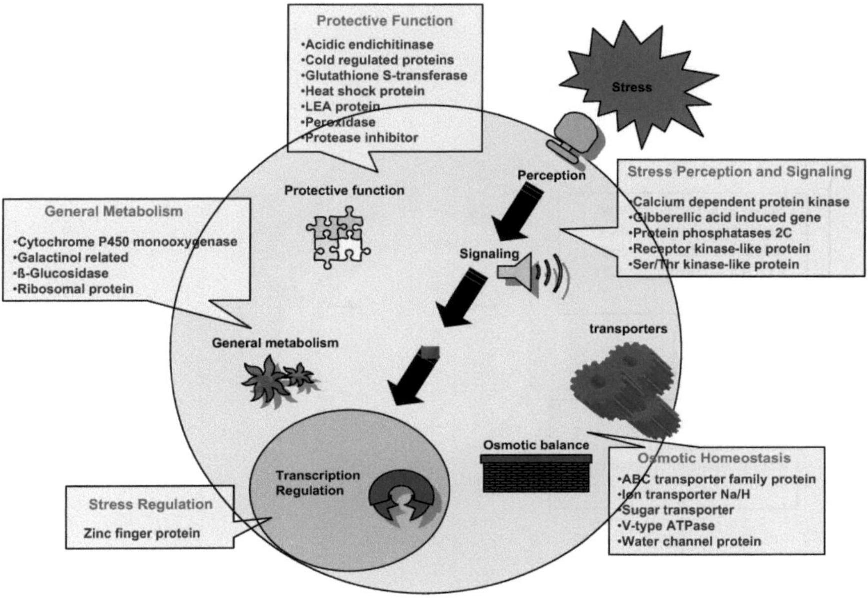

Plate 9. Cartoon depicting the salinity related transcriptome "fingerprints" conserved amongst the three model systems viz. *Arabidopsis*, rice and common ice plant *(see Fig. 1 on page 281)*

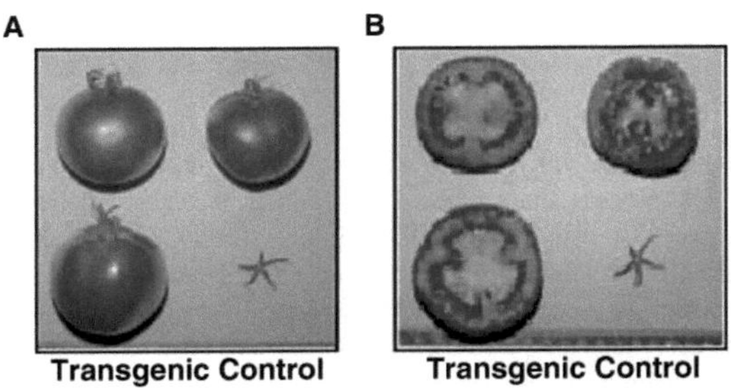

Plate 10. Induction of parthenocarpic tomato fruits by overproduction of auxin. (A) Fruits from pollinated (top) and unpollinated (bottom) flowers from transgenic (transformed with *DefH9::iaaM*) and control plants. (B) Cut fruits from pollinated (top) and unpollinated (bottom) flowers from transgenic (transformed with *DefH9::iaaM*) and control plants. (Adapted from Ficcadenti et al., 1999) *(see Fig. 5 on page 305)*

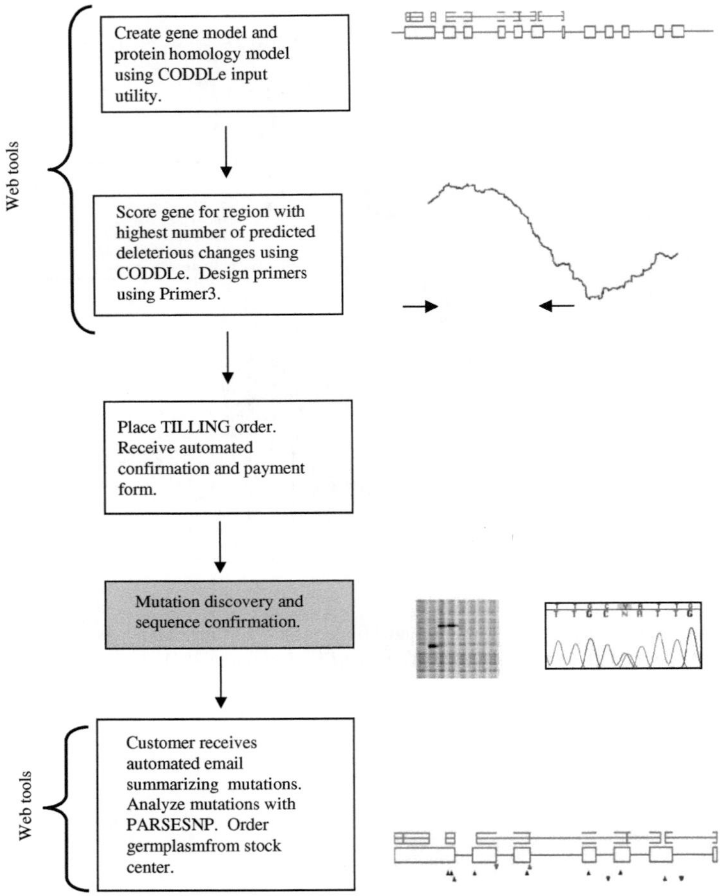

Plate 11. Outline of the steps involved in a public TILLING service. A series of web-based tools have been developed or adapted for the system. The process starts when a user creates a gene model and obtains and aligns homologous protein sequences by using the CODDLe input utility. CODDLe then identifies the region of the gene containing the highest density of potential nucleotide changes that could damage the protein when mutated. Primers design is accomplished with the program Primer3, and the researcher enters the selected primers. All of these steps are performed within the web browser window. The researcher received an automated email confirmation of the submitted order and a payment form. The primer order is automatically sent to the oligonucleotide supplier, and primers are shipped to the TILLING facility. Screening commences and mutations identified by TILLING are sequence-verified. The results are automatically emailed to the customer who placed the order. A link to PARSESNP output is provided in the report. PARSESNP graphically displays the location and type of mutations, predicts the severity of missense mutations, and provides restriction sites that are either gained or lost by the induced mutation (Taylor and Greene, 2003) *(see Fig. 4 on page 344)*

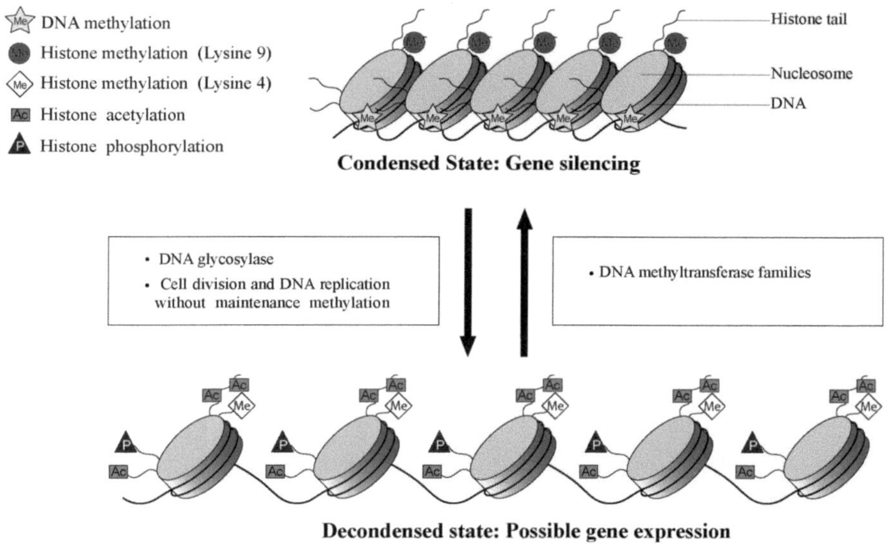

Plate 12. Model for the regulation of chromatin structure in plants. Only the processes controlling DNA methylation status are indicated *(see Fig. 1 on page 353)*

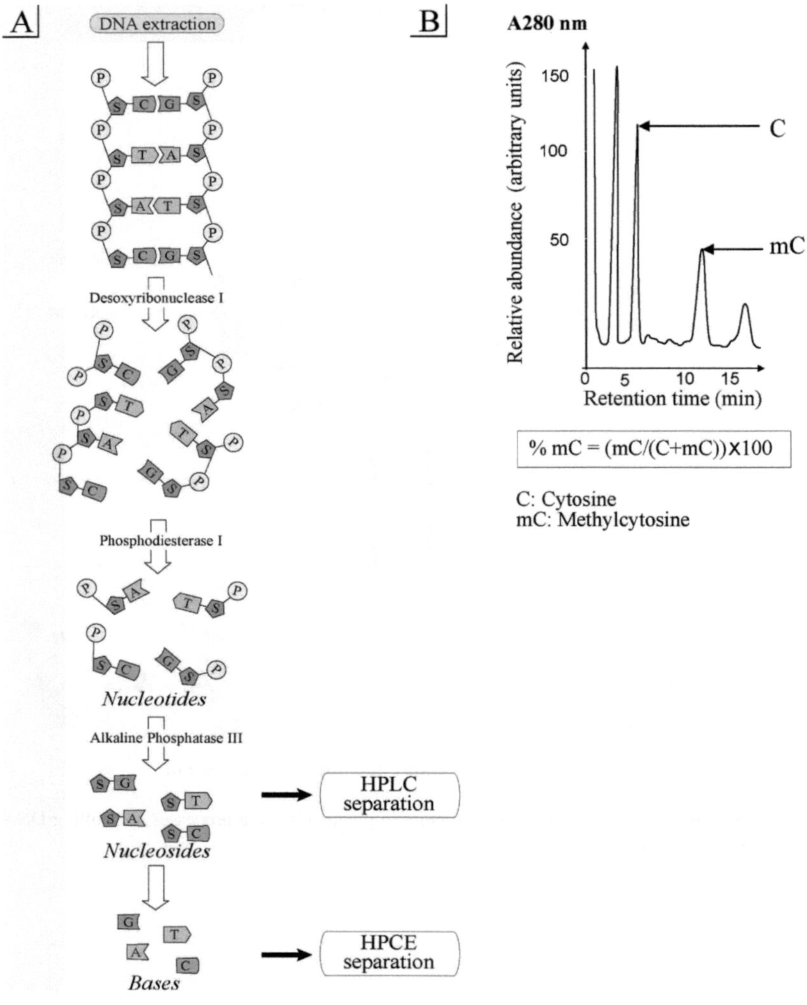

Plate 13. Determination of global genomic DNA methylation levels: A, Enzymatic DNA hydrolysis. B, HPLC chromatogram for the determination of methylcytosine percentage. P: Phosphate group. S: Sugar. A, T, C and G: Bases *(see Fig. 2 on page 357)*

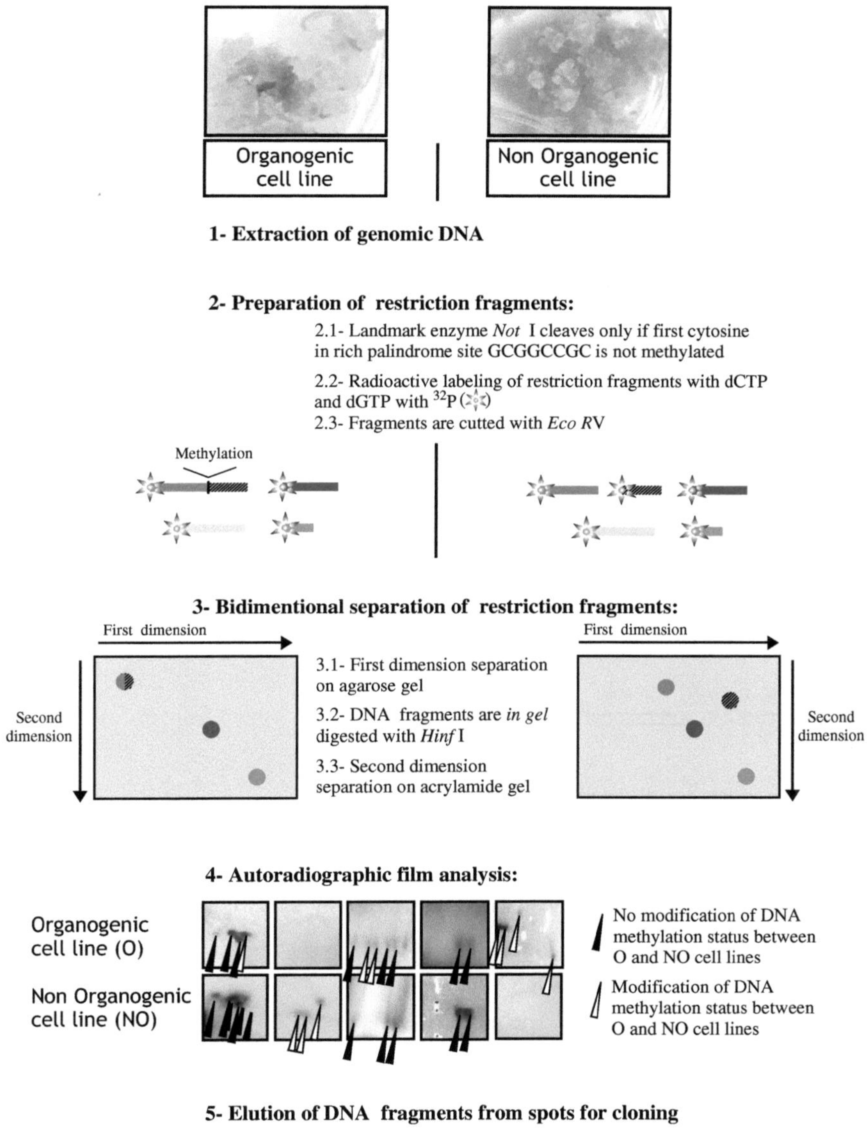

Plate 14. Principle of Restriction Landmark Genome Scanning (RLGS) method for the discovery of methylation biomarkers. RLGS sections were obtained with DNA extracted from organogenic or non-organogenic sugarbeet lines. Spots indicated by arrows correspond to fragments that can be superposed (black) or not (white) on the RLGS sections obtained with both lines. (Adapted from Causevic et al., 2006) *(see Fig. 3 on page 359)*

1- Ligation with ***Not*** **I** and ***Hinf*** **I** adaptaters. First PCR using primers designed on ***Not*** **I** adaptaters allow amplification ***Not*** **I / *Not*** **I** fragments.

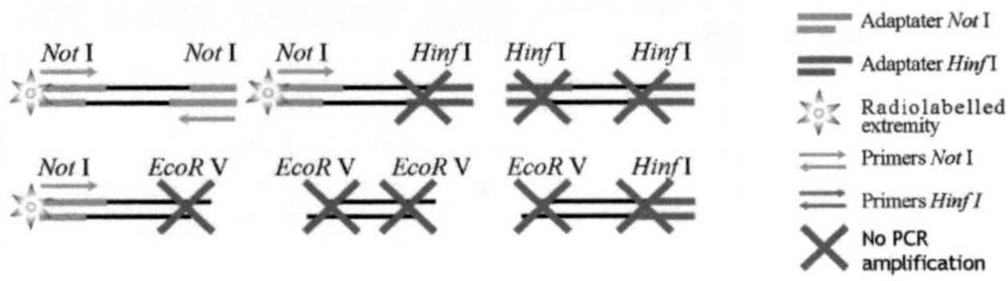

2- Second PCR using primers designed on ***Not*** **I** and ***Hinf*** **I** adaptaters allow amplification of ***Not*** **I / *Not*** **I** and ***Not*** **I / *Hinf*** **I** fragments.

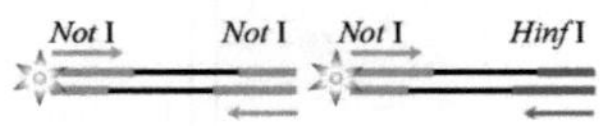

3- Amplified fragments are subcloned in adapted vector.

4- Sequencing and analysis.

Plate 15. Cloning strategy for epigenetic biomarkers screened by RLGS using adaptaters and PCR amplifications *(see Fig. 4 on page 361)*

A

Bisulfite sequencing

- Extraction of genomic DNA.

-Treatment by hydroxyquinone/bisulfite in order to deaminate unmethylated cytosine into uracile.

- PCR amplification with specific primers on genomic DNA treated or not.

- Subcloning of PCR products in a vector.

- Sequencing of about 10 clones by sequence.

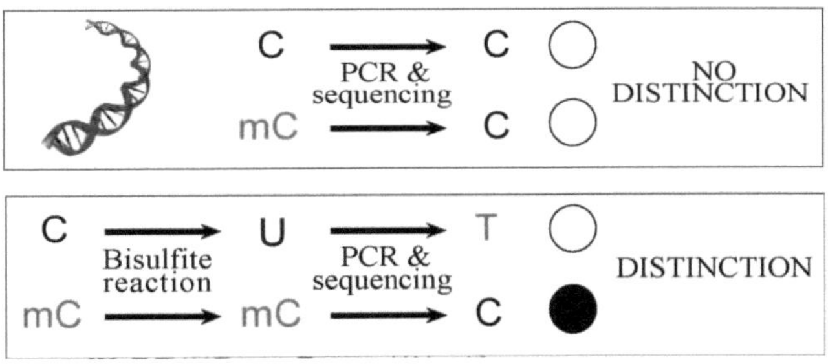

B

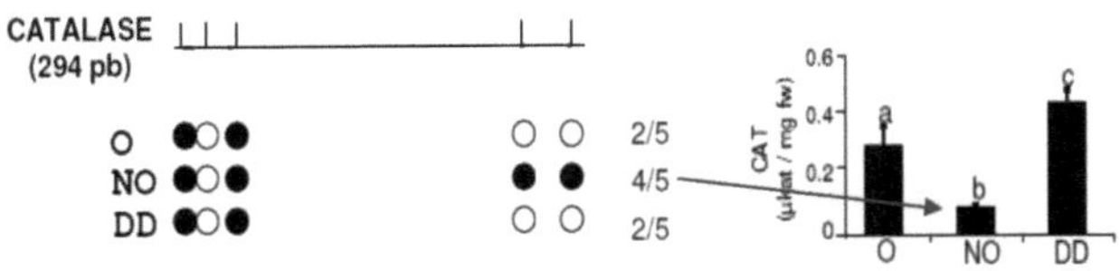

Plate 16. A, Principle of bisulfite-PCR sequencing method for the determination of the methylation status of gene candidates. B, Results of the methylation analysis of 5' regions of sugarbeet catalase gene by bisulfite sequencing. The potential methylated CpG sites in the sequence are indicated by perpendicular lines. For the three cell lines organogenic (O), non organogenic (NO) and dedifferentiated (DD), 6 to 10 PCR products were subcloned and sequenced. Five CpG sites were considered to be methylated when more than half the clones retained an unmodified cytosine at that position. Methylated CpG sites (Filled circles) and unmethylated CpG sites (open circles) are shown. The proportions of methylated CpG sites are indicated on the right for catalase activity as measured in the O, NO and DD sugarbeet cell lines. Data are means ± SE from three independent replicates. Values marked with different letters are significantly different between cell lines ($P \leq 0.05$) as determined by one-way ANOVA. *fw* fresh weight. (Adapted from Causevic et al., 2006) *(see Fig. 5 on page 365)*

CHAPTER 1

GENOMICS-ASSISTED CROP IMPROVEMENT: AN OVERVIEW

RAJEEV K. VARSHNEY[1,*] AND ROBERTO TUBEROSA[2,*]

[1]*International Crops Research Institute for the Semi-Arid Tropic Crops (ICRISAT), Patancheru-502324, A.P. India*

[2]*Department of Agroenvironmental Sciences and Technology, University of Bologna, Viale Fanin 44, 40127 Bologna, Italy*

Abstract: In recent years, a truly impressive number of advances in genetics and genomics have greatly enhanced our understanding of structural and functional aspects of plant genomes but at the same time have challenged us with many compelling avenues of investigation. The complete genome sequences of *Arabidopsis*, rice, sorghum and poplar as well as an enormous number of plant expressed sequence tags (ESTs) have become available. In the next few years, the entire genomes or at least gene space will likely be sequenced for most major crops. However, improved varieties, not sequences *per se*, contribute to improved economic return to the farmer. Functional genomics and systems biology research are facilitating the identification of gene networks that are involved in controlling genetic variation for agronomically valuable traits in elite breeding populations. Furthermore, combining the new knowledge from genomic research with conventional breeding methods is essential for enhancing response to selection, hence crop improvement. Superior varieties can result from the discovery of novel genetic variation, improved selection techniques, and/or the identification of genotypes with improved attributes due to superior combinations of alleles at multiple loci assembled through marker-assisted selection. Although it is clear that genomics research has great potential to revolutionize the discipline of plant breeding, high costs invested in/associated with genomics research currently limit the implementation of genomics-assisted crop improvement, especially for inbreeding and/or minor crops. A critical assessment of the status and availability of genomic resources and genomics research in model and crop plants, and devising the strategies and approaches for effectively exploiting genomics research for crop improvement have been presented in two volumes of the book. While Volume 1, entitled "Genomics approaches and platforms", compiles chapters providing readers with an overview of the available genomics tools, approaches and platforms, Volume 2, entitled "Genomics applications in crop improvement", presents a timely and critical overview on applications of genomics in crop improvement. An overview on the highlights of the chapters of these two volumes has been presented in the present introductory chapter.

*Authors' email: r.k.varshney@cgiar.org/ roberto.tuberosa@unibo.it

R.K. Varshney and R. Tuberosa (eds.), Genomics-Assisted Crop Improvement: Vol. 1: Genomics Approaches and Platforms, 1–12.

1. INTRODUCTION

Since the beginning of recombinant DNA technology, it has been suggested that plant molecular biology has the potential to initiate a new Green Revolution for sustainable agriculture to meet the needs of a fast-growing human population world wide. Genetic engineering has already shown the potential of plant biotechnology for crop improvement. As a result, transgenic plants with high agronomic and environmental value have been developed for several crop species such as maize, soybean, cotton, tomato, potato, tobacco, papaya, etc. (http://www.isaaa.org/). In parallel, because of their ecological and evolutionary novelty, transgenic crops have also raised a number of questions and public awareness (Wolfenbarger and Phifer 2000). In addition, the costs and uncertainties that result from the rapidly proliferating national and international regulations covering transgenic crops have significantly hindered further development of additional crops and traits (Kalaitzandonakes 2004, Bradford et al. 2005, Jank et al. 2005).

Molecular genomics, in addition to genetic engineering, is another important area of plant biotechnology that provides tools and opportunities for plant breeding. Molecular markers are one of such tools that can be used in a variety of ways to assist plant breeding activities. Some of these applications include: (i) marker-assisted selection (MAS) strategies for germplasm improvement and cultivar development, (ii) gene pyramiding for accumulating multiple genes for resistance to specific pathogens and pests within the same cultivar, and (iii) examining allelic diversity in natural populations or breeding material to select the desired genotypes (Jain et al. 2002, Gupta and Varshney 2004).

In recent years, remarkable progress has been achieved in the area of plant genomics as large-scale genome or expressed sequence tag (EST) sequencing projects were initiated in several model as well as crop plant species. In some plant species, the genome sequence data have already become available (e.g. *Arabidopsis*, rice, poplar, sorghum) and similar efforts are underway to sequence either the full genome or gene space in several other plant species such as *Medicago*, *Lotus*, tomato, soybean, maize, wheat, etc. However, sequences *per se* do not improve yield or resistance to major crop pests and abiotic stress. In fact, functional genomics studies for various agronomic traits are underway in several plant species to study gene expression and protein interactions for identifying gene networks responsible for trait phenotypes variation in the elite breeding populations. Furthermore, advances in molecular genetics and genomics provide novel approaches that can be integrated into plant breeding programmes for a more "holistic" crop improvement approach, also referred to as "genomics-assisted breeding" (Varshney et al. 2005b). The present book, therefore, aims to present in a timely and critical fashion (i) the existing information on genomics research, (ii) the prospects and approaches to apply this information to crop breeding, and (iii) the achievements and constraints of applications of genomics in crop improvement.

For the sake of convenience, the chapters of the book have been classified in two volumes. Volume 1, entitled "Genomics approaches and platforms", provides an overview on the available genomics tools, approaches and platforms, while

Volume 2, entitled "Genomics applications in crop improvement", compiles chapters that provide a critical overview on the applications of genomics for the improvement individual crops. The present introductory chapter provides an overview of all the chapters of both volumes of the book.

2. VOLUME 1: GENOMICS APPROACHES AND PLATFORMS

Due to significant progress in the area of molecular genetics during the last two decades, enormous genomic resources have been developed for major crop plant species. As an example, for most crop species a large number of molecular markers and high-density genetic maps, large-insert libraries (e.g. BAC libraries), ESTs, etc., have become available (see Phillips and Vasil 2001, Tuberosa et al. 2002, Gupta and Varshney, 2004). These tools, have been used for MAS (Koebner 2004, Varshney et al. 2006), and have facilitated mapping and cloning of genes or quantitative trait loci (QTLs) leading to sequencing and annotation of large genomic DNA fragments in several plant species (Stein and Graner 2004, Salvi and Tuberosa 2005).

Because of the availability of the above-mentioned genetic resources and significant progress in automation, robotics and bioinformatics, complete genome sequences have already become available for *Arabidopsis* (AGI, 2000), rice (Goff et al. 2002, Yu et al. 2002, 2005), poplar (http://www.eurekalert.org/pub_releases/2006-09/dgi-tft090806.php) and sorghum (http://www.phytozome.net/sorghum); similar efforts are underway for several other plant species (www.jgi.doe.gov/sequencing/index.html). Additionally, the availability of gene sequence data has made it possible to develop molecular markers directly from the genes. Such molecular markers include microsatellites or simple sequence repeats (SSR, Varshney et al. 2005a), single nucleotide polymorphisms (SNPs, Rafalski 2002) and conserved orthologous sequences (COS, Rudd et al. 2005) and very often are referred to as genic or functional molecular markers. Development and applications of such molecular markers are discussed in Chapter 2 by Rajeev Varshney and colleagues. The genic molecular markers represent a useful resource for (i) identifying the perfect or diagnostic markers, if the marker is derived from the gene(s) responsible for expression of the trait, (ii) assaying the functional diversity in the germplasm collection and (iii) comparative mapping and synteny studies. Molecular markers can play a pivotal role to enhance breeding strategies in a variety of ways. In Chapter 3, Anker Sørensen and colleagues advocate for adaptation of selection methods in breeding, towards the integrated use of genetic knowledge based on DNA markers so that the potential of molecular breeding and available germplasm resources can be better exploited. One of the main applications of molecular markers in plant breeding is to identify and map QTLs to understand the gene-to-phenotype relationships for the traits. In fact, how best to use the results of the mapping studies to enhance response to selection by MAS has always been a great concern to plant geneticists and breeders (Lande and Thompson 1990, Openshaw and Frascaroli 1997, Moreau et al. 2004, Podlich et al. 2004, Crosbie et al. 2006). In recent years, several studies have been conducted on modelling the effects of QTLs and MAS in plant breeding programmes (Hammer et al. 2005). Chapter 4, authored by Mark Cooper and colleagues, deals with illustrative examples and simulation experiments on modelling QTL effects and MAS in plant breeding.

The majority of the marker-trait association studies conducted in the past were based on linkage mapping that aimed to detect non-random association between a genotype and a phenotype. In recent years, association mapping based on linkage disequilibrium (LD), extensively used in human genetics, is becoming a very popular approach in plant genetics and breeding (Thornsberry et al. 2001). The LD-based association studies, involving associations within populations of unrelated accessions, offer a potentially powerful and rewarding approach for mapping causal genes/QTLs with modest effects and validating the role of functional candidate sequences (Hirschhorn and Daly 2005). The methodology and applications of LD-based association mappings to crop improvement have been discussed in Chapter 5 by Ersoz and colleagues. Although LD mapping offers the possibility of utilizing the potential of exotic germplasm in crop improvement, "advanced backcross (AB) QTL analysis" and "exotic libraries" are other approaches that increase the efficiency of harnessing natural biodiversity to improve yield, adaptation and quality of elite germplasm (Zamir et al. 2001). In Chapter 6, Silvana Grandillo and colleagues discuss the optimal exploitation of the naturally available genetic resources to generate new traits and improve crop performance.

Availability of EST/genome sequencing data from several crop species are being used presently to analyze the transcriptome/genomes of a species by using advanced bioinformatics tools (Zhang et al. 2004). Development of functional molecular markers from gene sequence data is an added value to the existing repository of molecular markers for their use in plant breeding. The use of sequence data by using a variety of applications in crop improvement has been discussed by Lim and colleagues in Chapter 7. Indeed, the integration of sequence-based molecular markers to the genetic maps has enhanced the resolution of comparative maps in related plant species (see Sorrells et al. 2003, Salse et al. 2004). In the case of cereals, where ca. 500,000 ESTs for each of major cereal species (e.g. wheat, maize, rice, barley) and the complete genome sequence data for rice and draft sequence for sorghum are available, comparative genomics has provided important insights on genome evolution and how to best utilize marker/gene information from one species to another. A comprehensive overview on cereal comparative genomics is presented by Jerome Salse and Catherine Feuillet in Chapter 8. Furthermore, comparative genomics allows us to isolate QTLs of agronomic interest (Salvi and Tuberosa 2005). The present status and future directions on QTL cloning have been summarized by Silvio Salvi and Roberto Tuberosa in Chapter 9.

The increasing emphasis on functional genomics and the wide accessibility of transcripts profiling has led to the establishment of various high-throughput methodologies of gene expression analysis. These methodologies include (i) EST sequencing, (ii) differential display (DD) (Liang and Pardee 1992), (ii) serial analysis of gene expression (SAGE) (Velculescu et al. 1995), (iii) nucleic acid hybridization of mRNA or cDNA fragments e.g. oligo chips (Lockhart et al. 1996), cDNA microarrays (Schena et al. 1995), (iv) cDNA-amplified fragment length polymorphism analysis (cDNA-AFLP) (Bachem et al. 1996), and (v) massively parallel signature sequencing (MPSS) (Brenner et al. 2000). These methodologies differ in principle, convenience, costs involved, number of transcripts assayed and sensitivity (Kuhn 2001). One of these approaches, i.e. SAGE, has been dealt with in detail in Chapter 10 by Prakash Sharma and colleagues. Some other important functional genomics approaches to identify candidate genes have been reviewed by Ashwani Pareek et al. in Chapter 12. Such kind of functional genomics approaches often highlight a very large number of genes associated with a trait of interest (Sreenivasulu et al. 2004). Nonetheless, pinpointing the relevant candidate genes and deploying them effectively as diagnostic markers for predicting the phenotype in plant breeding applications remain difficult and challenging tasks. In this respect, the "genetical genomics" approach, first introduced by Jansen and Nap (2001), appears quite promising. In this approach, the transcript level data generated from microarray analysis are analyzed in quantitative fashion to identify gene expression QTLs (eQTL). This approach has been referred to as "expression genetics" by Varshney et al. (2005b). In Chapter 11, Matias Kirst and Qibin Yu review the principles of genetical genomics, the results of these studies in plants and the use of this approach to dissect the genetic control of phenotypic traits of biological and agricultural interest. Although the approach is still in its infancy and remains prohibitively expensive, pioneering studies on this aspect have demonstrated its value to unravel genetic networks involved in transcription regulation, and to identify genes and pathways controlling phenotypic variation for quantitative traits. Indeed, the genetical genomics approach can help to fish out additional genes underlying the eQTLs identified for corresponding candidate genes in those plant species where genome sequence data are available.

It is important to note that the comparative analysis of the genome sequence data for *Arabidopsis* and rice is already providing important insights and understanding on the evolutionary relationships among various classes of gene families, including those representing components of hormone (especially auxin and cytokinin) signaling critical for plant development and growth. Jiten Khurana and colleagues in Chapter 13 have underlined the value of some of the auxin and cytokinin signaling components as genetic tools for manipulating agronomic traits in crops.

As a result of genome sequencing and functional genomics studies, massive sequence and expression data are being generated. Therefore, it is essential that statistical and mathematical standards, as well as guidelines for the experimental design and analysis of biological studies are upheld. Chapter 14 by Rebecca Doerge deals with past statistical issues, discusses current statistical advances that pertain to understanding complex traits, and promotes ideas about the data and statistical genomic models of the future. Such advances in the area of statistical genomics will improve the efficiency of identifying and validating the function of the most promising candidate genes. The function of candidate genes can also be verified through reverse genetics. Under these approaches, specific genes can be disrupted, and hypotheses regarding gene function can thus be directly tested *in vivo*. At present, a number of reverse genetic methods exist, many of which are limited in application because they are organism-specific, expensive, transgenic or only transiently disrupt gene function. As an alternative, TILLING (Targeting Induced Local Lesions IN Genomes; McCallum et al. 2000) deploys traditional mutagenesis and SNP discovery methods for a high-throughput, reverse-genetic strategy that is low in cost and applicable to most organisms. In fact, during the past six years, TILLING has moved from proof-of-concept to production with the establishment of publicly available services for several crop species such as maize, rice and barley. The protocols developed for TILLING have been adapted for the discovery of natural nucleotide diversity in germplasm collections and the method has been termed EcoTILLING. In Chapter 15, Bradley Till and colleagues review current TILLING and EcoTILLING technologies and discuss the progress that has been made in applying these methods to many different plant species.

It is evident from the above that there are several approaches available that can be used for applications in plant breeding. Some examples of successful utilization of genomics for crop improvement at least in some cereal species have been recently reviewed by Varshney et al. (2006). However, the anticipated success of molecular breeding has not materialized as expected. The critical factors responsible for it include the poor understanding and consideration of important genetic phenomena such as epigenetics, genome imprinting, epistasis, and regulatory variation (Morgante and Salamini 2003, Varshney et al. 2005). It has been shown that polymorphism in DNA methylation status leads to differences in gene expression and confers phenotypic effects (Ronemus et al. 1996). All these alleles having the same DNA sequence but differing in their methylation status correspond to different phenotypes and are referred to as epialleles (Kalisz and Purugganan 2004). In recent years, the polymorphism associated to the epialleles has been exploited to constitute the biomarkers. A biomarker is a substance or a process that is indicative of a phenotype or a biological event (Laird 2003) and can be used for a variety of applications. For example, in human cancer epialleles are used as biomarkers for early detection of cancer types. In plants, epigenetic inheritance is a source of polymorphism that holds great potential for selection and plant breeding (Tsaftaris et al. 2005). In Chapter 16, Marie-Véronique Gentil and Stéphane Maury describe the characterization of biomarkers using new molecular approaches and discuss the future role of biomarkers in plant breeding.

3. VOLUME 2: GENOMICS APPLICATIONS IN CROP IMPROVEMENT

The approaches and platforms presented in Volume 1 have already been utilized for crop improvement. In particular, due to the importance of cereals in the human diet and the availability of excellent genomic resources, cereal crops have been the major target of genomics approaches (see Varshney et al. 2006). In wheat, rice, maize, barley and sorghum, genomics research and MAS have already yielded fruitful results. Efforts are underway to exploit the genomics research for the improvement of several other plant species such as legumes, fruit species and trees.

Among different types of molecular markers, microsatellites have been the markers of choice for plant breeding applications until recently (Gupta and Varshney 2000). However, because of the availability of sequence data in recent years from a large number of plant species, the development and applications of single nucleotide polymorphism (SNP) markers is gaining momentum and will become more prevalent. In Chapter 1, Martin Ganal and Marion Röder review the development and applications of microsatellite and SNP markers for wheat breeding. Wheats (including durum and bread wheat) are the major foods for majority of the human population and are mainly consumed as processed products because of the unique functional properties they confer to the derived foods. Therefore, improvement of end-use quality has been a major concern to wheat breeders with an emphasis on developing cultivars for specific applications such as bread (leavened, flat, steamed, etc.), other baked goods (cakes, cookies, crackers, etc.), pasta and noodles, and a wide range of other products of restricted local uses. Processing and end-use quality of wheat-based products are influenced by several factors such as protein content and composition, starch, kernel hardness and lipids. The use of molecular markers to identify and manipulate the QTLs for important grain quality traits in wheat have been summarized in Chapter 2 by Domenico Lafiandra and colleagues. Among the factors that can affect the quantity and quality of final yield in cereal as well as in other crops abiotic stresses play a pivotal role. Of all abiotic stresses, drought ranks first in terms of economic importance and recalcitrance to breeders' efforts. The use of molecular markers and functional genomics for identifying genes/QTLs conferring tolerance to drought and their use in breeding has been discussed by Michael Baum and colleagues in Chapter 3.

After identifying the molecular markers associated with gene(s)/QTLs for traits of interest, the next step is to use molecular markers in back-crossing programmes to improve selection efficiency and to implement gene pyramiding especially for disease resistances. Applications of molecular markers in barley for marker-assisted back-crossing and gene pyramiding for several disease resistance genes have been summarized in Chapter 4 by Wolfgang Friedt and Frank Ordon. In addition to the direct use of molecular markers in breeding, high density fine mapping of genes/QTLs of interest allows for the isolation of QTLs (Salvi and Tuberosa 2005). Indeed, in several cereal species as a result of long-term efforts, a number of disease resistance genes/QTLs have been isolated. An update on cloning of important disease resistance genes/QTLs is provided in Chapter 5 by Beat Keller and colleagues.

Compared to wheat and barley, maize is a more important crop for the private sector. Therefore, the majority of genomics efforts and applications in maize genomics have been undertaken by private industries. Chapter 6 authored by Michael Lee deals with a comprehensive review on the applications of genomics and genetic engineering in maize breeding. Furthermore, this chapter presents a perspective on the requirements, pros and cons of genomics applications in maize breeding.

As a model of cereal species, rice has benefited greatly from the advances in plant genomics. For example, dense genetic maps, genome-wide physical maps and four drafts of the rice genome are available (Vij et al. 2006). These genomic resources helped the rice community to identify and apply molecular markers for the improvement of grain quality traits, disease resistance and abiotic stress tolerance. Applications of molecular markers linked with several traits of interest in MAS, gene pyramiding, breeding nurseries, etc., in rice are summarized by David Mackill in Chapter 7. Additionally, the use of the rice genome sequence, novel approaches such as candidate gene sequencing, SNP markers for rice breeding have been reviewed in Chapter 8 by Nagendra Singh and Trilochan Mohapatra.

Sorghum is an important crop that is more tolerant than other cereals to many abiotic stresses, including heat, drought, and flooding, making it an ideal crop for growing on marginal lands as the demand for food, feed and energy increases. In recent years, the use of molecular markers has been initiated for the genetic analysis and manipulation of agronomic and stress-tolerance traits important for sorghum improvement. For example, molecular markers have been identified to be associated with QTLs for many complex traits, including pre-flowering and post-flowering drought tolerance, early-season cold tolerance and resistance to the parasitic weed *Striga*. In some cases (e.g. stay green and *Striga*), efforts have been initiated to use the corresponding QTLs for the development of improved sorghum cultivars through MAS and trait introgression. Chapter 9, authored by Gebisa Ejeta and Joseph Knoll, provides an overview on development and application of molecular markers for the development of improved sorghum cultivars.

Although genomics has already provided important contributions for the improvement of major cereal species, relatively less important cereal crops such as the millets and rye have received much less attention and consequently, often lack reasonably dense genetic maps (Varshney et al. 2006). Nevertheless, in recent years, species-specific genomic resources are being generated and genomic resources from related cereal species are being transferred through comparative genomics studies. A similar case applies to grain legume crops which are very important for both human diets and animal feeds. The grain legume crops contribute 33% of human protein intake and are a major source of lipids. However, except for soybean and common bean, other legume crops have not received much attention in terms of genomics approaches. Chapter 10, authored by Rajeev Varshney and colleagues, reviews the progress in the area of molecular genetics and applications in breeding in case of three important semi-arid tropic crops, i.e. chickpea, pigeonpea and groundnut. In contrast to these species, well-saturated genetic maps, physical maps and functional genomics resources are available in soybean. The advances in the area of genomic resources and their applications in soybean improvement have been discussed in Chapter 11 by Tri Vuong and other colleagues.

In addition to cereal and legume species, advances are being made in the area of genomics-assisted crop improvement in several other species. For example, in case of forage crops, molecular breeding for quality trait has been discussed in Chapter 12 by Thomas Lübberstedt; while updates on molecular mapping, MAS and map-based cloning in tomato have been provided in Chapter 13 by Majid Foolad. Additionally, Chapter 14 by Pere Arús and Susan Gardiner reviews the progress and potential of genomics for improving Rosaceae temperate tree fruit species. In Chapter 15, Prasad Hendre and Ramesh Aggarwal discuss the development and applications of molecular markers for the genetic improvement of coffee.

Two chapters have been devoted to understand two important processes, i.e. nodulation in legumes and domestication in cereals. By using soybean as a plant system, Kyujung Van and colleagues report (Chapter 16) on the identification of nodulation mutants (e.g. non-nodulation, ineffective nodulation and super-/hypernodulation) and the genetic loci that control nodulation. These authors suggest that molecular gene identification should be combined with biochemical pathways for nodulation in order to better understand the symbiotic interactions between legume and Rhizobia. In Chapter 17, Carlo Pozzi and Francesco Salamini review several issues concerning the state of molecular knowledge of the effects induced by domestication and breeding of wheat crops. Genetic bottlenecks which have been associated to wheat domestication and breeding have also been discussed in the chapter.

Volume 2 is concluded by Chapter 18, authored by Paulo Arruda and Thaís Rezende Silva, on sugarcane improvement by using information provided by transcriptome analysis and functional genomics approaches. These authors have demonstrated the identification of genes, by analyzing EST resources, involved in biotic and abiotic stress response, disease resistance and sucrose accumulation.

4. CONCLUDING REMARKS

The two volumes of the book provide up-to-date information on genomics research including platforms, approaches, as well as the achievements of application of genomics in breeding. Although genomics holds great potential for improving breeding efficiency, the high costs associated with genomics research are a critical factor hindering further applications of genomics to crop improvement particularly for inbreeding and/or minor crop species. Nevertheless, there are several success stories on the development of improved superior cultivars. In the coming years, it is anticipated that the decreasing cost of genotyping and sequencing coupled with further advances in molecular platforms and bioinformatics, will allow genomics to become an integral part of crop breeding and to improve selection efficiency. We hope that our effort in compiling these two volumes will help and stimulate those working in crop genomics as well as conventional plant breeding to better focus their research plans for crop improvement programmes. Hopefully, the book will also help graduate students to develop a better understanding of this important aspect of modern plant science research.

REFERENCES

Arabidopsis Genome Initiative (2000) Analysis of the genome sequence of the flowering plant *Arabidopsis thaliana*. Nature 408:796–815

Bachem CWB, van der Hoeven RS, de Bruijn SM, Vreugdenhil D, Zabeau M, Visser RGF (1996) Visualization of differential gene expression using a novel method of RNA fingerprinting based on AFLP: analysis of gene expression during potato tuber development. Plant J 9:745–753

Bradford KJ, Deynze AV, Gutterson N, Parrott W, Strauss SH (2005) Regulating transgenic crops sensibly: lessons from plant breeding, biotechnology and genomics. Nat Biotechnol 23:439–444

Brenner S, Johnson M, Bridgham J, Golda G, Lloyd DH, Johnson D, Luo S, McCurdy S, Foy M, Ewan M et al (2000) Gene expression analysis by massively parallel signature sequencing (MPSS) on microbead arrays. Biotechnology 18:630–634

Crosbie TM, Eathington SR, Johnson GR, Edwards M, Reiter R, Stark S, Mohanty RG, Oyervides M, Buehler RE, Walker AK et al (2006) Plant breeding: past, present, and future. In: Lamkey KR, Lee M (eds) Plant breeding: the Arnel R. Hallauer international symposium. Blackwell Publishing, Ames, Iowa, pp 3–50

Goff SA, Ricke D, Lang TH, Presting G, Wang R, Dunn M, Glazebrook J, Sessions A, Oeller P, Varma H et al (2002) Draft sequence of the rice genome (*Oryza sativa* L. ssp. *japonica*). Science 296:92–100

Gupta PK, Varshney RK (2000) The development and use of microsatellite markers for genetics and plant breeding with emphasis on bread wheat. Euphytica 113:163–185

Gupta PK, Varshney RK (2004) Cereal genomics: an overview. In: Gupta PK, Varshney RK (eds) Cereal genomics. Kluwer Academic Press, Dordrecht, The Netherlands, pp 639–643

Hammer GL, Chapman S, Oosterom E, Podlich DW (2005) Trait physiology and crop modelling as a framework to link phenotypic complexity to underlying genetic systems. Aust J Agric Res 56:947–960

Hirschhorn JN, Daly MJ (2005) Genome-wide association studies for common diseases and complex traits. Nat Genet 6:95–108

Jain SM, Brar DS, Ahloowalia BS (2002) Molecular techniques in crop improvement. Kluwer Academic Publishers, Dordrecht, The Netherlands, pp 433–444

Jank B, Rath J, Spok A (2005) Genetically modified organisms and the EU. Trends Biotechnol 23:222–224

Jansen RC, Nap JP (2001) Genetical genomics: the added value from segregation. Trends Genet 17:388–391

Kalaitzandonakes N (2004) The potential impacts of the biosafety protocol on agricultural commodity trade. IPC Technology Issue Brief, Washington DC

Kalisz S, Purugganan MD (2004) Epialleles via DNA methylation: consequences for plant evolution. Trends Ecol Evol 19:309–314

Koebner RMD (2004) Marker-assisted selection in the cereals: the dream and the reality. In: Gupta PK, Varshney RK (eds) Cereal genomics. Kluwer Academic Publishers, Dordrecht, The Netherlands, pp 199–252

Kuhn E (2001) From library screening to microarray technology: strategies to determine gene expression profiles and to identify differentially regulated genes in plants. Ann Bot 87:139–155

Laird PW (2003) The power and the promise of DNA methylation markers. Nature Rev Cancer 3: 253–266

Lande R, Thompson R (1990) Efficiency of marker-assisted selection in the improvement of quantitative traits. Genetics 124:743–756

Liang P, Pardee AB (1992) Differential display of eukaryotic messenger RNA by means of the polymerase chain reaction. Science 257:967–971

Lockhart DJ, Dong H, Byrne MC, Follettie MT, Gallo MV, Chee MS, Mittmann M, Wang C, Kobayashi M (1996) Expression monitoring by hybridization to high-density oligonucleotide arrays. Nat Biotechnol 14:1675–1680

McCallum CM, Comai L, Greene EA, Henikoff S (2000) Targeting induced local lesions in genomes (TILLING) for plant functional genomics. Plant Physiol 123:439–442

Moreau L, Charcosset A, Gallais A (2004) Experimental evaluation of several cycles of marker-assisted selection in maize. Euphytica 137:111–118

Morgante M, Salamini F (2003) From plant genomics to breeding practice. Curr Opin Biotechnol 14: 214–219

Openshaw S, Frascaroli E (1997) QTL detection and marker-assisted selection for complex traits in maize. Proceedings of the 52nd Annual Corn and Sorghum research conference american seed trade association, Washington DC, pp 44–53

Phillips RL, Vasil IK (2001) DNA-based markers in plants. Kluwer Academic Publishers, The Netherlands, pp 497–503

Podlich DW, Winkler CR, Cooper M (2004) Mapping as you go: an effective approach for marker-assisted selection of complex traits. Crop Sci 44:1560–1571

Rafalski JA (2002) Applications of single nucleotide polymorphism in crop genetics. Curr Opin Plant Biol 5:94–100

Ronemus MJ, Galbiati M, Ticknor C, Chen J, Dellaporta SL (1996) Demethylation-induced developmental pleiotropy in *Arabidopsis*. Science 273:654–657

Rudd S, Schoof H, Klaus M (2005) PlantMarkers-a database of predicted molecular markers from plants. Nucleic Acids Res 33:D628–D632

Salse J, Piegu B, Cooke R, Delseny M (2004) New *in silico* insight into the synteny between rice (*Oryza sativa* L.) and maize (*Zea mays* L.) highlights reshuffling and identifies new duplications in the rice genome. Plant J 38:396–409

Salvi S, Tuberosa R (2005) To clone or not to clone plant QTLs: present and future challenges. Trends Plant Sci 10:297–304

Schena M, Shalon D, Davis RW, Brown PO (1995) Quantitative monitoring of gene expression patterns with a complementary DNA microarray. Science 270:467–470

Sorrells ME, Rota ML, Bermudez-Kandianis CE, Greene RA, Kantety R, Munkvold JD, Miftahudin, Mahmoud A, Ma X, Gustafson PJ, Qi LL, Echalier B (2003) Comparative DNA sequence analysis of wheat and rice genomes. Genome Res 13:1818–1827

Sreenivasulu N, Varshney RK, Kavikishore PV, Weschke W (2004) Tolerance to abioitic stress-a functional genomics approach. In: Gupta PK, Varshney RK (eds) Cereal genomics. Kluwer Academic Publishers, Dordrecht, The Netherlands, pp 483–514

Stein N, Graner A (2004) Map-based gene isolation in cereal genomes In: Gupta PK, Varshney RK (eds) Cereal genomics. Kluwer Academic Publishers, Dordrecht, The Netherlands, pp 331–360

Thornsberry JM, Goodman MM, Doebley J, Kresovich S, Nielsen D, Buckler ES (2001) *Dwarf8* polymorphisms associate with variation in flowering time. Nat Genet 28:286–289

Tsaftaris AS, Polidoros AN, Koumproglou R, Tani A, Kovacevic N, Abatzizou E (2005) Epigenetic mechanisms in plants and their implications in plant breeding. In: Tuberosa R, Phillips RL, Gale MA (eds). In the wake of the Double Helix: From the Green Revolution to the gene revolution, Avenue-media, Bologna, Italy, pp 157–172

Tuberosa R, Gill BS, Salvi S (2002) Cereal genomics: ushering in a brave new world. Plant Mol Biol 48:445–449

Varshney RK, Graner A, Sorrells ME (2005a) Genic microsatellite markers in plants: features and applications. Trends Biotechnol 23:48–55

Varshney RK, Graner A, Sorrells ME (2005b) Genomic-Assisted breeding for crop improvement. Trends Plant Sci 10:621–630

Varshney RK, Hoisington DA, Tyagi AK (2006) Advances in cereal genomics and applications in crop breeding. Trends Biotechnol 24:490–499

Varshney RK, Sigmund R, Borner A, Korzun V, Stein N, Sorrells ME, Langridge P, Graner A (2005c) Interspecific transferability and comparative mapping of barley EST-SSR markers in wheat, rye and rice. Plant Sci 168:195–202

Velculescu VE, Zhang L, Vogelstein B, Kinzler KW (1995) Serial analysis of gene expression. Science 270:484–487

Vij S, Gupta V, Kumar D, Vydianathan R, Raghuvanshi S, Khurana P, Khurana JP, Tyagi AK (2006) Decoding the rice genome. Bioessays 28:421–432

Wolfenbarger LL, Phifer PR (2000) Biotechnology and ecology: the ecological risks and benefits of genetically engineered plants. Science 290:2088–2093

Yu J, Hu S, Wang J, Wang G, Li SG, Wong KSG, Liu B, Deng Y, Dai L, Zhou Y, Zhang X et al (2002) A draft sequence of the rice genome (*Oryza sativa* L. *indica*). Science 296:79–92

Yu J, Wang J, Lin W, Li S, Li H, Zhou J, Ni P, Dong W, Hu S, Zeng C et al (2005) The genomes of *Oryza sativa*: a history of duplications. PLoS Biol 3:e38

Zamir D (2001) Improving plant breeding with exotic genetic libraries. Nat Rev Genet 2:983–989

Zhang H, Sreenivasulu N, Weschke W, Stein N, Rudd S, Radchuk V, Potokina E, Scholz U, Schweizer P, Zierold U, Langridge P, Varshney RK, Wobus U, Graner A (2004) Large-scale analysis of the barley transcriptome based on expressed sequence tags. Plant J 40:276–290

CHAPTER 2

GENIC MOLECULAR MARKERS IN PLANTS: DEVELOPMENT AND APPLICATIONS

RAJEEV K. VARSHNEY[1,*], THUDI MAHENDAR[1], RAMESH K. AGGARWAL[2] AND ANDREAS BÖRNER[3]

[1]*International Crops Research Institute for the Semi-Arid Tropics (ICRISAT), Patancheru - 502324, India*

[2]*Centre for Cellular and Molecular Biology (CCMB), Uppal Road, Hyderabad- 500 007, India*

[3]*Leibniz Institute of Plant Genetics and Crop Plant Research (IPK), Correnstrasse 3, D-06466 Gatersleben, Germany*

Abstract: The current advancement in plant biology research encompassing: generation of huge amount of molecular-genetic data, development of impressive methodological skills in molecular biology experimentation, and systems analyses, has set the stage to search for ways/means to utilize the available resources to strengthen interdisciplinary efforts to find solutions to the challenging goals of plant breeding efforts (such as abiotic stress tolerance) ultimately leading to gainful applications in crop improvement. A positive fall out of such a realization and efforts has been the identification/development of a new class of very useful DNA markers called genic molecular markers (GMMs) utilizing the ever-increasing archives of gene sequence information being accumulated under the EST sequencing projects on a large number of plant species in the recent years. These markers being part of the cDNA/EST-sequences, are expected to represent the functional component of the genome i.e., gene(s), in contrast to all other random DNA-based markers (RDMs) that are developed/generated from the anonymous genomic DNA sequences/domains irrespective of their genic content/information. Therefore, identifying DNA sequences that demonstrate large effects on adaptive plant behavior remains fundamental to the development of GMMs. The few recent studies have now demonstrated the utility of these markers in genetic studies, and also shown that GMMs may be superior than RDMs for use in the marker-assisted selection, comparative mapping, and exploration of the functional genetic diversity in the germplasm adapted to different environments. The only constraint of GMMs is their low level of polymorphism as compared to the RDMs, which is expected of their origin from the relatively conserved functional portion of the genome. This chapter provides a critical review of the development and various applications of the GMMs.

*Corresponding Authors: r.k.varshney@cgiar.org

R.K. Varshney and R. Tuberosa (eds.), Genomics-Assisted Crop Improvement: Vol. 1: Genomics Approaches and Platforms, 13–29.

1. MOLECULAR MARKERS IN PLANT BREEDING

In agriculture, one of the main objectives of plant breeder is to improve the existing cultivars, which are deficient in one or more traits by crossing such cultivars with lines that possess the desired trait. A conventional breeding programme thus involves crossing whole genomes followed by selection of the superior recombinants from among the several segregation products. Indeed, such a procedure is laborious and time consuming, involving several crosses, several generations, and careful phenotypic selection, and the linkage drag (tight linkage of the undesired loci with the desired loci) may make it further difficult to achieve the desired objective. Advent of DNA marker technology, development of several types of molecular markers and molecular breeding strategies offered possibilities to plant breeders and geneticists to overcome many of the problems faced during conventional breeding.

Molecular markers are now widely used to track loci and genome regions in several crop-breeding programmes, as molecular markers tightly linked with a large number of agronomic and disease resistance traits are available in major crop species (Phillips and Vasil 2001, Jain et al. 2002, Gupta and Varshney 2004). These molecular markers include: (i) hybridization-based markers such as restriction fragment length polymorphism (RFLP), (ii) PCR-based markers: random amplification of polymorphic DNA (RAPD), amplified fragment length polymorphism (AFLP) and microsatellite or simple sequence repeat (SSR), and (iii) sequence-based markers: single nucleotide polymorphism (SNP). The majority of these molecular markers has been developed either from genomic DNA libraries (e.g. RFLPs and SSRs) or from random PCR amplification of genomic DNA (e.g. RAPDs) or both (e.g. AFLPs). These DNA markers can be generated in large numbers and can prove to be very useful for a variety of purposes relevant to crop improvement. For instance, these markers have been utilized extensively for the preparation of saturated molecular maps (genetical and physical). Their association with genes/QTLs controlling the traits of economic importance has also been utilized in some cases for indirect marker-assisted selection (MAS) (e.g. Koebner 2004, Korzun 2002). Other uses of molecular markers include gene introgression through backcrossing, germplasm characterization, genetic diagnostics, characterization of transformants, study of genome organization and phylogenetic analysis (see Jain et al. 2002). For plant breeding applications, SSR markers, among different classes of the existing markers, have been proven and recommended as markers of choice (Gupta and Varshney 2000). RFLP is not readily adapted to high sample throughput and RAPD assays are not sufficiently reproducible or transferable between laboratories. While both SSRs and AFLPs are efficient in identifying polymorphisms, SSRs are more readily automated (Shariflou et al. 2001). Although AFLPs can in principle be converted into simple PCR assays (e.g. STSs), this conversion can become cumbersome and complicated as individual bands are often composed of multiple fragments (Shan et al. 1999), particularly in large genome templates.

2. GENIC MOLECULAR MARKERS: INTRODUCTION AND DEVELOPMENTS

Due to emphasis on functional genomics, several gene discovery projects in the form of genome sequencing, transcriptome sequencing or gene expression studies have been established since last five years. As a result, a large number of genes have been identified through 'wet lab' as well as *in silico* studies and a wealth of sequence data have been accumulated in public databases (e.g. http://www.ncbi.nlm.nih.gov; http://www.ebi.ac.uk) in the form of BAC (bacterial artificial chromosome) clones, ESTs (expressed sequence tags), full length cDNA clones and genes. The availability of enormous amount of sequence data from complete or partial genes has made it possible to develop the molecular markers directly from the parts of genes. These markers are referred as "genic" molecular markers (GMM).

The majority of the markers, developed and used in the past as described above in section 1, are directly derived from the genomic DNA, and therefore could belong to either the transcribed or the non-transcribed part of the genome without any information available on their functions. In contrast, GMMs developed from coding sequences like ESTs or fully characterized genes frequently have been assigned known functions. Based on the site of polymorphism and later's effect on phenotypic variation, GMMs have been classified into two groups (Anderson and Luebberstedt 2003):

(i) Gene-targeted markers (GTMs): derived from polymorphisms within genes, however not necessarily involved in phenotypic trait variation, e.g. untranslated regions (UTRs) of EST sequences (Schmitt et al. 2006; Aggarwal et al 2007);

(ii) Functional markers (FMs): derived from polymorphic sequences or sites within genes and, thus, more likely to be causally involved in phenotypic trait variation (e.g. candidate gene-based molecular markers). The FMs, depending on the involvement in the phenotypic trait variation, are further classified into two subgroups: (a) indirect functional markers (IFMs), for which the role for phenotypic trait variation is indirectly known, and (b) direct functional markers (DFMs), for which the role for the phenotypic trait variation is well proven.

As per the above terminology, the molecular markers derived from anonymous regions of the genome are called random DNA markers (RDMs), which may or may not be developed from the polymorphic site in gene or may not be developed from a gene at all.

Although genic markers were developed earlier also, these were in the form of cDNA–RFLP (Graner et al. 1991, Causse et al. 1994) for which functions could not be predicted at that time. However, some efforts were made to sequence these early cDNA clones to determine the genes and their functions (Michalek et al. 1999). Compared to these earlier efforts, development of genic markers have become a reality only in recent years, because of accumulation of large ESTs or gene sequences resources resulting from EST and genome sequencing projects in several crop species and also due to the developments in the field of bioinformatics (Gupta and Rustgi 2004). For example, several transcriptome resources have become available (http://www.ncbi.nlm.nih.gov/dbEST/dbEST_summary.html), and software tools or pearl scripts have been developed to search for SSRs and SNPs from EST or gene sequences (Varshney et al. 2004, 2005a).

Although, whole genome sequencing and annotation is the way to identify the entire gene repository of a species, this has been possible only for a limited number of crop species involving large scale sequencing of their genome or gene space. On the other hand, ESTs represent a basic commodity within the analysis of genomes and their genes for a species (Rudd et al. 2003). Whereas the complete sequencing of a genome may utilize either a clone-by-clone approach or a whole genome shotgun approach to acquire adequate coverage to assemble a meaningful scaffold, EST sequencing is directed at the quick, cheap and simple sequencing of partial gene transcripts (Sreenivasulu et al. 2002). As a result, a significant redundancy can be observed in gene sequence data obtained from EST sequencing projects (see Varshney et al. 2004). Therefore before developing molecular markers from ESTs, it is essential to define the "unigenes" after cluster analysis of random ESTs using appropriate computer programmes such as stackPack (Miller et al. 1999).

Once the unigene sequence data from EST analysis or non-redundant set of genes are available, molecular markers can be developed using two main approaches:

(1) Direct mapping: Under this approach, either the cDNA clones corresponding to the ESTs of interest can be used as RFLP probe or the PCR primers can be

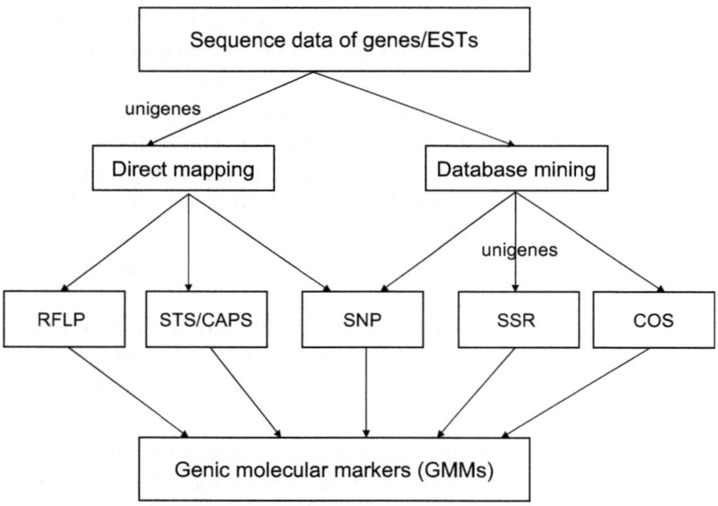

Figure 1. A scheme for development of genic molecular markers (GMMs). Two common ways to develop GMMs are shown in the figure. In the first method, the sequence data are used to define the unigenes and then the cDNA clones or genic clones corresponding to the unigenes can be assayed as RFLPs or the unigene sequence data can be used to design the primer pairs and assayed using STS/CAPS or SNP assays. In the second method, the sequence data can be mined by using some computer programmes or scripts to identify the SSRs, SNPs or COSs from given sequence data and then these markers, after defining the unigenes, can be assayed using appropriate genotyping platforms

designed for the EST/gene and used as STS or CAPS marker. Direct mapping approach should be undertaken with the unigene set of ESTs or genes only.

(2) *In silico* mining: In this approach, the SSR or SNP identification software tools are used to screen the sequence data for ESTs/genes. For identification of SNPs, the redundant set of EST data, generated from more than one genotype of a given species, are used. However, after identification of SNPs, only non-redundant set of ESTs should be considered for SNP mapping.

A scheme for development of GMMs has been shown in Figure 1. Development of FMs, however, requires: (i) functionally characterized genes, (ii) allele sequences from such genes, (iii) identification of polymorphic, functional motifs affecting plant phenotype within these genes, and (iv) validation of associations between DNA polymorphisms and trait variation. Therefore depending on the objective as well as available information or feasibility, the FMs, the special class of GMMs, can also be generated.

3. APPLICATIONS OF GENIC MOLECULAR MARKERS

Molecular markers have already shown their applications in a variety of ways in several plant species (see Gupta and Varshney 2004). The development of GMMs, now permits a targeted approach for detection of nucleotide diversity in genes controlling agronomic traits in plant populations. Some main areas of plant breeding and genetics, where the implementation of GMMs will prove quite useful, are discussed here.

3.1. Trait Mapping

One of the main applications of molecular markers in plant breeding is their use as diagnostic markers for the trait in the selection. However, use of random molecular markers (RDMs) as a diagnostic tool entails the risk of losing the linkage through genetic recombination. Even in case of GMMs, the gene-targeted markers (GTMs) where polymorphism was discovered through one allele analysis without any further specification of the polymorphic sequence motif are threatened by the same way (Rafalski and Tingey, 1993). In contrast to RDMs or GTMs, FMs (DFMs or IFMs) allow reliable application of markers in populations without prior mapping and the use of markers in mapped populations without risk of information loss owing to recombination.

The development of FMs is expensive and cannot be undertaken for all the traits and in all crop species, GMM have been developed and mapped in several plant species (Table 1). The genetic maps, developed after mapping/integration of GMM are called "transcript" or "gene" maps. For example, based on the candidate genes for drought tolerance, a comprehensive set of >200 gene-based markers have been developed for barley (Rostocks et al. 2005). Recently, a "transcript map" of barley after integrating more than 1000 gene-based markers (GTMs) has been developed, (Stein et al. 2007). A kind of transcriptome map based on deletion mapping of

Table 1. Some reports on development of genic molecular markers in important plant species

General name	Species	Type of markers developed	References
Cereals and grasses			
Barley	*Hordeum vulgare*	EST-SSR, EST-SNP, EST-RFLP, cDNA-RFLP	Thiel et al. 2003, Rostocks et al. 2005, Varshney et al. 2006, Willsmore et al. 2006, Stein et al. 2007, Varshney et al. 2007b
Maize	*Zea mays*	cDNA-RFLP, EST-SNP	Gardiner et al. 1993, Chao et al. 1994, Picoult-Newberg et al. 1999, Falque et al. 2005
Wheat	*Triticum aestivum*	EST-SSR, EST-SNP, cDNA-RFLP	Holton et al. 2002, Yu et al. 2004, Somers et al. 2003, Gao et al. 2004, Qi X. et al. 2004, Nicot et al. 2004
Rice	*Oriza sativa*	EST-SSR, EST-SNP, cDNA-RFLP, Intron Length Polymorphism (ILP)	Causse et al. 1994, Harushima et al. 1998, Temnykh et al. 2001, Feltus et al. 2004, Wang et al. 2005
Rye	*Secale cereale*	EST-SSR, EST-SNP	Hackauf and Wehling, 2002, Khlestkina et al. 2004, Varshney et al. 2007b
Sorghum	*Sorghum bicolor*	EST-SSR, cDNA-RFLP	Childs et al. 2001, Klein et al. 2003, Bowers et al. 2003, Ramu et al. 2006, Jayashree et al. 2006
Lolium	*Lolium perenne*	EST-SSR	Faville et al. 2004
Legumes			
White clover	*Trifolium repens*	EST-SSR	Barret et al. 2004
Soybean	*Glycine max*	EST-SSR	Song et al. 2004, Zhang et al. 2004
Fiber and oil seed crops			
Cotton	*Gossypium* sps.	EST-SSR	Zhang et al. 2005, Chee et al. 2004, Park et al. 2005-
Sunflower	*Helianthus* sps.	EST-SNP	Lai et al. 2005
Fruit and vegetables			
Grape	*Vitis vinifera*	EST-SSR	Chen et al. 2006
Kiwi fruit	*Actinidia chinensis*	EST-SSR	Fraser et al. 2004
Raspberry	*Rubus* spp.	EST-SSR	Graham et al. 2004
Tomato	*Lycopersicon esculentum*	EST-SSR	Frary et al. 2005
Strawberry	*Fragaria* spp.	EST-SSR	Sargent et al. 2006
Trees			
Pinus	*Pinus* ssp.	EST-SSR, ESTP	Cato et al. 2001
Coffee	*Coffea* ssp.	EST-SSR	Bhat et al. 2005, Aggarwal et al. 2007

more than 16,000 gene loci has been developed in wheat (Qi L-L et al. 2004). Such molecular maps, not only provide gene based molecular markers associated with the trait of interest after the QTL analysis, but also can be compared with those of the other related plant species in an efficient manner.

3.2. Functional Diversity

Characterization of genetic variation within natural populations and among breeding lines is crucial for effective conservation and exploitation of genetic resources for crop improvement programmes. Molecular markers have proven useful for assessment of genetic variation in germplasm collections (Hausmann et al. 2004; Maccaferri et al. 2006). Evaluation of germplam with GMMs might enhance the role of genetic markers by assaying the variation in transcribed and known function genes, although there may be a higher probability of bias owing to selection.

While using the genic SSR markers for diversity studies, the expansion and contraction of SSR repeats in genes of known function can be tested for association with phenotypic variation or, more desirably, biological function (Ayers et al. 1997). The presence of SSRs in the transcripts of genes suggests that they might have a role in gene expression or function; however, it is yet to be determined whether any unusual phenotypic variation might be associated with the length of SSRs in coding regions as was reported for several diseases in human (Cummings and Zoghbi 2000). Similarly, the use of SNP markers for diversity studies may correlate the SNPs of coding *vs*. non-coding regions of the gene with the trait variation. The variation associated with deleterious characters, however, is less likely to be represented in the germplasm collections of crop species than among natural populations because undesirable mutations are commonly culled from breeding populations (Cho et al. 2000).

Several studies involving GMMs, especially genic SSRs, have been found useful for estimating genetic relationship on one hand (see Gupta et al. 2003 Gupta and Rustgi 2004, Varshney et al. 2005a) while at the same time these have provided opportunities to examine functional diversity in relation to adaptive variation (Eujayl et al. 2001, Russell et al. 2004). It seems likely that with the development of more GMMs in major crop species, genetic diversity studies will become more meaningful by a shift in emphasis from the evaluation of anonymous diversity to functional genetic diversity in the near future. Nevertheless, use of the neutral RDM markers will remain useful in situations where: (i) GMMs would not be available, and (ii) to address some specific objectives e.g. neutral grouping of germplasm.

3.3. Interspecific or Intergeneric Transferability

Perhaps one of the most important features of the GMMs is that these markers provide high degree of transferability among distantly related species. In contrast, except RFLPs all other RDMs are generally constrained in this regard. Transferability of GMM markers to related species or genera has now been demonstrated in several studies (Table 2). For example, a computational study based on analysis

Table 2. Some examples of interspecific or intergeneric transferability of genic molecular markers

Plant species	Marker type	Species, recorded transferability	Reference
Cereals and grasses			
Barley (*Hordeum vulgare*)	EST-SSR, EST-SNP	Wheat, rice, rye	Thiel et al. 2003, Varshney et al. 2004, 2007b
Wheat (*Triticum aestivum*)	EST-SSR	*Aegilops* and *Triticum* species, barley, maize, rice, rye, oats, soybean, *Lophopyrum elongatum*	Holton et al. 2002, Gupta et al. 2003, Gao et al. 2003, Bandopadhyay et al. 2004, Yu et al. 2004, Mullan et al. 2005, Tang et al. 2006
Rice (*Oryza sativa*)	EST-SSR	wild species of rice	Cho et al. 2000
Sugarcane (*Saccharaum officinarum*)	EST-SSR	*Saccharum robustum*, *Erianthus* and Sorghum	Cordeiro et al. 2001
Sorghum (*Sorghum bicolor*)	EST-SSR	*Eleusine coracana*, *Seashore paspalum*, finger millet	Wang et al. 2005
Tall fescue (*Festuca*)	EST-SSR	subfamilies of Poaceae	Mian et al. 2005
Fiber and oilseed crops			
Cotton (*Gossypium hirsutum*)	EST-SSR	Cotton species	Saha et al. 2003
Sunflower (*Helianthus annus*)	EST-SSR	*Heliantus angustifolius*, *Helianthus verticillatus*	Pashley et al. 2006
Fruit and vegetables			
Strawberry (*Fragaria vesca*)	EST-SSR	*F. gracilis*, *F. iinumae*, *F. nilgerrensis*, *F. nipponica*	Bassil et al. 2006
Apricot (*Prunus armeniaca*)	EST-SSR	Vitaceae and Roseaceae family	Decroocq et al. 2003
Grape (*Vitis vinifera*)	EST-SSR	> 25 species from 5 Vitaceae and Roseaceae	Scott et al. 2000, Rossetto et al. 2002, Arnold et al. 2002, Decroocq et al. 2003
Tomato (*Solanum lycopersicum*)	EST-SSR	Solanaceous members	Frary et al. 2005
Ferns and trees			
Alpine lady-fern (*Atyrium distentifolium*)	EST-SSR	9 species from Woodsiaceae	Woodhead et al. 2003
Pinus (*Pinus taeda*)	EST-SSR	12 *Pinus* species	Komulainen et al. 2003, Changne et al. 2004, Liewlaksaneeyanawin et al. 2004
Spruce (*Picea glauca*)	EST-SSR	23 *Picea* species	Rungis et al. 2004
Citrus (*Citrus sinensis*)	EST-SSR	*Poncirus trifoliata*	Chen et al. 2006
Coffee (*Coffea arabica*, *Coffea canephora*)	EST-SSR	16 species of coffee and *Psilanthus*	Bhat et al. 2005, Poncet et al. 2006, Aggarwal et al. 2007

of ~1000 barley GMMs suggested a theoretical transferability of barley markers to wheat (95.2%), rice (70.3%), maize (69.3%), sorghum (65.9%), rye (38.1%) and even to dicot species (~16%). Infact, *in silico* analyses of GMMs of wheat, maize and sorghum with complete rice genome sequence data have provided a larger number of anchoring points among different cereal genomes as well as provided insights into cereal genome evolution (Sorrells et al. 2003, Salse et al. 2004).

In some studies, the use of GMMs of major crop species has been shown to enrich the genetic maps of related plant species for which little marker information is available. For example, barley EST-SSR as well as EST-SNP markers have been shown transferable as well as mappable in syntenic regions of rye (Varshney et al. 2004, 2005c, 2007a; Figure 2). Further, such kind of markers from the related plant species offers the possibility to develop anchor or conserved orthologous sets (COS) for genetic analysis and breeding in different species. In this direction, Rudd et al. (2005) identified a large repository of such COS markers and developed a database called "PlantMarker".

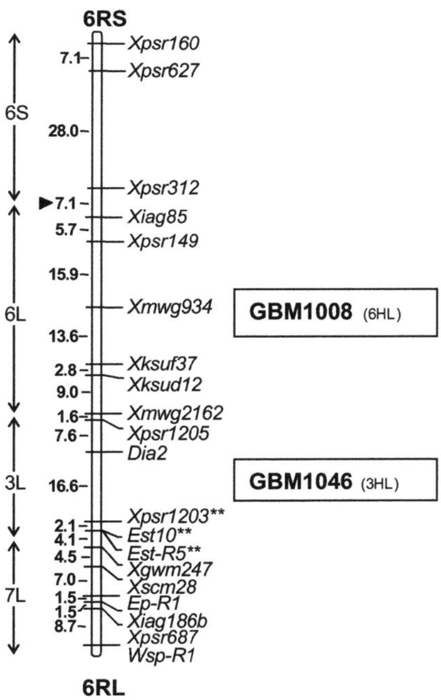

Figure 2. An example of integration of barley genic (EST-SSR) markers into syntenic regions of rye genetic map. Integrated barley markers (GBM1008, GBM1046) are shown in bold and capital font in boxes on right hand side. Details about other markers present on this linkage group are available in Korzun et al. (2001). Genetic distances are given in centimorgans (cM) on left hand side. The black triangle indicates the estimated centromere position. The relationship of the linkage group 6R in terms of Triticeae linkage group is shown on very left hand side (left to black triangle) as per Devos et al. (1993). Both barley genic markers from linkage group 3H and 6H are mapped into expected syntenic regions of the rye linkage group 6R. S = short arm, L = long arm

4. COMPARISON OF GMMs AND RDMs

Since the development of first molecular markers i.e. RFLPs in 1980 (Botstein et al. 1980), a diverse array of molecular marker technologies have come into being revolutionizing conventional plant breeding efforts for crop improvement. Significant strides have been made in crop improvement through conventional random molecular markers (RDMs). For instance, these molecular markers besides throwing light on organization, conservation and evolution of plant genomes, have also aided geneticists and plant breeders to tag genes, map QTLs for the traits of economic importance. Still, most of them are "anonymous" markers, that is to say their biological function is unknown. In comparison, a putative function for majority of the molecular markers, derived from the genes or ESTs, however can be deduced using some bioinformatics tools; such markers (GMMs) are commonly referred as functional markers (Varshney et al. 2005b). Although, in *stricto* sense, the functional markers are based on functionally defined genes underlying specific biochemical or physiological functions and therefore the FMs can be considered as a class of GMMs (Anderson and Luebberstedt 2003).

The GMMs, like RDMs, could detect both length and sequence polymorphisms in expressed regions of the genome but provide relatively stronger and robust marker assays. However, as compared to the RDMs the developmental costs of GMMs, depend on which specific class of GMMs is to be developed. Similarly the applied value of the GMMs as compared to the RDMs varies depending on the class of the GMMs. These relative costs and applications issues have been detailed in Table 3. In summary, if the GMMs based on the polymorphic site and verification are developed (i.e. FMs), these markers are superior to RDMs for using them as diagnostic tools in marker-assisted selection as they may owe the complete linkage with the trait locus alleles (Anderson and Luebberstedt 2003). In plant breeding, the GMMs are superior to RDMs for selection of, e.g., parent materials to build segregating populations, as well as subsequent selection of lines (line breeding) or inbreds (hybrid breeding). Depending on the mode of the GMM characterization, these can also be applied to the targeted combination of alleles in hybrid and synthetic breeding. In population breeding and recurrent selection programs, the GMMs can be employed to avoid genetic drift at characterized loci.

Being originated from the conserved proportion of the genome, the GMMs, as compared to the RDMs, are the candidate markers for interspecific/intergeneric transferability and comparative mapping/genomics studies in related plant species. Since the GMMs represent the expressed portion of the genome, they sample the variation in transcribed regions of the genome, and provide a more direct estimate of functional diversity while screening the markers on the germplasm adapted to different environments. Nevertheless, the GMMs, as compared to the RDMs are less polymorphic and provide less alleles and lower PIC values. Additionally, due to biased distribution in the genome, the GMMs are unsuitable for analyzing population structure.

Table 3. Comparison of genic molecular markers (GMMs) with random DNA markers (RDMs)

Feature	GMMs	RDMs		
		gSSRs, SNPs	RFLPs	RAPD/AFLP/ ISSR etc.
Need for sequence data	Genes/ESTs data Essential	Essential	Not required	Not required
Costs of generation	Low*	High	High	Low-moderate
Labour involved	Less	Much	Much	Less
Level of polymorphism	Low	High	Low	Low-moderate
Interspecific transferability and comparative mapping	High	Low-moderate	Moderate-High	Low-moderate
Function of markers	Known majority of times	Unknown majority of times	Unknown	Unknown
Utility in marker-assisted selection	Great, if the marker is derived from the gene, involved in expression of trait	High	Moderate	Low-moderate

*generally GMMs are by products of the available transcriptome resources being developed for functional genomic studies.

5. FUTURE DIRECTIONS OF GENIC MOLECULAR MARKERS

It is clear that the GMMs and especially the FMs are extremely useful source of markers in plant breeding for marker-assisted selection because these markers may represent the genes responsible for expression of target traits. If so, there will not be any recombination between the markers and the trait, thus representing perfect indirect selection tools. While low level of polymorphism is an inherent feature of the GMMs, it is compensated by their higher interspecific transferability as well as capacity to sample the functional diversity in the germplasm. These features make the development and application of the GMMs more attractive for plant breeding and genetics.

With more DNA sequence data being generated continuously, the trend is towards cross-referencing genes and genomes using sequence and map-based tools. Because polymorphism is a major limitation for many species, SSR- and SNP-based GMMs will be valuable tools for plant geneticists and breeders. In the longer term, development of allele-specific, functional markers (FMs) for the genes controlling agronomic traits will be important for advancing the science of plant breeding. In this context genic SSR and SNP markers together with other types of markers that target functional polymorphisms within genes will be developed in near future for major crop species. The choice of the most appropriate marker system, however, needs to be decided on a case-by-case basis and will depend on many issues including the availability of technology platforms, costs for marker development, species transferability, information content and ease of documentation.

REFERENCES

Aggarwal RK, Hendre PS, Varshney RK, Bhat PR, Krishna KV, Singh L (2007) Identification, characterization and utilization of EST-derived genic microsatellite markers for genome analyses of coffee and related species. Theor Appl Genet 114:359–372

Andersen JR, Lubberstedt T (2003) Functional markers in plants. Trends Plant Sci 8:554–560

Arnold C, Rossetto M, McNally J, Henry RJ (2002) The application of SSRs characterized for grape (*Vitis vinifera*) to conservation studies in *Vitaceae*. Am J Bot 89:22–28

Ayers NM, McClung AM, Larkin PD, Bligh HFJ, Jones CA, Park WD (1997) Microsatellites and a single nucleotide polymorphism differentiate apparent amylose classes in an extended pedigree of US rice germplasm. Theor Appl Genet 94:773–781

Bandopadhyay R, Sharma S, Rustgi S, Singh R, Kumar A, Balyan HS, Gupta PK (2004) DNA polymorphism among 18 species of *Triticum–Aegilops* complex using wheat EST-SSRs. Plant Sci 166:349–356

Barrett B, Griffiths A, Schreiber M, Ellison N, Mercer C, Bouton J, Ong B, Forster J, Sawbridge T, Spangenberg G, Bryan G, Woodfield D (2004) A microsatellite map of white clover. Theor Appl Genet 109:596–608

Bassil NV, Njuguna W, Slovin JP, (2006) EST-SSR markers from *Fragaria vesca* L. cv. yellow wonder. Mol Ecol Notes 6:806–809

Bhat PR, Kumar KV, Hendre PS, Kumar RP, Varshney RK, Aggarwal RK (2005) Identification and characterization of expressed sequence tags-derived simple sequence repeats, markers from robusta coffee variety 'C × R' (an interspecific hybrid of *Coffea canephora* × *Coffea congensis*). Mol Ecol Notes 5:80–83

Botstein D, White RI, Skolnick M, Davis RW (1980) Construction of a genetic linkage map in man using restriction fragment length polymorphisms. Am J Hum Genet 32:314–331

Bowers JE, Abbey C, Anderson S, Chang C, Draye X, Hoppe AH, Jessup R, Lemke C, Lennington J, Li Z (2003) A high-density genetic recombination map of sequence-tagged sites for sorghum, as a framework for comparative structural and evolutionary genomics of tropical grains and grasses. Genetics 165:367–386

Cato SA, Gardner RC, Kent J, Richardson TE (2001) A rapid PCR based method for genetically mapping ESTs. Theor Appl Genet 102:296–306

Causse MA, Fulton TM, Cho YG, Ahn SN, Chunwongse J, Wu K, Xiao J, Yu Z, Ronald PC, Harrington SE, Second G, McCouch SR, Tanksley SD (1994) Saturated molecular map of the rice genome based on an interspecific backcross population. Genetics 138:1251–1274

Chagne D, Chaumeil P, Ramboer A, Collada C, Guevara A, Cervera MT, Vendramin GG, Garcia V, Frigerio JMM, Echt C, Richardson T, Plomion C (2004) Cross-species transferability and mapping of genomic and cDNA SSRs in pines. Theor Appl Genet 109:1204–1214

Chao S, Baysdorfer C, Heredia-Diaz O, Musket T, Xu G, Coe Jr. EH (1994) RFLP mapping of partially sequenced leaf cDNA clones in maize. Theor Appl Genet 88:717–721

Chee PW, Rong JK, Williams-Coplin D, Schulze SR, Paterson AH (2004) EST derived PCR-based markers homologues in cotton. Genome 47:449–462

Chen C, Zhou P, YA, Huang S, Gmitter Jr. FG (2006) Mining and characterizing microsatellites from citrus ESTs. Theor Appl Genet 112:1248–1257

Childs KL, Klein RR, Klein PE, Morishige DT, Mullet JE (2001) Mapping genes on an integrated sorghum genetic and physical map using cDNA selection technology. Plant J 27:243–256

Cho YG, Ishii T, Temnykh S, Chen X, Lipovich L, McCouch SR, Parl WD, Ayres N, Cartinhour S (2000) Diversity of microsatellites derived from genomic libraries and GenBank sequences in rice (*Oryza sativa* L.). Theor Appl Genet 100:713–722

Cordeiro GM, Casu R, McIntyre L, Manners JM, Henry RJ (2001) Microsatellite markers from sugarcane (*Saccharum* spp.) ESTs cross-transferable to *Erianthus* and *Sorghum*. Plant Sci 160:1115–1123

Cummings CJ, Zoghbi HY (2000) Fourteen and counting: unraveling trinucleotide repeat diseases. Hum Mol Genet 9:909–916

Decroocq V, Fave MG, Hagen L, Bordenave L, Decroocq S (2003) Development and transferability of apricot and grape EST microsatellite markers across taxa. Theor Appl Genet 106:912–922

Devos KM, Atkinson MD, Chinoy CN, Francis HA, Harcourt RL, Koebner RMD, Liu CJ, Masojc P, Xie DX, Gale MD (1993) Chromosomal rearrangement in the rye genome relative to that of wheat. Theor Appl Genet 85:673–680

Eujayl I, Sorrells M, Baum M, Wolters P, Powell W (2001) Assessment of genotypic variation among cultivated durum wheat based on EST-SSRs and genomic SSRs. Euphytica 119:39–43

Falque M, Décousset L, Dervins D, Jacob AM, Joets J, Martinant JP, Raffoux X, Ribière N, Ridel C, Samson D, Charcosset A, Murigneux A (2005) Linkage mapping of 1454 new maize candidate gene l sequences or sites with in genes loci. Genetics 170:1957–1966

Faville MJ, Vecchies AC, Schreiber M, Drayton MC, Hughes LJ, Jones ES, Guthridge KM, Smith KF, Sawbridge T, Spangenberg GC, Bryan GT, Forster W (2004) Functionally associated molecular genetic marker map construction in perennial ryegrass (*Lolium perenne* L.). Theor Appl Genet 110:12–32

Feltus FA, Wan J, Schulze SR, Estill JC, Jiang N, Paterson (2004) An SNP resource for rice genetics and breeding based on subspecies *Indica* and *Japonica* genome alignments. Genetics 148:479–494

Frary A, Xu Y, Liu J, Tedeschi SME, Tanksley S (2005) Development of a set of PCR-based anchor markers encompassing the tomato genome and evaluation of their usefulness for genetics and breeding experiments. Theor Appl Genet 111:291–312

Fraser LG, Harvey CF, Crowhurst RN, De Silva HN (2004) EST-derived microstellites from *Actinidia* species and their potential for mapping. Theor Appl Genet 108:1010–1016

Gao L, Tang J, Li H, Jia J (2003) Analysis of microsatellites in major crops assessed by computational and experimental approaches. Mol Breed 12:245–261

Gao LF, Jing RL, Huo NX, Li Y, Li XP, Zhou RH, Chang XP, Tang JF, Ma ZY, Jia JZ (2004) One hundred and one new microsatellite loci derived from ESTs (EST-SSRs) in bread wheat. Theor Appl Genet 108:1392–1400

Gardiner J, Coe MEH, Melia-Hancock S, Hoisington DA, Chao S, (1993) Development of a core RFLP map in maize using an immortalized F_2 population. Genetics 134:917–930

Graham J, Smith K, Mac Kenzie K, Jorgenson L, Hackett C, Powell W (2004) The construction of a genetic linkage map of red raspberry (*Rubus idaeus* subsp. *idaeus*) based on AFLPs, genomic-SSR and EST-SSR markers. Theor Appl Genet 109:740–749

Graner A, Jahoor A, Schondelmaier J, Siedler H, Pillen K, Fischbeck G, Wenzel G, Herrmann RG (1991) Construction of an RFLP map of barley. Theor Appl Genet 83:250–256

Gupta PK, Rustgi S (2004) Molecular markers from the transcribed/expressed region of the genome in higher plants. Funct Integr Genomics 4:139–162

Gupta PK, Rustgi S, Sharma S, Singh R, Kumar N, Balyan HS (2003) Transferable EST-SSR markers for the study of polymorphism and genetic diversity in bread wheat. Mol Genet Genomics 270:315–323

Gupta PK, Varshney RK (2000) The development and use of microsatellite markers for genetics and plant breeding with emphasis on bread wheat. Euphytica 113:163–185

Gupta PK, Varshney RK (2004) Cereal genomics: an overview In: Gupta PK, Varshney RK (eds) Cereal genomics. Kluwer Academic Press, Dordrecht, The Netherlands, p 639

Hackauf B, Wehling P (2002) Identification of microsatellite polymorphisms in an expressed portion of the rye genome. Plant Breed 121:17–25

Harushima Y, Yano M, Shomura A, Sato M, Shimano T, Kuboki Y, Yamamoto T, Lin SY, Antonio BA, Parco A (1998) A high-density rice genetic linkage map with 2275 Markers using a single F_2 population. Genetics 148:479–494

Haussmann BI, Hess DE, Omanya GO, Folkertsma RT, Reddy BV, Kayentao M, Welz HG, Geiger HH, (2004) Genomic regions influencing resistance to the parasitic weed *Striga hermonthica* in two recombinant inbred populations of sorghum. Theor Appl Genet 109:1005–1016

Holton TA, Christopher JT, McClure L, Harker N, Henry RJ (2002) Identification and mapping of polymorphic SSR markers from expressed gene sequences of barley and wheat. Mol Breed 9:63–71

Jain SM, Brar DS, Ahloowalia BS (2002) Molecular techniques in crop improvement. Kluwer Academic Publishers, The Netherlands

Jayashree B, Ramu P, Prasad P, Bantte K, Hash CT, Chandra S, Hoisington DA, Varshney RK (2006) A database of simple sequence repeats from cereal and legume expressed sequence tags mined in silico: survey and evaluation. In Silico Biol 6:0054

Khlestkina EK, Than MHM, Pestsova EG, Roder MS, Malyshev SV, Korzun V, Borner A (2004) Mapping of 99 new microsatellite-derived loci in rye (*Secale cereale* L.) including 39 expressed sequence tags. Theor Appl Genet 109:725–732

Klein PE, Klein RR, Vrebalov J, Mullet JE (2003) Sequence-based alignment of sorghum chromosome 3 and rice chromosome 1 reveals extensive conservation of gene order and one major chromosomal rearrangement. Plant J 34:605–622

Koebner RMD (2004) Marker-assisted selection in the cereals: the dream and the reality. In: Gupta PK, Varshney RK (eds) Cereal genomics. Kluwer Academic Publishers, Netherlands, p 199

Komulainen P, Brown GR, Mikkonen M, Karhu A, Garcia-Gil MR, Malley DO, Lee B, Neale DB, Savolainen O (2003) Comparing EST-based genetic maps between *Pinus sylvestris* and *Pinus taeda*. Theor Appl Genet 107:667–678

Korzun V (2002) Use of molecular markers in cereal breeding. Cell Mol Biol Lett 7:811–820

Korzun V, Malyshev S, Voylokov AV, Börner A (2001) A genetic map of rye (*Secale cereale* L.) combining RFLP, isozyme, protein, microsatellite and gene loci. Theor Appl Genet 102:709–717

Lai Z, Livingstone K, Zou Y, Church SA, Knapp SJ, Andrews J, Rieseberg LH (2005) Identification and mapping of SNPs from ESTs in sunflower. Theor Appl Genet 111:1532–1544

Liewlaksaneeyanawin C, Ritland CE, El-Kassaby YA, Ritland K (2004) Single-copy, species-transferable microsatellite markers developed from loblolly pine ESTs. Theor Appl Genet 109:361–369

Maccaferri M, Sanguineti MC, Natoli E, Araus-Ortega JL, Ben Salem M, Bort J, Chenenaoui S, Deambrogio E, Garcia DML, De Montis A et al (2006) A panel of elite accessions of durum wheat (*Triticum durum* Desf.) Suitable for association mapping studies. Plant Genet Res 4:79–85

Mian MAR, Saha MC, Hopkins AA, Wang ZY (2005) Use of tall fescue EST-SSR markers in phylogenetic analysis of cool-season forage grasses. Genome 48:637–647

Michalek W, Kunzel G, Graner A (1999) Sequence analysis and gene identification in a set of mapped RFLP markers in barley (*Hordeum vulgare*). Genome 42:849–853

Miller RT, Christoffels AG, Gopalakrishnan C, Burke J, Ptitsyn AA, Broveak TR, Hide WA (1999) A comprehensive approach to clustering of expressed human gene Sequence: the sequence tag alignment and consensus knowledge base. Genome Res 9:1143–1155

Mullan DJ, Platteter A, Teakle NL, Appels R, Colmer TD, Anderson J, Francki M (2005) EST-derived SSR markers from defined regions of the wheat genome to identify *Lophopyrum elongatum* specific loci. Genome 48:811–822

Nicot N, Chiquet V, Gandon B, Amilhat L, Legeai F, Leroy P, Bernard M, Sourdille P (2004) Study of simple sequence repeat (SSR) markers from wheat expressed sequence tags (ESTs). Theor Appl Genet 109:800–805

Park YH, Alabady MS, Ulloa M, Sickler B, Wilkins TA, Yu J, Stelly DM, Kohel RJ, El-Shihy OM, Cantrell RG (2005) Genetic mapping of new cotton fiber loci using EST-derived microsatellites in an interspecific recombinant inbred line cotton population. Mol Genet Genomics 274:428–441

Pashley CH, Ellis JR, Mccauley DE, Burke JM (2006) EST Databases as a source for molecular markers: lessons from helianthus. J Hered 97:381–388

Phillips RL, Vasil IK (2001) DNA-based markers in plants. In: Phillips RL, Vasil IK (eds) DNA-based markers in plants. Kluwer Academic Publishers, Dordrecht, The Netherlands, p 497

Picoult-Newberg L, Ideker TE, Pohl MG, Taylor SL, Donaldson MA, Nickerson DA, Boyce-Jacino M (1999) Mining SNPs from EST databases. Genome Res 9:167–174

Poncet V, Rondeau M, Tranchant C, Cayrel A, Hamon S, de Kochko A, Hamon P (2006) SSR mining in coffee tree EST databases: potential use of EST–SSRs as markers for the *Coffea* genus. Mol Genet Genomics 276:436–449

Qi L-L, Echalier B, Chao S, Lazo GR, Butler GE, Anderson OD, Akhunov ED, Dvorak J, Linkiewicz AM, Ratnasiri et al (2004) A chromosome bin map of 16,000 expressed sequence tag loci and distribution of genes among the three genomes of polyploid wheat. Genetics 168:701–712

Qi X, Pittaway TS, Lindup TS, Liu S, Liu H, Waterman E, Padi FK, Hash CT, Zhu J, Gale, MD, Devos KM (2004) An integrated genetic map and a new set of simple sequence repeat markers for pearl millet, *Pennisetum glaucum*. Theor Appl Genet 109:1485–1493

Rafalski JA, Tingey SV (1993) Genetic diagnostics in plant breeding: RAPDs, microsatellites and machines. Trends Genet 9:275–280

Ramu P, Bantte K, Ashok CK, Jayashree B, Rolf TF, Senthilvel S, Mahalakshmi V, Reddy AL, Fakrudin B, Hash CT (2006) Development and mapping of EST-SSR markers in sorghum for comparative mapping with rice. In: Plant and animal genome XIV conference, San Diego, CA, USA, P 200 (http://www.intl-pag.org/14/abstracts/PAG14_P200.html)

Rossetto M, McNally J, Henry RJ (2002) Evaluating the potential of SSR flanking regions for examining taxonomic relationships in the *Vitaceae*. Theor Appl Genet 104:61–66

Rostoks N, Mudie S, Cardle L, Russell J, Ramsay L, Booth A, Svensson J, Wanamaker S, Walia H, Rodriguez E, Hedley P, Liu H, Morris J, Close T, Marshall D, Waugh R (2005) Genome-wide SNP discovery and linkage analysis in barley based on genes responsive to abiotic stress. Mol Genet Genomics 274:515–527

Rudd S, Mewes HW, Mayer KF (2003) Sputnik: a database platform for comparative plant genomics. Nucleic Acids Res 31:128–132

Rudd S, Schoof H, Klaus M (2005) PlantMarkers-A database of predicted molecular markers from plants. Nucleic Acids Res 33:D628–D632

Rungis D, Berube Y, Zhang J, Ralph S, Ritland CE, Ellis BE, Douglas C, Bohlmann J, Ritland K (2004) Robust simple sequence repeat markers for spruce (*Picea* spp.) from expressed sequence tags. Theor Appl Genet 109:1283–1294

Russell J, Booth A, Fuller J, Harrower B, Hedley P (2004) A comparison of sequence-based polymorphism and haplotype content in transcribed and anonymous regions of the barley genome. Genome 47:389–398

Saha S, Karaca M, Jenkins JN, Zipf AE, Reddy OUK, Kantety RV (2003) Simple sequence repeats as useful resources to study transcribed genes of cotton. Euphytica 130:355–364

Salse J, Piegu B, Cooke R, Delseny M (2004) New *in silico* insight into the synteny between rice (*Oryza sativa* L.) and maize (*Zea mays* L.) Highlights reshuffling and identifies new duplications in the rice genome. Plant J 38:396–409

Sargent DJ, Clarke J, Simpson DW, Tobutt KR, Aru´s P, Monfort A, Vilanova S, Denoyes-Rothan B, Rousseau M, Folta KM, Bassil NV, Battey NH (2006) An enhanced microsatellite map of diploid *Fragaria*. Theor Appl Genet 112:1349–1359

Schmitt BA, Costa JH, Dirce Fernandes de Melo (2006) AOX –A functional marker for efficient cell reprogramming under stress? Trends Plant Sci 11:281–287

Scott KD, Eggler P, Seaton G, Rossetto M, Ablett EM, Lee LS, Henry RJ (2000) Analysis of SSRs derived from grape ESTs. Theor Appl Genet 100:723–726

Shan X, Blake TK, Talbert LE (1999) Conversion of AFLP markers to sequence-specific PCR markers in barley and wheat. Theor Appl Genet 98:1072–1078

Shariflou MR, Hassani ME, Sharp PJ (2001) A PCR-based DNA marker for detection of mutant and normal alleles of the *Wx-D1* gene of wheat. Plant Breed 120:121–124

Somers DJ, Robert K, Moniwa M, Walsh A (2003) Mining single-nucleotide polymorphisms from hexaploid wheat ESTs. Genome 46:431–437

Song QJ, Marek LF, Shoemaker RC, Lark KG, Concibido VC, Delannay X, Specht JE, Cregan PB (2004) A new integrated genetic linkage map of the soybean. Theor Appl Genet 109: 122–128

Sorrells ME, Rota ML, Bermudez-Kandianis CE, Greene RA, Kantety R, Munkvold JD, Miftahudin, Mahmoud A, Ma X, Gustafson PJ, Qi LL, Echalier B, Gill BS, Matthews DE, Lazo GR, Chao S, Anderson OD, Edwards H, Linkiewicz AM, Dubcovsky J, Akhunov ED, Dvorak J, Zhang D, Nguyen HT, Peng J, Lapitan NLV, Gonzalez-Hernandez JL, Anderson JA, Hossain K, Kalavacharla V, Kianian SF, Dong-Woog C, Close TJ, Dilbirligi M, Gill KS, Steber C, Walker-Simmons MK, McGuire PE, Qualset CO (2003) Comparative DNA sequence analysis of wheat and rice genomes. Genome Res 13:1818–1827

Sreenivasulu N, Kavi Kishor PB, Varshney RK, Altschmied L (2002) Mining functional information from cereal genomes – the utility of expressed sequence tags (ESTs). Curr Sci 83:965–973

Stein N, Prasad M, Scholz U, Thiel T, Zhang H, Wolf M, Kota R, Varshney R, Perovic D, Grosse I, Graner A (2007) A 1,000-loci transcript map of the barley genome: new anchoring points for integrative grass genomics. Theor Appl Genet 114:823–839

Tang J, Gao L, Cao Y, Jia J (2006) Homologous analysis of SSR-ESTs and transferability of wheat SSR-EST markers across barley, rice and maize. Euphytica 151:87–93

Temnykh S, DeClerck G, Lukashova A, Lipovich L, Cartinhour S, McCouch S (2001) Computational and experimental analysis of microsatellites in rice (*Oryza sativa* L.): frequency, length variation, transposon associations, and genetic marker potential. Genome 14:1812–1819

Thiel T, Michalek W, Varshney RK, Graner A (2003) Exploiting EST databases for the development and characterization of gene-derived SSR-markers in barley (*Hordeum vulgare* L.). Theor Appl Genet 106:411–422

Varshney RK, Beier U, Khlestkina EK, Kota R, Korzun V, Graner A, Börner A (2007a) Single nucleotide polymorphisms in rye (*Secale cereale* L.): discovery, frequency and applications for genome mapping and diversity studies. Theor Appl Genet 114:1105–1116 (doi 10.1007/s00122–007–0504–7)

Varshney RK, Graner A, Sorrells ME (2005a) Genic microsatellite markers in plants: features and applications. Trends Biotechnol 23:48–55

Varshney RK, Graner A, Sorrells ME (2005b) Genomic-assisted breeding for crop improvement. Trends Plant Sci 10:621–630

Varshney RK, Grosse I, Hahnel U, Siefken R, Prasad M, Stein N, Langridge P, Altschmied L, Graner A (2006) Genetic mapping and BAC assignment of EST-derived SSR markers shows non-uniform distribution of genes in the barley genome. Theor Appl Genet 113:239–250

Varshney RK, Korzun V, Börner A (2004) Molecular maps in cereals: Methodology and progress. In: Gupta PK, Varshney RK (eds) Cereal genomics. Kluwer Academic Publishers, The Netherlands, p 35

Varshney RK, Marcel TC, Ramsay L, Russell J, Röder M, Stein N, Waugh R, Langridge P, Niks RE, Graner A (2007b) A high density barley microsatellite consensus map with 775 SSR loci. Theor Appl Genet (DOI 10.1007/S00122-007-0503-7)

Varshney RK, Sigmund R, Borner A, Korzun V, Stein N, Sorrells ME, Langridge P, Graner A (2005c) Interspecific transferability and comparative mapping of barley EST-SSR markers in wheat, rye and rice. Plant Sci 168:195–202

Wang HY, Liu DC, Yan ZH, Wei YM, Zheng YL (2005) Cytological characteristics of hybrid F2 population between *Triticum aestivum* L. and *T. durum* with reference to wheat breeding. J Appl Genet 46:365–369

Willsmore KL, Eckermann P, Varshney RK, Graner A, Langridge P, Pallotta M, Cheong J, Williams KJ (2006) New eSSR and gSSR markers added to Australian barley maps. Aust J Agric Res 57:953–959

Woodhead M, Russell J, Squirrell J, Hollingsworth PM, Cardle L, Ramsay L, Gibby M, Powell W (2003) Development of EST-SSRs from the alpine lady-fern, *Athyrium distentifolium*. Mol Eco Notes 3:287–290

Yu JK, La Rota M, Kantety RV, Sorrells ME (2004) EST derived SSR markers for comparative mapping in wheat and rice. Mol Genet Genomics 271:742–751

Zhang LY, Bernard M, Leroy P, Feuillet C, Sourdille P (2005) High transferability of bread wheat EST-derived SSRs to other cereals. Theor Appl Genet 111:677–687

Zhang WK, Wang YJ, Luo GZ, Zhang JS, He CY, Wu XL, Gai JY, Chen SY (2004) QTL mapping of ten agronomic traits on the soybean (*Glycine max* L. Merr.) genetic map and their association with EST markers. Theor Appl Genet 108:1131–1139

CHAPTER 3

MOLECULAR BREEDING: MAXIMIZING THE EXPLOITATION OF GENETIC DIVERSITY

ANKER P. SØRENSEN[1,*], JEROEN STUURMAN[1], JEROEN ROUPPE VAN DER VOORT[2] AND JOHAN PELEMAN[3]

[1]*Keygene N.V., P.O. Box 216, 6700 AE, Wageningen, The Netherlands*
[2]*Enza Zaden B.V., P.O. Box 7, 1600 AA Enkhuizen, The Netherlands*
[3]*Nunhems B.V., P.O. Box 4005, 6080 AA Haelen, The Netherlands*

Abstract: The use of molecular markers is gradually expanding from the field of scientific genetic analysis towards the implementation and application in breeding programs. Applications of DNA markers in breeding are based on the knowledge of the relation between genotypic and phenotypic variation. This overview of the field of molecular breeding describes current and future methods for establishing these relations through the combined use of modern DNA technologies and the laws of inheritance. The modern molecular breeder has the opportunity to control an increasing amount of traits in the breeding process through efficient application of DNA markers. Traits with different level of complexity require different approaches for discovery and molecular control. These approaches include control of genotypes and traits, at the level of linked markers, haplotypes, genes and gene alleles. In order to fully exploit the potential of molecular breeding as well as the potential of available germplasm resources, the selection methods in breeding will have to be adapted, towards the integrated use of genetic knowledge based on DNA markers.

1. INTRODUCTION

The purpose of plant breeding has always been to adapt the growth and production of the plant to the needs of man. The earliest advances to meet this purpose are now called crop domestication and have for cereals and pulses taken place ca. 10.000 years ago (Lev-Yadun et al. 2000). The first selections in natural populations were

*Authors' email: anker.sorensen@keygene.com

R.K. Varshney and R. Tuberosa (eds.), Genomics-Assisted Crop Improvement: Vol. 1: Genomics Approaches and Platforms, 31–56.

aimed at preventing seed dispersal at maturity, leading to the selection of the non-brittle spikes in cereals. As a result of the domestication, many crops of today must rely on human intervention for their reproduction.

After the discovery of the Mendelian principles of heredity, the methodologies of modern plant breeding through directed crosses and selection for the desired recombinants have been developed, leading to an enormous and steady increase of crop productivity and quality. The progress in plant breeding is entirely based on the availability of genetic variation. In principle the task of the modern plant breeders is to exploit this genetic variability. In conventional breeding schemes, the genetic variation of breeding populations is estimated (and selected) by measuring the phenotypic performance only. Even though this process has proven to be effective, it is commonly accepted that selection directly at the genotype level, would greatly increase the efficiency of the breeding efforts. This is due to the environmental influence on the phenotypic measurements, resulting in a biased measure of the true genetic potential of an individual. Prerequisite for the use of selection based on genotype is that the relative value of the different genotypes is well known and predictable.

After the discovery of the DNA molecule as the carrier of genetic and heritable information (Watson & Crick 1953) the possibility of factually describing the genotype of individuals, and thus using this information through selection, became feasible. The first molecular technique to address this challenge in plants, the RFLP technique, was reviewed (Tanksley et al. 1989). However it was only after the development of PCR based molecular marker technologies such as RAPD (Williams J. et al. 1990) and AFLP (Vos et al. 1995), that the technologies could be applied at an acceptable cost for marker applications in plant breeding. The challenge since then has been to develop methodologies needed to discover molecular markers, which can link the genotypic scores to phenotypic performance in a repeatable, robust and affordable manner (Young 1999). This challenge has been a major focus of Keygene over the last 16 years.

The following chapters will summarize the knowledge and experience that has been generated through many different projects using DNA markers. The last chapter, will present our vision on how the molecular plant breeder of the future will integrate the use of technologies en methodologies with the ultimate aim of maximizing the exploitation of genetic variation in plants.

2. MOLECULAR MARKERS FOR THE CONTROL OF BREEDING OBJECTIVES

DNA markers have greatly enhanced the ability of the modern plant breeder to efficiently meet different selection objectives in the breeding program. DNA markers are highly reliable selection tools as they are abundant, stable, not influenced by environmental conditions and relatively easy to score in an experienced laboratory. Compared to phenotypic assays, DNA markers offer great advantages to accelerate the variety development time as a result of:

1. increased reliability: the outcome of phenotypic assays is affected, among others, by environmental factors, the heritability of the trait, the number of genes involved, the magnitude of their effects and the way these loci interact. Hence, error margins on the measurement of phenotypes tend to be significantly larger than those of genotyping scores based on DNA markers.
2. increased efficiency: DNA markers can be scored at the seedling stage. This is especially advantageous when selecting for traits which are expressed only at later stages of development, such as flower, fruit and seed characteristics. By selecting at the seedling stage, considerable amounts of time and space can be saved.
3. reducing cost: there are ample traits where the determination of the phenotype costs more than a PCR assay. In high throughput setting, the total cost for a PCR assay will typically not exceed 1 Euro. In comparison, the growth of a tomato or pepper plant to full maturity in a heated greenhouse will cost approximately 20 Euro. Every plant that can be rejected before planting will in such settings save a considerable amount of money.

Before deciding to follow DNA marker assisted approaches, practical concerns and cost-benefit analysis need to be addressed. Leaders of breeding programs must address a multifaceted evaluation of DNA marker-assisted approaches before committing to such endeavors.

The most straightforward applications of DNA markers in breeding programs are; genetic distance analysis, variety identification and purity control of seed lots and marker assisted backcrossing. These applications (currently) make use of what can be called genome wide polymorphisms. We define genome wide polymorphisms as molecular markers detectable between two defined genotypes (lines), but without knowing any linkage of the markers to traits or genes. Using this definition, genome wide polymorphisms only have a relative value in relation to the score of the marker in other individuals. Nevertheless such markers, when used in multiplex, have proven to be applicable for a number of objectives aimed at controlling the genome constitution of lines in plant breeding as described below.

A different group of applications of DNA markers in plant breeding is the identification and use of markers tightly linked to specific genes, monogenic traits and QTL. These markers can be used directly for indirect selection purposes. The different applications are discussed in more detail in the following paragraphs.

2.1. Controlling Genomes

2.1.1. Variety identification and seed purity analysis

Genotyping using (genome wide DNA markers can be considered as the most reliable method for the identification of fixed lines and varieties. Therefore DNA fingerprinting methods have been implemented in many breeding companies to analyze the purity of seed lots of inbred lines as well as hybrid seed lots Roldán-Ruiz et al. 2000). The immediate benefit of using molecular markers for this purpose is the fact that a marker score on a pooled seed-lot sample is much cheaper then a grow-out and visual inspection of the seedlings. The largest benefit of using markers for seed-lot purity assessment however is the fact that the assay is fast, giving the opportunity to move seed-lots to commercial valorization much quicker and saving storing costs of the expensively produced seeds. Any multiplex DNA fingerprinting technique (RAPD, AFLP®, SSR) can be used for this purpose, as long as sufficient polymorphisms can be sampled at low cost to unambiguously identify varieties or determine hybrid purity.

2.1.2. Genetic distance analysis

A general measure of the genetic diversity and inter-relatedness of a germplasm sample can most easily be obtained by performing a genetic distance analysis. In principle all individuals of the germplasm samples are fingerprinted, with preferably a multiplex fingerprinting technique. Subsequently a pair-wise distance matrix is calculated for each pair of individuals, followed by a clustering algorithm or a Principal Component Analysis (PCA). The result is a grouping of individuals within the germplasm according to their inter-relatedness (Figure 1). Different methods of calculating and interpreting similarity matrices have been reviewed (Reif et al. 2005). Genetic distance analysis is very powerful in assorting unknown genotypes to groups within the germplasm, such as heterotic groups in maize (Lübberstedt et al. 2000) without knowledge of the pedigree of the individuals. In addition the pair-wise similarities are being used for designing crosses aimed at maximizing the genetic segregation, either for use in optimizing the efficiency of genetic mapping populations or for optimizing the expected segregation of traits. The use of genetic distances based on molecular marker scores to predict heterosis has been investigated (Syed et al. 2004; Vuylsteke et al. 2000). A general verdict on the usefulness of genome wide markers for this purpose is still open, but it seems that an intelligent filter on the polymorphism sampling will need to be applied in order to reliably correlate marker diversity to hybrid performance. In that case the term genome wide markers however would no longer apply.

2.1.3. Marker assisted backcross breeding

Marker Assisted Backcross (MABC) breeding has now become a standard application in modern plant breeding and the optimization of MABC strategies was already reviewed (Frisch et al. 1999; Reyes-Vadez 2000). The reason for the success of this strategy is two-fold. Firstly, the time required for conversion of a Recurrent Line into it's Near Isogenic Line through MABC can be reduced from 6 to 3 generations. Such a significant reduction in time to market can significantly influence the success of a new variety. Secondly, a good performing variety represents a very valuable fixed combination of alleles and keeping this combination intact, when adding simply inherited traits, is very attractive for many variety improvement programs.

Two different aspects can be distinguished in MABC breeding:

(i) selection for recurrent parent genomic content: In this application, the DNA fingerprints are used to calculate the % recurrent parent genome in each

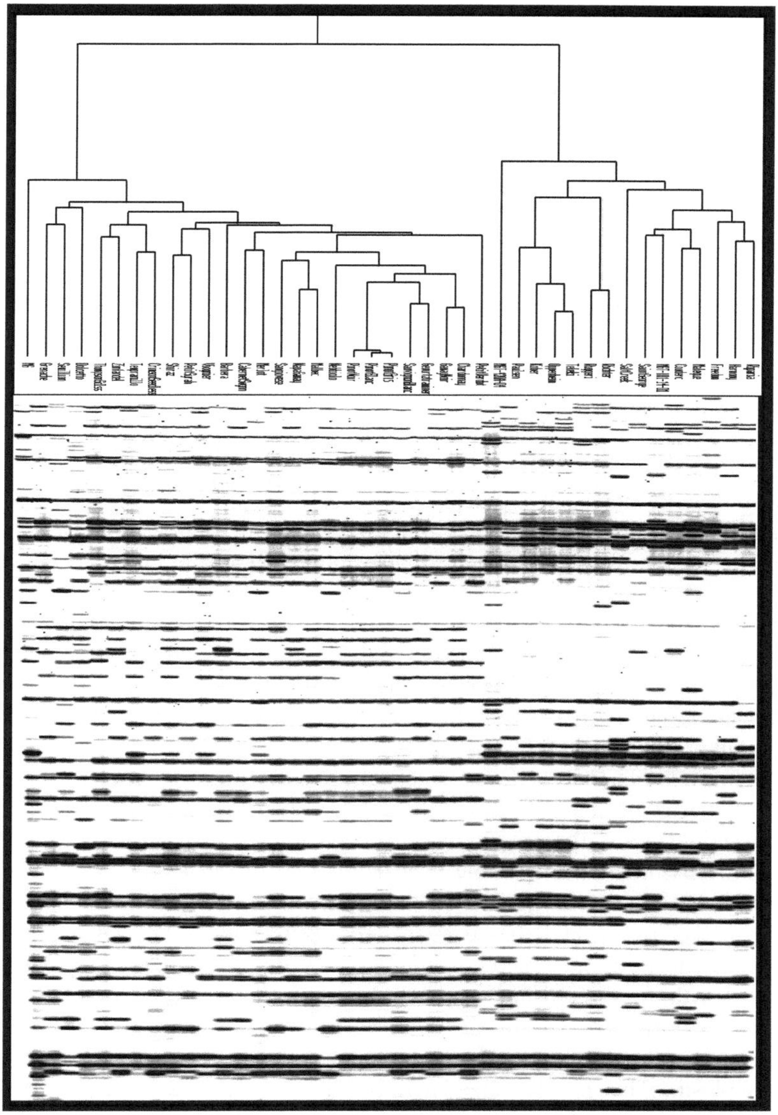

Figure 1. Combined AFLP fingerprint (below) and dendrogram (above) of a panel of grape varieties and rootstocks. Based on the AFLP fingerprint data, the genetic relatedness of all the varieties was determined and displayed by using a dendrogram. Subsequently, the fingerprints of the varieties were re-ordered according to the order of the dendrogram

backcross individual, hereby taking the genome representation of the markers into account. Furthermore, the methodology of analysis includes the identification of donor and recurrent parent segments remaining in the Backcross individuals.

(ii) selection against linkage drag: When negative characteristics are linked with the trait that needs to be introgressed, molecular markers can be used to select for recombinants or double recombinants in the genomic region surrounding the trait of interest. After phenotypic testing of these recombinants, individuals may be selected in which the region responsible for the linkage drag has been removed from the locus of interest. This particular application has proven to be very valuable, especially when introgressing traits from wild relatives into elite lines (Peleman et al. 2003).

2.2. Controlling Traits

2.2.1. Indirect selection

Indirect selection with molecular markers has proven to be a powerful method of selection in plant breeding. Especially for traits for which the phenotypic tests are unreliable or expensive, markers offer a great solution. Before indirect selection can be applied, the genetic basis of the trait of interest needs to be elucidated and markers linked to the gene(s) of interest have to be identified. The methods for identifying linked markers are described below.

The AFLP marker technology has been used for this purpose in a great variety of species for numerous traits (Paran et al. 2003; Dekkers et al. 2002; Koornneef et al. 2001). Once linked markers have been identified, the AFLP markers can be converted into simple PCR assays, which allow screening of large numbers of plants for the trait of interest in a cost effective manner. A suitable linked DNA marker should allow the prediction of the phenotype in a broad range of the germplasm. It is therefore advisable to test multiple linked markers for their level of Linkage Disequilibrium (LD) with the trait of interest before conversion of the linked marker into a PCR based assay. To ensure reliable implementation of the marker in the practical breeding schemes, the identified marker typically must be located within a 1-2 cM interval from the trait of interest. The occurrence of multiple alleles in the germplasm for a desired locus may often complicate the identification of one single marker that will predict the phenotype in the entire breeding germplasm. This hurdle can generally be overcome by using a string of multiple linked markers, 'marker haplotypes', describing the diversity of the targeted locus across the breeding germplasm.

2.2.2. Development of markers for monogenic (qualitative) traits

For the identification of markers linked with monogenic traits, different approaches can be followed. The preferred approaches are all based on knowledge of the mendelian segregation of the trait combined with screening a limited number of samples with a relatively large number of markers. This way, many (initially random) marker loci can be screened with a limited effort. The number of samples that need to be fingerprinted can be limited by screening on set(s) of Near Isogenic Lines (NIL's), if these are available. Candidate markers that are identified this way, are then screened on a panel of phenotypically well characterised germplasm lines to confirm their linkage and to determine the predictive value of the markers on the germplasm.

An extremely powerful approach to identify linked markers consists of the 'Bulked Segregant Analysis' (B.S.A.) method (Michelmore et al. 1991). For this type of screening, individuals from a segregating population are pooled on the basis of their phenotype, and the pools are then fingerprinted until a sufficient number of markers emerge. This method can be used for both dominant and recessive monogenic traits. For dominant genes, 'cis' markers (linked in coupling phase with the trait of interest) will emerge from the screening, whereas 'trans' markers (linked in repulsion phase with the opposite allele) will be identified for recessive traits. The B.S.A. approach has proven to be useful for the identification of linked markers for di-genic traits as well. The initial B.S.A. screen will then reveal markers that will turn out to be unlinked when screening individuals of the mapping population. A secondary pool design based on combined markers scores and phenotype class will enable the screening for additional linked markers to each of the two loci.

This method has been applied in numerous cases and has delivered many very reliable markers which have been implemented for high throughput screening and indirect selection in breeding programs (http://www.fao.org/BIOTECH/docs/Barone.pdf). Moreover, the method forms the basis for map based cloning of genes responsible for a specific trait expression. In fact the limitation of the use is on the number of recombination events that can be generated during the meiosis, and not the genotyping method.

2.2.3. Development of markers for polygenic (quantitative) traits

A majority of agronomically important traits like flowering time, fruit quality, reproductive behavior, stress tolerance and yield exhibit a continuous phenotypical variation (Paterson et al. 1988; Mackay 2001; Morgante et al. 2003). Such traits are determined by a number of genes, collectively termed quantitative trait loci (QTL), each contributing partially to the phenotype in interaction with additional genetical and environmental factors. The polygenic nature of such complex traits has hindered marker development for indirect selection as well as gene isolation projects, mainly due to the lack of discrete phenotypic segregations. A reliable phenotypic evaluation of a quantitative trait is affected by environmental factors, the number of replicates, the number of genes involved, the magnitude of their effect and the way in which these loci interact. For example, the phenotypic effect of a QTL may easily remain undetected as a result of epistatic interactions with other genetic factors. Therefore the predictive use of DNA markers for complex traits is not straightforward and we consider this aspect to be the main challenge for the molecular plant breeder for the coming decade, even though some studies have shown that traits displaying a continuous distribution might have a relatively simple inheritance (Frary et al. 2000; Thornsberry et al. 2001; Rouppe van der Voort et al. 2000).

2.2.3.1. Bi-parental mapping populations Current QTL mapping strategies are using segregating populations of two parent lines and generally lead to the assignment of a QTL to a region of 10–20 cM. In the case of molecular breeding applications, such a rough localization leads to inefficient indirect selection; the association between the linked markers and the trait may become lost during the breeding process, negative traits may be closely linked with the QTL and will not be separated by selecting a large region, and identification of different alleles through haplotyping is cumbersome and expensive for large genomic regions. Furthermore, the current family based QTL mapping methods, will estimate the breeding value of only two alleles (the parents of the mapping population) and the interaction effects of only two genomes. For the complete breeding germplasm used by modern plant breeders, we can safely assume that multiple alleles exist for which the breeding value "*per se*" as well as in interaction with other loci influencing the trait, will remain unknown. Hence, there is a need for efficient methods that allow the precise mapping of QTL and establishing the breeding value of all allelic variants at each QTL locus.

Key factors in high-resolution QTL mapping strategies are the number of identified recombination events, the marker density, and the trait complexity. Sufficient recombination events in QTL intervals can be identified for species where large progenies can be generated easily and cheap (summarized in Darvasi 1998). In general, all these methods require large phenotyped populations to reduce the trait complexity (Darvasi et al. 1993; Darvasi 1998), which renders the cost for these applications relatively high. In plants, QTL have been fine mapped by applying a mapping strategy based on the analysis of large progenies derived from near-isogenic lines (NILs) (Frary et al. 2000; Fridman et al. 2000; El-Din El-Assal et al. 2001; Takahashi et al. 2001; Liu et al. 2002; Salvi et al. 2002; Bentsink et al. 2003). This approach requires the construction of highly inbred lines involving many generations prior to generating the cross needed for fine mapping.

QIR analysis has been used successfully by us for the mapping of oligo-genic traits. Instead of homogenizing the complete genetic background, as in the NIL approach, the QIR analysis approach focuses specifically on the loci involved in expression of the phenotype (Peleman et al. 2005). This strategy involves simultaneous fine mapping of QTL already at the F2/F3 stage rather than producing inbred lines prior to fine mapping (Figure 2). The main principle of the approach is the selective genotyping and phenotyping of only those plants that yield information on the map position of the QTL. Such plants are selected after a first rough-scale mapping by standard methods (e.g., 200 F2 individuals). After identification of the QTL for the trait of interest, a larger part of the population (e.g., 1000 F2 plants) is screened with markers flanking the QTL to identify sets of QTL isogenic recombinants (QIRs). QIR plants carrying a recombination event in one QTL while e homozygous at all other QTL are most informative. The trait complexity can thus be reduced to a monogenic trait as plants with all but one QTL having an identical homozygous genotype are selected. These QIRs are subsequently genotyped with sufficient markers at the recombinant QTL region to

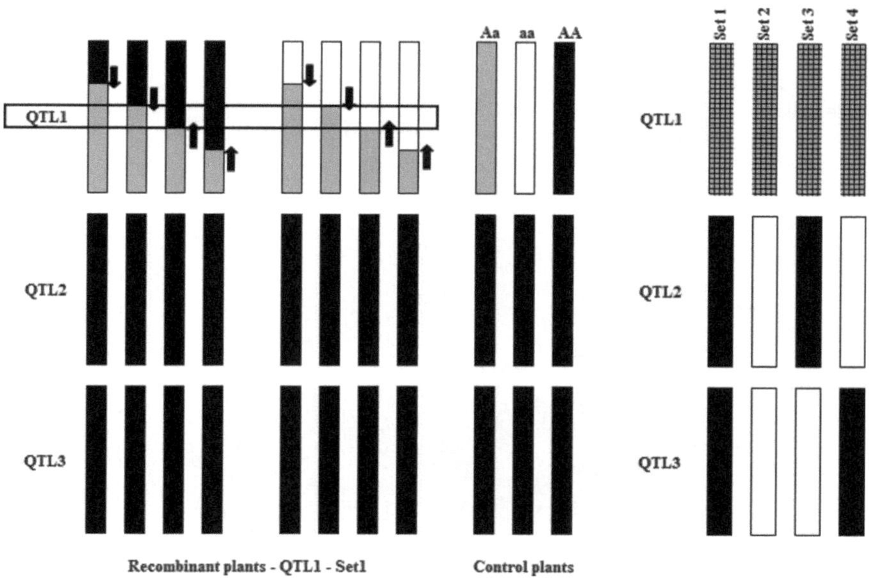

Figure 2. The principle of QTL fine mapping based on selection of QIR sets. Black segments indicate homozygous for parent 1 alleles (AA), white segments indicate homozygous for parent 2 alleles (aa), grey segments indicate heterozygosity. The left panel shows QIR set 1 with recombinations within the QTL1 interval while the other QTL intervals are homozygous for the parent 1 alleles. The position of the gene is determined based on the phenotypic values of the recombinants relative to the values of the control plants (indicated by arrows). The right panel shows the different QIR sets that can be constructed for one locus in a three QTL system. Set 1 and 2 will be the most informative

precisely map the recombination event within the QTL-bearing interval. Phenotyping the QIRs becomes more reliable by reducing the trait complexity as these plants are nearly isogenic for all QTL that affect the trait. The downside of the QIR analysis approach is that the method is limited to traits that are determined by a maximum of five significant QTL.

2.2.3.2. Introgression line libraries (ILLs) A preferred method for mapping many agronomically important quantitative traits segregating in the offspring of a bi-parental cross is the use of Introgression Line Libraries. An Introgression line (IL) library consists of a series of lines harboring a single homozygous donor segment introgressed into a uniform, cultivated background (Figure 3) (reviewed in Zamir 2001; Eshed et al. 1995).

The great advantages of IL libraries in comparison with other mapping approaches are:

- IL Libraries consist of homozygous 'immortal' lines and therefore can be phenotyped repeatedly in multiple environments and used for the simultaneous mapping of many traits;

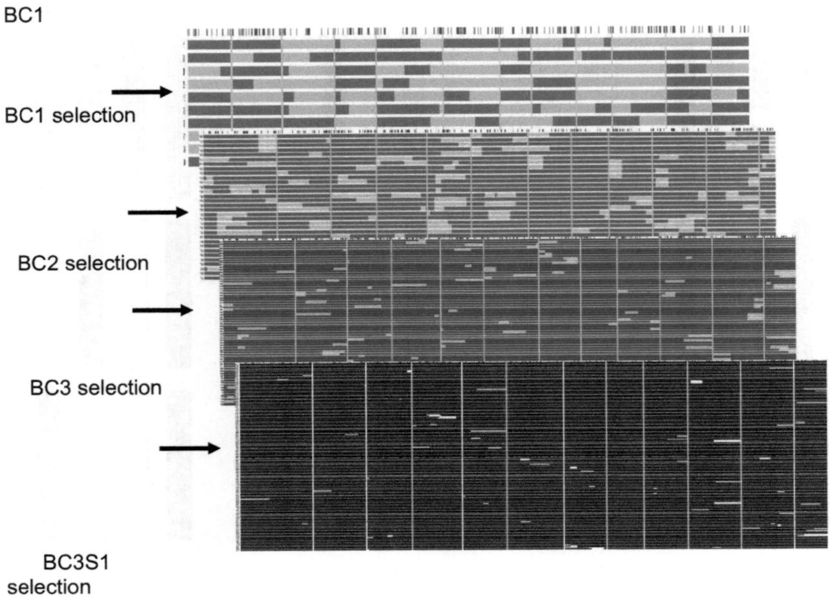

Figure 3. Example of an IL library construction process in tomato (12 chromosomes each separated by a grey bar) by Marker Assisted BackCrossing. Each horizontal bar represents an individual (best visible in the BC1 selection). In grey, homozygous recurrent parent segments are shown. In black the homozygous donor segments are indicated (see BC3S1). Light bars represent heterozygous segments. The black bars on top of the figure show the relative positions of the markers used to assist the selection of the appropriate introgression lines

(Note: Reprinted from: Trends Plant Sci. 8, Peleman J-D, Rouppe van der Voort J, Breeding by Design, 330-334 © (2003), with permission from Elsevier)

- IL Libraries contain homogenous genetic backgrounds, only differing from one another by the introgressed donor segment. Thus, epistatic effects from the donor parent are eliminated.
- QTL are dissected into separate monogenic components which increases the reliability of measuring phenotypic traits
- ILs containing interesting QTL can be backcrossed to various lines to investigate interactive effects.
- Although dependent on the resolution of the IL library (= average introgression segment size), QTLs are typically mapped into smaller intervals than by classical QTL mapping; IL libraries provide optimal starting material for the fine mapping of the mapped loci.
- A secondary bonus from using IL libraries is that often new 'exotic' alleles can be found that have a positive effect in the culture crop germplasm (providing the cross is composed of a elite x exotic cross).

To study interaction (epistasis) between loci, reciprocal IL Libraries can be constructed. In such case, IL libraries from line A into B and vice versa are constructed. By doing so, phenotypes which can not be detected because they are mediated by interacting loci in the AxB library will be measured as a knocked-out phenotype in the BxA library. Subsequently, crosses between individual introgression lines each bearing one of the interacting alleles can be made to investigate the extent of the interaction (Eshed et al. 1996).

To map loci contributing to heterosis, the IL library can be crossed to a tester parent. This will create an F1 IL library in which each introgression segment is present in the heterozygous state. This F1 IL library is then phenotyped to detect heterotic effects caused by specific introgression segments. The past five years, considerable progress has been made for a number of different crops in the construction of Introgression Line Libraries (Zamir 2001). IL libraries also provide perfect starting material for fine mapping of an interesting locus or even isolation of the causal gene: each line containing a locus of interest can be backcrossed to the recurrent parent (and if necessary, selfed) to create a large segregating population. This population can be used to identify recombinants within the introgression segment using flanking markers. Phenotyping these recombinants will enable the mapping of the locus at high resolution.

2.2.3.3. Multi-parental mapping populations A novel method for plant geneticists that in theory will be able to circumvent some of the limitations of the bi-parental mapping method has been used in animal breeding. Recently a specific population design called "heterogeneous stocks", based on multiple parent individuals, has been proposed as a central resource for mouse genetic mapping (The Complex Trait Consortium 2004; Mott et al. 2000) We refer to these methods as Multi-parental mapping reviewed by Flint (Flint et al. 2005). The mapping populations are produced by multiple rounds of intercrossing a group of lines, generating the heterogeneous stock and subsequently drawing RIL populations from this stock through subsequent selfing generations. The choice of the initial lines is essential for the success of the strategy, since the aim is to be able to cover a large part of the genetic diversity within the breeding germplasm. A Genetic Distance Analysis with molecular markers will provide a good support for the optimal choice of complementary lines to combine. The strategy will require an increased amount of markers fingerprinting as compared to a bi-parental mapping population. This is because the number of recombination is highly increased and the number of alleles per locus will be larger then two. The benefit however will be a higher precision of the mapping (resolution) as well as a higher power of detecting QTL alleles and QTL interactions. When multiplex marker fingerprinting cost will be further reduced, this method can be an attractive addition to the efforts of unraveling the genetics of complex traits.

2.2.3.4. Association mapping An alternative approach which has attracted increasing attention from plant geneticist over the last years in the use of association mapping, often referred to as Linkage Disequilibrium (LD) mapping. This method is increasingly recognized as a valuable addition to the toolbox for identifying loci contributing to quantitative traits (Peleman et al. 2003; Thornsberry et al. 2001; Kraakman et al. 2004; Hagenblad et al. 2004). In association mapping, statistical association between genotypes and phenotypes is analysed in large germplasm sets, thereby eliminating the need for population development for the purpose of QTL mapping. Furthermore the method holds the potential of simultaneous detection of loci and estimation of the phenotypic value of all different allelic versions of the QTL that are present in the germplasm sets tested. The identification of loci influencing the expression of the trait is based on the assumption that a historical relationship exists between alleles at two closely situated loci, and this original co-occurrence will gradually decay in the population by recombinations between the loci during meioses. Consequently, the relative allele distributions of an unknown gene and that of a very nearby situated marker will be non-random, or in other words, the two are in dis-equilibrium. In plant breeding germplasm sets, we can expect the presence of population structure, which will significantly influence the results of an association study and cause spurious trait-marker associations. Algorithms and methods are being developed to correct for these effects (Pritchard et al. 2000; Zöllner et al. 2005; Caldwell et al. 2006). A conceptual advantage of association mapping is that the linkage is evaluated over the large pool of historic meioses, allowing gene localization with a higher resolution than when using linkage mapping (Ranalla et al. 2000). The power of the method will therefore depend on the extent of LD present around each QTL in the germplasm set used for the analysis in combination with the resolution of the genotypic scores on the germplasm set as well as the correct ordering (mapping) of all markers scored. (Buckler et al. 2002; Rafalski et al. 2004; Smid et al. 2006). Association mapping can be performed with either single markers or with haplotypes. Haplotypes can be defined as common sets of (marker) alleles in linkage phase in adjacent loci. When using haplotypes in association studies, the information of several linked bi-allelic markers is combined to emulate a single, multi-allelic informative marker. Haplotypes can be generated from physical map sequences or re-sequenced loci (sequence haplotypes) or genetic maps (marker haplotypes). We believe that haplotype based association mapping will provide more power of detection when applied for whole genome scanning for QTLs then when using single marker association tests (see Figure 4). (Buntjer et al. 2005; JoséAranzana et al. 2005; The International HapMap Consortium 2005; Niu 2004).

2.3. Controlling Genes

Whatever methods are employed for parental selection and genetic mapping, most approaches in molecular breeding start with the assumption that allelic variants at one or more genes underlie phenotypic variation. This does not always need to be true, as supra-genic phenomena as genomic imprinting may also cause phenotypic variation. However, it is correct to assume that if traits can be fine mapped on genetic linkage maps, they are ultimately caused by alleles of single genes, either

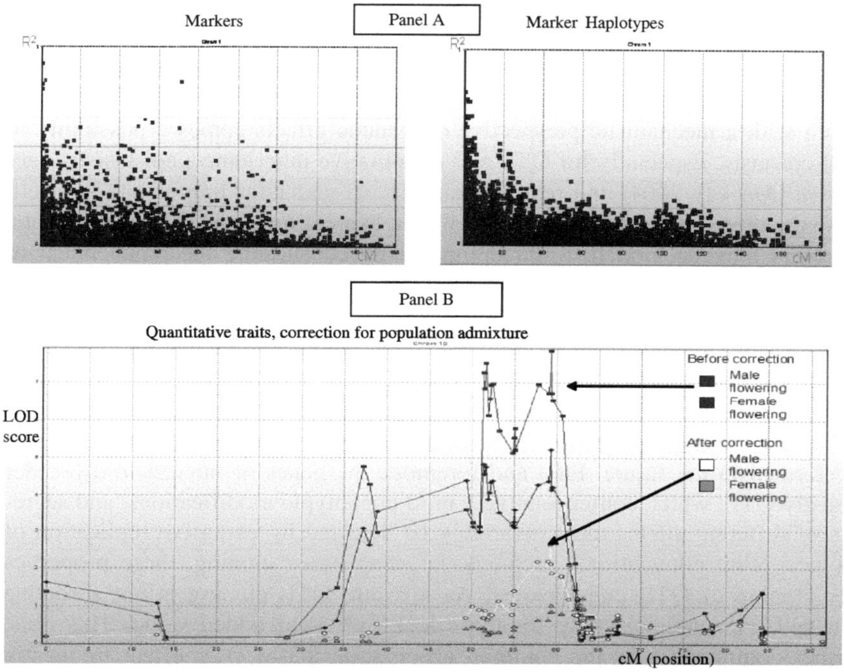

Figure 4. Analysis of LD and LD mapping in a maize line panel. AFLP® marker scores of 115 Dent and 23 Flint lines were aligned to the Keygene integrated maize genetic linkage map, for haplotype based whole genome LD analysis. **Panel A** displays LD expressed as R^2 values of single markers (left) and marker haplotypes (right, haplotype length = 3) of chromosome 1. A slight increase in the extent of LD, an improved relation between cM distance and LD and a clear reduction of spurious LD, can be observed when using marker haplotypes as compared to single markers. **Panel B** display association LOD peaks found for the traits male and female flowering. The influence of the population stratification was assumed to be the cause of some spurious association peaks for these traits. This was confirmed for a number of the LD peaks found. High LOD scores disappeared after correction for the subpopulation stratification

Mendelian major genes, or QTL. The highest resolution of molecular breeding is therefore the cloning of individual genes.

2.3.1. Development of markers through positional cloning

There are several reasons why the identification of individual genes would be beneficial for breeding purposes. The most obvious is that a causal gene would allow the development of perfect allele-specific molecular markers for indirect selection. However, in crops with high levels of LD, such fine resolution may not be required, and a breeder may be satisfied with a haplotype of the locus of interest. More important is the fact that gene cloning results in mechanistic models as to why a particular allelic variant performs better or worse than another, rendering it a functional marker. This allows a breeder to design intelligent selection criteria to discover novel alleles based on pre-defined molecular features, such as a specific nucleotide sequence, expression level or expression pattern. Molecular selection can be done before any actual breeding, allowing phenotypic assessment to be restricted to a few promising alleles from a wider germplasm.

Along with a mechanistic perspective on genetic effects comes a possibility to control epistasis. Especially for QTL and quantitative inheritance, epistasis is very pervasive (Mackay 2004) and requires not only an understanding of a QTL itself, but also the effect of genetic background on the expression of the QTL. The genetic background may be modelled and optimised by considering the biochemical or physiological pathways in which a QTL functions. Finally, a benefit of gene cloning is one of legal protection. If specific alleles lead to improved varieties, breeders can protect their investments by describing molecular features, allowing highly specific genotype tests for variety identification.

Presently, gene cloning has not been of major impact for molecular breeding. This has a number of reasons, which dictate the way in which gene cloning will be performed in the future. First and foremost, the genes of interest to a breeder are mostly QTL, which confer relatively mild phenotypic modifications, and segregation of different alleles must invariably be detected by extensive replication of genotypes. With conventional methods, i.e. map based cloning, these properties render cloning of QTL cumbersome. As a result, QTL cloning is not a routine option and is economical only for those loci with clear added value. This issue is further complicated by the common observation that a QTL effect frequently fails to replicate well in different genetic backgrounds, leading many breeders to suggest that single QTL genes have relatively little to add except in very specific cases. It is imperative, therefore, to improve the technical toolkit to such extent that cloning QTL, whether of major or minor effect, becomes a cost effective and simple routine. Developments in this area have been reviewed in detail elsewhere (Salvi et al. 2005; Morgante et al. 2003). We believe that, with the conventional hierarchical map based cloning approaches, this goal will for many traits be difficult to achieve, creating a need to explore alternative methods for higher throughput cloning of genes with breeding value.

2.3.2. Development of markers through the candidate gene approach

One line of progress comes from emerging genomic tools. Currently, it seems that evaluation of candidate genes in high throughput association mapping procedures is one of the most promising routes to identifying genes with QTL effects. This has been demonstrated in several cases and in several crops (Li et al. 2005; Olsen et al. 2004; Szalma et al. 2005; Thornsberry et al. 2001; Thumma et al. 2005; Wilson et al. 2004). Biochemical and physiological predictions about the functions of particular plant genes has led to the identification of subtle quantitative effects of specific alleles on traits of interest. Once good association mapping panels are established and characterized for phenotype and population structure (Flint-Garcia et al. 2005), they can be used indefinitely to assess the effects on selected genes and polymorphisms within them. A limiting factor becomes the speed and cost of resequencing and polymorphism discovery in a large number of genes. With many possible candidate genes and their gene families, effective association analysis will become a large scale resequencing endeavour.

The selection of candidate genes can follow from various sources of information. Among these are pre-established roles from knock-out phenotypes or biochemical function, but also from genomic location, expression studies, etc. It is important to note that associations are always statistical, and must be followed up by confirmatory studies in which specific alleles are tested in segregating populations, or ultimately by direct genetic modification.

2.3.3. Development of markers through QTL tagging

A second avenue towards identifying genes with useful phenotypic effects is to reconsider what actually constitutes a QTL. With quantitative inheritance, it is generally accepted that many genes contribute to phenotype, and that these genes can be quite different in different crosses. If so, almost any gene can be a QTL, a notion that opens up at least one novel approach. As has been demonstrated amply in Drosophila (Norga et al. 2003; Harbison et al. 2004; Mackay 2004), assessing the quantitative phenotypic effects of a collection of random homozygous insertions (such as transposons) is very effective in de novo induction and simultaneous cloning of QTL by tagging. Interestingly, a fair proportion of insertion alleles appear to improve the phenotype of Drosophila strains that carry them, or correspond to positionally cloned QTL (Norga et al. 2003; Harbison et al. 2004). Several excellent transposition systems are available in several species of model plants (Droc et al. 2006; http://www.arabidopsis.org; McCarty et al. 2005; Stuurman et al. 2005), which would allow a similar approach as followed in Drosophila. With this method implemented, QTL identification can get a significant new avenue. Although large scale insertional mutagenesis is not available for most crop plants, knowledge about quantitative trait genes in model species should have significant impact on understanding corresponding genetic systems in crops.

In summary, we believe that cloning genes is a necessary extension of molecular breeding. With new tools and concepts in hands, the catalogue of useful genes and their associated phenotypic variation is expected to expand exponentially, providing many novel opportunities for creating designer genotypes of superior performance. For QTL of large effect, enhanced procedures for map based cloning should expedite their isolation. For isolation of QTL with small effects, it is likely that the future will see a greater emphasis on evaluation of candidate genes and on mutagenesis based approaches such as tagging with insertion mutagens.

3. OPTIMAL EXPLOITATION OF (BREEDING) GERMPLASM

The application of DNA markers for indirect genome and trait selection, based on developments as described above, has improved the gain of selection in recurrent breeding programs by increasing heritability and selection intensities simultaneously. Moreover specific applications of DNA markers can create substantially more added value in the variety development process. By applying markers in a creative manner, new traits can be introduced which either could not or could only be obtained with great difficulty by classical breeding. Therefore, the application of markers in breeding has created a competitive advantage to those breeders / companies which have successfully integrated DNA markers as part of their working tools. Below, a number of examples are provided where creative applications of markers clearly provide a major benefit over classical breeding.

3.1. Removal of Linkage Drag

One of the earliest creative applications has been called removal of linkage drag. Many valuable traits are (or have been) introduced into elite germplasm through crosses with wild relatives and backcrosses with elite lines as the recurrent parent (RP). Breeders have experienced that this strategy often is hampered by the fact that the desirable trait from the wild relative is linked to a trait that influenced the performance of the elite line in a negative way. In order to remove this linkage drag, one (or more) recombination event(s) has to take place within the initially introgressed genomic segment from the wild relative. As a consequence of relatively large sequence diversity between the introgressed segment from the wild relative and the orthologue genomic segment of the elite RP line, recombination frequency will be depressed in the introgressed segment (Roger et al. 2000). Only the screening of a very large (thousands) segregating population, will enable the identification of the desired recombinant individual. This effort is enormous if the screening must be based on phenotypic evaluations, but with current high throughput marker screening platforms, thousands of individuals can be screened with markers flanking the introgression segment in a day. Only recombinant individuals will need to be phenotyped in order to confirm the removal of the linkage drag. As an example of a successful application of this strategy, which has delivered varieties with novel traits we refer to the development of lettuce varieties resistant to the aphid *Nasonovia ribisnigri* (Jansen, 1996) and the development of fertile Rye hybrid varieties (http://www.pollenplus.de/pollenplus_62.php.).

3.2. Pyramiding Favourable Alleles and QTL

A very powerful example of using markers for the creation of novel varieties is by pyramiding favourable alleles of genes in one variety (genotype). This approach can offer great financial rewards through extending the life span of new varieties. Pyramiding in combination with molecular marker selection is a widely used term for many different applications. The basis for using efficient marker assisted pyramiding lies in the precise knowledge of the genetic positions of the genes and the availability of markers which are in very tightly linked to these genes. Combining different pathogen resistance genes using this approach has been reported among others by Gebhardt et al (2006). (Gebhardt et al. 2006; Witcombe et al. 2000). A different application of the same principle relates to the combination of multiple race-specific resistance genes, which reside in a resistance gene cluster. Favorable alleles of homologous resistance genes may be located in tandem, but present in different accessions. In such case it is of paramount importance to precisely fine map the alleles of the different genes with respect to one another. This goal can only be achieved using DNA markers. Subsequently, the linked markers can be utilized to select for the rare recombinants that combine the favorable alleles in tandem.

An increasing amount of knowledge about the molecular basis of the expression of traits is available to the modern plant breeder. Extending the principle of pyramiding favorable alleles to multiple traits, quantitative and qualitative, will increase the possibility for the plant breeder to create favorable and novel genotypes by monitoring the segregation of traits during the breeding process, and focusing the selection on novel combinations of favorable (QTL) alleles (Servin et al. 2004). For a successful implementation of such an approach, in our opinion, a re-definition of traits is necessary. A separation of traits into trait-components that can be mapped separately will be necessary. For example, an extremely important phenotype like yield is determined by a vast array of component characters, such as root size, plant size, number of fruit, size of fruit, fruit contents, etc. Mapping the genes involved in these separate components provides a better understanding of the complex trait and a higher chance of success. This approach will significantly aid in unraveling the complexity of agronomically important traits. It is with such traits that, in the long term, the biggest benefits of MAS can be obtained.

3.3. Effective Exploitation of (Exotic) Germplasm

There are several strategies for exploiting the variation in germplasm collections. For some traits, it might be necessary to use wild ancestors of crop plants and to introgress some of the diversity that was lost during domestication in order to improve agricultural performance under optimal as well as stress conditions. Most of the genetic variation that is present in wild species and unadapted germplasm in gene banks, has a negative effect on the adaptation of plants to the agricultural high performing environments; hence, the challenge is to identify and introduce into the breeding germplasm, only the advantageous alleles. This is particularly the case for quantitative traits because the value of a wild or exotic accession for contributing useful alleles cannot be determined a priori with certainty, either because the "*per se*" contribution of the advantageous alleles is not detectable in the phenotyping range of the wild accession, or because epistatic effects render the beneficial allele undetected in the background of the wild genome. Breeders have traditionally been limited in the use of wild germplasm in their breeding programs due to complex, long-term and unpredictable outcomes, particularly in crops where quality traits are important market criteria. This is a pity because in most crops, the cultured germplasm only represents a small section of the vast diversity available in the species. Tanksley and co-workers have clearly demonstrated that wild relatives of tomato contain genes contributing to interesting culture characteristics which are generally not expected to reside in those species (Tanksley et al. 1997; Frary et al. 2000). Marker assisted backcrossing now enables the breeders to precisely introgress small sectors of wild/exotic accessions thereby providing breeders with the tools to effectively unleash the vast resources held in germplasm collections.

DNA marker based diversity analysis enables gene banks to define core collections, which will provide a user friendly entry point for breeders to access large and varied germplasm collections. A large scale genetic distance analysis of the complete CGN genebank of lettuce in The Netherlands has been performed using AFLP markers. The analysis involved more than 6.800 samples and a enormous data set of more than 1,35 million datapoints was produced in this study (Jansen et al. 2006). This type of analyses will greatly aid selection of genotypes for an effective broadening the genetic base of the elite breeding germplasm.

Using markers tightly linked to a gene of interest, so called locus haplotyping can be performed on accessions of germplasm to identify those samples that bear different alleles at the locus of interest (Peleman et al. 2003). It enables identification of accessions/lines bearing different alleles at a single locus, which can then be evaluated into further detail with respect to performance. This enables the breeders to efficiently identify new traits or better versions of existing traits which then can be quickly introgressed into their breeding lines. Even more precision can be obtained when the sequence of the gene underlying the trait of interest is known. Current costs of sequencing allow for re-sequencing of a gene throughout a germplasm collection to identify all different sequence alleles that can be found in the wide germplasm collection. This approach allows the effective exploitation of germplasm without the enormous task of having to phenotype all accessions (Sicard et al. 1999; Huang et al. 1992).

A method for the transfer of QTLs of agronomically important traits from a wild species into a crop variety called 'advanced backcross QTL analysis' (AB-QTL) was first proposed by Steven Tanksley (Tanksley et al. 1996). In this approach, a wild species is crossed to an elite line from the breeding germplasm and the progeny of backcross families is selected for the quantitative trait. Typically BC2F2 or BC2F3 populations are evaluated for retention of the traits of interest and genotyped with polymorphic DNA markers. The data are used for QTL mapping and analysis of recurrent parent genome recovery simultaneously, thus selecting for an elite line carrying genomic segments from the wild donor line, which are responsible for an increased performance of the trait of interest.

The AB-QTL approach has been evaluated in many crops (Septiningish et al. 2003; Zhi-Kang et al. 2005; Pillen et al. 2004). A disadvantage of the method is that one can not be certain at the beginning of an AB-QTL program, that the wild accession will contribute useful QTL alleles and thus justify the substantial investment that needs to be made.

4. SUMMARY AND OUTLOOK

The concept of Breeding by Design (Peleman et al. 2003) is basically simple: combine all favourable alleles at all loci of agronomical importance, by controlling the appropriate traits at the molecular level. To achieve this objective, a number of technical, methodological as well as psychological barriers have to be overcome. These barriers and the degree in which these have been resolved so far will be summarized in this section.

Both the dissection of the genetic basis of traits and the efficient application of the DNA markers in breeding, require reliable, efficient and cheap genotyping technology. In the past 15 years, major progress has been reached in reducing the cost per data point at least 100 fold. This evolution of developing new and cost reducing marker technologies is still ongoing. Recent reviews provide ample overview on novel high throughput genotyping technologies (Syvanen 2001; Syvanen 2005; Chen et al. 2003; Borewitz et al. 2003). It can be expected that with the current technological progress the cost per datapoint will be reduced to such an extent that its cost will not be limiting for application in breeding at all. This will render a number of the applications mentioned in the previous chapters, such as genome wide LD mapping, multi-parental mapping and high throughput gene cloning, affordable to be applied at a broad scale. Major breakthroughs in this field can especially be expected from advances in novel sequencing technology (Margulies et al. 2005; Chan 2005; Shendure et al. 2004). If we consider complete gene sequences as the ultimate resolution for genotyping, a cost effective multi-locus sequencing technology would be the ultimate genotyping technology to describe the genetic variability in the germplasm of the plant breeder. It is noted that the wealth of DNA sequence in model (crop) species will allow the exploitation of syntenic relations of orthologous gene function across species. Such syntenic analyses will play an important role in speeding up molecular breeding, especially in a large number of agricultural and horticultural crops, for which the investments needed for whole genome genetic analysis with DNA markers and DNA sequences will be too high in comparison with the margins that are made in these crops. Parallel to developing crop specific molecular tools like high dense integrated molecular marker maps, these crops may benefit from the sequence knowledge that is gathered through research on their crop-relatives or the model species for development of "functional markers" (Anderson et al. 2003)

Not only technology has been rate limiting in the application of genetic mapping studies. Originally, genetic mapping in plants was typically performed in segregating F2, BC, or at best in DH populations. Often these populations were only segregating for a limited number of phenotypic characters. Mapping traits this way was inefficient and costly, since many populations needed to be set-up and genotyped and only 2 alleles were typically mapped per locus.

The design and use of novel population types have significantly attributed to the unravelling of genotype-phenotype relations. As we have demonstrated in the chapters above, this trend is still continuing and novel ways of developing mapping populations and families are being created and tested (multi-parental populations, LD mapping panels). In general we can identify a trend towards the utilization of populations generated as part of the breeding program as opposed to the specific development of mapping populations purely for the purpose of the mapping project. Those breeding programs that will implement these integrated strategies in the most efficient way will benefit directly from the molecular marker work during the process of variety development.

If we extrapolate on the technical possibilities and the cost reduction of genotyping, the key factor of the future for determining the exact relation between the genotypic and the phenotypic variability will be the precision of phenotyping. Knowledge on how to dissect complex phenotypes into separate components helps to objectively determining the values of each of these components and to be able to determine the genetic factors which have a causal effect on the variation of the trait component. A method which aims at reducing the phenotype complexity is to measure metabolites instead of the trait as such (Schauer et al. 2006). Other methods are based on image analysis of growing plants, which objectively captures quantitative differences of many characteristics simultaneously (http://www.cropdesign.com/general.php; http://www.lemnatech.com). Irrespective of the method used for dissecting the phenotype into measurable components, the strategy will require the re-assembling of complex relations between the components, with respect to the agronomically important trait under selection in the breeding program. This will present a challenge in itself and novel algorithms will need to be developed that not only will detect reliable relations, but also will be able to present the complex relations to the user (the plant breeder) in a way that will actually help the breeder in making decisions during the breeding process. Furthermore we know that the expression of many quantitative traits is strongly influenced by the environment and that genotype by environmental interaction is more the rule than the exception. Modern plant breeders have adapted their phenotypic selection methods by taking the environmental effects on the performance of candidate varieties into account (Cooper et al. 1994; Van Eeuwijk et al 2005). The challenge for the molecular breeder will be to understand the QTL by environment interaction as well, which again can only be realized by intelligent visualisation of such complex relations.

The large scale mapping of traits and genotype by environment interaction effects will yield an increasing amount of genetic knowledge to be exploited by the plant breeder. So how will the molecular breeder be able to efficiently exploit the vast amount of genetic knowledge that will gradually become available? Based on the developments described in the chapters above, we believe that their will be two major changes in the way the molecular breeder will operate in the future. First of all we expect that the design process of ideal genotypes, and the crosses that lead thereto, will gain more importance in the total breeding program. We have described this process as Breeding by Design ™ (Peleman et al. 2003). This is basically an extension of the principle of pyramiding favourable alleles, but also removing negative alleles from the population (Figure 5). When considering this principle for multiple QTL of complex traits, displaying environmental and epistatic interactions, positive selection is not *a priori* straightforward. In these cases the breeder will have to adopt a selection strategy combining positive and negative selection. Positive selection can be performed on alleles and allele combinations, which are absolutely required for the performance of the variety and a simultaneous negative selection

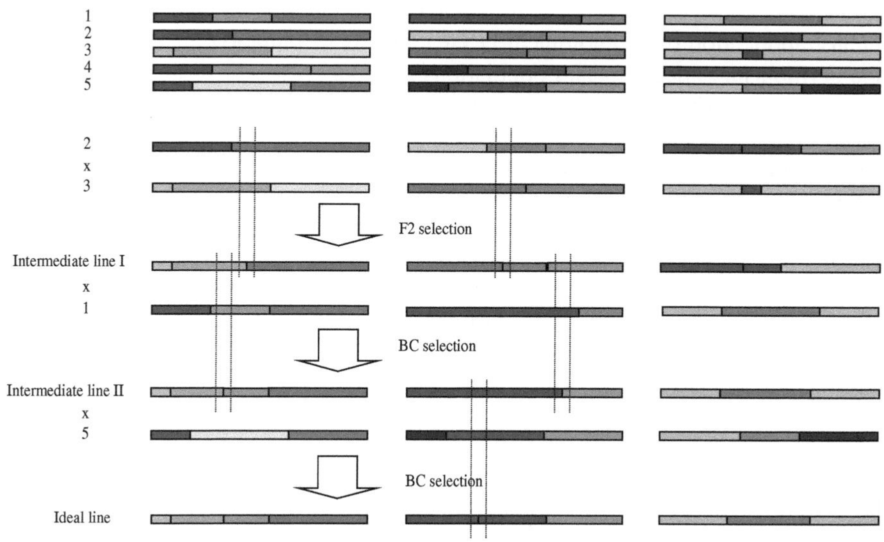

Figure 5. **The principle of Breeding by Design.** Subsequent crosses and selections using markers lead to the desired superior elite line genotype starting from a collection of 5 parental lines. Dotted lines indicate marker positions used to select for the desired recombinants *(see plate 1)*
(Note: Reprinted from: Trends Plant Sci. 8, Peleman J-D, Rouppe van der Voort J, Breeding by Design, 330-334 © (2003), with permission from Elsevier)

can be performed on all allele combinations which in previous candidate varieties have proven not to produce varieties suitable for market introduction. Such marker assisted selection methods can be applied in the development of inbred lines as well as the development of test-hybrids and have the potential of reducing the amount of test-varieties to be phenotyped significantly. In this way the phenotyping cost can be reduced thus allowing to intensify the breeding effort by increasing the chance of generating test varieties harbouring novel allele combinations which are likely to provide a better performance.

The second major change which we consider a prerequisite for implementation of these strategies of marker assisted breeding will be the availability of integrated data-management systems, combining phenotypic, genotypic, pedigree and environmental information. These systems must be able to register the phenotypic "value" of all genotypes tested in previous breeding circles and environments. The data must be presented to the breeder in a manner that will support the selection process as well as the design of new selection circles. With the increasing amount of data and complex relations known, this can only be achieved if such data management systems will be developed as "self-learning" systems (McKay et al. 2003; McKay et al. 2002) that can propose selection and design decisions to the breeder, based on prior knowledge present in the data-system.

Ideally, the molecular breeder of the future will be steering an integrated, complementary application of technological DNA tools, exact phenotyping capacities and wide germplasm collections and populations in order to develop superior varieties. During this process, an enormous resource of knowledge is generated which, with the aid of data management and decision support systems will enable the breeders to deploy more rational and refined breeding strategies and selection choices. The recent technological developments are bringing this strategy within reach. However, there still remain bottle necks at the level of phenotyping precision, the understanding of genotype-environment interactions, and the assimilation capacity of the molecular breeder himself, which needs resolving. Removing these last obstacles will enable the optimal exploitation of the naturally available genetic resources and will create unsurpassed possibilities to generate new traits and superior crop performance.

ACKNOWLEDGEMENTS

The authors gratefully acknowledge Gera van Kalkeren for assisting with the manuscript preparation. J. Antoni Rafalski of Dupont Crop Genetics and Albrecht E. Melchinger from the University of Hohenheim are gratefully acknowledged for very fruitful discussions concerning the possibilities of association mapping in plants. We would like to thank Milena Ouzunova & Carsten Knaak from KWS SAAT AG for providing datasets for the analysis of LD characteristics in maize germplasm sets. Last but not least we gratefully acknowledge the Bioseeds companies Takii Seed, Vilmorin Clause & Cie, De Ruiter Seeds, Enza Zaden and Rijk Zwaan for their continuous support of our research efforts towards the optimal exploitation of marker technologies and applications in plant breeding. *The AFLP® technology is covered by patents and/or patent applications of Keygene N.V.; AFLP® is a registered trademark of Keygene N.V.; Breeding by Design ™ is a trademark of Keygene N.V.*

REFERENCES

Anderson J-R, Lübberstedt T (2003) Functinal markers in plants. Trends Plant Sci 8:554–560

Aranzana M-J, Kim S, Zhao K, Bakker E, Horton M, Jakob K, Lister C, Molitor J, Shindo C, Tang C, Toomajian C, Traw B, Zheng H, Bergelson J, Dean C, Marjoram P, Nordborg M (2005) Genome-wide Association mapping in Arabidopsis identifies previously known flowering time and pathogen resistance genes. PLoS Genet 1:e60:0531–0539

Bentsink L, Yuan K, Koornneef M, Vreugdenhil D (2003) The genetics of phytate and phosphate accumulation in seeds and leaves of Arabidopsis thaliana, using natural variation. Theor Appl Genet 106:1234–1243

Borewitz J-O, Liang D, Ploiffe D, Chang H-S, Zhu T, Weigel D, Berry C-C, Winzler E, Chory J (2003) Large-scale identification of single-feature polymorphisms in complex genomes. Genome Res 13:513–523

Buckler E-S, Thornsberry J-M (2002) Plant molecular diversity and applications to genomics. Curr Opin Plant Biol 5:107–111

Buntjer J-B, Sørensen A-P, Peleman J-D (2005) Haplotype diversity: The link between statistical and biological associations. Trends Plant Sci 10:466–471

Caldwell K-S, Russel J, Langridge P, Powel W (2006) Extreme population-dependent linkage disequilibrioum detected in an inbreeding plant species, hordeum vulgare. Genetics 172:557–567

Chan E-Y (2005) Advances in sequencing technology. Mutat Res 573:13–40, www.sciencedirect.com.

Chen X, Sullivan P-F (2003) Single nucleotide polymorphism genotyping: Biochemistry, protocol, cost and throughput. Pharmacogenomics J 3:77–96

Chetelata R-T, Meglicb V, Cisnerosa P (2000) A genetic map of tomato based on BC1 lycopersicon esculentum x solanum lycopersicoides reveals overall synteny but suppressed recombination between these homeologous genomes. Genetics 154:857–867

Cooper M, DeLacy I-H (1994) Relationships among analytical methods used to study genotypic variation and genotype-by-environment interaction in plant breeding multi-environment experiments. Theor Appl Genet 88:561–572

Darvasi A (1998) Experimental strategies for the genetic dissection of complex traits in animal models. Nat Genet 18:19–24

Darvasi A, Weinreb A, Minke V, Weller J-I, Soller M (1993) Detecting marker-QTL linkage and estimating QTL gene effect and map location using a saturated genetic map. Genetics 134:943–951

Dekkers J-C, Hospital F (2002) The use of molecular genetics in the improvement of agricultural populations. Nat Rev Genet 3:22–32

Droc G, Ruiz M, Larmande P, Pereira A, Piffanelli P, Morel J-B, Dievart A, Courtois B, Guiderdoni E, Perin C, OryGenes DB (2006) A database for rice reverse genetics. Nucleic Acids Res 1:34

El-Din El-Assal S, Alonso-Blanco C, Peeters A-J, Raz V, Koornneef M (2001) A QTL for flowering time in Arabidopsis reveals a novel allele at CRY2. Nat Genet 29:435–440

Eshed Y, Zamir D (1995) An introgression line population of *Lycopersicon pennellii* in the cultivated tomato enables the identification and fine mapping of yield-associated QTL. Genetics 141:1147–1162

Eshed Y, Zamir D (1996) Less than additive epistatic interactions of QTL in tomato. Genetics 143:1807–1817

Flint J, Valder W, Shifman S, Mott R (2005) Strategies for mapping and cloning quantitative trait genes in rodents. Nat Rev genet 6:271–285

Flint-Garcia S-A, Thuillet A-C, Yu J, Pressoir G, Romero S-M, Mitchell S-E, Doebley J, Kresovich S, Goodman M-M, Buckler E-S (2005) Maize association population: a high-resolution platform for quantitative trait locus dissection. Plant J 44:1054–1064

Frary A, Clint NT, Frary A, Grandillo S, van der Knaap E, Cong B, Liu J, Meller J, Elber R, Alpert K-B, Tanksley S-D (2000) *fw2.2*: A quanititative trait locus key to the evolution of tomato fruit size. Science 289:85–88

Fridman E, Pleban T, Zamir D (2000) A recombination hotspot delimits a wild-species quantitative trait locus for tomato sugar content to 484 bp within an invertase gene. Proc Natl Acad Sci USA 97:4718–4723

Frisch M, Bohn M, Melchinger A-E (1999) Comparison of selection strategies for marker-assisted backcrossing of a gene. Crop Sci 39:1295–1301

Gebhardt C, Bellin D, Henselewski H, Lehmann W, Schwarzfischer J, Valkonen J-P-T (2006) Marker-assisted combination of major genes for pathogen resistance in potato. Theor Appl Genet 112: 1458–1464

Hagenblad J, Tang C, Molitor J, Werner J, Zhao K, Zheng H, Marjoram P, Weigel D, Nordborg M (2004) Haplotype structure and phenotypic associations in the chromosomal regions surrounding two Arabidopsis thaliana flowering time loci. Genetics 168:1627–1638

Harbison S-T, Yamamoto A-H, Fanara J-J, Norga K-K, Mackay T-F (2004) Quantitative trait loci affecting starvation resistance in Drosophila melanogaster. Genetics. 166:1807–1823 http://www.arabidopsis.org; http://www.cropdesign.com/general.php; http://www.fao.org/BIOTECH/docs/Barone.pdf; http://www.lemnatec.com/; http://www.pollenplus.de/pollenplus_62.php/

Huang N, Stebbins G-L, Rodriguez R-L (1992) Classification and evolution of *A-amylase* genes in plants. Proc Natl Acad Sci USA 89:7526–7530

Jansen J-P-A (1996) Aphid resistance in composites. International application published under the patent cooperation treaty (PCT) No. WO 97/46080.

Jansen J, Verbakel H, Peleman J, van Hintum Th-J-L (2006) A note on the measurement of genetic diversity within genebank accessions of lettuce (*Lactuca sativa* L.) using AFLP markers, Theor Appl Genet 112:554–561

Koornneef M, Stam P (2001) Changing paradigms in plant breeding. Plant Physiol 125:156–159

Kraakman A-T-W, Niks R-E, Van den Berg P-M-M-M, Stam P, Van Eeuwijk F-A (2004) Linkage disequilibrium mapping of yield and yield stability in modern spring barley cultivars. Genetics 168:435–446

Lev-Yadun S, Gopher A, Abbo S (2000) The cradle of agriculture Archeology. Science 288:1602–1603

Li Z-K, Fu B-Y, Gao Y-M, Xu J-L, Ali J, Lafitte H-R, Jiang Y-Z, Domingo Rey J, Vijayakumar C-H-M, Maghirang R, Zheng T-Q, Zhu L-H (2005) Genome-wide introgression lines and their use in genetic and molecular dissection of complex phenotypes in rice (*Oryza sativa* L.). Plant Mol Biol 59:33–52

Li L, Strahwald J, Hofferbert H-R, Lubeck J, Tacke E, Junghans H, Wunder J, Gebhardt C (2005) DNA variation at the invertase locus invGE/GF is associated with tuber quality traits in populations of potato breeding clones. Genetics. 170:813–821

Liu J, van Eck J, Cong B, Tanksley S-D (2002) A new class of regulatory genes underlying the cause of pear-shaped fruit. Proc Natl Acad Sci USA 99:13302–13306

Lübberstedt T, Melchinger A-E, Dussle C, Vuylsteke M, Kuiper M (2000) Relationship among early European maize inbreds: IV genetic diversity revealed with AFLP markers and comparison with RFLP, RAPD, and pedigree data. Crop Sci 40:783–791

Mackay T-F-C (2001) The genetic architecture of quantitative traits. Annu Rev Genet 35:303–339

Mackay T-F-C (2004) The genetic architecture of quantitative traits: lessons from Drosophila. Curr Opin Genet Dev 14:253–257

Margulies M, Michael M, Altman W-E, Attiya S, Bader J-S, Bemben L-A, Berka J, Braverman M-S, Chen Y-J, Chen Z, Dewell S-B, Du L, Fierro J-M, Gomes X-V, Godwin B-C, He W, Helgesen S, Ho C-H, Irzyk G-P, Jando S-C, Alenquer M-L-I, Jarvie T-P, Jirage K-B, Kim J-B, Knight J-R, Lanza J-R, Leamon J-H, Lefkowitz S-M, Lei M, Li J, Lohman K-L, Lu H, Makhijani V-B, McDade K-E, McKenna M-P, Myers E-W, Nickerson E, Nobile J-R, Plant R, Puc B-P, Ronan M-T, Roth G-T, Sarkis G-J, Simons J-F, Simpson J-W, Srinivasan M, Tartaro K-R, Tomasz A, Vogt K-A, Volkmer G-A, Wang S-H, Wang Y, Weiner M-P, Yu P, Begley R-F, Rothberg J-M (2005) Genome sequencing in microfabricated high-density picolitre reactors. Nature 437:376–380

McCarty D-R, Settles A-M, Suzuki M, Tan B-C, Latshaw S, Porch T, Robin K, Baier J, Avigne W, Lai J, Messing J, Koch K-E, Hannah L-C (2005) Steady-state transposon mutagenesis in inbred maize. Plant J 44:52–61

McKay B, Slaney J-K (2002) Advances in artificial intelligence, 15th Australian joint conference on artificial intelligence, Canberra, Australia, December 2–6, 2002, Proceedings Springer 2002

McKay R-I, Abbass H-A (2003) Artificial life: an introduction. Intern J Computat Intellig and Appl 3:143–144

Michelmore R-W, Paran I, Kesseli R-V (1991) Identification of markers linked to disease-resistance genes by bulked segregant analysis: a rapid method to detect markers in specific genomic regions by using segregating populations. Proc Natl Acad Sci USA 88:9828–9832

Morgante M, Salamini F (2003) From plant genomics to breeding practice. Curr Opin Biotech 14:214–219

Mott R, Talbot C-J, Turri M-G, Collins A-C, Flint J (2000) A method for fine mapping quantitative trait loci in outbred animal stocks. Proc Natl Acad Sci USA 97:12649–12654

Niu T (2004) Algorithms for inferring haplotypes. Genet Epidemiol 27:334–347

Norga K-K, Gurganus M-C, Dilda C-L, Yamamoto A, Lyman R-F, Patel P-H, Rubin G-M, Hoskins R-A, Mackay T-F, Bellen H-J (2003) Quantitative analysis of bristle number in *Drosophila* mutants identifies genes involved in neural development. Curr Biol 19;13:1388–1396

Olsen K-M, Halldorsdottir S-S, Stinchcombe J-R, Weinig C, Schmitt J, Purugganan M-D (2004) Linkage disequilibrium mapping of Arabidopsis CRY2 flowering time alleles. Genetics 167:1361–1369

Paran I, Zamir D (2003) Quantitative traits in plants: Beyond the QTL. Trends Genet 19:303–306

Paterson A-H, Lander E-S, Hewitt J-D, Peterson S, Lincoln S-E et al (1988) Resolution of quantitative traits into Mendelian factors using a complete linkage map of restriction fragment length polymorphisms. Nature 335:721–726

Peleman J-D, Rouppe van der Voort J (2003) Breeding by design. Trends Plant Sci 8:330–334

Peleman J-D, Wye C.L, Zethof J, Sørensen A-P, Verbakel H, van Oeveren J, Gerats T, Rouppe van der Voort J (2005) Quantitative trait locus (QTL) Isogenic recombinant analysis: a method for High-resolution mapping of QTL within a single population. Genetics 171:1341–1352

Pillen K, Zacharias A, Léon J (2004) Comparative AB-QTL analysis in barley using a single exotic donor of *Hordeum vulgare* ssp. spontaneum. Theor Appl Genet 108:1591–1601

Pritchard J-K, Stephens M, Rosenberg N-A, Donnelly P (2000) Association mapping in structured populations. Am J Hum Genet 67:1070–181

Rafalski A, Morgante M (2004) Corn and humans: Recombination and linkage disequilibrium in two genomes of similar size. Trends Genet 20:103–111

Rannala B, Slatkin M (2000) Methods for multipoint disease mapping using linkage disequilibrium. Genet Epidemiol 9:S71–77

Reif J-C, Melchinger A-E, Frisch M (2005) Genetical and mathematical properties of similarity and dissimilarity coefficients applied in plant breeding and seed bank management. Crop Sci 45:1–7

Reyes-Valdes M-H (2000) A model for Marker-based selection in gene introgression breeding programs. Crop Sci 40:91–98

Roldán-Ruiz I, Calsyn E, Gilliland T-J, Coll R, van Eijk M-J-T, De Loose M (2000) Estimating genetic conformity between related ryegrass (*Lolium*) varieties. 2. AFLP characterization. Mol Breed 6: 593–602

Rouppe van der Voort J, van der Vossen E, Bakker E, Overmars H, van Zandvoort P, Hutten R, Klein Lankhorst R, Bakker J (2000) Two additive QTLs conferring broad-spectrum resistance in potato to *Globodera pallida* are localized on resistance gene clusters. Theor Appl Genet 101:1122–1130

Salvi S, Tuberosa R (2005) To clone or not to clone plant QTLs: present and future challenges. Trends Plant Sci 10:297–304

Salvi S, Tuberosa R, Chiapparino E, Maccaferri M, Veillet S et al (2002) Toward positional cloning of Vgt1, a QTL controlling the transition from the vegetative to the reproductive phase in maize. Plant Mol Biol 48:601–613

Schauer N, Semel Y, Roessner U, Gur A, Balbo I, Carrari F, Pleban T, Perez-Melis A, Bruedigam C, Kopka J, Willmitzer L, Zamir D, Fernie A-R (2006) Comprehensive metabolic profiling and phenotyping of interspecific introgression lines for tomato improvement. Nat Biotechnol 24:447–454

Septiningish E-M, Trijatmiko K-R, Moeljoparwiro S, McCouch.S-R (2003) Identification of quantitative trait loci for grain quality in an advanced backcross population derived from the *Oryza sativa* variety IR64 and the wild relative *O rufipogon* Theor Appl Genet 107:1433–1441

Servin B, Martin O-C, Mézard M, Hospital, F (2004) Toward a theory of marker-assisted gene pyramiding. Genetics 168:513–523

Shendure J, Mitra R-D, Varma C, Church G-M (2004) Advanced sequencing technologies: methods and goals. Nat Rev Genet 5:335–344

Sicard D, Woo S-S, Arroyo-Garcia R, Ochoa O, Nguyen D, Korol A, Nevo E, Michelmore R (1999) Molecular diversity at the major cluster of disease resistance genes in cultivated and wild *Lactuca* spp. Theor Appl Genet 99:405–418

Smid K-J, Tórjèk O, Meyer R, Schmuths H, Hoffman M-H, Altmann T (2006) Evidence for large-scale population structure of *Arabidopsis thaliana* from genome-wide single nucleotide polymorphis markers. Theor Appl Genet 112:1104–1114

Stuurman J, Kuhlemeier C (2005) Stable two-element control of dTph1 transposition in mutator strains of *Petunia* by an inactive ACT1 introgression from a wild species. Plant J 41:945–55

Syed N-H, Chen Z-J (2004) Molecular marker genotypes, heterozygosity and genetic interaction explain heterosis in *Arabidopsis thaliana*. Heredity 94:295–304

Syvanen A-C (2001) Accessing genetic variation: genotyping single nucleotide polymorphisms. Nat Rev Genet 2:930–942

Syvanen A-C (2005) Toward genome-wide SNP genotyping. Nat Genet 37:10

Szalma S-J, Buckler E-S, Snook M-E, McMullen M-D (2005) Association analysis of candidate genes for maysin and chlorogenic acid accumulation in maize silks. Theor Appl Genet 110:1324–1333

Takahashi Y, Shomura A, Sasaki T, Yano M (2001) Hd6, a rice quantitative trait locus involved in photoperiod sensitivity, encodes the alpha subunit of protein kinase CK2. Proc Natl Acad Sci USA 98:7922–7927

Tanksley S-D, Grandillo S, Fulton T-M, Zamir, D, Eshed Y, Petiard V, Lopez J, Beck-Bunn T (1996) Advanced backcross QTL analysis in a cross between an elite processing line of tomato and its wild relative *L. pimpinellifolium*. Theor Appl Genet 92:213–224

Tanksley S-D, McCouch S-R (1997) Seed banks and molecular maps: unlocking genetic potential from the wild. Science 277:1063–1066

Tanksley S-D, Nelson J-C (1996) Advanced backcross QTL analysis: A method for the simultaneous discovery and transfer of valuable QTLs from unadapted germplasm into elite breeding lines. Theor Appl Genet 92:191–203

Tanksley S-D, Young N-D, Paterson A-H, Bonierbale M-W (1989) RFLP mapping in plant breeding: new tools for an old science. Biotechnology 7:257–264

The Complex Trait Consortium (2004) The collaborative cross, a community resource for the genetic analysis of complex traits. Nat Genet 36:1133–1137

The International HapMap Consortium (2005) A haplotype map of the human genome. Nature 437:27

Thornsberry J-M, Goodman M-M, Doebley J, Kresovich S, Nielsen D, Buckler E-S (2001) IV, Dwarf8 polymorphisms associate with variation in flowering time. Nat Genet 28:286–289

Thumma B-R, Nolan M-F, Evans R, Moran G-F (2005) Polymorphisms in cinnamoyl CoA reductase (CCR) are associated with variation in microfibril angle in *Eucalyptus*. Genet 171:1257–1265

van Eeuwijk F-A, Malosetti M, Yin X, Struik P-C, Stam P (2005) Statistical models for genotype by environment data: from conventional ANOVA models to eco-physiological QTL models. Aust J Agric Res 56:883–894

Vos P, Hogers R, Bleeker M , Reijans M, van der Lee T, Hornes M, Frijters A, Pot J, Peleman J, Kuiper M, Zabeau M (1995) AFLP: a new technique for DNA fingerprinting. Nucleic Acid Res 23: 4407–4414

Vuylsteke M, Kuiper M, Stam P (2000) Chromosomal regions involved in hybrid performance and heterosis: thir AFLP ®-based identification and practical use in prediction models. Heredity 85: 208–218

Watson & Crick (1953) Molecular structure of nucelic acids. Nature 4356:737–738

Williams J, Kubelik A, Livak K, Rafalski J, Tingey S (1990) DNA polymorphisms amplified by arbitrary primers are useful as genetic markers. Nucleic Acids Res 18:6531–6535

Wilson L-M, Whitt S-R, Ibanez-Carranza A-M, Goodman M-M, Rocheford T-R, Buckler E-S (2004) Dissection of maize kernel composition and starch production by candidate gene association, Plant Cell 16:2719–2733

Witcombe J-R, Hash C-T (2000) Resistance gene deployment strategies in cereal hybrids using maker-assisted selection: gene pyramiding, three-way hybrids and synthetic parent populations. Euphytica 112:175–186

Young N-D (1999) A cautiously optimistic vision for marker-assisted breeding. Mol breed 5:505–510

Zamir D (2001) Improving plant breeding with exotic genetic libraries. Nat Rev Genet 2:993–989

Zöllner S, Pritchard J-K (2005) Coalescent-based association mapping and fine mapping of complex trait loci. Genetics 169:1071–1092

CHAPTER 4

MODELING QTL EFFECTS AND MAS IN PLANT BREEDING

MARK COOPER*, DEAN W. PODLICH AND LANG LUO
Pioneer Hi-Bred International. 7250 NW 62nd Ave, Johnston, IA 50131-0552, USA

Abstract: The empirical evidence accumulated to date indicates that the genetic architecture of the different traits of organisms, emphasizing here those relevant to plant breeding, should be viewed as a genetic complexity continuum. This concept is not new to plant breeders. What is new is that geneticists and plant breeders can now apply high throughput molecular technologies to identify and study the genes and alleles responsible for the standing genetic and phenotypic variation for traits in elite breeding populations. Plant breeders undertake research to develop robust breeding strategies that take advantage of this growing body of trait genetics knowledge and seek breeding methods that can be practically applied to improve multiple traits to achieve defined breeding objectives. While experimental and quantitative methods are developed to detect quantitative trait loci (QTL) and to implement marker-assisted selection (MAS) for the detected trait QTL as components of a comprehensive plant breeding strategy, simulation modeling methods can be applied to quantify the robustness of the chosen QTL analysis and MAS methods for the trait genetics complexity continuum. We review methods that can be applied to model the effects of QTL and outcomes from MAS in plant breeding as our view of the trait genetic complexity continuum unfolds. Some key lessons from this body of research are discussed.

1. INTRODUCTION

Today there is widespread interest in the field of genetics and much ongoing research aimed at mapping and studying genetic variation for the regions of the genome that influence the phenotypic variation for many different traits of organisms. A motivation for these efforts is an anticipated predictive capability associated with achieving some level of characterization of the genetic architecture of the standing variation for traits in natural and constructed reference populations by identifying

*Corresponding Author: mark.cooper@pioneer.com

R.K. Varshney and R. Tuberosa (eds.), Genomics-Assisted Crop Improvement: Vol. 1: Genomics Approaches and Platforms, 57–95.

quantitative trait loci (QTL) for the standing variation within the genomes of the organisms. The QTL so discovered can be useful entry points for many other types of genetic investigations conducted to understand the fine detail of the DNA sequence polymorphisms in the region of the QTL and to understand the gene-to-phenotype relationships for the traits. In plant breeding these broad areas of interest have progressed to the stage of questioning how best to use the results of these mapping studies to enhance the success rate in breeding programs by marker-assisted selection (MAS; e.g., Lande and Thompson 1990, Openshaw and Frascaroli 1997, Podlich et al. 2004, Johnson 2004, Niebur et al. 2004, Moreau et al. 2004a, Crosbie et al. 2006). Given these interests in the use of QTL as targets for MAS as part of a comprehensive breeding program strategy, we consider motivations for modeling the effects of QTL and MAS in plant breeding programs. Some recent illustrative examples are considered and two simulation experiments are included to complement the literature review and demonstrate some key points (Figure 1). The simulation experiments are used mainly as examples to demonstrate that relevant questions and situations can be tackled in a comprehensive way by applying appropriate modeling approaches within the context of the key components of a breeding program: i.e., the germplasm and elite reference population of the breeding program (Rasmusson 1996), the genetic architecture of the standing variation for the traits in the reference genotype-environment system (Cooper et al. 2005) and the breeding strategy (Hallauer and Miranda 1988, Comstock 1996). Today, simulation modeling methods can be applied as powerful complementary approaches when comprehensive investigation by closed-form theoretical and empirical methods is not feasible for the questions and genetic systems under consideration (Kempthorne 1988, Podlich and Cooper 1998, 1999, Cooper et al. 2005, Walsh 2005).

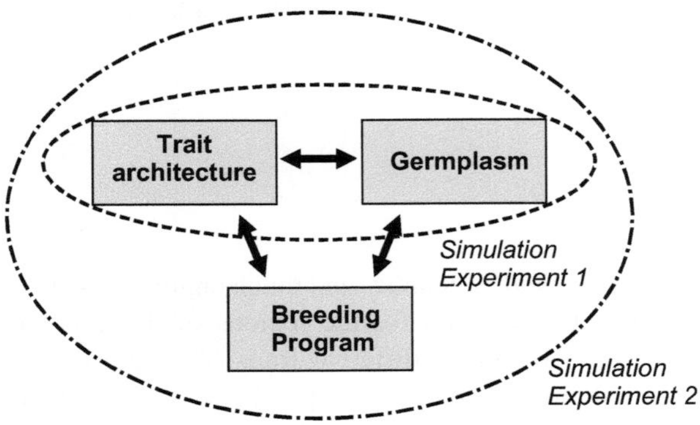

Figure 1. Focus of two simulation experiments conducted to investigate aspects of Quantitative Trait Locus (QTL) detection and Marker Assisted Selection (MAS) within the context of the germplasm reference population and trait genetic architecture components of a breeding program: (1) Simulation Experiment 1 QTL detection, (2) Simulation Experiment 2 QTL detection and application within a MAS strategy. The dashed lines indicate the areas of emphasis for Simulation Experiments 1 and 2

Each day we learn more about the gene-to-phenotype architecture of many of the traits of a diverse array of organisms and also the combined impact that genetic and environmental variation has on the phenotypic variation for these traits. This growing and developing bank of knowledge is created from diverse studies that span microorganisms in laboratory and natural environments, plants and animals in their natural ecology, plants and animals in agricultural systems and humans in different societies. However, while we have learnt much we still know a lot less than we would like to know about many of the genetic and environmental properties of these genotype-environment systems in order to predict key aspects of their behavior. While our knowledge of some systems is comprehensive and can be used effectively for prediction in particular situations our views for complex traits such as yield and stress tolerance are still often context dependent for many of the reasons that are discussed below. These context dependencies will always set limits on the predictive power of the gene-to-phenotype models we construct and therefore should be examined and understood. We are in the same situation for the agricultural systems within which plant breeding programs operate. Further, agricultural systems continually change, sometimes gradually and other times rapidly relative to the timeframe of a breeding program cycle. As our understanding of the components and dynamics of these agricultural systems advances, breeding programs continue to operate with objectives to improve the level and stability of yield, abiotic and biotic stress tolerance and the important end-use quality traits of the varieties of plants used by farmers for agricultural production (e.g., Figure 2; Duvick et al. 2004). In systems where formal breeding and selection programs have had a long history there are many opportunities to study the genetic basis of the progress that has already been made and the types of genetic changes that have contributed to the realized changes and improvements that have been achieved for the phenotypes of different traits (Figure 2; Rajaram and van Ginkel 2001, Campos et al. 2004, Duvick et al. 2004, Janick 2004a,b). Knowledge of the genetic changes that have contributed and those that did not contribute to genetic gain for the stated breeding objectives in the past can be used as a foundation for understanding the properties of the current genotype-environment systems and as a basis for developing predictions of viable experimental and applied paths to explore further opportunities for genetic improvements in the productivity and sustainability of crops within the target agricultural systems.

The design of comprehensive breeding strategies that utilize MAS for multiple traits is an extremely interesting and challenging scientific problem relevant to many aspects of applied plant breeding today. The common result observed from experimental investigations designed to map traits is a partial picture of the genetic architecture of the standing variation for the traits under investigation (Openshaw and Fascaroli 1997, Schön et al. 2004, Cooper et al. 2005). Expectations of the genetic changes that can be realized from MAS can be defined based on the assumption of the additive effects of the QTL identified by trait mapping (Lande and Thompson 1990, Walsh 2005). In combination with the additive effects of QTL alleles and the associated additive genetic variation, footprints of the influences

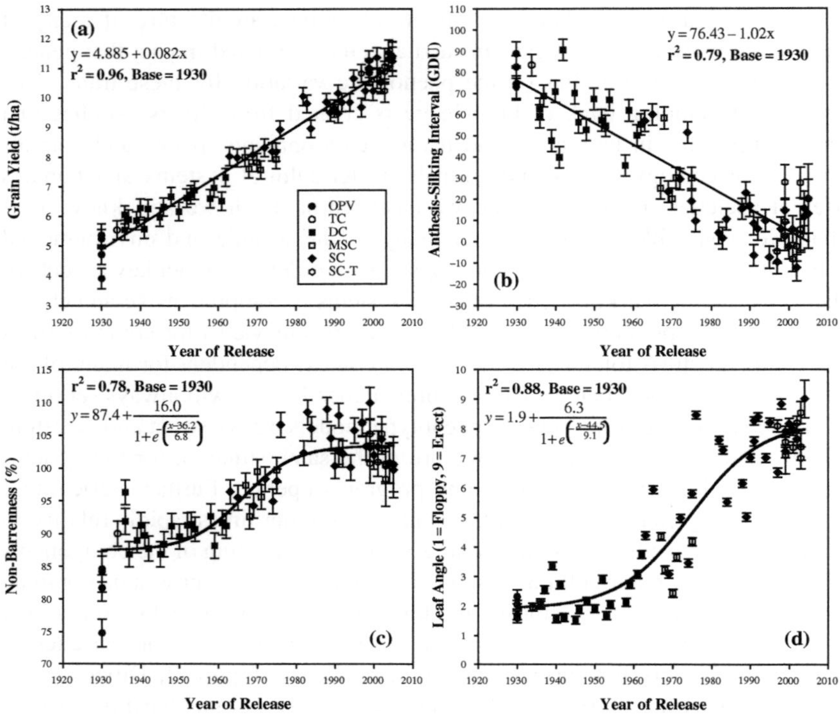

Figure 2. Changes in elite hybrid trait performance over the history of the Pioneer breeding program; (a) grain yield, (b) Anthesis-Silking Interval, (c) Non-Barrenness measured as the percentage of non-barren plants, (d) Leaf angle measured on a 1 to 9 rating scale with 1 = floppy leaves and 9 = erect leaves. Hybrid trait means are computed as Best Linear Unbiased Predictors (BLUPs) across multiple environments sampled in side-by-side experiments conducted from 1990 to 2005. For further details see Duvick et al. (2004)

of a range of non-additive effects, attributed to epistasis, gene-by-environment interactions and pleiotropy, can also be observed (e.g., Moreau et al. 2004a,b, 2006, van Eeuwijk et al. 2005, Blanc et al. 2006, Carlborg et al. 2006, Holland 2006, Li et al. 2006). Thus, daily the breeder deals with a trait genetic complexity continuum. It is within the context of this trait genetic complexity continuum that we can consider the concept of modeling QTL effects and MAS and evaluating the merits of MAS breeding strategies.

2. MODELING METHODS

2.1. Theory to Application

The concept of modeling the properties of a system to enable prediction of its behavior is well accepted in many fields of science (Casti 1997, Kauffman 1993,

Wolfram 2002) and business (Axelrod and Cohen 1999, Bass 1999, Schrage 2000). In association with the evolution of faster computer hardware and software there have been advances in the theory and the range of modeling methods that can be applied to practical problems in genetics and plant breeding (Fraser and Burnell 1970, Kempthorne 1988, Lynch and Walsh 1998, Podlich and Cooper 1998, Cooper et al. 2002). The application of statistical modeling methods in population and quantitative genetics has a long history (Fisher 1918, Wright 1932, Crow and Kimura 1970, Kempthorne 1988, Falconer and Mackay 1996, Lynch and Walsh 1998). The range of statistical methods available to the geneticist has progressed; expanding from the familiar least squares methods applied to linear models to the uses of maximum likelihood and Bayesian approaches within various linear and non-linear modeling frameworks. With advances in technologies for high throughput genomic, environmental and trait phenotype measurements the availability of large multidimensional data sets has broadened the range of experimental investigations that can be conducted to study genetic variation for traits. As the genetic questions have become more comprehensive and complex the computational requirements have become more demanding and necessitated further advances in computing infrastructure and algorithms. With the availability of the required computing infrastructure, fast and comprehensive simulation capabilities have been developed to assist investigation of many of the challenging questions that are relevant to genetics and plant breeding (Podlich and Cooper 1998, 1999). These enhanced computing tools have opened up many new opportunities to study the properties of "real-world" complex systems (Casti 1997, Kauffman 1993, Williams 1997, Podlich and Cooper 1998, 1999, Micallef et al. 2001, Wolfram 2002, Chapman et al. 2003, Crutchfield and Schuster 2003, Wagner 2005, Newman et al. 2006). The main objective of this paper is to review and discuss some potential applications of simulation modeling methodologies as they have and can be used to identify and understand QTL effects of traits and to evaluate the potential for augmenting breeding by MAS. A secondary objective is to relate the status of the empirical trait mapping results that have accumulated to date to some of the expectations that emerge from modeling the properties of genotype-environment systems whenever interactions among the genetic and environmental components are present; the interactions emphasized here are inspired by the genetic concepts of epistasis, gene-by-environment interactions and pleiotropy. Our intent with this second objective is not to make firm conclusions, these fields of investigation are still in their early stages, but to encourage careful scientific consideration of the different mapping results that have accumulated to date.

2.2. Mapping Traits

Here we define a QTL to be any region of the genome that is associated with the standing variation for a trait phenotype in a relevant reference population that can be identified by one or more sequence-based DNA markers when they are applied in combination with a suitable experimental design and statistical analysis method. Similarly MAS is defined to be any application of the associated DNA markers to select for combinations of QTL alleles to create and test genotypes with a predicted trait phenotype in a target genotype-environment system. It is assumed that the markers can be arranged in the form of a genetic map that represents their linear order on chromosomes. It is also assumed that within the region identified as the QTL there is a functional polymorphism in the DNA sequence that influences the differential realization of the trait phenotype for different genotypes of the QTL. Thus, it is expected that mapping traits to identify QTL reveals information about the genomic positions of important functional polymorphisms in the DNA sequence of the organism that are present within the reference mapping population. The genetic and functional bases of QTL variation for traits and their influences on trait phenotypes are considered further below and elsewhere in this book. Typically MAS strategies used in applied plant breeding will involve some combined index utilizing QTL marker and trait phenotypic information.

The mapping resolution that can be achieved in defining the relevant regions of the genome by QTL analysis methods depends in part on the extent of linkage disequilibrium in the reference mapping population. In the perfect situation the marker sequence polymorphism and the functional DNA sequence polymorphism contributing to the trait phenotypic variation would be the same. This may result when a QTL has been previously cloned or when a candidate gene is used to identify markers and a polymorphism at the gene contributes to the standing quantitative trait variation. However, more typically the QTL are identified by a genome scan using many markers selected to cover the genome. In these cases the marker polymorphisms themselves are likely to be neutral and are associated with the functional polymorphism only by the linkage disequilibrium that exists for the DNA molecule within the reference population of genotypes. When such a genome scan is used to identify the QTL that are to be targets for MAS in applied breeding it is important to understand the extent of linkage disequilibrium in the reference population used for identification of QTL and that which exists in the elite breeding populations targeted for application of a MAS strategy. Linkage disequilibrium between marker and functional polymorphisms in the DNA sequence is necessary within a reference population to identify an association between a marker and a QTL. The extent of linkage disequilibrium in mapping populations can be manipulated by controlling the structure and number of cycles of inter-mating of individuals (e.g., Winkler et al. 2003). With relatively sparse genetic maps it will be necessary for the linkage disequilibrium to extend for large segments of the chromosomes to detect the associations. Alternatively with dense genetic maps it is an advantage to have less extensive linkage disequilibrium to enable finer mapping of the QTL. In parallel with the mapping studies that are used to identify QTL, the individuals within the reference population of a breeding program are continually inter-mated in designed crossing schemes and subjected to selection. Thus, in these breeding crosses recombination can continually operate to both break up current and create new physical linkage associations between alleles of different

QTL on the same chromosome and between markers and the functional polymorphisms of the QTL. Therefore, care must be taken to ensure that the linkage phases in the mapping study are relevant to those in the breeding reference population that will be the target for MAS. The two simulation experiments considered below are designed to take into consideration these influences of linkage disequilibrium (Figure 1).

Following the arguments given above, here we consider mapping traits within the context of an ongoing breeding program (e.g., Figures 1 and 2). Thus, any genetic gain from MAS will need to build on the progress that has been achieved by conventional breeding strategies. The realized genetic gain for quantitative traits that has been achieved by breeding can be understood as a long-term outcome from the application of open recurrent selection strategies that are designed to manage genetic diversity and manipulate multiple traits over multiple cycles of selection to improve and stabilize the yield and quality traits for the sets of genotypes grown by farmers (e.g., Figure 2; Rajaram and van Ginkel 2001, Duvick et al. 2004, Barker et al. 2004). Much of the breeding progress to date for complex traits such as yield has been achieved by pedigree and recurrent selection strategies (e.g., Hallauer and Miranda 1988, Comstock 1996, Duvick et al. 2004) applied to select for the desired trait phenotypes within relevant pools of genetic diversity, rather than by molecular enhanced approaches such as MAS. However, as with previous changes in core breeding methodology in the 20th Century there will be an exploratory transition phase and we can expect this situation to change as the 21st Century unfolds. The availability of large numbers of polymorphic markers that can be assayed rapidly and economically by high throughput technologies in large populations of genotypes has created interest and provides opportunity for augmenting the breeding process by MAS for the QTL polymorphisms at specific regions of the genome that are indicated as being responsible for the standing trait genetic variation (Cahill and Schmidt 2004, Niebur et al. 2004, Crosbie et al. 2006). This is a proven method for traits under the control of a few major genes or major QTL (e.g., Cahill and Schmidt 2004, Crosbie et al. 2006). The extension of this methodology to traits that are genetically more complex is feasible but requires consideration of the relative importance of the additive and non-additive components of genetic variation for the traits within the elite populations used for breeding and importantly an understanding of the genetic bases of these sources of variation and the potential influences of different trait genetics on the outcomes of the chosen MAS strategy (Cooper and Podlich 2002, Niebur et al. 2004, Podlich et al. 2004, Walsh 2005, van Eeuwijk et al. 2005, Cooper et al. 2005, Welch et al. 2005, Tardieu et al. 2005, Hammer et al. 2005). Walsh (2005) reminds us that to exploit the additive effects of alleles requires only identification of the desirable allele of a QTL and selection of that allele, regardless of the allele combinations at other loci. However, to exploit non-additive effects requires methods for identification of desirable combinations of alleles (for dominance, allele combinations at a single locus; and for epistasis, allele combinations at multiple loci) and selection of these allele combinations and their consistent transmission to subsequent generations. This effort becomes even more challenging in the presence of QTL with pleiotropic effects and QTL-by-Environment interactions (QEI). The empirical challenge that these complexities present in applied breeding is the need to conduct more comprehensive mapping studies to both discover the QTL and the desirable allele combinations and to evaluate their practical selection by MAS.

While we emphasize modeling methods in this paper, empirical evaluations of MAS strategies will always be necessary (e.g. Bouchez et al. 2002, Moreau et al. 2004a, Crosbie et al. 2006). However, there are many potential implementations of MAS and many details of the genetic architecture of traits to consider, making experimental evaluation of all possibilities impractical. Therefore, a combined empirical modeling evaluation of the potential of MAS strategies by focusing on some of the common genetic issues is likely to be a more feasible approach (Wang et al. 2003, 2004, Chapman et al. 2003, Cooper et al. 2005, Hammer et al. 2005). A common feature of the many alternatives that have been proposed is that marker alleles associated with favorable QTL alleles by coupling phase linkage are used to manipulate trait phenotypes by selecting for designated favorable combinations of the QTL alleles at one or more QTL. The trait phenotypes for the target QTL genotypes can be predicted based on experimentally determined effects of the QTL alleles. By defining and constructing some of the different target genotypes based on predictions from the multi-QTL models, validation experiments can be conducted to compare the predicted and realized phenotypes and to estimate the realized genetic gain from MAS. Even for restricted cases such comparisons of MAS with other breeding strategies is costly, they take considerable time and it is questionable whether the design of such experiments with adequate power is feasible for complex traits that are typically improved over multiple cycles of selection within a breeding program; the breeding program does not wait for the results of such studies. Given this non-stationary situation within applied breeding programs there is merit in modeling QTL detection methods and MAS strategies as part of a comprehensive research program organized to design, refine and optimize a breeding program if the results are to positively impact the outcomes of the breeding program.

3. QUANTITATIVE TRAITS

3.1. Phenotyping

Attention to relevant phenotyping is a critical component of any trait mapping experiment (Campos et al. 2004). Capacity for phenotyping should be developed in combination with efforts to develop the genetic resources and mapping tools. Phenotyping requires definition of the correct plant traits and environmental measurements, in addition to consideration of experimental plot design and the observational units to be used to take the measurements, equipment needs, throughput and skill required to take the measurements, experimental design and analysis methods, and an understanding of the characteristics of the target population of environments (TPE) within which the experiments are to be conducted. Choice of environments depends on the traits in question. For traits where prior evidence indicates relatively simple genetic architecture and no evidence of confounding genotype-by-environment interactions (GEI), replication to achieve a moderate to high heritability within a single environment may be sufficient for many coarse mapping objectives. When the genetics are more complex and GEI are known to be important, greater attention must be given to the number and types of environments used in the experiment. Options that have been considered range from managed environments chosen to impose specific conditions (e.g., Ribaut et al. 1996, 1997, 2004, Crossa et al. 1999, Vargas et al. 2006) to relatively large samples taken to represent different locations and years (e.g., Openshaw and Fascaroli 1997, Schön et al. 2004, Moreau et al. 2004b). In all cases the interpretation of the QTL effects and the determination of their importance for MAS can be enhanced by characterizing the relationship between the sample of environments used for QTL detection and the frequency of occurrence of these environmental conditions in the TPE (Chapman et al. 2000a,b,c, Moreau et al. 2004b, Löffler et al. 2005, Hammer et al. 2005).

For complex traits, such as yield and tolerance to abiotic and biotic stresses, consideration is often given to the scope for dissection of the primary trait into components that are expected to be simpler to work with than the ultimate trait of interest (e.g., Nguyen and Blum 2004, Ribaut et al. 1996, 1997, 2004, Campos et al. 2004). This approach has elements that can be viewed as advantages and disadvantages. It is often argued that trait dissection provides an important advantage when the component traits themselves have a higher heritability in the reference mapping population than the ultimate target trait and they have a genetic correlation with the target trait. For example, in the case of maize grown in environments where a water deficit during flowering is a common occurrence, the synchrony of the development of the male and female inflorescences (ASI; Anthesis to Silking Interval) on a plant and the variability of this synchrony among plants within an experimental plot is considered to be an appropriate target for investigation of genetic variation for drought tolerance that ultimately influences yield under the water deficit (Ribaut et al. 1996, 1997, 2004, Campos et al. 2004). This approach in maize is strongly motivated by the associated long-term trends in breeding programs towards higher grain yield and reduced ASI (e.g., Figure 2, Duvick 1977, Chapman et al. 1997, Duvick et al. 2004). However, a challenge that is associated with such trait dissection approaches is the difficulty of determining the impact that the component traits have on the target trait once they are integrated back into the whole plant response within the context of a TPE (Chapman et al. 2003, Campos et al. 2004, Hammer et al. 2005). Hammer et al. (2006) have argued and demonstrated a case for the use of crop growth models as appropriate quantitative frameworks for focusing these trait dissection and integration efforts. They reviewed three case studies where models with predictive success were developed for traits within plants. Their key point was that with any trait dissection strategy for complex traits a complementary trait integration strategy is necessary to determine the importance of and to deal with any effects of epistasis, GEI and pleiotropy and achieve impact from MAS in the target genotype-environment system. They demonstrated an application of this approach for the design of breeding strategies aimed to improve the grain yield of sorghum for drought prone environments (Chapman et al. 2003, Hammer et al. 2005).

3.2. Genetic Architecture

Thus, clearly the widespread availability of molecular markers and their organization in the form of genetic maps has opened up new opportunities to directly study the genetic architecture of standing genetic variation for quantitative traits by enabling mapping of the traits to identify QTL in the elite germplasm of breeding programs (Niebur et al. 2004, Crosbie et al. 2006). Further, with the availability of the complete genome sequence of the target organisms we can investigate the gene-to-phenotype properties of the DNA sequence polymorphisms contributing to the detected QTL within the target organism and other organisms through syntenic relationships. Similar efforts are underway in model organisms and these can inform the investigations conducted in agricultural plants (Mackay 2001, 2004, Welch et al. 2005). With identification of QTL for traits it becomes feasible to study their number, distribution across the genome, the number and frequency of alleles for each QTL and the influence of the alleles on trait phenotypes (e.g., Openshaw and Fascaroli 1997, Li et al. 2006). In turn, knowledge of the number and effects of QTL for quantitative traits presents the breeder with new possibilities to apply this genetic knowledge to extend the range of selection strategies used in breeding. It is important to recall that mapping will identify only those components of the traits where there is standing genetic variation within the chosen reference population of genotypes. The QTL and their effects are conditional on the reference population of genotypes and the environments within which the mapping studies were conducted and the rigor with which these inference spaces are sampled. Therefore, the views we obtain are likely to be always partial and context dependent and as such conditional on the reference germplasm and the details of the environments sampled from the TPE. While any one mapping approach is unlikely to reveal information for all of the genes responsible for the relevant trait biology, the identified QTL provide logical entry points for further studies of the gene-to-phenotype architecture of the traits and their contributions to genotypic variation for plant performance in agricultural environments (e.g., Laurie et al. 2004, Cooper et al. 2005, Li et al. 2006). Further, by combining the results from multiple mapping studies a more comprehensive view of the genetic architecture of the standing variation for traits can be obtained (e.g., Jansen et al. 2003, Chardon et al. 2004, Blanc et al. 2006).

Experience from applied breeding indicates that the genetic architecture of the standing variation for the traits that are targeted for improvement in the elite germplasm of a breeding program is most likely to be a genetic complexity continuum. The continuum extends from "simple" traits that are under the control of one (Mendelian) or a few (oligogenic) additive genes to more "complex" performance traits, such as grain yield and tolerance of abiotic stresses, that are the outcome of multiple component traits, each in turn oligogenic or under the control of multiple genes (polygenic). The appropriate QTL detection and MAS strategies will differ for traits along the genetic complexity continuum. The majority of the molecular evidence suggests that many of the genes involved in the variation for the traits are potentially under a combination of additive and non-additive influences arising from the effects of epistasis, pleiotropy and gene-by-environment interactions (Moreau et al. 2004b, Welch et al. 2005, Holland 2006, Carlborg et al. 2006, Li et al. 2006). Empirical evidence from map-based cloning of a few of the trait QTL identified to date indicates that the functional bases of the genetic polymorphisms associated with QTL can be diverse and may be in the regulatory or coding sequences of the genes that reside within the region of the chromosome indicated as the QTL (Doebley et al. 1995, Frary et al. 2000, Mackay 2004, Salvi and Tuberosa 2005). Thus, the breeder that is working with many traits simultaneously is continually dealing with all aspects of this genetic complexity continuum, from simple to complex trait genetics and from predominantly additive allele effects to strongly context dependent non-additive allele effects. This situation is not new to plant breeders, but what is new is that we can now study this trait complexity continuum at the genetic level by applying appropriate QTL mapping strategies. At the genetic level useful models of the trait genetic complexity continuum can be formally defined and quantified within the *E(NK)* modeling framework (Kauffman 1993, Cooper and Podlich 2002). Therefore, some of the gene-to-phenotype properties of traits can be modeled as a continuum extending from simple finite locus models based on one or a few genes to complex networks of interacting genes (Kauffman 1993, Clark 2000, Cooper and Podlich 2002, Cooper et al. 2005, Hammer et al. 2006). The finite locus and mixed finite locus and polygenic models studied within the *E(NK)* framework can be compared and combined with the more classical statistical models (van Eeuwijk et al. 2005, Walsh 2005, Cooper et al. 2005). From a quantitative genetics and an applied breeding perspective the concept of trait genetic complexity can be combined with the concept of trait heritability to examine the discovery power of different trait QTL mapping methods and to evaluate the relative strengths and weaknesses of different MAS breeding strategies. Two illustrative examples demonstrating some of these applications of the *E(NK)* modeling framework are considered below (Figure 1).

3.3. Statistical Modeling Methods

The regions of interest that are studied as QTL are usually identified by the presence of a significant statistical association between the sequence polymorphisms of the markers and phenotypic variation for the traits of interest. Both marker genotype and trait phenotype are measured on individuals sampled from a relevant reference population of genotypes. Recommended statistical methods have been developed, but this is still an area of ongoing research. Attention to the impact of population size on the distribution of estimated QTL effects is necessary to avoid inflation of the estimates of QTL effects (Beavis 1998, Openshaw and Fascaroli 1997, Utz et al. 2000, Schön et al. 2004).

If additive QTL effects are the predominant component of the genetic variation for traits then appropriate closed-form expressions can be used to model the expected response to selection from alternative breeding strategies as a function of the proportion of the additive genetic variation accounted for by the QTL. Lande and Thompson (1990) provided examples of such expressions for a range of breeding strategies. Alternatively, where non-additive effects predominate or the standing additive genetic variation is strongly conditioned by the non-additive effects these expressions may not capture important properties of the genetic architecture of the traits that can influence both the interpretation of QTL mapping studies and the outcomes of MAS. In this more complex situation, simulation modeling methods in combination with comprehensive mapping efforts can be used to evaluate many detailed aspects of MAS strategies (Niebur et al. 2004, Podlich et al. 2004, Cooper et al. 2005).

3.4. Gene Networks

As the results from large numbers of trait mapping studies accumulate our view of the genetic architecture of the standing variation for many traits is broadening. Summaries from multiple studies indicate distributions of QTL effects that are suggestive of a genetic architecture where there are a few QTL with large effects and many QTL with small effects (e.g., Kearsey and Farquhar 1998). The early mapping studies were conducted in relatively small mapping populations (<200 individuals) with the ambition to identify a few major QTL with broad relevance. For some simple traits this model of the genetic architecture of traits seems to be applicable. For many of the complex yield and stress tolerance traits of relevance to breeding in elite populations this does not appear to be a general result (Openshaw and Frascaroli 1997, Schön et al. 2004, Carlborg et al. 2006, Li et al. 2006). The synoptic view obtained from analysis of multiple mapping experiments strongly suggests that in addition to some consistent additive QTL effects with broad applicability across genetic backgrounds and environments the genetic architecture of traits in elite breeding populations includes important components of epistasis, gene-by-environment interaction and pleiotropy. Therefore, not unexpectedly the relative importance of these different components of the genetic architecture changes with trait, reference population of genotypes and environments. Collectively these observations indicate we require approaches for studying the genetic architecture of traits that allow diagnosis of the relative contributions and importance of the different additive and non-additive components that contribute to the standing genetic variation in elite breeding populations. We have tackled this as a genetic modeling problem by developing a flexible quantitative framework for studying the genetic architecture of traits in terms of gene networks (Kauffman 1993, Omholt et al. 2000, Cooper and Podlich 2002, Peccoud et al. 2004, Cooper et al. 2005, Holland 2006). Therefore, the classical Mendelian and additive polygenic genetic models are viewed as special cases within the more general framework. This framework builds on the study of properties of networks that has its origins in graph and complex systems theory (e.g., Kauffman 1993, Williams 1997, Crutchfield and Schuster 2003, Dorogovtsev and Mendes 2003, Wagner 2005, Newman et al. 2006). Importantly, the theoretical framework can be applied to design and analyze trait mapping experiments to estimate important properties of the gene-to-phenotype architecture of complex traits (Cooper et al. 2002, van Eeuwijk et al. 2005, Welch et al. 2005, Tardieu et al. 2005, Hammer et al. 2005) and to study the impact of these properties on the expected outcomes of breeding strategies (Cooper et al. 2002, Chapman et al. 2003, Podlich et al. 2004, Peccoud et al. 2004, Wang et al. 2003, 2004, Cooper et al. 2005).

3.5. Crop Growth and Development Models

Crop growth and development models have been proposed as a natural quantitative framework that serves the dual roles of guiding trait dissection and trait integration for mapping complex traits; (1) trait dissection to study the genetic architecture of the complex traits, and (2) integration of the information generated from the trait dissection investigations to study and quantify the effects of trait QTL on whole plant response in the TPE (Hammer et al. 2005, 2006). The challenge of predicting effects at the level of the trait phenotype from changes at the level of the DNA sequence can be viewed as a scaling problem in biology, where the objective is to scale knowledge of DNA sequence variants from the level of the molecular polymorphism to predict the relevant trait phenotype of the selected plant genotypes replicated across the environments of the TPE. For the effective application of molecular genetics to plant breeding this interface between trait dissection and integration appears to be fertile territory for productive interactions between experimental and theoretical trait genetics research (e.g., Bower and Bolouri 2001, Hammer et al. 2006). Hammer et al. (2006) considered desirable properties of models for such applications. They focused on the levels of detail that are likely to be required to adequately capture the key features of an interacting set of genes to enable prediction of their collective influence on plant growth and development processes. Three case studies were reviewed; (1) the developmental transition from the vegetative to the reproductive stage in *Arabidopsis thaliana* (Welch et al. 2005), (2) leaf expansion by maize under variable environmental conditions (Tardieu et al. 2005), and (3) multi-trait drought tolerance strategies for improvement of grain yield of sorghum in a diverse set of dryland environments (Hammer et al. 2005). In each of the case studies genetic variation was related to a model of the target physiological process that responded to key environmental inputs. The physiological models they considered did not capture all of the molecular details of the interactions among the genes influencing the traits. Instead the models captured the integrated behavior of sets of genes by way of coefficients that quantified the influence of environmental variables on the key plant growth and development processes that influenced the traits studied; Tardieu et al. (2005)

referred to these key growth and development relationships as meta-processes. Van Eeuwijk et al. (2005) demonstrated how statistical mixed models can be defined to detect and predict the effects of QTL for the coefficients of the plant growth and development processes targeted by Tardieu et al. (2005) and Hammer et al. (2005).

4. QUANTITATIVE TRAIT MODELING: QTL ANALYSIS AND MAS STRATEGIES

Combining the modeling concepts discussed above it is possible to apply simulation and statistical methods to model QTL detection methods and MAS breeding strategies (Figure 1). Some aspects are demonstrated below. A recurrent population improvement scenario (Figure 3) is used as the breeding strategy example and trait genetic architecture is simulated as a complexity continuum applying the *E(NK)* modeling framework (Kauffman 1993, Cooper and Podlich 2002).

At all stages of a breeding program experiments are designed to measure trait phenotypes for the sampled genotypes in a sample of test environments. We refer to these experiments in general as multi-environment trials (METs; Figure 3). In combination with measuring trait phenotypes the genotypes can be fingerprinted to characterize genetic polymorphism and the environments can be characterized to understand the relationship of the MET to the TPE (Chapman et al. 2003, Moreau et al. 2004b, Löffler et al. 2005). The trait data generated from METs can be organized as a two-way table with rows indexed for different genotypes and columns indexed for different environments. Each cell in the table represents a phenotypic observation on a genotype in an environment. This basic structure can be extended for multiple observations per trait and multiple traits. Thus, the phenotypes

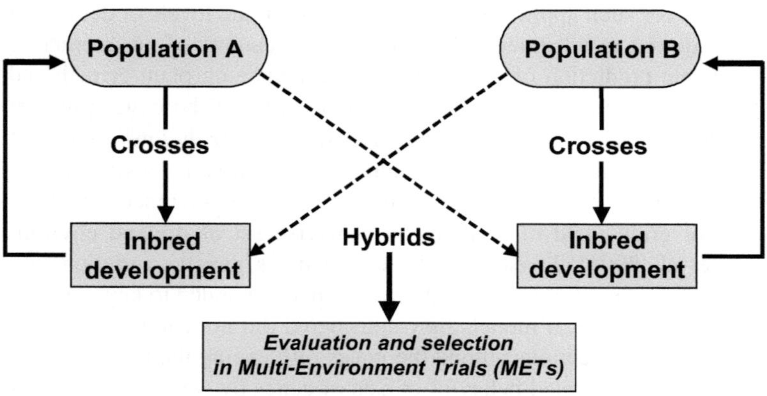

Figure 3. Schematic representation of a reciprocal recurrent selection breeding strategy used to simulate the breeding strategy in Simulation Experiment 2

for traits measured on the sample of genotypes in the sample of environments ($\underline{P}_{ijr}$) can be modeled within a statistical framework as:

$$(1) \qquad \underline{P}_{ijr} = \mu + \underline{G}_i + \underline{E}_j + \underline{(GE)}_{ij} + \underline{\varepsilon}_{ijr}$$

Where μ is the grand mean, $\underline{G}_i$ is a genotypic main-effect of individual i, i=1,..., I, $\underline{E}_j$ is the environmental main-effect of environment j, j=1,..., J, $\underline{(GE)}_{ij}$ is the genotype-by-environment interaction effect for the combination of individual i and environment j, and $\underline{\varepsilon}_{ijr}$ is the residual effect for phenotypic observation r on individual i in environment j, r=1,...R. The underlining of the terms in the model indicates that these are treated as random variables sampled from the reference genotype-environment system being explored by the breeding program. A common statistical assumption is to assume that the random terms are variables distributed as Normal(0,σ^2). A range of different specifications and extensions of this statistical model with associated biological interpretations of the terms can be given within a mixed model framework (e.g., van Eeuwijk et al. 2001, 2005, Smith et al. 2002a,b). To provide a complementary framework for a genetic interpretation of the observed trait variation we can also consider components of the trait phenotypic variation arising from the combination of a "*genetic signal*" component ($G_i + (GE)_{ij}$), an "*environmental context*" component (E_j) and an "*environmental noise*" ($\underline{\varepsilon}_{ijr}$) component. The genetic signal component can be viewed as a genetically determined trait value for individual i with genotype $(NK)_i$ as a physiological outcome of the combined action of the alleles of the N genes influencing the trait for individual i, within the context of environment E_j. To emphasize that the N genes can interact to determine an observable trait outcome for a genotype the K parameter is used to indicate the interaction topology of a gene network influencing the trait (Kauffman 1993, Cooper and Podlich 2002, Cooper et al. 2005); K=0 indicates the N genes act independently in the model and increasing levels of K indicate increasing levels of interaction among the N genes. Thus, the genetic signal component ($G_i + (GE)_{ij}$) can be written as $E_j(NK)_i$; read as the NK genotypic value for genotype i within environmental context j. This form of the $E(NK)$ model provides a basis for relating the gene-to-phenotype biology dimension of the trait to the genetic variation dimension that exists within a particular reference population of genotypes, such as the elite genetics of a breeding program, and its reference genotype-environment system. Here the noise component $\underline{\varepsilon}_{ijr}$ is taken to be an outcome of the systematic and random sources of environmental variation associated with growing and measuring the trait phenotypes of the sampled genotypes in an experimental sample of environments. Taking the genetic signal, environmental context and noise components together we can rewrite equation (1) as:

$$(2) \qquad \underline{P}_{ijr} = E_j(NK)_i + \underline{\varepsilon}_{ijr}$$

Some of the relationships between the form of the *E(NK)* model given in equation (2) to the extensions of model (1) used for QTL mapping of traits have been discussed previously (Cooper et al. 2005, van Eeuwijk et al. 2005). Applying equations (1) and (2) together we can consider QTL mapping from the perspective of extracting the genetic signal component associated with the *N* genes contributing to the trait phenotypic variation and the interpretation of the characterized genetic variation within an *E* environmental and *NK* genetic background context. Understanding the genetic and environmental contexts of the detected genetic signal provides a sound basis for interpretation of the QTL effects and evaluating the opportunities for MAS. In the two *in silico* examples considered below we will use aspects of the relationship between equations (1) and (2) to simulate the process of QTL detection in mapping studies and the application of the detected QTL for enabling MAS within a breeding strategy. Equation (2) is used to simulate an ensemble of different genotype-environment systems, ranging from simple to complex, and QTL mapping methods are applied within the statistical framework indicated by equation (1) to identify the *N* genes and their effects on the trait phenotype.

To simulate an ensemble of genotype-environment systems, for any *E(NK)* gene-to-phenotype model for a trait we define a set of *N* genes and distribute these across a genome of 10 chromosomes (Figure 4). For the applications considered here the *N* genes have positions in the genome that are defined in terms of a genetic map. For the selected levels of *E* and *K* each of the *N* genes is characterized for any gene-by-environment interaction effects and epistatic effects by identifying the other *K* genes with which it interacts (Figure 4) and the number of environment specific effects. Pleiotropic effects are not explicitly considered in the two examples below. The *E(NK)* model effects for the *N* genes can be derived in a number of ways. Approaches we have considered include, sampling from an underlying statistical distribution of gene effects (e.g., Kauffman 1993, Cooper and Podlich 2002), direct specification of genotypic values where knowledge of physiological epistasis is available (*e.g.*, Podlich and Cooper 1998, Cooper and Podlich 2002), defining appropriate sets of differential equations to capture the dynamics of biochemical pathways (*e.g.*, Peccoud et al. 2004), defining coefficients for appropriate processes within crop growth and development models (Chapman et al. 2003, Hammer et al. 2005). From any of these different parameterizations of the *E(NK)* model genotypic values can be determined for any genotype-environment combination within the reference system by applying equation (2). Clearly the investigator will select an approach relevant to their objectives. Given a definition of the *E(NK)* model we can simulate a MET within the context of a breeding program by sampling a set of *I* genotypes from the relevant stage in the breeding program and determining their genotypic values in a sample of *J* environments applying the chosen *E(NK)* model. The $\underline{\varepsilon}_{ijr}$ noise component can be added to the genotypic values by defining the appropriate trait heritability for the reference population of genotypes and applying this to the sample of *I* genotypes to determine an appropriate empirical error variance component. Given a noise component Normal($0,\sigma_{\varepsilon}^2$)

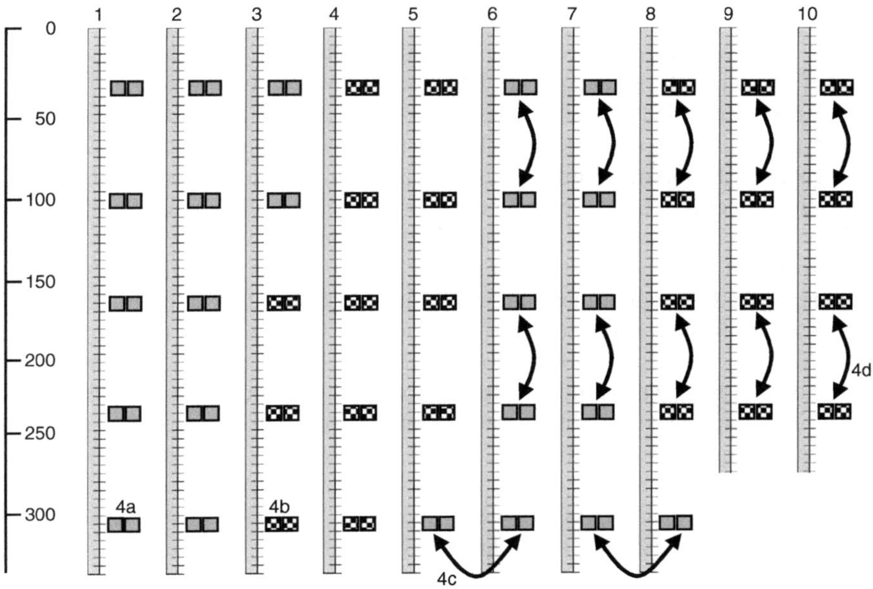

Figure 4. Representation of an example genome used in the simulation experiment. Genome is shown for an *E(NK)*=2(48:1) genetic model (i.e. two environment types, 48 genes influencing trait performance, additive and epistatic gene interactions; labeled as *K*=1 for simplicity). Molecular markers are shown as lines. Genes are shown as squares. Genes that have the same performance across the two environment types are shown in grey. Genes that have different performances across the two environment types (i.e. gene-by-environment interaction) are shown in black. Genes involved in di-genic epistatic networks are highlighted by arrows. Example gene effects for the different types of gene action are shown in Figure 5; Labels 4a=additive, 4b=gene-by-environment, 4c=epistatic, 4d=combination of epistatic and gene-by-environment interaction

random $\underline{\varepsilon}_{ijr}$ variables can be added to the $E_j(NK)_i$ genotypic value to give a phenotypic $\underline{P}_{ijr}$ value for each genotype. Once the trait phenotypes are defined for the simulated MET (Figure 3) any appropriate QTL analysis method can be applied to the simulated MET data set (e.g., van Eeuwijk et al. 2005). For the purposes of this study we define the objective of the QTL mapping step in the simulation experiments as the identification of as many of the *N* genes as possible and the correct estimation of the global additive effects of the identified subset of the *N* genes.

5. DEFINING THE BREEDING PROGRAM AND GERMPLASM CONTEXT

To investigate QTL detection and MAS for the trait genetic complexity continuum within a breeding program context the QU-GENE software (Podlich and Cooper 1998) was applied to simulate population improvement over five cycles of Reciprocal

Recurrent Selection (RRS; Figure 3). Many of the detailed steps involved in the breeding cycle were omitted from the simulation of the breeding program to focus on specific aspects of modeling QTL detection and MAS. Additional details of the breeding cycle can be included when considered necessary. Following the definition of an appropriate ensemble of *E(NK)* models to be used to represent the putative properties of the gene-to-phenotype architecture of a trait a suitable reference population of founder genotypes is defined for the breeding program. Two related simulation experiments are considered below; (1) Simulation Experiment 1 was designed to investigate aspects of the power of QTL detection and definition of QTL effects within the context of a bi-parental mapping study, and (2) Simulation Experiment 2 was designed to investigate applications of the results of QTL analyses to enable MAS over multiple cycles of a breeding program (Figure 1). Here we do not attempt to simulate or investigate any particular breeding program or exemplify a particular MAS strategy. The emphasis is on demonstrating that a comprehensive modeling approach can be applied to investigate the QTL effects that are likely to be detected from a QTL analysis approach and to consider the impact of a MAS strategy that utilizes the detected QTL information. As discussed above the scenario chosen here can be replaced and parameterized for the situations of interest to the investigator.

5.1. Simulation Experiment 1

5.1.1. Modeling QTL detection

In Simulation Experiment 1 (Figure 1) the results from QTL analysis of a bi-parental cross were examined for a range of simple to complex polygenic gene-to-phenotype models (Table 1).

Applying equation (2) an ensemble of *E(NK)* genetic models and associated bi-parental mapping populations was simulated for a single trait for N=48, E=1, 2, 5, K=0, 1, 2. As discussed above the objective is to demonstrate the modeling process rather than conduct a comprehensive analysis of a particular scenario. Therefore, the ensemble considered here is small and restricted in range of complexity compared to other investigations we have undertaken previously, but is sufficient to demonstrate the modeling process. The number of genes was held constant to limit the size of the simulation experiment; N=48 was chosen to represent a moderate sized

Table 1. E(NK) genetic model combinations, heritability levels and replication applied to simulate a trait genetic complexity continuum for Simulation Experiments 1 and 2

Parameter type	Levels	Level Details
Number of genes	1	N=48
Number of environments	3	E=1,2,5
Level of epistasis	3	K=0,1,2
Heritability	5	H=0.1,0.3,0.5,0.7,0.9
Model parameterizations	50	Replicates 1 to 50

polygenic system. For E=1 there is one environment-type for the TPE. For E=2 and E=5 there are multiple environment-types and gene-by-environment interactions are introduced. For the E=2 and E=5 models the different environment-types were assumed to occur with equal frequency in their respective TPE; E=2 frequency $E1=E2=0.5$ and for E=5 frequency $E1=E2=E3=E4=E5=0.2$. Further, for the E=2 cases, a subset of 24 genes from the total of N=48 genes was defined to have no gene-by-environment interaction and the remaining 24 genes were defined to have gene-by-environment interactions. For the E=5 cases, there was an increase in the number of genes defined to have gene-by-environment interactions relative to environment-type 1, ranging from 10, 19, 29 and 38 genes across environment-types 2 through 5, respectively. For the K=0 cases, all 48 genes were defined as additive (i.e., no epistatic interactions). For the K=1 cases, there were 24 additive genes and 12 di-genic epistatic networks. For the K=2 cases, there were 12 additive genes and 12 tri-genic epistatic networks. For each of the 9 factorial combinations of E and K, samples of 50 parameterizations of the *E(NK)* model were created by sampling the genetic effects from a Uniform(0,1) distribution (Kauffman 1993, Cooper and Podlich 2002). Typical examples of the types of gene effects that can be generated are shown for a subset of the N genes in Figure 5; i.e., (a) additive, (b) gene-by-environment, (c) epistatic, and (d) combined epistasis and gene-by-environment effects).

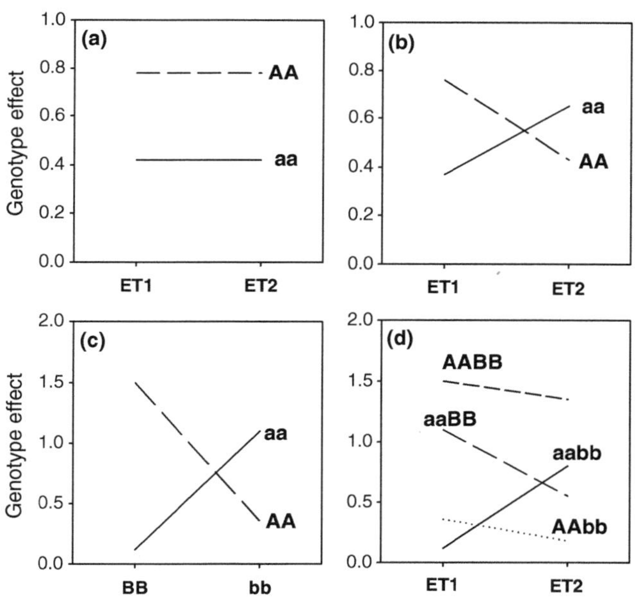

Figure 5. Example set of gene effects generated by the *E(NK)* model. The types of effects shown are: (a) additive, (b) gene-by-environment, (c) epistatic, and (d) combination of epistatic and gene-by-environment. Only the homozygous genotypes are shown. The simulated genome locations of the genes responsible for the simulated genotype responses are shown in Figure 4

Five levels of environmental noise (i.e., error in equations (1) and (2); $\underline{\varepsilon}_{ijr} \sim N(0, \sigma_\varepsilon^2)$) were considered to represent different levels of heritability ranging from high to low (Table 1). Environmental noise across the experiment was defined in relation to the *E(NK)* = 1(48:0) case (i.e., additive genetics case). Error variance component estimates ($\hat{\sigma}_\varepsilon^2$) were computed from $\hat{\sigma}_\varepsilon^2 = \hat{\sigma}_g^2 \left(\frac{1}{H} - 1\right)$ for five levels of broad sense heritability $H = 0.1, 0.3, 0.5, 0.7, 0.9$, based on the estimated genetic variance $\hat{\sigma}_g^2$ present in the *E(NK)* = 1(48:0) case. For a defined level of H the estimates of the error variances were applied across all genetic models (i.e., all levels of E and K). Thus, the five levels of H applied in the simulation experiment represent a constant $\hat{\sigma}_\varepsilon^2$ that is referenced to the *E(NK)* = 1(48:0) model. In total 2,250 combinations of *E(NK)* model, parameterizations and heritability levels were considered (9×50×5).

To simulate a bi-parental mapping population two individuals contrasting for all N=48 genes were defined as parents. The two individuals were crossed to generate an F1 from which a random set of doubled-haploids (DHs) was generated for mapping. In all cases a mapping population size of 500 DHs was used. This population sample size is large in comparison to that used in many published mapping studies, but is not large enough to avoid the Beavis effect for polygenic systems such as those considered here (Beavis 1998, Openshaw and Fascaroli 1997, Schön et al. 2004). The phenotypes of the 500 DH lines used to map the QTL of the trait were measured in a single environment sampled at random from the TPE. QTL mapping in a single environment was chosen to demonstrate within the simulation experiments some influences of context dependent effects due to gene-by-environment interactions. For each of the 2,250 genetic model combinations, 20 bi-parental mapping population replicates were considered. The 20 replicates represented different samples of the 500 DHs. Thus, a total of 45,000 DH populations were created for QTL mapping. The QTL mapping of the trait was conducted using composite interval mapping as implemented in QTL Cartographer V1.16 available for the Linux operating system (Basten et al. 1995). The QTL Cartographer software and the composite interval mapping method were chosen because of their wide use as reported in the literature. Other mapping methods can be applied. All QTL tests were conducted to test for additive QTL effects and no explicit tests were conducted for QTL-by-QTL effects or QTL-by-Environment effects. These are potential and useful extensions that we recommend in order to generalize the analyses considered here, but were considered beyond the illustrative scope of this investigation. Forward and backward regression (SRmapqtl with method FB) was used to search for cofactors. QTL scanning was conducted every 2 cM with two additional parameters, where genetic background was set to 10 and window size was 10 cM. QTL were deemed to be detected if the peak of the LOD profile exceeded the defined empirical significance threshold and fell within 15 cM of either side of the true QTL position. For each analysis the number of QTL identified and the estimated additive effects of the QTL were recorded. The number of QTL detected was compared with the total number of simulated QTL (N=48) as a measure of the power of QTL detection. The estimated QTL effects were compared with the true QTL effects graphically to obtain measures of the Type I (false positives), Type II (false negatives) and Type III (correct detection of QTL position with incorrect ranking of the alleles for additive effects relative to the global additive effects) error rates associated with detection and definition of QTL effects. True QTL effects for *E(NK)* genetic models with gene-by-environment interaction (*E*>1) were computed by averaging allele effects across individual environment-types in the TPE. True QTL additive effects for *E(NK)* genetic models with epistasis (*K*>0) were computed by contrasting allele effects at a gene, averaged across the rest of the genetic background in a given epistatic network; i.e., these average effects were intended to define 'global' additive effects for each of the *N* genes within the complete system of *N*=48 genes segregating in the mapping population.

5.1.2. Results

For all *E(NK)* models, the number of QTL detected decreased with heritability level (Figure 6). The largest percentage of true QTL detected was for the gene-to-phenotype models with no epistatic interactions (*K*=0, Figure 6a–c). There was a decrease in the percentage of QTL detected when epistatic interactions were included in the gene-to-phenotype models (Figure 6d–f, *K*=1; Figure 6g–i, *K*=2). In terms of QTL detection, there was little difference in the results for increasing levels of *E*. The limited influence of gene-by-environment interaction on the total number of QTL detected in this case was due to the fact that the mapping populations were evaluated in a single environment-type sampled from the TPE and the crossover types of gene-by-environment interactions simulated (e.g., Figure 5b).

For the additive *E(NK)*=1(48:0) models the effect of decreasing heritability was a reduction in the ability to detect the QTL with small effects (Figure 7). While there were fewer large-effect QTL than small-effect QTL present to be detected (Figure 7b,e,h) a greater percentage of the large-effect QTL that were present were consistently detected (Figure 7c,f,i). In general there was a tendency to over-estimate the true effects of the subset of the *N* genes detected as QTL. The tendency to over-estimate the true effects increased with decreasing heritability (Figures 7a *cf.* 7b, 7d *cf.* 7e, 7g *cf.* 7h).

The QTL effects were most accurately estimated for the additive *E(NK)*=1(48:0) models at *H*=0.9 (Figure 8a). For this combination of *E(NK)* model and heritability 30.4% of the QTL were detected. The overall number of Type I, II and III errors was low relative to the number of correct detections of QTL and determination of the additive effects of the alleles. The majority of the Type II errors were associated with genes of small effect. Increasing the complexity of the gene-to-phenotype model by introducing epistasis (*K*>0) and gene-by-environment interactions (*E*>1) created greater uncertainty in the correct detection and determination of the true additive gene effects by QTL detection (Figure 8). With the inclusion of gene-by-environment and epistatic interactions, there was a tendency to over-estimate the additive effect relative to the true additive effect. Notably there was also an increase in the number of Type III errors relative to the additive model at all levels of heritability (Figure 8; e.g., 8a *cf.* 8b,c). For all *E(NK)* genetic models, the capacity

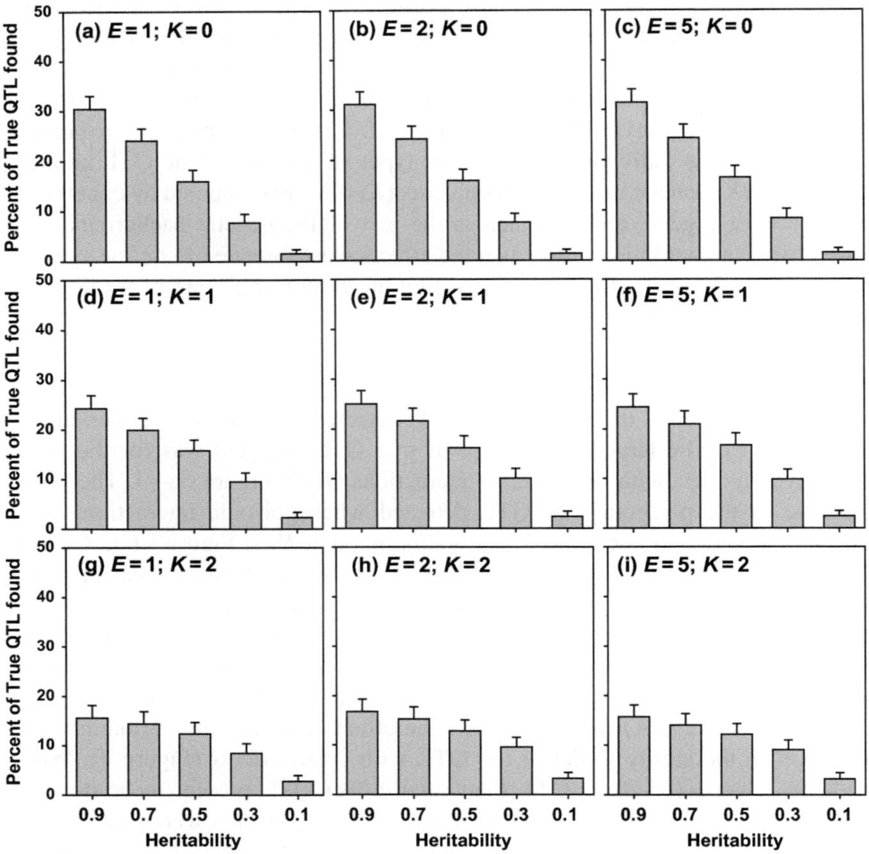

Figure 6. Percent of true QTL detected by Composite Interval Mapping (CIM) in the bi-parental mapping populations. Results are shown for nine genetic models (i.e. factorial combinations of E and K) and five levels of heritability (H=0.9,0.7,0.5,0.3,0.1). For each combination, the results are the average across the 50 genetic parameterizations and 20 bi-parental replications (i.e. 1000 data sets)

to detect and correctly estimate additive gene effects decreased with heritability (Figure 8; e.g., 8a,d,g). With lower heritability, the effects of the correctly detected QTL tended to be over estimated. Further, genes with small true effects could still be declared as significant QTL and in these cases the effect of the gene was greatly overestimated by the declared QTL effect. While the number of QTL detected decreased with the presence of epistasis and gene-by-environment interaction effects, some of the N=48 genes were still detected as QTL, even though the correct definition of their global additive effects was more difficult. The results obtained from simulating QTL mapping for different genetic model scenarios emphasize some of the challenges associated with QTL mapping. For example, the over-estimation of QTL effects in genetic models with gene-by-environment

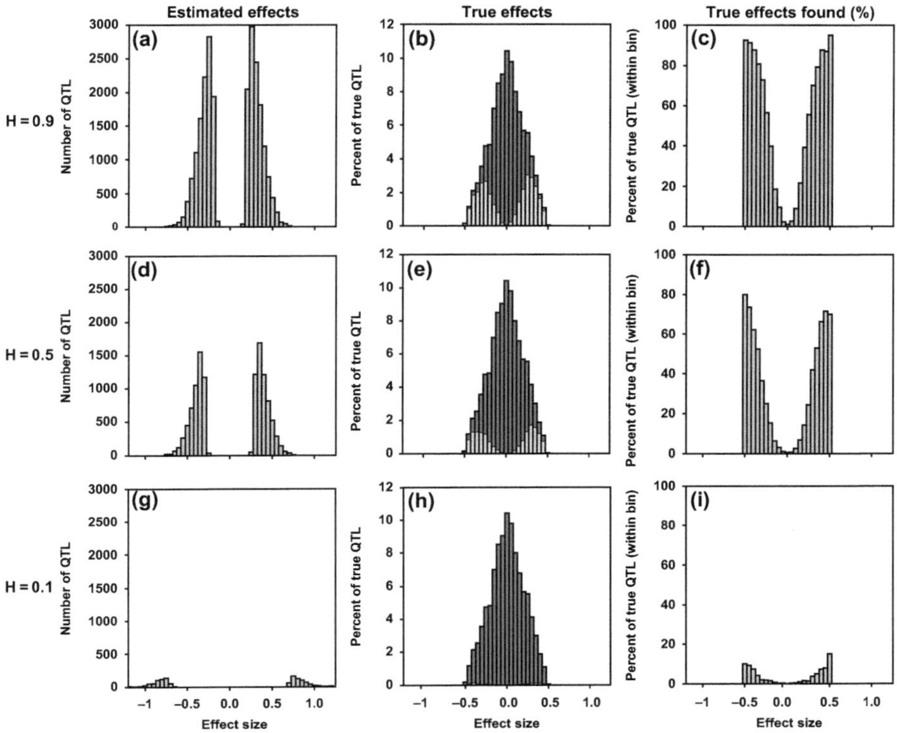

Figure 7. Distribution of estimated and true QTL effects detected for the *E(NK)*=1(48:0) model at three heritability levels. Panels (a), (d), (g) show the distribution of estimated QTL effects. Panels (b), (e), (h) show the distribution of true QTL effects. The QTL detected in the mapping studies are shown in light grey. The QTL not detected in the mapping studies are shown in dark grey. Panels (c), (f) and (i) show the distribution of true QTL effects for the set of true QTL detected in the mapping studies

and epistatic interactions are indicative of the fact that QTL effects from mapping studies are often estimated based on a partial picture of the genotype-environment system (e.g., in this experiment the mapping populations were evaluated in a single environment-type from the TPE). The implications of these results for MAS are considered further in Simulation Experiment 2.

5.2. Simulation Experiment 2

5.2.1. QTL detection and MAS

In Simulation Experiment 2 (Figure 1), the results of QTL mapping studies, as described in Simulation Experiment 1, were used as inputs into a MAS breeding strategy for comparison with the genetic gain achieved by phenotypic selection (PS). The MAS and PS selection schemes were investigated within a RRS

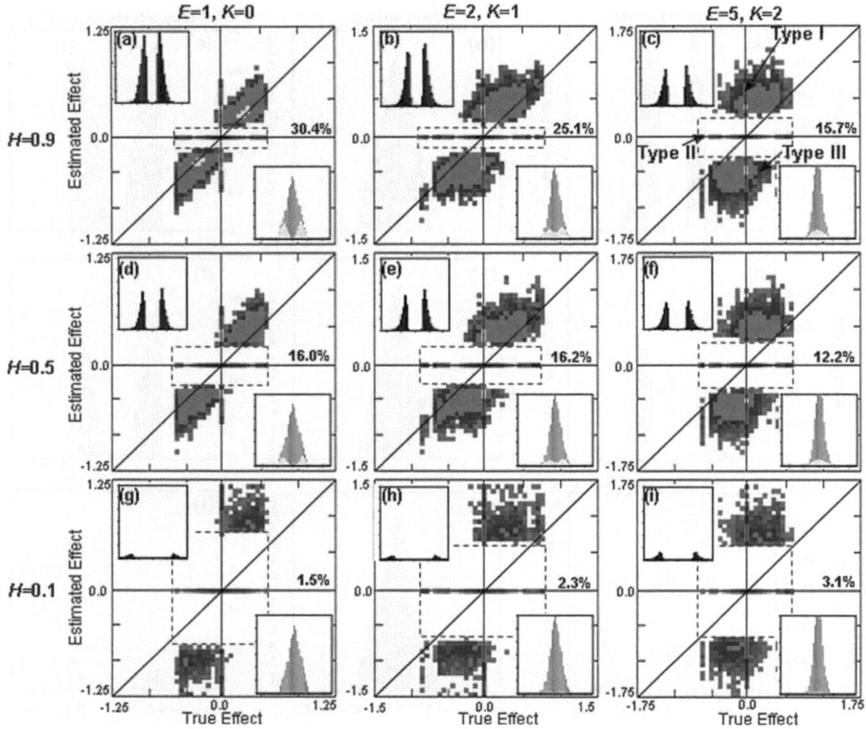

Figure 8. True and estimated additive QTL effects for three genetic models (*E(NK)*=1(48:0); *E(NK)*=2(48:1); *E(NK)*=5(48:2)) and three levels of heritability (*H*=0.9,0.5,0.1). Results are shown as a heat plot, using true and estimated QTL detected by Composite Interval Mapping (CIM) in the bi-parental mapping populations. For each sub-panel, the results are displayed for the 50 genetic parameterizations and 20 bi-parental replications (i.e. 1000 data sets). Colors range from cyan through dark red. Type I, II and III errors are highlighted by arrows. Type I errors represent cases where QTL were falsely detected in a given map region (i.e. false positives), Type II errors represent cases where the true QTL were not detected by CIM, (i.e false negatives) and Type III errors represent cases where the QTL were correctly detected but the estimated favorable allele was incorrectly defined. The percentage of true QTL detected is listed in each sub-panel *(see plate 2)*

breeding strategy conducted for five cycles (Figure 3). The objective of the RRS breeding strategy was defined as increasing trait value for single-cross hybrids; e.g. to simulate the case of selecting for higher grain yield of hybrids (Figure 2). The reference genotype-environment systems, genetic models and QTL mapping methods for Experiment 2 were the same as those defined for Experiment 1 (Table 1). To implement a MAS strategy the results of the QTL analysis were applied to define a Target Genotype (TG). Given the breeding objective was to achieve increased trait value the TG within any cycle of the breeding program was based on the definition of the favorable additive allele effects for the identified QTL. Selection was then implemented by using the identified QTL allele profiles of the individuals to identify desirable bi-parental crosses that would enable an increase in the number of favorable QTL allele effects in progeny of the cross. In combination with QTL-based cross selection, phenotypic selection for increased trait value was conducted within each cross based on the results of a simulated MET. Consistent with the RRS strategy all individuals were tested for trait phenotypic performance in testcross combination, with testers selected from the complementary heterotic group (Figure 3). The selection was conducted for higher trait performance within both of the heterotic groups and genetic progress was measured in terms of improvements in hybrid performance, where the hybrids were single-cross F1 hybrids based on combinations of elite inbreds, with one parent taken from each of the two heterotic groups.

Following Podlich et al. (2004) a Mapping-As-You-Go strategy was applied over the five cycles of selection. Therefore, a QTL analysis was conducted for the founding reference population of genotypes and for both heterotic groups for each cycle of selection as in Simulation Experiment 1. The parents used in each QTL mapping study were defined in conjunction with the germplasm present at any given cycle of the breeding program. The genotypes were evaluated for their trait phenotype in a single environment-type sampled at random from the TPE. For each mapping reference population at each cycle of selection extreme individuals were identified, one with a low and one with a high trait phenotype. The extreme individuals were crossed to generate an F1 from which a random set of 500 DHs was generated for mapping as in Simulation Experiment 1. However, in contrast to Simulation Experiment 1, not all genes were guaranteed to be segregating in the mapping study in any given cycle of selection. In total, QTLCartograher was run 225,000 times in Simulation Experiment 2 (9 genetic models × 50 genetic parameterizations × 5 heritability levels × 20 breeding replications × 5 cycles of selection). Thus, the QTL analysis and interpretation procedures described for Simulation Experiment 1 were completed 225,000 times in Simulation Experiment 2 to implement the MAS strategy. In parallel with the MAS strategy a comparable PS strategy was conducted for the same reference populations. The response for MAS was then measured in terms of the advantage demonstrated over phenotypic selection (MAS-PS) for increase in trait value in the TPE.

5.2.2. Results

The number of detected QTL segregating in the mapping population was typically much less than the total $N = 48$ polymorphic genes in the reference population of the breeding program (Figure 9a–c). The largest number of QTL detected was usually observed in cycle 1 and the number of QTL detected decreased over the five cycles of selection (Figure 9a–c). For all genetic models the number of QTL detected decreased with heritability. For the genetic models with increasing levels of epistasis (Figure 9b,e $K = 1$ and 9c,f $K = 2$) there was the potential for an increased genetic signal component relative to the noise component, compared to that defined for the additive genetic model (Figures 9a,d $K = 0$), and there was an

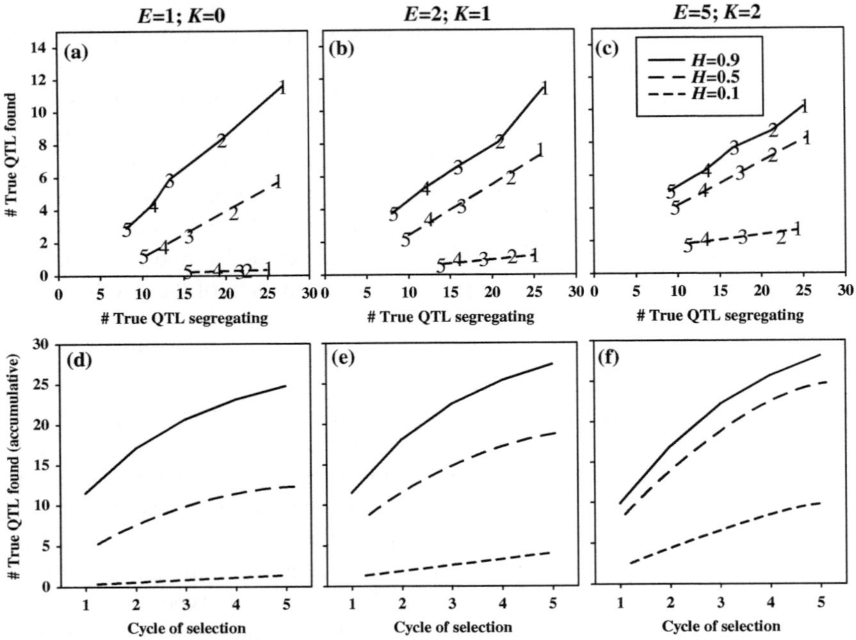

Figure 9. The average number of true QTL segregating against the average number of true QTL detected for five cycles of selection (a-c) and the average cumulative number of true QTL detected over five cycles of selection (d-f). The results are shown for three genetic models ($E(NK) = 1(48:0)$; $E(NK) = 2(48:1)$; $E(NK) = 5(48:2)$) and three levels of heritability ($H = 0.9, 0.5, 0.1$). The numbers on panel (a)-(c) represent the five cycles of selection. The average cumulative number of true QTL was computed taking into account the number of "new" QTL detected each cycle of selection (i.e. QTL that were not detected in any previous cycle of selection). For each combination, the results were computed from the 50 genetic parameterizations and 20 breeding replications (i.e. 1000 simulations)

associated increase in the number of QTL identified (Figure 9b,c) compared to the additive model at the same starting level of heritability (Figure 9a).

For the additive model (Figure 9a), there was a relatively large decrease in the number of QTL detected from cycle to cycle over the five cycles of selection. The genetic models with gene-by-environment and epistatic interactions (Figure 9 b,c) showed different patterns of decrease in QTL detection over the cycles of selection. With high heritability (H=0.9) the genetic models with higher levels of gene-by-environment interactions and epistasis had an initially lower number of QTL detected than for the additive genetic model (e.g. Figure 9c *cf.* 9a). However, the cumulative number of QTL detected was higher over cycles of selection compared to the additive model (Figure 9f *cf.* 9d). The cumulative QTL discovery results suggest that when gene-by-environment and epistatic interactions were a component of the genetic architecture of the trait, new components of genetic variation were more consistently identified by QTL mapping over cycles of selection compared to the additive model (Cheverud and Routman 1996, Podlich et al. 2004).

For the RRS MAS strategy considered there was genetic improvement in the population trait mean over the five cycles of selection (Figure 10). For all genetic models considered response to selection decreased with lower heritability. The rate of improvement of the trait, normalized relative to the global target genotype with the highest possible trait performance, decreased with the increased levels of genetic complexity introduced by including and increasing the levels of epistasis and gene-by-environment interactions.

The genetic change over cycles of breeding, as measured in terms of genetic distance (Hamming Distance) from the global target genotype, differed among the *E(NK)* genetic models (Figure 11). For the additive model (Figure 11a) progress towards the global target genotype was achieved over the five cycles of selection and was more consistent and less variable (Figure 11b) when compared to the genetic models with epistasis and gene-by-environment interactions (Figures 11a *cf.* 11c,e mean genetic distance; Figures 11b *cf.* 11d,f variance of genetic distance). There are

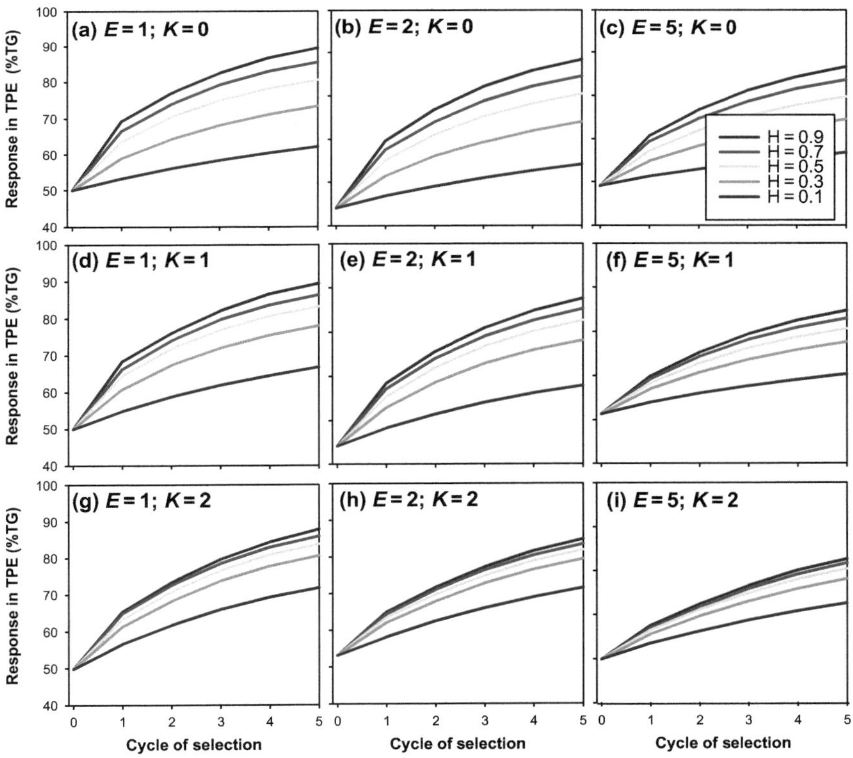

Figure 10. Average response in the TPE for the nine genetic models (factorial combinations of E and K) and five levels of heritability over five cycles of selection. For each combination, the results were computed from the 50 genetic parameterizations and 20 breeding replications (i.e. 1000 simulations) *(see plate 3)*

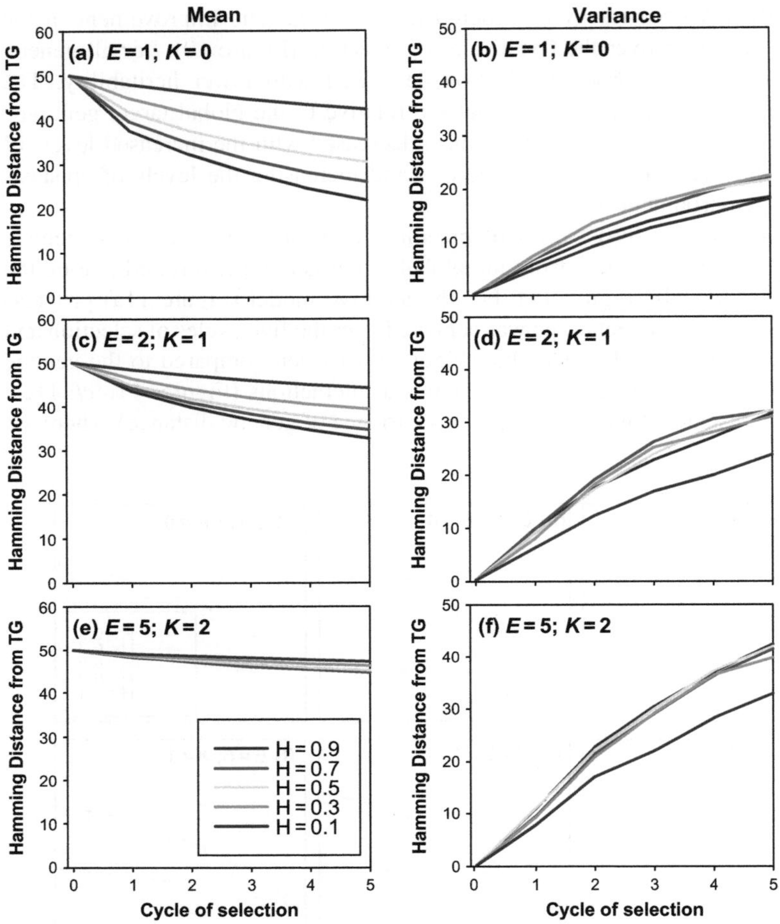

Figure 11. Mean and Variance of Hamming Distances (HD) of hybrid combinations from the target genotype for three genetic models (*E(NK)* = 1(48:0); *E(NK)* = 2(48:1); *E(NK)* = 5(48:2)) and five levels of heritability (*H* = 0.9,0.7,0.5,0.3,0.1), over five cycles of selection. For each combination, the results were computed from the 50 genetic parameterizations and 20 breeding replications (i.e. 1000 simulations) *(see plate 4)*

important differences in the paths towards genetic improvement of the trait between the additive $E(NK) = 1(48{:}0)$ genetic models and those with epistasis ($K > 0$) and gene-by-environment ($E > 1$) interactions. For the additive model all paths to genetic improvement lead towards the global target genotype and a more restricted number of paths towards genetic improvement were exploited among the different replicates. Thus, there was a decrease in the genetic distance between the breeding population and the global target genotype over cycles of selection and the rate of approach to the target genotype was greater with higher heritability (Figure 11a).

When epistasis was a component of the genetic model there were multiple workable paths towards genetic improvement of the trait from the initial reference population. Many of these paths to higher trait performance resulted in the creation of genotypes with higher trait performance that were different from the global target genotype that was defined as a reference point. A number of these different genetic paths towards trait improvement were identified among the different replicates of the breeding programs starting from the same reference population conditions. Thus, when epistasis was a component of the genetic model, genetic improvement in trait performance was still achieved over the cycles of selection (Figure 10) while on average there was less progress towards the global target genotype (Figures 11c,e cf. 11a) and greater variation among replicates (Figures 11d,f *cf.* 11b).

On average the MAS selection strategy resulted in a greater rate of increase in trait performance compared to the PS strategy for the five cycles of selection (Figure 12). Thus, augmenting phenotypic selection with selection based on QTL

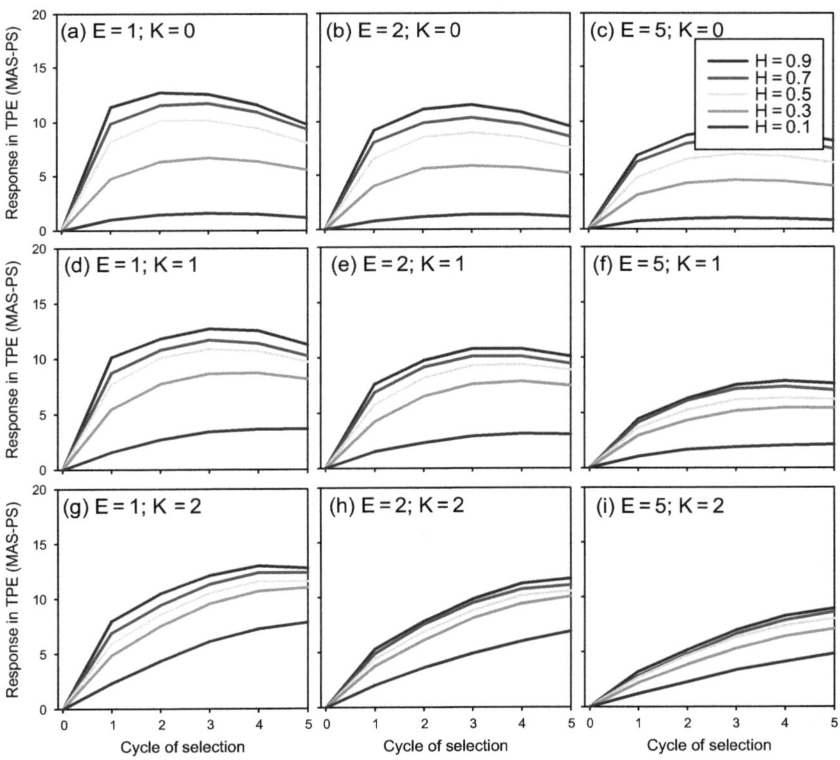

Figure 12. Difference in the average response (Marker-assisted selection – Phenotypic selection) for the nine genetic models (factorial combinations of *E* and *K*) and five levels of heritability over five cycles of selection. For each combination, the results were computed from the 50 genetic parameterizations and 20 breeding replications (i.e. 1000 simulations of each breeding strategy) *(see plate 5)*

model predictions contributed to greater rates of genetic gain over multiple cycles of breeding. While the magnitude differed the advantage of MAS over PS was observed for all genetic models and heritability levels considered. For the additive model, following an initial increase in the advantage of MAS for the first two cycles the advantage over PS had begun to decrease by cycle 5. For the more complex genetic models with epistasis as a component of the genetic architecture of the trait the increase in the advantage of MAS over PS persisted for more cycles of selection (e.g. Figures 12d,g *cf.* 12a) and in some cases continued to increase over the five cycles of selection (Figures 12h,i). For all genetic models the advantage of MAS over PS decreased with lower heritability (Figure 12). This was consistent with the result that the number of QTL detected decreased with decreasing heritability (Figure 9).

6. IMPLICATIONS OF MODELING

A comprehensive modeling approach can be applied to assist a breeder in understanding the likely outcomes of a QTL analysis method, the properties of the detected QTL effects given the selected analysis strategy and also their potential influence on the outcomes of a MAS strategy. Importantly with advances in simulation modeling methods these methods can be applied to many relevant contexts that cover the likely trait genetic complexity continuum relevant to the breeder and need not be restricted to considering only the additive effects of polygenic systems. These opportunities have previously been emphasized by Kempthorne (1988) and Lande (1991). We advocate the application of comprehensive modeling approaches in combination with a relevant program of empirical investigations to map the genetic architecture of the traits. The *in silico* experiments can be harmonized with and parameterized for the questions relevant to MAS within the target breeding program and the genotype-environment system context.

The two illustrative simulation experiments that were considered here were designed to be simple relative to the complexity of the trait genetics that can be encountered within real genotype-environment systems and also in terms of the specific details involved in conducting multiple cycles of a breeding program. Our intention was not to model any specific "real-world" scenario but to remove as much of the detail as possible to focus on a few key points relevant to QTL mapping of traits and MAS in breeding in general. Some results that were emphasized by the simulation experiments were:

1. Improvements in trait phenotyping that lead to either higher heritability by reducing sources of experimental error or refined definitions of the key traits to be measured and the environment-types within which they should be measured will lead to improved power to detect and characterize QTL and enhance the opportunities to achieve positive results from MAS.
2. The results from trait mapping studies are expected to change over the course of the cycles of a plant breeding program as selection changes the composition of the standing genetic variation. Thus, trait mapping that is intended to impact

MAS in breeding programs must be done within the context of the reference germplasm of the breeding program to obtain reliable indications of the potential benefits of MAS.

3. Selection is effective at achieving genetic improvements in trait phenotypes along the genetic complexity continuum. This result is consistent for both theoretical (Cooper and Podlich 2002, Podlich et al. 2004) and importantly with empirical studies (Figure 2; Duvick et al. 2004, Crosbie et al. 2006) of genetic gain from breeding. With appropriate attention to trait heritability, QTL will be identified for traits along the genetic complexity continuum, from the simplest to most complex traits. Moving beyond classical Mendelian systems in specific reference populations it would seem appropriate to consider the results of any trait mapping study as entry points into understanding the genetic architecture of the complex traits. These entry points can be built on with subsequent studies.
4. Theoretical and empirical treatments of QTL analysis and MAS indicate that it is easier to demonstrate impact from MAS for traits under more simple genetic control but that the major opportunities for long-term improvements over conventional breeding strategies are actually for the more complex traits. This experimental to application gap requires us to extend our current quantitative genetics modeling framework to more explicitly deal with the complexities of the genetic architecture of traits, such as epistasis, gene-by-environment interactions and pleiotropy, if we are to realize the potential advantages from MAS for complex traits in applied breeding (Cooper et al. 2005, Holland 2006, Hammer et al. 2006).

While we demonstrated some aspects of the points listed above in two relatively simple simulation experiments our experience is that they apply equally to more comprehensive situations involving higher levels of genetic complexity and that are more inclusive of the details of the genotype-environment system and the breeding programs under consideration (Cooper and Podlich 2002, Wang et al. 2003, 2004, Cooper et al. 2005). Below we consider further details of these points and their implications for MAS in plant breeding.

What have plant breeders been able to learn to date from the results of the different trait mapping experiments? The current empirical evidence obtained from mapping traits in elite breeding populations indicates that the genetic architecture of traits is a complexity continuum that extends from simply inherited traits controlled by one or a few additive genes to complex traits under the control of many interacting genes that frequently influence more than one trait. The view of the genetic complexity continuum that is achieved by any one study will likely be context dependent and depend on the traits considered and the germplasm and environments examined (Cooper et al. 2005; see also comments by Sing et al. 2003 and McClearn 2006). This is not a negative result but is an important consideration that should caution investigators against broad generalizations from the context dependent perspectives of their individual experiments. Further, in addition to mapping the more traditional plant traits we can now map molecular phenotypes and study many of the details that are involved in scaling effects that traverse from the level of expression and translation of DNA sequence variation, biochemical pathway and molecular signaling components to the levels of multiple trait phenotypes of plants grown across multiple environments. Much of this molecular detail suggests there is potential for complex context dependent effects of the allelic variation for genes with the potential to introduce non-additivity into the gene-to-phenotype architecture of traits (e.g., Holland 2006, Li et al. 2006, Carlborg et al. 2006). However, the presence of a complex genetic architecture including epistasis, pleiotropy and gene-by-environment interactions does not preclude the identification of QTL that exhibit properties of additive genetic variation within a given reference population, as shown in the above two simulation experiments. Thus, even with complex trait genetics both the currently available theory and empirical evidence are consistent in indicating that many forms of genetic architecture can be observed depending on the choice of trait and reference population of genotypes and environments for QTL mapping. The empirical evidence that demonstrates the presence of important interactions in the gene-to-phenotype architecture of traits often forces us to reject the global applicability of the naïve additive model and at least be circumspect of the level of predictability that can be achieved from studies that show predominantly additive effects. The reference system may well behave additively in its present context. However, breeders work with germplasm and manage utilization of populations to achieve levels of predictability over multiple cycles of selection (Bubeck et al. 2006). Therefore, the important question for the breeder is whether the current additivity can be used to project into the new genetic states of the system that will be created by selection. An important second result that is observed both empirically and in the gene network theory we apply is that even in the presence of complex genetic interactions selection can still operate to identify paths towards genetic changes that contribute to improved trait phenotypes (Cooper and Podlich 2002). Thus, an important distinction between the simple and complex genetic models of trait genetic architecture that we have considered here is in the cumulative nature of the additive genetic variation exploited by selection. Theoretical investigations have demonstrated that new sources of additive genetic variance can be released within a reference population by restricting the dimensions of an epistatic genetic system contributing to the standing genetic variance (Cheverud and Routman 1996, Podlich et al. 2004). From an analysis perspective this potential source of new genetic variation can be viewed as conditional genetic variation within appropriate reference populations (e.g., Carlborg et al. 2006). The modeling of different forms of trait genetic architecture inspired by the empirical evidence provides some views of how effective paths towards genetic improvement of multiple traits can be identified along the trait genetic complexity continuum (Hammer et al. 2006).

Given the arguments above, why model QTL effects and MAS for plant breeding applications? An answer we give is that with our current knowledge of the genetic architecture of the traits the process of designing and optimizing a MAS breeding strategy that outperforms conventional selection (i.e., one based on selection on pedigree and phenotypic information) to create novel genotypes with improved phenotypes is a complex problem beyond the limits of complete specification and long-term prediction. As with other complex scientific problems of this nature, one reason that motivates the use of modeling methods is to accelerate and enable the design and implementation of practical breeding strategies that deal with multiple traits of varying genetic complexity. Formal modeling methods provide a quantitative framework for integrating the multiple streams of phenotypic, genetic, pedigree and environmental information that must be combined to detect QTL and apply the information used by the breeder to implement forward selection in an applied breeding program. Such a quantitative modeling framework can be encoded within an appropriate high performance computing infrastructure and designed to deal in real-time with the large volumes of data generated by breeding programs (Cooper et al. 2006). As research continues to advance our understanding of the genetic architecture of traits new ideas for mapping genetic variation for traits and applying the results within MAS strategies will continue to emerge. The modeling framework we outlined above can be used to extend the prerequisite empirical studies and accelerate an understanding of the strengths and weaknesses of the proposed breeding strategies and quantify their potential benefits. Thus, as new genetic evidence accumulates we can more rapidly advance from ideas to practical application by breeders whenever benefits are identified.

Theory suggests potential benefits from MAS for simple and complex traits (Lande 1991, Lande and Thompson 1990, Podlich et al. 2004, Cooper et al. 2005). However, for quantitative traits it is often difficult to empirically demonstrate the advantages over conventional selection on phenotypes alone (Moreau et al. 2004a). A comprehensive approach to modeling QTL effects to evaluate the opportunities for MAS in applied breeding requires recognition that we are dealing with multiple traits and a genetic complexity continuum. Further, to evaluate the merits of breeding strategies, such as MAS, we require a quantitative genetics framework that enables investigation of the performance of breeding strategies for all parts of the continuum. Classical quantitative genetics theory (Lynch and Walsh 1998, Walsh 2005) provides such a framework when we are interested in the expected genetic gain that can be attributed to selection for the additive components of the genetic architecture of the traits. The breeder's prediction equation can be extended to indicate the proportion of the additive genetic variation accounted for by QTL (Lande 1991, Lande and Thompson 1990) and to indicate the contributions of non-additive sources of variation to short and long-term gain (Walsh 2005). However, for many of the practical issues related to implementing MAS in a breeding program and where the non-additive components of genetic variance are important components of the genetic architecture of the traits the classical framework does not provide answers to many of the breeders' questions. Thus, we need an equivalent of the breeder's equation that can deal explicitly with both the additive and non-additive components of the genetic architecture of traits (Holland 2006). For these situations we can extend the quantitative framework using stochastic simulation methods based on genetic models that are defined to incorporate the additive and non-additive components of the genetic architecture of traits (Cooper et al. 2005). The *E(NK)* model is an example of a framework that can be applied for this purpose and customized for the genetic information available on the architecture of traits within defined reference populations of genotypes. Combined with access to the growing body of knowledge on the molecular basis of the genetic architecture of traits many "real-world" scenarios can be incorporated and used to guide the definition of the relevant genetic models (Cooper et al. 2005, Walsh 2005, Hammer et al. 2006). Implementation of the quantitative framework applied here within an appropriate computing infrastructure enables the breeder to deal with many of the complexities that arise from the intersection of theoretical models and the empirical evidence describing components of the genetic architecture of traits.

7. SUMMARY AND OUTLOOK

Today the trait changes that have been brought about by plant breeding (e.g., Figure 2) can be described at the genotype and phenotype levels. Even though the molecular and functional bases for most of the genetic changes are not fully understood genetic progress is still made by breeding programs (Figure 2). MAS strategies offer opportunities for accelerating the rates of genetic progress that can be achieved. Currently some of the early proposed applications of MAS are under empirical evaluation and others are being applied (e.g., Moreau et al. 2004a, Cahill and Schmidt 2004, Niebur et al. 2004, Podlich et al. 2004, Hammer et al. 2005, Crosbie et al. 2006). Beyond the theoretical considerations there are many issues that require detailed consideration when applying MAS in applied breeding. Given that mapping studies will identify only a component of the standing genetic variation for traits in a sample of the reference genotype-environment system at a point in time, theory and experience suggests that these studies should be viewed as entry points into the study of the genetic architecture of traits that will need to be continually refined (Podlich et al. 2004). Further, it is important to realize that modeling the effects of QTL is one component of a larger modeling effort aimed at understanding the nature and role of genetic variation in the sustainability of the genotype-environment systems that have been adopted in agriculture. Modeling QTL effects and MAS is an evaluation of our capacity to understand and manipulate the standing genetic variation to create and evaluate new genotype-management-environment states within agricultural systems by selecting on identified components of the genetic variation. This should be viewed as a potential refinement and enhancement of how breeders have operated in the past. The "Green Revolution" for wheat and rice in the second half of the 20th Century is an example of such a change that was realized by conventional breeding. The "Green Revolution" was enabled when suitable genes for reduced plant height were incorporated into the reference populations of the wheat and rice breeding programs to decrease the incidence of plant lodging and enable farmers in suitable ecogeographical regions to increase grain yield by applying increased levels of fertilizer and water inputs to the system (e.g., Perkins 1979, Rajaram and van Ginkel 2001, Borlaug and Dowswell 2005, Khush 2005, Duvick 2006). As was the case in the past sustainable progress from breeding will result from a clear definition and execution of breeding objectives within a comprehensive breeding program strategy that combines knowledge of germplasm, trait genetics, selection strategies and the TPE (Hallauer and Miranda 1988, Comstock 1996, Podlich and Cooper 1999, Cooper et al. 2005, Duvick 2006, Hallauer and Pandey 2006). In association with our expanding body of knowledge of both the gene-to-phenotype architecture of traits and the nature of the standing genetic variation in elite breeding populations, trait mapping and MAS provides the breeder with many new opportunities for genetic improvement of multiple traits to enhance the performance and sustainability of agricultural systems.

ACKNOWLEDGEMENTS

We thank Chris Winkler for his support in running the simulation experiments.

REFERENCES

Axelrod R, Cohen MD (1999). Harnessing complexity: organizational implications of a scientific frontier. The Free Press, Sydney, Australia

Barker T, Campos H, Cooper M, Dolan D, Edmeades G, Habben J, Schussler J, Wright D, Zinselmeier C (2005) Improving drought tolerance in maize. Plant Breed Rev 25: 173–253

Bass TA (1999) The predictors. Henry Holt and Company, New York

Basten CJ, Weir BS, Zeng ZB (1995) QTL cartographer: a reference manual and tutorial for QTL mapping. Center for Quantitative Genetics, NC St U

Beavis WD (1998) QTL analyses: power, precision and accuracy. In: Paterson AH (ed) Molecular dissection of complex traits. CRC Press, Boca Raton, pp 145–162

Blanc G, Charcosset A, Mangin B, Gallais A, Moreau L (2006) Connected populations for detecting quantitative trait loci and testing for epistasis: an application in maize. Theor Appl Genet 113:206–224

Borlaug NE, Dowswell CR (2005) Feeding a world of ten billion people: a 21st century challenge. In: Tuberosa R, Phillips RL, Gale M (eds) In the wake of the double helix: from the green revolution to the gene revolution. Avenue media, Bologna, Italy, pp 3–23

Bouchez A, Hospital F, Causse M, Gallais A, Charcosset A (2002) Marker-assisted introgression of favorable alleles at quantitative trait loci between maize elite lines. Genetics 162:1945–1959

Bower JM, Bolouri H (eds) (2001) Computational modeling of genetic and biochemical networks. A Bradford Book, The MIT Press, Cambridge, Massachusetts

Bubeck, DM, Carlone MR, Fox RL, Hoffbeck MD, Segebart RL, Stucker DS (2006) Breeding progress measured in eight elite inbred families. Maydica 51:141–149

Cahill DJ, Schmidt DH (2004) Use of marker assisted selection in a product development breeding program. In: Fischer T, Turner N, Angus J, McIntyre L, Robertson M, Borrell A, Lloyd D (eds) New directions for a diverse planet: proceedings of the 4th international crop science congress. (Brisbane, Australia, 26 September to 1 October, 2004, Online Proceedings www.cropscience.org.au)

Campos H, Cooper M, Habben JH, Edmeades GO, Schussler JR (2004) Improving drought tolerance in maize: a view from industry. Field Crop Res 90:19–34

Carlborg Ö, Jacobsson L, Åhgren P, Siegel P, Andersson L (2006) Epistasis and the release of genetic variation during long-term selection. Nat Gene 38:418–420

Casti JL (1997) Would-be worlds: how simulation is changing the frontiers of science. John Wiley & Sons, Inc., Brisbane, Australia

Chardon F, Virlon B, Moreau L, Falque M, Joets J, Decousset L, Murigneux A, Charcosset A (2004) Genetic architecture of flowering time in maize as inferred from quantitative trait loci meta-analysis and synteny conservation with the rice genome. Genetics 168:2169–2185

Chapman SC, Crossa J, Edmeades GO (1997) Genotype by environment effects and selection for drought tolerance in tropical maize. I. Two mode pattern analysis of yield. Euphytica 95:1–9

Chapman SC, Cooper M, Butler DG, Henzell RG (2000a) Genotype by environment interactions affecting grain sorghum. I. Characteristics that confound interpretation of hybrid yield. Aus J Agr Sci 51:197–207

Chapman SC, Cooper M, Hammer GL, Butler DG (2000b) Genotype by environment interactions affecting grain sorghum. II. Frequencies of different seasonal patterns of drought stress are related to location effects on hybrid yields. Aus J Agr Sci 51:209–221

Chapman SC, Hammer GL, Butler DG, Cooper M (2000c) Genotype by environment interactions affecting grain sorghum. III. Temporal sequences and spatial patterns in the target population of environments. Aus J Agr Sci 51:223–233

Chapman S, Cooper M, Podlich D, Hammer G (2003) Evaluating plant breeding strategies by simulating gene action and dryland environment effects. Agrono J 95:99–113

Cheverud J, Routman E (1996) Epistasis as a source of increased additive genetic variance at population bottlenecks. Evolution 50:1042–1051

Clark AG (2000) Limits to prediction of phenotypes from knowledge of genotypes. Evolut Bio 32:205–224

Comstock, RE (1996) Quantitative genetics with special reference to plant and animal breeding. Iowa State University Press Ames, Iowa

Cooper M, Podlich DW (2002) The *E(NK)* model: extending the *NK* model to incorporate gene-by-environment interactions and epistasis for diploid genomes. Complexity 7:31–47

Cooper M, Podlich DW, Smith OS (2005) Gene-to-phenotype models and complex trait genetics. Aus J Agr Res 56:895–918

Cooper M, Chapman SC, Podlich DW, Hammer GL (2002) The GP problem: quantifying gene-to-phenotype relationships. In Silico Bio 2:151–164

Cooper M, Smith OS, Merrill RE, Arthur L, Podlich DW, Löffler CM (2006) Integrating breeding tools to generate information for efficient breeding: past, present, and future. In: Lamkey KR, Lee M (eds) Plant breeding: The Arnel R. Hallauer International Symposium. Blackwell Publishing, Ames, Iowa, pp 141–154

Crosbie TM, Eathington SR, Johnson GR, Edwards M, Reiter R, Stark S, Mohanty RG, Oyervides M, Buehler RE, Walker AK, Dobert R, Delannay X, Pershing JC, Hall MA, Lamkey KR (2006) Plant breeding: past, present and future. In: Lamkey KR, Lee M (eds) Plant breeding: The Arnel R. Hallauer International Symposium. Blackwell Publishing, Ames, Iowa, pp 3–50

Crossa J, Vargas M, van Eeuwijk FA, Jiang C, Edmeades GO, Hoisington D (1999) Interpreting genotype × environment interaction in tropical maize using linked molecular markers and environmental covariables. Theor Appl Genet 99:611–625

Crow JF, Kimura M (1970) An introduction to population genetics theory. Burgess Publishing Company, Minneapolis, Minnesota

Crutchfield JP, Schuster P (Eds) (2003) evolutionary dynamics: exploring the interplay of selection, accident, neutrality, and function. Oxford University Press, Oxford, UK

Doebley J, Stec A, Gustus C (1995) *Teosinte branched1* and the origin of maize: evidence for epistasis and the evolution of dominance. Genetics 141:333–346

Dorogovtsev SN, Mendes JFF (2003) Evolution of networks: from biological nets to the internet and WWW. Oxford University Press, Oxford

Duvick, DN (1977). Genetic rates of gain in hybrid maize yields during the past 40 years. Maydica 22:187–196

Duvick, DN (2006) Social and environmental benefits of plant breeding. In: Lamkey KR, Lee M (eds) Plant breeding: The Arnel R. Hallauer International Symposium. Blackwell Publishing, Ames, Iowa, pp 61–72

Duvick DN, Smith JSC, Cooper M (2004) Long-term selection in a commercial hybrid maize breeding program. Plant Breed Rev 24(2):109–151

Falconer DS, Mackay TFC (1996) Introduction to quantitative genetics, 4th edn. Longman, Burnt Mill, Harlow, Essex, England

Fisher RA (1918) The correlation between relatives on the supposition of mendelian inheritance. T Roy Soc Edin 52:399–433

Frary A, Nesbitt TC, Frary A, Grandillo S, van der Knaap E, Cong B, Liu J, Meller J, Elber R, Alpert KB, Tanksley SD (2000) *fw2.2*: a quantitative trait locus key to the evolution of tomato fruit size. Science 289:85–88

Fraser A, Burnell D (1970) Computer models in genetics. McGraw Hill Book Company, New York

Hallauer AR, Miranda FO JB (1988). Quantitative genetics in maize breeding 2nd edn. Iowa State University Press, Ames

Hallauer AR, Pandey S (2006) Defining and achieving plant-breeding goals. In: Lamkey KR, Lee M (eds) Plant breeding: the Arnel R. Hallauer International Symposium. Blackwell Publishing, Ames, Iowa, pp 73–89

Hammer GL, Chapman S, van Oosterom E, Podlich DW (2005) Trait physiology and crop modelling as a framework to link phenotypic complexity to underlying genetic systems. Aus J Agr Res 56:947–960

Hammer G, Cooper M, Tardieu F, Welch S, Walsh B, van Eeuwijk F, Chapman S, Podlich D (2006) Models for navigating biological complexity in breeding improved crop plants. Trends in Plant Sci 11:587–593

Holland JB (2006) Theoretical and biological foundations of plant breeding. In: Lamkey KR, Lee M (eds) Plant breeding: the Arnel R. Hallauer International Symposium. Blackwell Publishing, Ames, Iowa, pp 127–140

Janick J (ed) (2004a) Plant breeding reviews 24, Part 1: long-term selection: maize. John Wiley & Sons, Inc., Hoboken, New Jersey, USA

Janick J (ed) (2004b) Plant breeding reviews 24, Part 2: long-term selection: crops, animals, and bacteria. John Wiley & Sons, Inc., Hoboken, New Jersey, USA

Jansen RC, Jannink J-L, Beavis WD (2003) Mapping quantitative trait loci in plant breeding populations: use of parental haplotype sharing. Crop Sci 43:829–834

Kauffman, SA (1993) The origins of order: self-rganization and selection in evolution. Oxford University Press, New York

Kearsey MJ, Farquhar GL (1998) QTL analysis in plants; where are we now? Heredity 80, 137–142

Kempthorne O (1988) An overview of the field of quantitative genetics. In: Weir BS, Eisen EJ, Goodman MM, Namkoong G (eds) Proceedings of the second international conference on quantitative genetics, Sunderland, Massachusetts: Sinauer Associates, Inc., pp 47–56

Khush GS (2005) Green revolution: challenges ahead. In: Tuberosa R, Phillips RL, Gale M (eds) In the wake of the double helix: from the green revolution to the gene revolution. Avenue media, Bologna, Italy, pp 37–51

Lande R (1991) Marker-assisted selection in relation to traditional methods of plant breeding. In: Stalker HT, Murphy JP (eds) Plant breeding in the 1990s, CAB International, Wallingford, UK. pp 437–451

Lande R, Thompson R (1990) Efficiency of marker-assisted selection in the improvement of quantitative traits. Genetics 124:743–756

Laurie CC, Chasalow SD, LeDeaux JR, McCarroll R, Bush D, Hauge B, Lai C, Clark D, Rocheford TR, Dudley JW (2004) The genetic architecture of response to long-term artificial selection for oil concentration in the maize kernel. Genetics 168:2141–2155

Li ZK, Arif M, Zhong DB, Fu BY, Xu JL, Domingo-Rey J, Ali J, Vijayakumar CHM, Yu SB, Khush GS (2006). Complex genetic networks underlying the defensive system of rice (*Oryza sativa* L.) to *Xanthomonas oryzae* pv. *oryzae*. Proc Nat Acad Sci 103:7994–7999

Löffler CM, Wei J, Fast T, Gogerty J, Langton S, Bergman M, Merrill B, Cooper M (2005) Classification of maize environments using crop simulation and geographic information systems. *Crop Sci* 45:1708–1716

Lynch M, Walsh B (1998) Genetics and analysis of quantitative traits. Sinauer Associates, Inc. Sunderland Massachusetts, USA

Mackay TFC (2001) The genetic architecture of quantitative traits. Ann Rev Genet 35:303–339

Mackay TFC (2004) The genetic architecture of quantitative traits: lessons from *Drosophila. Curr Opin Genet Devel* 14:1–5

McClearn GE (2006) Contextual genetics. Trends Genet 22:314–319

Micallef KP, Cooper, M. and Podlich DW (2001) Using clusters of computers for large QU-GENE simulation experiments. Bioinformatics 17:194–195

Moreau L, Charcosset A, Gallais A (2004a) Experimental evaluation of several cycles of marker-assisted selection in maize. Euphytica 137:111–118

Moreau L, Charcosset A, Gallais A (2004b) Use of trial clustering to study QTL × environment effects for grain yield and related traits in maize. Theor Appl Genet 110:92–105

Nguyen HT, Blum A (eds) (2004) Physiology and biotechnology integration for plant breeding. Marcel Dekker, Inc. New York

Niebur WS, Rafalski JA, Smith OS, Cooper M (2004) Applications of genomics technologies to enhance rate of genetic progress for yield of maize within a commercial breeding program. In: Fischer T, Turner N, Angus J, McIntyre L, Robertson M, Borrell A, Lloyd D (eds) New directions for a diverse planet: proceedings of the 4th international crop science congress. Brisbane, Australia, 26 September to 1 October, 2004, Online Proceedings www.cropscience.org.au

Newman M, Barabási A-L, Watts DJ (eds) (2006) The structure and dynamics of networks. Princeton University Press, Princeton

Omholt SW, Plahte E, Øyehaug L, Xiang K (2000) Gene regulatory networks generating the phenomena of additivity, dominance and epistasis. Genetics 155:969–980

Openshaw SJ, Frascaroli E (1997) QTL detection and marker-assisted selection for complex traits in maize. Proceedings of the 52nd annual corn and sorghum research conference. American Seed Trade Association, Washington DC, USA pp 44–53

Peccoud J, Vander VK, Podlich DW, Winkler C, Arthur L, Cooper M (2004) The selective values of alleles in a molecular network model are context-dependent. Genetics 166:1715–1725

Perkins JH (1979) Geopolitics and the green revolution: wheat, genes, and the cold war. Oxford University Press, Oxford

Podlich DW, Cooper M (1998) QU-GENE: a platform for quantitative analysis of genetic models. Bioinformatics 14:632–653

Podlich DW, Cooper M (1999) Modelling plant breeding programs as search strategies on a complex response surface. Lect Notes Comput Sci 1585:171–178

Podlich DW, Winkler CR, Cooper M (2004) Mapping as you go: an effective approach for marker-assisted selection of complex traits. Crop Sci 44:1560–1571

Rajaram S, van Ginkel M (2001) Mexico: 50 years of international wheat breeding. In: Bonjean AP, Angus WJ (eds) The world wheat book; A history of wheat breeding. Lavoisier Publishing, Paris, France, pp 579–608

Rasmusson DC (1996) Germplasm is paramount. In: Reynolds MP, Rajaram S, McNab A (eds) Increasing yield potential in wheat: breaking the barriers. Mexico, DF, CIMMYT, pp 28–35

Ribaut J-M, Hoisington DA, Deutsch JA, Jiang C, Gonzalez-de-Leon D (1996) Identification of quantitative trait loci under drought conditions in tropical maize. 1. Flowering parameters and the anthesis-silking interval. Theor Appl Genet 92:905–914

Ribaut J-M, Jiang C, Gonzalez-de-Leon D, Edmeades GO, Hoisington DA (1997) Identification of quantitative trait loci under drought conditions in tropical maize. 2. Yield components and marker-assisted selection strategies. Theor Appl Genet 94:887–896

Ribaut J-M, Hoisington D, Bänziger M, Setter TL, Edmeades GO (2004) Genetic dissection of drought tolerance in maize: a case study. In: Nguyen HT, Blum A (eds) Physiology and biotechnology integration for plant breeding. Marcel Dekker, Inc., New York, pp 571–609

Salvi S, Tuberosa R (2005) To clone or not to clone plant QTLs : present and future challenges. Trends in Plant Sci 10:297–304

Schön CC, Utz HF, Groh S, Truberg B, Openshaw S, Melchinger AE (2004) Quantitative trait locus mapping based on resampling in a vast maize testcross experiment and its relevance to quantitative genetics for complex traits. Genetics 167:485–498

Schrage M (2000) Serious play: how the world's best companies simulate to innovate. Harvard Business School Press, Boston, Massachusetts

Sing CF, Stengård JH, Kardia SLR (2003) Genes, environment, and cardiovascular disease. Arterioscler, Thromb, Vasc Biol 23:1190–1196

Smith A, Cullis B, Thompson R (2002) Exploring variety-environment data using random effects AMMI models with adjustments for spatial field trend: Part 1: Theory. In: Kang MS (ed) Quantitative genetics, genomics and plant breeding. CAB International, Wallingford, UK, pp 323–335

Smith A, Cullis B, Luckett D, Hollamby G, Thompson R (2002) Exploring variety-environment data using random effects AMMI models with adjustments for spatial field trend: Part 2: applications. In: Kang MS (ed) Quantitative genetics, genomics and plant breeding. CAB International, Wallingford, UK, pp 337–351

Tardieu F, Reymond M, Muller B, Granier C, Simonneau T, Sadok W, Welcker C (2005) Linking physiological and genetic analyses of the control of leaf growth under changing environmental conditions. Aus J Agr Res 56:937–946

Utz HF, Melchinger AE, Schön CC (2000) Bias and sampling error of the estimated proportion of genotypic variance explained by quantitative trait loci determined from experimental data in maize using cross validation and validation with independent samples. Genetics 154:1839–1849

van Eeuwijk FA, Malosetti M, Yin X, Struik PC, Stam P (2005) Statistical models for genotype by environment data: from conventional ANOVA models to eco-physiological QTL models. Aus J Agr Res 56:883–894

Vargas M, van Eeuwijk FA, Crossa J, Ribaut J-M (2006) Mapping QTLs and QTL × environment interaction for CIMMYT maize drought stress program using factorial regression and partial least squares methods. Theor Appl Genet 112:1009–1023

Wagner A (2005) Robustness and evolvability in living systems. Princeton University Press, Princeton.

Walsh B (2005) The struggle to exploit non-additive variation. Aus J Agr Res 56:873–881

Wang J, van Ginkel M, Trethowan R, Ye G, DeLacy I, Podlich D, Cooper M (2004) Simulating the effects of dominance and epistasis on selection response in the CIMMYT wheat breeding program using QuCim. Crop Sci 44:2006–2018

Wang J, van Ginkel M, Podlich D, Ye G, Trethowan R, Pfeiffer W, DeLacy IH, Cooper M, Rajaram S (2003) Comparison of two breeding strategies by computer simulation. Crop Sci 43:1764–1773.

Welch SM, Dong Z, Roe JL, Das S (2005) Flowering time control: gene network modelling and the link to quantitative genetics. Aus J Agr Res 56:919–936

Williams GP (1997) Chaos theory tamed. Joseph Henry Press, Washington DC

Winkler CR, Jensen NM, Cooper M, Podlich DW, Smith OS (2003) On the determination of recombination rates in intermated recombinant inbred population. Genetics 164:741–745

Wolfram S (2002) A new kind of science. Wolfram Media, Inc, Champaign, Illinois, USA

CHAPTER 5

APPLICATIONS OF LINKAGE DISEQUILIBRIUM AND ASSOCIATION MAPPING IN CROP PLANTS

ELHAN S. ERSOZ[1], JIANMING YU[1] AND EDWARD S. BUCKLER[1,2,*]

[1]*Institute for Genomic Diversity, Cornell University, Ithaca, NY, 14853, USA*

[2]*USDA-ARS, US Plant, Soil and Nutrition Laboratory, Ithaca, NY, 14853–2901*

Abstract: The investigations of patterns of linkage disequilibrium for designing association-mapping studies are fast becoming a method of interest for complex trait dissection and improvement practices in many crop plants. The methodology and its applications to crop improvement, to date are discussed.

1. INTRODUCTION

Association mapping, also known as linkage disequilibrium mapping, is a relatively new and promising genetic method for complex trait dissection. Association mapping has the promise of higher mapping resolution through exploitation of historical recombination events at the population level, that may enable gene level mapping on non-model organisms where linkage based approaches would not be feasible (Nordborg and Tavare 2002; Risch and Merikangas 1996).

Association mapping utilizes ancestral recombinations and natural genetic diversity within a population to dissect quantitative traits and is built on the basis of linkage disequilibrium concept (Geiringer 1944; Lewontin and Kojima 1960). One of the working definitions of linkage disequilibrium (here on will be referred to as LD) is the non-random co-segregation of alleles at two loci.

In contrast to linkage based studies, linkage disequilibrium based genetic association studies offer a potentially powerful approach for mapping causal genes with modest effects (Hirschhorn and Daly 2005). While linkage analysis is based upon

*Corresponding Author: esb33@cornell.edu

R.K. Varshney and R. Tuberosa (eds.), Genomics-Assisted Crop Improvement: Vol. 1: Genomics Approaches and Platforms, 97–119.

detection of non-random association between a genotype and a phenotype in well-characterized pedigrees, association mapping focuses on associations within populations of *unrelated* individuals. In general, chromosomes sampled from *unrelated* individuals in a population will be much more distantly related than those sampled from members of traditional pedigrees. In other words, the time to most recent common ancestor (MRCA) of any given two individuals from a population of unrelated individuals would be greater than that of a pedigree population. This is what makes LD mapping suitable for fine-scale mapping: there will have been more opportunities for recombination to take place over several generations, between many alleles, in a species, while there can be only a few generations of recombination present in pedigree populations. Increase in the rate of recombination will lead to reshuffling of the chromosomal segments into smaller pieces. This will lead to reduction of the LD in short distances around loci, and lead to significant co-occurrence (i.e. LD) between only loci physically close, allowing high resolution. Whereas pedigree studies work with recombination events in few generations that enable exchange between chromosomes at the order of megabases, association studies deal with segmental exchanges measured in kilobases (Paterson et al. 1990; Stuber et al. 1992; Thornsberry et al. 2001).

2. WHAT IS LINKAGE DISEQUILLIBRIUM AND HOW IS IT RELATED TO ASSOCIATION MAPPING STUDIES?

The term *linkage disequilibrium* was first introduced back in late 1940's to describe the degree of non-random association between pairs of loci. In the absence of demographic effects that might confound the linkage disequilibrium patterns, LD summary statistics such as r^2 can be used to define the level of co-occurrence of alleles at two loci (Hill and Robertson 1968). When r^2 is zero, alleles at two loci do not co-occur more frequently than would be expected under random sampling. r^2 approaches its maximum of 1 as alleles at two loci show more frequent co-occurrence within the population sample examined. There are various other linkage disequilibrium statistics that can be used for this purpose (Hedrick 1987) all of which aim to estimate the predictive value of a marker locus on another locus that is displaying non-zero LD with it (if LD statistic is zero, two loci examined have zero predictive value for each other).

Association mapping uses these properties of the measures of pairwise LD statistics to infer the predictive value of a marker locus for the association of the chromosomal region it resides with the phenotype. The high-LD chromosomal region around a marker locus defines the predictive range of a certain genetic marker. If LD within this genomic range is complete, any polymorphism within this range will have the same predictive value for the association with the phenotype. Hence, as a result of a significant marker-phenotype association, it can be concluded that the causative polymorphism resides within this high LD region around the marker locus.

With respect to association mapping, the most significant aspect of LD is its predictive properties over the haplotype it resides in. However, the extent of LD (in base pairs) within species and even within individual genomes are highly variable, and therefore most reliably estimated empirically (Long and Langley 1999). Theoretical estimation of the levels of LD for realistic population models that does not satisfy the assumptions of Wright-Fisher model is complex. The hardship is mostly due to the large number of interrelated factors involved in the formation of patterns of LD, including but not limited to genetic drift, population admixture, and natural selection (Pritchard and Przeworski 2001; Wall and Pritchard 2003).

The statistical power of associations is determined by the extent of LD with the causative polymorphism, as well as sample size used for the study (Long and Langley 1999; Wang and Rannala 2005). If LD decays too fast within a region, large number of markers would be required to scan target regions of a genome. On the other hand, if LD decays too slowly, the size of the haplotype blocks would be too large to unambiguously reveal underlying causative locus. In other words, the decay of LD over physical distance in the study population determines the marker density required and the level of resolution that may be obtained in an association study.

2.1. How to Estimate LD

There are several summary statistics proposed for estimation of linkage disequilibrium (Hedrick 1987), however the most commonly used summary statistic within the association study framework is known as r^2 (Hill and Robertson 1968; Lewontin 1988). Conceptually and mathematically r is the Pearson's (product moment) *correlation coefficient* of the correlation that describes the predictive value of the allelic state at one polymorphic locus on the allelic state at another polymorphic locus, where r^2 is the squared value of correlation coefficient that is also called *coefficient of determination*. r^2 explains the proportion of a sample variance of a response variable that is *explained* by the predictor variables when a linear regression is performed.

Lewontin's D, is another summary statistic for LD that is commonly used. *D* describes the difference between the coupling gamete frequencies and repulsion gamete frequencies at two loci. From *D* a second measure of linkage disequilibrium, that is normalized *D'* can also be estimated. Even in samples taken from populations at equilibrium under neutrality, variances of linkage disequilibrium summary statistics are typically large but *D'* has the lowest variance (Hedrick 1987). However, estimation using *D'* may generate erratic and unreliable results when low frequency alleles or small sample sizes are used for the analysis. It is advised to collapse the alleles using an allele frequency cut-off prior to estimation of linkage disequilibrium statistics *D* and *D'*.

Other than these commonly used summary statistics for LD, there are also likelihood-based methods that investigate probability of independence between pairs of sites using two-locus sampling distributions, rather than calculating a summary statistic for LD. These methods, usually referred to as model-based LD estimators, also provide means of estimating population recombination parameter $4Nc$ under neutral equilibrium model from nucleotide sequence data (Golding 1984; Hudson 1985; Hudson 2001) or generating other model-based estimates of LD for comparisons with observed patterns (Mueller 2004) under various population structure and demographic history scenarios. Although the estimation of LD through these methods are more computationally intensive compared to the pairwise-LD estimation methods, they are extensively used for evolutionary and population genetic studies as well as investigations on the domestication of various crop plant species (Wright et al. 2005; Wright and Gaut 2005).

2.2. Interpretation of LD Data

Estimating LD from empirical data is a straightforward procedure; however interpretation of results of LD analysis and extrapolation of this information to the genome may be more complex. It is important to estimate the rate of decay of LD with physical distance to be able to extrapolate information gathered from a small collection of sampled loci to the whole genome investigated. This extrapolation is essential for association mapping study design since it may be used for determining the marker density required for scanning previously unexplored regions of the genome as well as determining the maximum resolution that can be achieved for genotype phenotype associations for the study population.

The levels of LD are expected to be highly variable across the genome, due to several factors such as variation in recombination rate and selection. For reliable results, this variation needs to be taken into account when designing experiments to exploit LD. Variation in rate of recombination across the genome is a key factor that contributes to the variance observed in patterns of LD. A number of researchers have focused on the distance at which average r^2 is reduced to 0.10, as a reasonable point to conclude there is minimal LD to detect associations with complex traits. The reasoning for this r^2cut-off is as follows: in a complex trait a large quantitative trait locus (QTL) may only explain approximately 10% of the phenotypic variation. If a marker only explains 10% of the total QTL variation, then the marker will only explain one percent of the phenotypic variation. Detection of locus effects that cause smaller than 1% phenotypic variation requires exponentially increasing population sizes therefore such small effects would be considered undetectable in a moderate size study population.

Sufficient power for association studies of complex traits requires LD blocks to be defined more strictly for greater LD as well as larger population sizes. Current human genetic studies focus on genome scans aiming for much higher LD (e.g. $r^2 > 0.80$) (Barrett 2006), and are developing haplotype based approaches that can help capture more variants (Pe'er et al. 2006).

2.3. LD in Plants

Studies on rates of decay of linkage disequilibrium in various plant taxa (Flint-Garcia et al. 2003) such as maize (*Zea mays* ssp. *mays*) (Ching et al. 2002;

Palaisa et al. 2003; Remington et al. 2001a; Tenaillon et al. 2002), barley (*Hordeum vulgarae*) (Caldwell et al. 2004; Caldwell et al. 2006), *Arabidopsis thaliana* (Nordborg et al. 2002; Nordborg et al. 2005), and sorghum (*Sorghum bicolor*) (Hamblin et al. 2005) and durum wheat (*Triticum durum*) (Maccaferri et al. 2005), indicate tremendous variation in the extent of linkage disequilibrium. This variation is mostly due to founder effect followed by genetic drift that leads to unequal number of effective recombinations in species sub-populations. Furthermore, selfing also plays an important role (Nordborg 2000).

The population sample effect is clearly observed in maize, where LD decays within 1kb in land races (Tenaillon et al. 2001), in approximately 2 kb in diverse inbred lines (Remington et al. 2001a) and can extend up to 100 kb in commercial elite inbred lines (Ching et al. 2002). In barley, in a study of four loci Caldwell et al. (2006) shows that LD might extend up to 212 Kb in elite lines while it might decay below $r^2 = 0.2$ within 0.4 kb for the same region in wild lines. In wild barley (*Hordeum spontaneum*) the results on analysis of LD over 18 loci suggests that LD decay displays a pattern quite similar to that of maize at some loci, that decays below significant levels within 2 kb (Morrell et al. 2005). However, there are a proportion of the loci that show more extensive LD, which may be the result of admixture. In European Aspen (*Populus tremula*), Ingvarsson (2005) shows that there is substantial variation not only across populations but also across loci, and estimates the range of decay of LD to an expected value of r^2 to less than 0.05 within a few hundred basepairs. In a comparison of nine loci across two population samples of loblolly pine (*Pinus taeda* L.), Gonzalez-Martinez et al. (2006a) shows that the rates of decay of linkage disequilibrium are fast; decays below the level of $r^2 < 0.2$ within 2 kb but is variable and not significantly different for the independent population samples investigated for loblolly pine.

In predominantly selfing *Arabidopsis*, LD at a key flowering time locus (*FRI*) extends beyond 250 Kb (Nordborg et al. 2002). However, in large genomic surveys, the decay of LD was reported to be much faster genome-wide: below the level of $r^2 < 0.2$ within about 30 Kb (Nordborg et al. 2005). In another selfing species, soybean (*Glycine max*), Zhu et al. (2003) studied the patterns of LD in 143 short amplicons that spans approximately 12.5 cM of the genome. The study reports that significant decay of LD was detectable within approximately 2–2.5 cM that roughly equals to 1–1.5 Mb. There are few studies that investigate LD in rice (*Orzya sativa*) to date; at a disease resistance locus it was reported that substantial LD extends beyond 100 kb (Garris et al. 2003) and even further at the *waxy* domestication locus (Olsen et al. 2006). For the rice genome, more comprehensive studies are underway.

3. ASSOCIATION POPULATIONS AND STATISTICS

There are five main stages for association studies: (1) Selection of population samples, (2) Determination of the level and influence of population structure on the sample, (3) Phenotyping the population sample for traits of interest (4) Genotyping

Preliminary analysis & Feasibility Study

Population: Small sized diversity sample(s) to be used as a *Discovery Panel.*

Data: Nucleotide sequence, from locus samples with genome-wide coverage from the *Discovery Panel.*

Analysis: Nucleotide diversity(θ), decay of linkage disequilibrium with physical distance (r^2), population recombination rate (ρ), population structure and demography.

Results: Range of diversity to be sampled for association population, marker density required for sufficient coverage of target genomic regions (or the genome) for association, level of population structure that exists within the species, evaluation of genome-wide influence of demography, determination of genomic regions targeted by natural selection and domestication, and number and density of the neutral markers required to evaluate background associations.

Data Collection

Genotype: Select, species wise informative, and high throughput genotyping amenable markers. Choice of genotyping platform is dependent on the size of the population to be studied as well as the number of available markers, thereby per marker per individual experimental cost is optimized. In addition genotypes from the candidate regions that are trait dependent, at least as many neutral markers should be genotyped as well, in order to test the levels of background-stochastic associations.

Phenotype: Phenotypes of interest should be replicated temporally and spatially to increase accuracy and precision of the phenotypic measurements. Quantitative measures of the traits of interest are preferable over categorical phenotyping. Evaluation of the heritability help define the expectation for the genetic component of the phenotypic variance.

Statistical Association

- Build statistical model(s) for the expectation of phenotypic correlation with environmental and genetic variability ($V_P = V_G + V_E$).
- Evaluate the level of co-variance between the phenotypes, and combine the highly correlated traits in the same model.
- Evaluate co-variance between the neutral marker genotypes and candidate gene genotypes.
- Determine the Type I error thresholds according to the number of tests performed, and the level of flexibility in the study.
- Determine power and false positive rate expectations for the study.
- Run statistical association tests.

Post-Association Follow-up

Evaluation: The genotypic value of the associated allele should be evaluated on several different genetic backgrounds, for its overall phenotype as well as biochemical and molecular genetic studies for elucidation of structure and function.
Verification: The association reported should be verified either through re-evaluation in an independent population sample or through allelic silencing/knock-outs.
Breeding : The best alleles obtained through the study should be incorporated into breeding programs for integration into elite varieties.

Box 1. The steps employed during an association study

the population, for either candidate genes/regions or as a genome-wide scan and (5) Testing the genotypes and phenotypes for their associations (Box 1).

The choice of association test is the last step of the study and is mostly dependent on the previous steps according to the characteristics of the population that was used to collect the genotypic and phenotypic data (Breseghello and Sorrells 2006a; Breseghello and Sorrells 2006b; Lewis 2002). Furthermore, possible complications due to population structure in the study sample may adversely affect the association test results. The influence of population structure on each association study depends on the relatedness between sampled individuals in the studied population. Therefore, the populations amenable for association studies may be classified according to the level of relatedness between the individuals forming the association population.

In the following subsections, we will first discuss the influences of population structure on various association study designs, followed by examples of control for its influences by accounting for the relatedness between individuals forming the association population.

3.1. Population Structure

Most important constraint for the use of association mapping for crop plants is unidentified population substructuring and admixture due to factors such as adaptation or domestication (Thornsberry et al. 2001; Wright and Gaut 2005). Population structure creates genome-wide linkage disequilibrium between unlinked loci. When the allele frequencies between sub-populations of a species is significantly different, due to factors such as genetic drift, domestication or background selection, genetic loci that do not have any effect whatsoever on the trait may demonstrate statistical significance for their co-segregations with a trait of interest. Provided that a large number of neutral markers are available for estimation of genome wide effects of structure, it is possible to statistically account for such effects in association data analysis (Yu et al. 2006b).

In cases where the population structuring is mostly due to population stratification (Bamshad et al. 2004; Pritchard 2001) three methods are often acknowledged to be suitable for statistically controlling the effects of population stratification on association tests: (1) genomic control (GC) (Devlin et al. 2004; Devlin and Roeder 1999; Devlin et al. 2001), (2) structured association (SA) method including two extensions that are modified for the type of association study as case-control (SA-model) (Pritchard et al. 2000b) or quantitative trait association study (Q-model) (Camus-Kulandaivelu et al. 2006; Thornsberry et al. 2001), (3) unified mixed model approach (Q+K) (Yu et al. 2006b).

First method suggested for statistically controlling population structure was GC that assumes population structuring has equivalent effects on all loci genome-wide. In GC method, a small random set of markers (e.g., polymorphisms unlikely to affect the trait of interest) are used to estimate influence of population structure on the association test statistics *(inflation factor)*, such that the significance of the association statistic (P value) estimated is adjusted to account for population structure. The general principle of GC is to use individual genomes from the sample, to estimate levels of confounding due to substructure and more direct relatedness such as familial relationship in the study and scale the final significance level of the association reported accordingly (Devlin et al. 2001).

Structured association methodology, utilizes marker loci unlinked to the candidate genes under investigation to infer *subpopulation membership*. The application of structured association to qualitative and quantitative traits is done using the appropriate model depending on the trait and population type, with either SA or Q models respectively. In application of SA for quantitative trait association (Q-model), a two stage procedure is constructed where for the first stage each subject's probability of membership in each subpopulation is estimated (Pritchard et al. 2000a; Pritchard et al. 2000b) and then in the next stage, a test of association is conducted using subpopulation membership as a variable for the anociation model tested (Pritchard et al. 2000b). In case-control studies, the probability of the SNP frequency distribution based on population structure is compared between the case and control samples. For quantitative traits, the population structure estimates are used as co-variates in the regression model that defines the correlation of the genotype with the phenotype (Camus-Kulandaivelu et al. 2006; Thornsberry et al. 2001).

In unified mixed model approach (aka Q+K model) of Yu and Pressoir et al.(2006b), a large set of random markers that can provide genome-wide coverage are used to estimate population structure (Q) and relative kinship matrix (K), which are fit into a mixed-model framework to test for marker-trait association. In the unified mixed-model approach, each of the factors that may confound association analysis, that is, familial relatedness between individuals (K) and relatedness due to population structure (Q) are considered as independent variables within the species population. In order to account for the combined affects of such relatedness factors, they are included as covariates into the regression model that defines the correlation between genotype and the phenotype during association testing.

The genetic makeup of the study population that was used to collect genotype and phenotype data defines the model and type of association statistics to be used for association tests. This will be discussed further in the next section.

3.2. Classic Association Populations

If the individuals forming the study population are *effectively* unrelated, the study population may be considered a random sample of individuals from species population and is therefore equivalent to any natural population. The relatedness amongst the individuals forming the population can be either estimated using pedigrees (Emik and Terrill 1949) or inferred using molecular markers (Blouin 2003; Lynch and Ritland 1999; Oliehoek et al. 2006; Wang 2002). These individuals can either be selected from originally natural populations, or subselected from selections included in breeding programs, to form a classic association population. Selecting individuals from breeding programs offers the advantage of easy incorporation into future breeding programs, however the number of lineages incorporated in the association study becomes limited (Breseghello and Sorrells 2006a; Breseghello and Sorrells 2006b).

All the previously mentioned statistical methods for population structure inferences are applicable to the classic association populations; however Q+K model has the widest base of applicability across all structured association study designs in natural populations.

In plants, so far the focus has been on quantitative traits in natural populations. In maize, using diverse inbred lines it was possible to select a sample of 102 lines with relatively few closely related individuals by sampling across the world's breeding programs (Remington et al. 2001a; Thornsberry et al. 2001). However, as larger samples were gathered to increase statistical power to over 300 maize lines it became extremely difficult to find samples that match the structure expected in natural populations (Flint-Garcia et al. 2005). These are the cases where the combined natural and family based approaches are most powerful (Yu et al. 2006a). In *Arabidopsis* (Nordborg et al. 2005), natural samples were collected from around the world but because of strong population structure and selfing, these samples in many respects behave more like families for association mapping purposes (Aranzana et al. 2005). Association studies with some tree species are more likely to fall into the model of effectively unrelated individuals (Gonzaléz-Martinéz et al. 2006b; Thumma et al. 2005). Most crop plant studies will probably fall on a continuum between natural and family-based association populations.

3.3. Family Based Association Populations

If the association population is a collection of unrelated families, instead of single unrelated individuals, it is possible to perform a joint linkage and association analysis on the population, that potentially can be more informative on the trait of interest than either approach alone (Holte et al. 1997; Karayiorgou et al. 1999). For instance, in human genetics, where the association populations are collections of parent-offspring trios, two types of study design is considered: transmission disequilibrium tests (TDTs) (Allison 1997; Fulker et al. 1999; Monks et al. 1998; Rabinowitz 1997; Spielman et al. 1993) , family based association tests (FBATs) (Herbert et al. 2006; Horvath et al. 2001; Laird et al. 2000; Laird and Lange 2006; Lake et al. 2000; Lange et al. 2003). Stich et al. (2006) modified the QTDT algorithm to test its applicability to inbred plant populations, and developed a model named Quantitative Inbred Pedigree Disequilibrium Test (QIPDT), for analysis of joint linkage and association data from crop plant populations. Another family based population design that was essentially developed for crop and livestock breeding is the Henderson's Mixed Model Approach (Henderson 1975), generally known for its applications in Best Linear Unbiased Predictors (BLUPs). Family based association study design investigates co-segregation and linkage simultaneously (Spielman et al. 1994).

A long standing mixed model method has been used by animal scientists to analyze the data from extended pedigree in dairy or cattle breeding programs (Henderson 1975; Henderson 1976; Henderson 1984). The superiority of the mixed model lies in its incorporation of the phenotypic observations from relatives of an individual into the estimation of the breeding value of that individual. The amount of information that is incorporated depends on the heritability of the trait and the genetic relationships (traditionally defined by pedigree information) among individuals. Naturally, this method has been extended to quantify the single gene effect while accounting for the pedigree relationship (Kennedy et al. 1992) and is applicable to association mapping with family based association populations. Taking this mixed model framework, Yu et al. (2006b) suggested to replace the pedigree-based co-ancestry with a marker-based relative kinship (K) to account for the relatedness among individuals.

This unified mixed model approach is demonstrated to be the most powerful statistic compared to all the rest of the statistics, for the family based association studies and those studies falling between classical and family-based designs. The flexibility and generality of this approach allow association studies to be carried out on any population without the restriction on the specific family structure.

3.4. Special Association Populations

Recently, the field of plant association genetics pioneered the use of a new type of association population, designed to incorporate advantages of both linkage based and linkage disequilibrium based quantitative trait dissection approaches in association studies, in a stronger design than Transmission-Disequilibrium Test (TDT) design. This builds off of some of the joint linkage-association approaches encountered in cattle breeding (Blott et al. 2003; Meuwissen and Goddard 1997). The *Nested Association Populations* (NAM) are developed through controlled crosses between a diverse selection of unrelated individuals according to a breeding scheme that aims shuffling of alleles in diverse samples either across backgrounds or against a reference background while keeping track of number and locations of the recombination events that shuffle the parental chromosomes (Yu et al. 2006a). The subsequent generations of progeny of the crosses can then used as association populations. A population generated according to this described scheme not only provides tremendous power to the statistical tests of association, but also enables the projection of genotype information from the parents to the progeny optimizing genotyping cost for large studies. The cross design is expected to effectively reduce many of the affects of admixture and population structure on the association population. For such populations, a two step procedure for associations is suggested.

The two stage study design of nested association mapping requires deep sequencing or genotyping of the parents for SNP identification across the genome followed by lower density genotyping in the progeny in order to infer the locations of the recombination breakpoints during the crosses. Once the recombination breakpoints are localized and the recombination blocks are traced back to the contributing parent, the haplotype information from the parents can be directly projected on the progeny genome, without further need for genotyping within these blocks.

This design scheme enables the researcher to utilize the advantages of both linkage based and linkage disequilibrium based genetic mapping approaches. It provides genome wide coverage, with high resolution and is performed on an experimental cross that is robust to genetic heterogeneity with representation of several alleles per loci in a large population.

Because of the balanced design, straightforward multiple regression approaches can be applied (Yu et al. 2006a) for association testing. Currently, availability of such nested association populations are reported for maize (Yu et al. 2006a) and loblolly pine (Baltunis 2005; Ersoz 2006; Kayihan et al. 2005). Further statistical methods that are going to utilize and combine information from both parent and progeny generations for NAM type populations are currently under development.

These mentioned association population structures represent the continuum of LD levels from low in classic association populations towards high in biparental breeding populations. Nested association populations that are similar to heterogenous intermated populations (Niebur et al. 2004) fall in the mid-range of this continuum with moderate levels of LD and linkage.

4. FALSE POSITIVES AND POWER OF ASSOCIATION

One of the major concerns in the association mapping studies is the statistical power of the association testing, since as it stands, there is a trade off between the power and accuracy for reporting associations due to false positives. The major determinant of the levels of false positives and power of associations is the level of population structure in the association population.

A false positive (Type I error) occurs when a test incorrectly reports that it has found a positive result where none really exists. The classical definition of Type I error is an incorrect rejection of the null hypothesis - accepting the alternative hypothesis even though the null hypothesis was true. The second functional biological definition of false positives is also used in association studies. In this framework, false positives do not only arise due to the failure of the statistical test performed, but also in cases where the statistical test is valid and the association exists but it is an association with population structure instead of the trait of interest. Population structure can lead to identification of loci that generate statistically significant but biologically invalid associations solely due to their tight correlation with population structure. However, if the population structure in an association study is properly dealt with, this is not expected to be a source of false positives.

Traditionally, Type I error rate (α) for multiple testing is controlled with the Bonferroni correction. The Bonferroni correction in general is conservative and leads to power loss for detection if the polymorphisms are in linkage disequilibrium and/or the traits are correlated with one another.

Another statistical method suggested for control for multiple testing is False Discovery Rate (FDR) procedure. The FDR is the proportion of positive results that are actually false positives to the whole set of positive results obtained from a statistical test. The procedure can be used to estimate a cutoff for a particular FDR (Benjamini and Hochberg 1995), or estimate an FDR for a particular cutoff (Storey 2002; Storey and Tibshirani 2003). The FDR approaches may be most appropriate when multiple traits are being compared or when the markers are not in extensive LD (Chen and Storey 2006). Essentially based on the relative costs of false positives on further follow-up research, appropriate false discovery rates should be determined and be used.

A third procedure that can be applied for multiple testing correction is the permutation test (Churchill and Doerge 1994; Doerge and Churchill 1996), which controls for the genome-wide error rate (GWER). The permutation test has the ability to estimate effects on significance levels caused by the use of correlated markers as well as correlated traits. In this approach, the trait values are permuted relative to the genotypic data. These permutation approaches are appropriate ways to control the GWER, however, they can be quite conservative if one expects numerous QTLs. Recently, the $GWER_k$ approach of Chen and Storey (2006) incorporates a method for a more liberal balance of true and false positives provides a reasonable avenue.

Other than these statistical methods proposed, it is also possible to non-parametrically estimate the false discovery rate through comparison of distributions of P values, against a set of markers of known influence and a set of random markers scored on the same association population, with simulations. The probability of false associations is simply the ratio of the proportion of significant associations detected in the random set to the proportion of significant associations detected in the simulated set of known influence loci. This method provides a fast and rigorous way of estimating FDR, if a set of random markers has been scored on the association population. Since random markers are required to estimate population structure, this method should be applicable for association testing in most cases.

The power of a statistical test is the probability that the test will reject a false null hypothesis. Some of the relevant parameters that can effect the power of association studies are, but not limited to (1) The type of association test, single marker or haplotype based, (2) The multiplicity control method, (3) Population-Structure control method, (4) Genetic architecture of the trait, (5) Population size, (6) Marker density, (7) Type of populations used for associations, family based or effectively unrelated (Long and Langley 1999).

Simulation studies that investigate the power of the association tests for candidate gene association approach report that 300 individuals in a natural population provide enough power to detect *repeatable* associations when population structure is controlled properly (Camus-Kulandaivelu et al. 2006; Long and Langley 1999; Thornsberry et al. 2001; Yu et al. 2006a). These power estimates are based on candidate gene studies, where there are few SNPs being evaluated relative to the entire genome. Genome scan type association studies rapidly becoming feasible, but for such studies the population sample size required to obtain sufficient power will be larger. The exact population size required will depend on the LD structure for the population. Population sizes of 1000 to 5000 genotypes will likely be sufficient in most cases.

The power of association will be low, if the trait is highly correlated with population structure. Statistical controls for population structure, under such circumstances would result in false negatives. An example of such a case is demonstrated for maize and *Arabidopsis* flowering time traits (Aranzana et al. 2005; Flint-Garcia et al. 2005). The reason for flowering time and population structure to be correlated is that flowering time is an adaptive trait that largely defines the structure. The Q+K model can produce somewhat better results in these situations (Yu et al. 2006b), but in general a different sample or genetic design is required to work with traits that are tightly correlated with population structure. From a study of 60 traits on a maize diversity panel of 302 inbred lines, the only traits that showed strong relationship with structure were two flowering time related traits.

Three studies using different germplasm have analyzed maize flowering time and the *dwarf8* (*d8*) gene (Andersen et al. 2005; Camus-Kulandaivelu et al. 2006; Thornsberry et al. 2001). These studies highlight the difficulties of studying traits related to population structure. In all three studies, when population structure is ignored; highly significant associations between the traits and polymorphisms in *d8* are detected that are often much more significant than any of the random markers. It is clear that the putatively functional allele is segregating with a very high allele frequency in some populations while it is represented at very low frequencies in other populations. This is exactly what would be expected if flowering time is under diversifying selection between the various sub-populations. Furthermore, upon application of standard corrections for managing population structure (Q) the *d8*-flowering time association is significant for some samples but not for others in all three studies.

Essentially, there is low statistical power to evaluate candidate genes that are involved in the clinal adaptation and/or creation of population structure. While empirical significance estimates obtained through contrasting the significances of the candidates with large numbers of random markers, the most effective approach for this type of trait may be specially constructed association populations, with balanced designs.

5. PHENOTYPING AND GENOTYPING STRATEGIES FOR ASSOCIATION TESTING

As in all other quantitative genetic studies, the success of an association study is heavily dependent on the accurate evaluation of the phenotype of interest. The within population variation observed for genotypes and phenotypes for an association is much greater than that found in most bi-parental mapping populations. While greater variation is preferable while aiming for higher resolution and allele mining, it can pose problems for accurate evaluation of this variation in a meaningful way in a single environment.

The inherent variation observed in phenotypic trait measurement, when combined with the substantial genetic variation included in some association studies, requires careful experimental design to acquire quality data. In addition, evaluations in multiple environments with controls and unbalanced designs may be required. In our experience with maize, we found that evaluating the germplasm in short day environments has facilitated some trait evaluation by reducing photoperiod effects between lines. Additionally, we found that evaluating the germplasm in testcrosses (F1 hybrids) has reduced the phenotypic range into a manageable level. Since each of these approaches interact with the genetic architectures of the traits, future studies will be needed to fully understand the tradeoffs of various study design approaches.

In the association study design, genotyping is required for both inferences on the genotype/phenotype associations and on the population structure and demography. The first aim of querying candidate regions for polymorphisms is best achieved by genotyping SNPs within these candidate regions. The second aim of gathering information on population specific phenomenon like structure, linkage, demography, and kinship can be achieved through genotyping neutral background markers, such as SNPs on non-coding regions or SSRs (simple sequence repeats) distributed evenly throughout the genome.

All genetic markers can be used for investigating association; however, SNPs potentially have the most utility compared to rest of the genetic markers. Various assays were developed for detection of known and unknown SNPs. Some are relatively easy to implement and low in cost, others are developed for high volume screening at substantial cost. As the cost of genotyping reduces, genome-wide scans of all available polymorphisms in a species genome are becoming rapidly feasible and preferable over targeted SNP genotyping approaches. SSR markers have historically been useful in association studies and do have high information content, but they may be difficult to find in candidate gene regions and they are several fold more expensive to score than SNPs.

For the purposes of inferences on the population history, genotype information from a large number of neutral marker loci is required. We are using the term neutral marker loosely here, to indicate the non-candidate loci, i.e. the loci that were *not* designated as candidate loci that can putatively influence a trait of interest. The density of the markers required should be scaled to provide genome-wide coverage. Simulation studies suggest 100 SSR or 200 SNP markers would suffice to get a reasonable estimate of population structure and relatedness for most crop plants (Yu and Buckler unpublished results).

When targeting candidate loci for association studies, the greatest statistical power is achieved when the marker and QTL have equal allele frequencies (Abecasis et al. 2001) in the study population. This is due to opportunity created for maximal linkage and LD since robust detection of associations requires the marker and trait loci are in phase. If there is no knowledge of the QTL frequency distribution *a priori*, the best alternative is to choose markers with a wide range of allele frequencies that are likely to mimic the QTL mutation rate. Some SSRs probably mutate faster and have a different frequency distribution than QTL, which may make them less useful for association mapping. SNPs with a wide range of allele frequencies are most likely to be informative. In order to maximize the information content of SNPs, a large number of them can be chosen to scan a particular genomic region, and this can be achieved with numerous algorithms available for choosing SNPs. (Ackerman et al. 2003; Daly et al. 2001; Forton et al. 2005; Gabriel et al. 2002; Halldorsson et al. 2004; Johnson et al. 2001; Ke and Cardon 2003; Patil et al. 2001; Sebastiani et al. 2003; Zhang and Jin 2003).

Wether the phenotype of interest has a binary or quantitative phenotype is also of interest for the association study design. When a binary trait is being investigated, case-control type populations are required for association analysis, where equivalent sized sub-populations of individuals that display the phenotype of interest (cases) and do not display the phenotype of interest (controls) are querried for allelic association of genetic loci with the case and control phenotypes in a statistically significant manner. The statistical test performed is simply a hypothesis test, that asks weather or not the allelic frequency distribution of a locus is the same or different for a given locus between the two sub-populations. Bulk Segregant Analysis (BSA) type (Michelmore et al. 1991) bulked sample genotype screening methods for all the available marker loci may facilitate the candidate gene and association discovery, for binary traits (Shaw et al. 1998). The challange of case-control type studies is to make sure that the case and control groups are comparable in terms of their genetic makeup. Most of the statistical methods aim to detect and correct for the affects of population statification and ancestry differences between the case and control groups (Price et al. 2006; Pritchard et al. 2000b).

6. ASSOCIATION MAPPING IN CROP PLANTS

The motivations for attempting association mapping in different crop plants are highly variable. For historically well studied crop plants, such as maize and rice, the major motivation for association approach is dissection of complex traits at very high-level resolution, as well as allele mining from natural genetic diversity resources. For other organisms where there is insufficient or little genetic resources the major motivation is functional marker development and identification of molecular markers tightly linked to the trait locus for marker assisted selection and breeding practices. Thus, each association study stands alone for their own motivations and should be evaluated for its utility and success based on their initial motivations and aims.

Association mapping approach requires extensive infrastructure development and preliminary studies to determine population structure and LD (Box 1). Once the preliminary data and infrastructure for association mapping for a species is available, several association studies on various plant taxa report successful results for tests of associations between candidate locus genotypes and various complex phenotypes (Table 1).

In model organism *Arabidopsis*, the association mapping practice is mostly motivated by generating proof of concept, identification of QTL involved in adaptation, and additional alleles to supplement other mutagenesis approaches. The candidate-gene association study at the *CRY2-Cryptochrome2* locus reported diverse functional alleles (Olsen et al. 2004). In their first attempt for a genome-wide association study in *Arabidopsis*, Aranzana et al. (2005) reports identification of previously known flowering time (*FRI* locus) and three known pathogen resistance genes.

In maize, all reported association studies so far have targeted candidate genes with known mutant phenotypes and are motivated by high resolution mapping and allele mining purposes. For instance, *d8* locus with flowering time (Andersen et al. 2005; Camus-Kulandaivelu et al. 2006; Thornsberry et al. 2001), *bt2(brittle2)*, *sh1(shrunken1)* and *sh2(shrunken2)* with kernel composition, *ae1(amylose extender1)* and *sh2(shrunken2)* with starch pasting properties (Wilson et al. 2004) and sweet taste (Tracy et al. 2006), *a1(anthocyaninless1)* and *whp1(whitepollen1)* genes with maysin synthesis (Szalma et al. 2005), *lyc-e (lycopene epsilon cyclase)* gene with carotenoid content (Harjes et al. 2006) are studies that report very high resolution associations, as well as localizing the causative polymorphism within 1–2 Kb of the marker loci reported. In maize, very little is known about association mapping from a genomic scale, mostly due to incomplete genomic sequence and very rapid decay of LD. At the *Y1* locus a relatively large genomic context was examined. *Y1* is a key gene in carotenoid production in maize (Buckner et al. 1990; Buckner et al. 1996), and through an association study (Palaisa et al. 2003) the allelic variation was traced down to multiple independent insertions in the *Y1* promoter region that cause up regulation of the downstream *Y1* gene. At this locus, associations were also shown to extend to neighboring genes (Palaisa et al. 2004) albeit with weaker significances. This extended LD is mostly the result of breeding efforts in the 20th century that specifically targeted this simple Mendelian inherited trait. The extended LD at *Y1* locus is likely to be one of the most extensive in the maize genome; effective over 100s of kb, while other domestication loci *tb1* (*teosinte branched 1*) (Lukens and Doebley 2001) and *tga (teosinte glume architechture)* (Wang et al. 2005) show LD that extends over 10s of kb. However, it should be emphasized that *tb1* and *tga* domestication loci demonstrate patterns of reduced diversity as well as extended LD,

Table 1. Association studies that report significant results. SA: Structured Association, GLM: General Linear Model, MLM: Mixed Linear Model, DRR: Double Round Robin, FR: Fusiform Rust, PC: Pitch Canker. PB-AM: Pedigree Based Association Mapping and FB-AM: Family Based Association Mapping are two special applications of Nested Association Mapping (NAM) applications described in the text

Species	Population type	Association method	Trait	Reference(s)
Zea mays	Diverse Inbred Lines	SA(Q model)	Flowering Time	**Thornsberry et al. 2001 Andersen et al. 2005 Camus-Kulandaivelu et al. 2006**
		SA(Q model)	Kernel Composition Starch Pasting properties	**Wilson et al. 2005**
		SA(Q model)	Maysin Synthesis	**Szalma et al. 2005**
		Case-Control	Carotenoid Content	**Palaisa et al. 2004**
		MLM (Q+K model)	Caretenoid Content	**Harjes et al. 2006**
Zea mays	Diverse Inbred lines	Haplotype TreeScanning	Sweet Taste	**Tracy et al. 2006**
Arabidopsis thaliana	Diverse Ecotypes	SA(Q Model)	Flowering Time	**Olsen et al. 2004**
			Disease Resistance Flowering Time	**Aranzana et al. 2005**
Eucalyptus spp.	Unstructured Natural Population	Regression (GLM)	Microfibril Angle	**Thumma et al. 2005**
Triticum aestivum	Diverse Cultivars	PB-AM (Q Model)	Kernel Size Milling Quality	**Breseghello and Sorrells 2006b**
Oryza sativa	Diverse Land Races	Haplotype Tree Scanning	Glutinous Phenotype	**Olsen and Purugganan 2002**
Pinus taeda	DRR-Cross of Diverse Parents	Case-Control FB-AM (GLM)	FR Resistance PC Tolerance	**Ersoz 2006**
Pinus taeda	Unstructured Natural Population	Regression(GLM) & MLM (K model)	Wood Specific Gravity% Late Wood Microfibril Angle Cellulose Content	**Gonzaléz-Martinéz et al. 2006a**
Lolium perenne	Diverse Natural Populations	SA(GLM)	Heading Date	**Skøt et al. 2005**

indicating that the estimates of LD is not as efficient as they are at *Y1*. Furthermore it is plausible to assume that not all of the selection events may have similar LD patterns to that of *Y1* locus.

Rice is another crop plant that was extensively studied and has whole genome sequence available. Association studies in rice are mostly motivated by allele mining for economically important traits. An example of such a study is the associations reported between *WAXY* locus of and glutinous phenotype that is commonly known as the *sticky rice* (Olsen and Purugganan 2002).

In many important plant species such as forest trees, the generation time of the organism presents a tribulation for the complex trait dissection through genetic analysis. In these species, association-mapping approach offers the opportunity to overcome the limitations of organismal systems, and enables fast trait improvement. Several successful results in candidate gene based association studies have recently been reported from forestry crop species eucalyptus (*Eucalyptus nitens* and *Eucalyptus globulus*) and loblolly pine (*Pinus taeda* L.). For instance, a study by Thumma et al. (2005) reports an association between the Cinnamoyl-CoA-Reductase (*CCR*) gene and microfibril angle in *Eucalyptus spp*. In a loblolly pine candidate gene joint-linkage and association study, associations of several candidate regions with fungal disease resistance traits are reported (Ersoz 2006). Also in loblolly pine wood quality candidate gene association study for association of chemical and physical wood property traits *cad* and *sams2* genes with early wood formation, *lp3-1* gene with percent late wood, *4CL* with juvenile and mature wood, *α-tubulin* with microfibril angle, and *CesA3* with cellulose content are reported (Gonzaléz-Martinéz et al. 2006b).

Another motivation for association approach is the opportunity to unify the elite germplasm resource of an organism through investigation of the breeding material. In an association study, Breseghello and Sorrells (2006b) investigate the wheat kernel size and milling quality in an elite germplasm collection of soft-winter wheat from eastern US. It identifies, three candidate regions on chromosomes 2D, 5A and 5B that are significantly associated with traits (Breseghello and Sorrells 2006b). This study clearly demonstrates the utility of association mapping as a powerful method that can provide a bridge for closing the gap between the implementation of the genetic trait dissection results to marker-assisted selection.

Several AFLP based genome scan studies have also been successful in discovering associations in germplasm samples with high LD. In perennial rye grass *Lolium perenne* (Skøt 2005) successful associations for underlying major flowering time (heading date) QTL were identified. In sea beet (*Beta vulgaris* ssp. *maritime*) (Hansen et al. 2001) identification of several AFLP markers that show significant associations with another flowering time trait (bolting date) is also reported.

7. CONCLUSIONS

So far, map based cloning approaches are reported to successfully clone 12 major effect QTL and nine small effect QTL (Price 2006). The time scale from QTL mapping to positional cloning practice is estimated to be between 5 to 10 years, while sufficient resolution for QTL cloning through association mapping can be achieved within 2–3 years. Furthermore, there is a substantial lag between the QTL discovery to marker assisted crop improvement practices, dedicated to verification of the presence and stability of QTL, in the traditional linkage based studies. In a well-designed association study, some of the results can be immediately applied to marker-assisted improvement.

The true large scale applications of association mapping will become apparent as multiple species began to have marker densities sufficiently high for whole genome scan by association mapping. Currently, several research groups are working on whole genome scan approaches in half a dozen species that have whole genome sequences available, and there are at least 50 more species whose genome sequences are being completed in the near future.

The goal of association mapping in many crop plants is to identify key genes controlling various traits and mine the best alleles from diverse germplasm to be incorporated in elite breeding material. Traditionally genetic markers were mostly used for trait improvement through several breeding based approaches such as Marker Assisted Selection (MAS), Marker Assisted Breeding (MAB) and Mapping As You Go (MAYG) (Podlich 2004) as well as QTL cloning/transformation based approaches (Remington et al. 2001b). Association mapping has the potential to provide numerous useful alleles to these marker assisted breeding programs. These markers assisted breeding programs using association data are now underway in numerous plant breeding companies. In the next few years, we will also witness the applications of association mapping and MAS for public breeding programs.

Association mapping holds an important and rapidly expanding niche in quantitative trait mapping studies along with linkage mapping and positional cloning, and it is likely that this niche will continue to expand over the next decade.

REFERENCES

Abecasis GR, Cookson WO, Cardon LR (2001) The power to detect linkage disequilibrium with quantitative traits in selected samples. Am J Hum Genet 68:1463–1474

Ackerman H, Usen S, Mott R, Richardson A, Sisay-Joof F et al (2003) Haplotypic analysis of the TNF locus by association efficiency and entropy. Genome Biol 4:R24

Allison DB (1997) Transmission-disequilibrium tests for quantitative traits. Am J Hum Genet 60:676–690

Andersen JR, Schrag T, Melchinger AE, Zein I, Lubberstedt T (2005) Validation of Dwarf8 polymorphisms associated with flowering time in elite European inbred lines of maize (*Zea mays* L.). Theor Appl Genet 111:206–217

Aranzana MJ, Kim S, Zhao K, Bakker E, Horton M et al (2005) Genome-wide association mapping in *Arabidopsis* identifies previously known flowering time and pathogen resistance genes. PLoS Genet 1:e60

Baltunis BS, Huber DA, White TL, Golfard B, Stelzer HE (2005) Genetic effects of rooting loblolly pine stem cuttings from a partial diallel mating design. Can J Forest 35:1098–1108

Bamshad M, Wooding S, Salisbury BA, Stephens JC (2004) Deconstructing the relationship between genetics and race. Nat Rev Genet 5:598–609

Barrett JC, Cardon LR (2006) Evaluating coverage of genome-wide association studies. Nat Genet 38:659–662

Benjamini Y, Hochberg Y (1995) Controlling the false discovery rate – a practical and powerful approach to multiple testing. J Roy Stat Soc Ser B Methodol 57:289–300

Blott S, Kim JJ, Moisio S, Schmidt-Kuntzel A, Cornet A et al (2003) Molecular dissection of a quantitative trait locus: a phenylalanine-to-tyrosine substitution in the transmembrane domain of the bovine growth hormone receptor is associated with a major effect on milk yield and composition. Genetics 163:253–266

Blouin JD (2003) DNA-based methods for pedigree reconstruction and kinship analysis in natural populations. Trends Ecol Evol 18:503–511

Breseghello F, Sorrells M (2006a) Association analysis as a strategy for improvement of quantitative traits in plants. Crop Sci 46:1323–1330

Breseghello F, Sorrells ME (2006b) Association mapping of kernel size and milling quality in wheat (*Triticum aestivum* L.) cultivars. Genetics 172:1165–1177

Buckner B, Kelson TL, Robertson DS (1990) Cloning of the y1 locus of maize, a gene involved in the biosynthesis of carotenoids. Plant Cell 2:867–876

Buckner B, Miguel PS, Janick-Buckner D, Bennetzen JL (1996) The y1 gene of maize codes for phytoene synthase. Genetics 143:479–488

Caldwell KS, Langridge P, Powell W (2004) Comparative sequence analysis of the region harboring the hardness locus in barley and its colinear region in rice. Plant Physiol 136:3177–3190

Caldwell KS, Russell J, Langridge P, Powell W (2006) Extreme population-dependent linkage disequilibrium detected in an inbreeding plant species, *Hordeum vulgare*. Genetics 172:557–567

Camus-Kulandaivelu L, Veyrieras JB, Madur D, Combes V, Fourmann M et al (2006) Maize adaptation to temperate climate: relationship between population structure and polymorphism in the Dwarf8 gene. Genetics 172:2449–2463

Chen L, Storey JD (2006) Relaxed significance criteria for linkage analysis. Genetics 173:2371–2381

Ching A, Caldwell KS, Jung M, Dolan M, Smith OS et al (2002) SNP frequency, haplotype structure and linkage disequilibrium in elite maize inbred lines. BMC Genet 3:19

Churchill GA, Doerge RW (1994) Empirical threshold values for quantitative trait mapping. Genetics 138:963–971

Daly MJ, Rioux JD, Schaffner SF, Hudson TJ, Lander ES (2001) High-resolution haplotype structure in the human genome. Nat Genet 29:229–232

Devlin B, Bacanu SA, Roeder K (2004) Genomic control to the extreme. Nat Genet 36:1129–1130; author reply 1131

Devlin B, Roeder K (1999) Genomic control for association studies. Biometrics 55:997–1004

Devlin B, Roeder K, Wasserman L (2001) Genomic control, a new approach to genetic-based association studies. Theor Popul Biol 60:155–166

Doerge RW, Churchill GA (1996) Permutation tests for multiple loci affecting a quantitative character. Genetics 142:285–294

Emik LO, Terrill CE (1949) Systematic procedures for calculating inbreeding coefficients. J Hered 40:51–55

Ersoz ES (2006) Candidate gene-association mapping for dissecting fungal disease resistance in loblolly pine. PhD Dissertation in Genetics, University of California, Davis

Flint-Garcia SA, Thornsberry JM, Buckler ES (2003) Structure of linkage disequilibrium in plants. Ann Rev Plant Biol 54:357–374

Flint-Garcia SA, Thuillet AC, Yu J, Pressoir G, Romero SM et al (2005) Maize association population: a high-resolution platform for quantitative trait locus dissection. Plant J 44:1054–1064

Forton J, Kwiatkowski D, Rockett K, Luoni G, Kimber M et al (2005) Accuracy of haplotype reconstruction from haplotype-tagging single-nucleotide polymorphisms. Am J Hum Genet 76:438–448

Fulker DW, Cherny SS, Sham PC, Hewitt JK (1999) Combined linkage and association sib-pair analysis for quantitative traits. Am J Hum Genet 64:259–267

Gabriel SB, Schaffner SF, Nguyen H, Moore JM, Roy J et al (2002) The structure of haplotype blocks in the human genome. Science 296:2225–2229

Garris AJ, McCouch SR, Kresovich S (2003) Population structure and its effects on haplotype diversity and linkage disequilibrium surrounding the xa5 locus of rice (*Oryza sativa* L.). Genetics 165:759–769

Geiringer H (1944) On the probability theory of linkage in Mendelian heredity. Ann Math Stat 15:25–57

Golding GB (1984) The sampling distribution of linkage disequilibrium. Genetics 108:257–274

Gonzaléz-Martinéz SC, Ersoz E, Brown GR, Wheeler NC, Neale DB (2006a) DNA sequence variation and selection of tag single-nucleotide polymorphisms at candidate genes for drought-stress response in *Pinus taeda* L. Genetics 172:1915–1926

Gonzaléz-Martinéz SC, Wheeler N, Ersoz ES, Neale DB (2006b) Association genetics in *Pinus taeda* L.I. wood property traits. Genetics 2007 175:399–409

Halldorsson BV, Bafna V, Lippert R, Schwartz R, De La Vega FM et al (2004) Optimal haplotype block-free selection of tagging SNPs for genome-wide association studies. Genome Res 14:1633–1640

Hamblin MT, Salas Fernandez MG, Casa AM Mitchell SE, Paterson AH et al (2005) Equilibrium processes cannot explain high levels of short- and medium-range linkage disequilibrium in the domesticated grass Sorghum bicolor. Genetics 171:1247–1256

Hansen M, Kraft T, Ganestam S, Sall T, Nilsson NO (2001) Linkage disequilibrium mapping of the bolting gene in sea beet using AFLP markers. Genet Res 77:61–66

Harjes CE, Yates H, Torbert R, Wurtzel E, Buckler ES (2007) Characterization of maize kernel carotenoid diversity and identification of functionally distinct alleles by association mapping- **In preparation**

Hedrick PW (1987) Gametic disequilibrium measures: proceed with caution. Genetics 117:331–341

Henderson CR (1975) Best linear unbiased estimation and prediction under a selection model. Biometrics 31:423–447

Henderson CR (1976) Simple method for computing inverse of a numerator relationship matrix used in prediction of breeding values. Biometrics 32:69–83

Henderson CR (1984) Application of linear models in animal breeding. University of Guelph, Ontario

Herbert A, Gerry NP, McQueen MB, Heid IM, Pfeufer A et al (2006) A common genetic variant is associated with adult and childhood obesity. Science 312:279–283

Hill WG, Robertson, A (1968) Linkage disequilibrium in finite populations. Theor Appl Genet 38:226–231

Hirschhorn JN, Daly MJ (2005) Genome-wide association studies for common diseases and complex traits. Nat Rev Genet 6:95–108

Holte S, Quiaoit F, Hsu L, Davidov O, Zhao LP (1997) A population based family study of a common oligogenic disease – Part I: association/aggregation analysis. Genet Epidemiol 14:803–807

Horvath S, Xu X, Laird NM (2001) The family based association test method: strategies for studying general genotype–phenotype associations. Eur J Hum Genet 9:301–306

Hudson RR (1985) The sampling distribution of linkage disequilibrium under an infinite allele model without selection. Genetics 109:611–631

Hudson RR (2001) Two-locus sampling distributions and their application. Genetics 159:1805–1817

Ingvarsson PK (2005) Nucleotide polymorphism and linkage disequilbrium within and among natural populations of European Aspen (*Populus tremula* L., Salicaceae). Genetics 169:945–953

Johnson GC, Esposito L, Barratt BJ, Smith AN, Heward J et al (2001) Haplotype tagging for the identification of common disease genes. Nat Genet 29:233–237

Karayiorgou M, Sobin C, Blundell ML, Galke BL, Malinova L et al (1999) Family-based association studies support a sexually dimorphic effect of COMT and MAOA on genetic susceptibility to obsessive-compulsive disorder. Biol Psychiat 45:1178–1189

Kayihan GC, Huber DA, Morse AM, White TL, Davis JM (2005) Genetic dissection of fusiform rust and pitch canker disease traits in loblolly pine. Theor Appl Genet 110:948–958

Ke X, Cardon LR (2003) Efficient selective screening of haplotype tag SNPs. Bioinformatics 19:287–288

Kennedy B, Quinton M, Vanarendonk J (1992) Estimation of effects of single genes on quantitative traits. J Anim Sci 70:2000–2012

Laird NM, Horvath S, Xu X (2000) Implementing a unified approach to family-based tests of association. Genet Epidemiol 19:S36–42

Laird NM, Lange C (2006) Family-based designs in the age of large-scale gene-association studies. Nat Rev Genet 7:385–394

Lake SL, Blacker D, Laird NM (2000) Family-based tests of association in the presence of linkage. Am J Hum Genet 67:1515–1525

Lange C, Lyon H, DeMeo D, Raby B, Silverman EK et al (2003) A new powerful non-parametric two-stage approach for testing multiple phenotypes in family-based association studies. Hum Hered 56:10–17

Lewis CM (2002) Genetic association studies: design, analysis and interpretation. Brief Bioinform 3:146–153

Lewontin RC (1988) On measures of gametic disequilibrium. Genetics 120:849–852

Lewontin RC, Kojima K (1960) The evolutionary dynamics of complex polymorphisms. Evolution 14:458–472

Long AD, Langley CH (1999) The power of association studies to detect the contribution of candidate genetic loci to variation in complex traits. Genome Res 9:720–731

Lukens L, Doebley J (2001) Molecular evolution of the teosinte branched gene among maize and related grasses. Mol Biol Evol 18:627–638

Lynch M, Ritland K (1999) Estimation of pairwise relatedness with molecular markers. Genetics 152:1753–1766

Maccaferri M, Sanguineti MC, Noli E, Tuberosa R (2005) Population structure and longrange disequilibrium in a durum wheat elite collection. Mol Breed 15:271–290

Meuwissen TH, Goddard ME (1997) Estimation of effects of quantitative trait loci in large complex pedigrees. Genetics 146:409–416

Michelmore RW, Paran I, Kesseli RV (1991) Identification of markers linked to disease-resistance genes by bulked segregant analysis: a rapid method to detect markers in specific genomic regions by using segregating populations. Proc Natl Acad Sci USA 88:9828–9832

Monks SA, Kaplan NL, Weir BS (1998) A comparative study of sibship tests of linkage and/or association. Am J Hum Genet 63:1507–1516

Morrell PL, Toleno DM, Lundy KE, Clegg MT (2005) Low levels of linkage disequilibrium in wild barley (*Hordeum vulgare* ssp. spontaneum) despite high rates of self-fertilization. Proc Natl Acad Sci USA 102:2442–2447

Mueller J (2004) Linkage disequilibrium for different scales and applications. Brief Bioinform 5:355–364

Niebur W, Rafalski JA, Smith OS, Cooper M (2004) New directions for a diverse planet. Proceedings of the 4th International Crop Science Congress. Brisbane, Australia, http://www.cropscience.org.au

Nordborg M (2000) Linkage disequilibrium, gene trees, and selfing: an ancestral recombination graph with partial self-fertilization. Genetics 154:923–929

Nordborg M, Borevitz JO, Bergelson J, Berry CC, Chory J et al (2002) The extent of linkage disequilibrium in *Arabidopsis thaliana*. Nat Genet 30:190–193

Nordborg M, Hu TT, Ishino Y, Jhaveri J, Toomajian C et al (2005) The pattern of polymorphism in Arabidopsis thaliana. PLoS Biol 3:e196

Nordborg M, Tavare S (2002) Linkage disequilibrium: what history has to tell us. Trends Genet 18:83–90

Oliehoek PA, Windig JJ, van Arendonk JA, Bijma P (2006) Estimating relatedness between individuals in general populations with a focus on their use in conservation programs. Genetics 173:483–496

Olsen K, Caicedo A, Polato N, McClung A, McCouch S et al (2006) Selection under domestication: evidence for a sweep in the rice waxy genomic region. Genetics 173:975–983

Olsen KM, Halldorsdottir SS, Stinchcombe JR, Weinig C, Schmitt J et al (2004) Linkage disequilibrium mapping of *Arabidopsis* CRY2 flowering time alleles. Genetics 167:1361–1369

Olsen KM, Purugganan MD (2002) Molecular evidence on the origin and evolution of glutinous rice. Genetics 162:941–950

Palaisa K, Morgante M, Tingey S, Rafalski A (2004) Long-range patterns of diversity and linkage disequilibrium surrounding the maize Y1 gene are indicative of an asymmetric selective sweep. Proc Natl Acad Sci USA 101:9885–9890

Palaisa KA, Morgante M, Williams M, Rafalski A (2003) Contrasting effects of selection on sequence diversity and linkage disequilibrium at two phytoene synthase loci. Plant Cell 15:1795–1806

Paterson AH, DeVerna JW, Lanini B, Tanksley SD (1990) Fine mapping of quantitative trait loci using selected overlapping recombinant chromosomes, in an interspecies cross of tomato. Genetics 124:735–742

Patil N, Berno AJ, Hinds DA, Barrett WA, Doshi JM et al (2001) Blocks of limited haplotype diversity revealed by high-resolution scanning of human chromosome 21. Science 294: 1719–1723

Pe'er I, Chretien YR, de Bakker PI, Barrett JC, Daly MJ et al (2006) Biases and reconciliation in estimates of linkage disequilibrium in the human genome. Am J Hum Genet 78:588–603

Podlich DW, Winkler CR, Cooper M (2004) Mapping as you go: an effective approach for marker-assisted selection of complex traits. Crop Sci 44:1560–1571

Price AH (2006) Believe it or not, QTLs are accurate! Trends Plant Sci 11:213–216

Price AL, Patterson NJ, Plenge RM, Weinblatt ME, Shadick NA et al (2006) Principal components analysis corrects for stratification in genome-wide association studies. Nat Genet 38:904–909

Pritchard JK (2001) Deconstructing maize population structure. Nat Genet 28:203–204

Pritchard JK, Przeworski M (2001) Linkage disequilibrium in humans: models and data. Am J Hum Genet 69:1–14

Pritchard JK, Stephens M, Donnelly P (2000a) Inference of population structure using multilocus genotype data. Genetics 155:945–959

Pritchard JK, Stephens M, Rosenberg NA, Donnelly P (2000b) Association mapping in structured populations. Am J Hum Genet 67:170–181

Rabinowitz D (1997) A transmission disequilibrium test for quantitative trait loci. Hum Hered 47:342–350

Remington DL, Thornsberry JM, Matsuoka Y, Wilson LM, Whitt SR et al (2001a) Structure of linkage disequilibrium and phenotypic associations in the maize genome. Proc Natl Acad Sci USA 98: 11479–11484

Remington DL, Ungerer MC, Purugganan MD (2001b) Map-based cloning of quantitative trait loci: progress and prospects. Genet Res 78:213–218

Risch N, Merikangas K (1996) The future of genetic studies of complex human diseases. Science 273:1516–1517

Sebastiani P, Lazarus R, Weiss ST, Kunkel LM, Kohane IS et al (2003) Minimal haplotype tagging. Proc Natl Acad Sci USA 100:9900–9905

Shaw SH, Carrasquillo MM, Kashuk C, Puffenberger EG, Chakravarti A (1998) Allele frequency distributions in pooled DNA samples: applications to mapping complex disease genes. Genome Res 8(2):111–123

Skøt L, Humpreys MO, Armstead I, Heywood S, Skøt K et al (2005) An association mapping approach to identify flowering time genes in natural populations of Lolium perenne. Mol Breed 15: 233–245

Spielman RS, McGinnis RE, Ewens WJ (1993) Transmission test for linkage disequilibrium: the insulin gene region and insulin-dependent diabetes mellitus (IDDM). Am J Hum Genet 52:506–516

Spielman RS, McGinnis RE, Ewens WJ (1994) The transmission/disequilibrium test detects cosegregation and linkage. Am J Hum Genet 54:559–560; author reply 560–553

Stich B, Melchinger AE, Piepho H-P, Heckenberger M, Maurer HP, Reif JC (2006) A new test for family based association mapping with inbred lines from plant breeding programs. Theor Appl Genet 113:1121–1130

Storey JD (2002) A direct approach to false discovery rates. J Roy Stat Soc Ser B Stat Methodol 64:479–498

Storey JD, Tibshirani R (2003) Statistical significance for genomewide studies. Proc Natl Acad Sci USA 100:9440–9445

Stuber CW, Lincoln SE, Wolff DW, Helentjaris T, Lander ES (1992) Identification of genetic factors contributing to heterosis in a hybrid from two elite maize inbred lines using molecular markers. Genetics 132:823–839

Szalma SJ, Buckler EST, Snook ME, McMullen MD (2005) Association analysis of candidate genes for maysin and chlorogenic acid accumulation in maize silks. Theor Appl Genet 110:1324–1333

Tenaillon MI, Sawkins MC, Anderson LK, Stack SM, Doebley J et al (2002) Patterns of diversity and recombination along chromosome 1 of maize (*Zea mays* ssp. *mays* L.). Genetics 162:1401–1413

Tenaillon MI, Sawkins MC, Long AD, Gaut RL, Doebley JF et al (2001) Patterns of DNA sequence polymorphism along chromosome 1 of maize (*Zea mays* ssp. *mays* L.). Proc Natl Acad Sci USA 98:9161–9166

Thornsberry JM, Goodman MM, Doebley J, Kresovich S, Nielsen D et al (2001) Dwarf8 polymorphisms associate with variation in flowering time. Nat Genet 28:286–289

Thumma BR, Nolan MF, Evans R, Moran GF (2005) Polymorphisms in cinnamoyl CoA reductase (CCR) are associated with variation in microfibril angle in *Eucalyptus* spp. Genetics 171:1257–1265

Tracy WF, Whitt SR, Buckler ES (2006) Recurrent mutation and genome evolution: example of *Sugary1* and the origin of sweet maize. Crop Sci 46:1–7

Wall JD, Pritchard JK (2003) Haplotype blocks and linkage disequilibrium in the human genome. Nat Rev Genet 4:587–597

Wang H, Nussbaum-Wagler T, Li B, Zhao Q, Vigouroux Y et al (2005) The origin of the naked grains of maize. Nature 436:714–719

Wang J (2002) An estimator for pairwise relatedness using molecular markers. Genetics 160: 1203–1215

Wang Y, Rannala B (2005) In silico analysis of disease-association mapping strategies using the coalescent process and incorporating ascertainment and selection. Am J Hum Genet 76:1066–1073

Wilson LM, Whitt SR, Ibanez AM, Rocheford TR, Goodman MM et al (2004) Dissection of maize kernel composition and starch production by candidate gene association. Plant Cell 16:2719–2733

Wright SI, Bi IV, Schroeder SG, Yamasaki M, Doebley JF et al (2005) The effects of artificial selection on the maize genome. Science 308:1310–1314

Wright SI, Gaut BS (2005) Molecular population genetics and the search for adaptive evolution in plants. Mol Biol Evol 22:506–519

Yu J, Holland JB, McMullen MD, Buckler ES (2006a) Genome-wide complex trait dissection through nested association mapping -in review

Yu J, Pressoir G, Briggs WH, Vroh Bi I, Yamasaki M et al (2006b) A unified mixed-model method for association mapping that accounts for multiple levels of relatedness. Nat Genet 38:203–208

Zhang K, Jin L (2003) Haplo block finder: haplotype block analyses. Bioinformatics 19:1300–1301

Zhu YL, Song QJ, Hyten DL, Van Tassell CP, Matukumalli LK et al (2003) Single-nucleotide polymorphisms in soybean. Genetics 163:1123–1134

CHAPTER 6

EXPLOITATION OF NATURAL BIODIVERSITY THROUGH GENOMICS

SILVANA GRANDILLO[1,*], STEVE D. TANKSLEY[2] AND DANI ZAMIR[3]

[1]*CNR-IGV, Institute of Plant Genetics, Portici, Via Università 133, 80055 - Portici (NA), Italy*

[2]*Department of Plant Breeding and Department of Plant Biology, Cornell University, Ithaca, NY 14853, USA*

[3]*Faculty of Agriculture, The Hebrew University of Jerusalem, PO Box 12, Rehovot 76100, Israel*

Abstract: The genetic improvement of crop plants is the most viable approach to meeting the increasing demand for agricultural output. This goal may be achieved by using the wealth of genetic variation provided by nature. Until now, scientists have been unable to exploit the genetic potential warehoused in plant germplasm repositories for quantitative traits associated with agricultural yield. Here we review the development and application of the advanced-backcross and introgression-line breeding populations for the identification of wild species derived chromosome segments that improve agricultural performance of elite germplasm. The results of studies in a wide range of crops indicate that, unlike their domestic relatives, which are often depleted in genetic variation, wild populations of plants carry a tremendous wealth of potentially valuable alleles, many of which would not have been predicted from the phenotype of the wild plants. The results from these studies may help open up new sources of genetic variation for plant breeding and biotechnology and shed light on the nature of quantitative trait variation.

1. THE RATIONALE FOR RE-DOMESTICATION OF NATURAL BIODIVERSITY

Today modern agriculture – and, for that matter, human existence – is dependent on the cultivation of a few highly productive crop species. All these species were originally domesticated by humans from wild relatives about 10,000 years ago (Ladizinsky 1998; Simmonds 1976). Although the exact mechanisms by which domestication was carried out are not known, there is evidence suggesting that

*Corresponding Author: grandill@unina.it

R.K. Varshney and R. Tuberosa (eds.), Genomics-Assisted Crop Improvement: Vol. 1: Genomics Approaches and Platforms, 121–150.

domestication of most crop plants occurred in specific 'centers of origin' throughout the world, and has generally involved only a few founding genotypes. The 'founder effect' principle in crop evolution is responsible for the fact that many crop plants contain only a small fraction of the genetic variation that is present in their wild ancestors. In the case of tomato, for example, new-world founder cultivars were introduced into Europe in the sixteenth century, and these few introductions represented the starting point for the development of the improved germplasm that was then disseminated to many areas of the world. Similarly, the majority of modern U.S. hard red winter wheat varieties originated from only two lines imported from Russia and Poland, while almost all soybean varieties in the United States derive from a dozen of introductions from China, and cotton varieties trace back to a few Mexican lines. By contrast, maize, a naturally outcrossing species, has experienced more gene flow between cultivated and wild species, which has given rise to a highly polymorphic genome (reviewed by Zamir 2001).

The first outcome of the domestication process were landrace varieties that can be considered the earliest form of cultivars. They are the result of a slow breeding process that farmers have conducted over the centuries by selecting improved plant types in their fields, which have arisen through naturally occurring mutations, recombination, and spontaneous outcrossing events. Since landraces have been selected for subsistence agricultural environments, they produced low, but stable yields. As a result of the selection exerted by humans during domestication in favor of desired traits including large fruit and seed size, sweet flavor and pleasant aroma, or against unfavorable ones such as seed shattering or unpleasant aroma, cultivated germplasm often shows a wide range of extreme phenotypes, which can frequently be more diverse than those observed in the original wild germplasm. However, this phenotypic variation may not always correspond to a proportionally wide underlying genetic variation as single gene mutations can exert wide pleiotropic effects.

Domestication, therefore, represents the first genetic bottleneck that was imposed by humans on wild germplasm, and the derived early domesticates carry only a subset of the genetic variation found in the wild ancestors (Ladizinsky 1998; Simmonds 1976). After domestication, intensive breeding of crop varieties by modern science has further eroded the genetic base in many crops. Due to the overall superior performance of elite crops over their related wild species, most modern plant breeding programs are often based on repeated intercrossing of a limited number of genetically closely related elite lines, which leaves most of the genetic variation contained in the unadapted germplasm basically unexploited. Moreover, as farmers throughout the world shifted to growing high-yield varieties, many landraces were lost. The problem of a reduced gene pool of cultivated germplasm is particularly relevant in self-pollinated crops, such as tomato and rice, where the level of genetic variation in cultivated varieties can be lower than 5% of that available in nature (Miller and Tanksley 1990; Wang et al. 1992).

The reduced genetic base which characterizes many modern crop varieties not only makes them more susceptible to disease epidemics, but it also reduces the chances for plant breeders to identify useful new combinations of genes, thus causing in the long term a slower rate of crop improvement. On the other hand, exotic germplasm, including wild relatives and early landrace varieties, offer a vast genetic resource that can potentially broaden the genetic base of modern varieties (Tanksley and McCouch 1997). The potential value of exotic germplasm was recognized early at the beginning of the past century (Bessey 1906; Burbank 1914), Nikolai Vavilov (1887– 1943) and Jack Harlan (1917–1998) being among the first to set up plant collections. These examples, combined with the alarming rate at which locally adapted landraces are being lost and at which natural habitats are being damaged, have led the international community to invest efforts and resources in large plant collections and preservation in the form of seed banks, focusing primarily on "exotics". Worldwide, there are more than 700 documented seed collections holding an estimated 2.5 million entries including many exotics, and for a staple crop like rice, more than 20,000 wild accessions are stored in seed banks (Plunknett et al. 1987)

In spite of the wealth of genetic potential preserved in germplasm collections, breeders have so far been unable to fully exploit it, especially for the improvement of complex traits important to agriculture, including yield, nutritional quality and stress tolerance. Such traits often show a polygenic inheritance pattern resulting from the segregation of numerous interacting quantitative trait loci (QTL), with varying magnitude of effect, whose expression is modified by the genetic background and the environment. Exotic germplasm has been commonly used as a source for major genes for disease and insect resistances (Plunknett et al. 1987), as shown by the high number of resistance genes derived from wild species which can be found in elite germplasm. For example, at present, commercial tomato hybrids include different combinations of 15 independently introgressed disease-resistance genes that originate from various wild resources; in rice, the genes for resistance to more than seven pathogens have been introgressed into cultivated rice germplasm from wild species, and some lines containing the wild introgressions are in commercial cultivation; and in wheat, wild relatives have been used as sources for approximately 30 independent resistance genes (as reviewed by Zamir 2001).

The limited use of exotic genetic resources for the improvement of quantitative traits can be explained by the fact that the transfer of traits from unadapted germplasm that carries many undesirable genes into elite lines is a time-consuming, laborious process which requires an efficient selection procedure and many generations of backcrossing to the adapted parent in order to recover most of the desirable agronomic traits, without always ending in a successful product. Moreover, several inherent problems are often associated with crosses involving wild and domesticated species, including unilateral incompatibility, hybrid inviability or sterility, infertility of the segregating generations, suppressed recombination between the chromosomes of the two species, and 'linkage drag'– the transfer of tightly-linked undesirable loci with the traits of interest. Furthermore, much of the wild germplasm is phenotypically inferior to modern cultivars for many of the quantitative traits that breeders would like to improve.

With the availability of co-dominant DNA markers, it has become possible to construct saturated genetic maps which have provided the necessary tools to overcome some of the above-mentioned problems associated with the use of exotic germplasm, and to allow its more systematic and efficient use as a source of valuable alleles for the improvement of quantitative traits. Marker-based estimates of genetic variability within and between accessions permit a more rational and efficient sampling of genebanks. Molecular maps have allowed the genetic dissection of the loci underlying quantitative traits, and by fine mapping QTL it is possible to distinguish pleiotropy from close linkage. Moreover, recombinants can be more efficiently identified in which close linkages are broken, thus reducing the negative effects of linkage drag (Tanksley 1993). Once tightly-linked markers to the target QTL are identified, marker assisted selection (MAS) can be used to transfer the QTL more precisely and efficiently into the desired genetic background. Finally, QTL mapping studies have also provided stronger evidence that despite the inferior phenotype, unadapted germplasm is likely to be a source of agronomically favorable QTL alleles associated with transgressive segregation observed in several interspecific crosses (de Vicente and Tanksley 1993; Eshed and Zamir 1995; Tanksley et al. 1996; Tanksley and McCouch 1997). These results suggest that there are many favorable alleles that were "left behind" by the domestication process and that these alleles can now be more efficiently "recovered" using innovative genomic-assisted breeding strategies such as molecular maps and the integrative power of QTL analysis (Tanksley and McCouch 1997; Zamir 2001; McCouch 2004).

Despite the numerous QTL-mapping studies conducted and reported for many crops, the contribution of QTL analysis to breeding new varieties has so far been low. This may in part be due to the fact that QTL-mapping efforts and plant breeding programs have generally been independent processes. Moreover, almost all QTL studies have used early segregating generations (F_2, F_3 and BC_1) for mapping and QTL detection. Favorable QTL alleles identified in these early generations often lose their effects once they are introgressed into the genetic background of elite lines. This can be explained with the relatively high level of epistatic interactions that occur between donor QTL alleles and other donor genes in early mapping generations.

New tools and concepts must therefore be developed that would allow us to use more efficiently the genetic potential stored in seed banks and in exotic germplasm, for the improvement of elite genotypes, thereby enriching the genetic base of crop species and accelerating the rate of genetic improvement. Here we review two related molecular breeding strategies, the "advanced backcross (AB) QTL method" and "exotic libraries" that have been developed and tested in several crops with the purpose of increasing the efficiency with which natural biodiversity can be exploited to improve yield, adaptation and quality of elite germplasm. These approaches, along with the recent developments in technology and statistical methodology, have laid the premises for the 'Breeding by Design' concept, which aims to design superior genotypes '*in silico*'. This is pursued by understanding the genetic basis of agronomically important traits and allelic variation at the target loci through a combination of precise genetic mapping, high-resolution chromosome haplotyping and extensive phenotyping (Peleman and van der Voort 2003). The final goal of this new concept is the optimal exploitation of the naturally available genetic resources to generate new traits and improve crop performance.

2. THE ADVANCED BACKCROSS QTL MAPPING STRATEGY

Advanced backcross QTL analysis (AB-QTL) was proposed by Tanksley and Nelson (1996) as a new breeding method that integrates the process of QTL discovery with variety development, by simultaneously identifying and transferring useful QTL alleles from unadapted (e.g., land races, wild species) to elite germplasm, thus broadening the genetic diversity available for breeding (Figure 1).

The strategy differs from other QTL mapping methods because the molecular-marker and phenotypic analyses are delayed until advanced generations, like BC_2 or BC_3, when the frequency of the donor-parent genome is reduced and therefore the segregating population resembles the recurrent parent of the cross. Moreover, during the development of these advanced populations a negative genotypic and/or phenotypic selection is exerted against unfavorable alleles originating from the unadapted parent. This avoids the masking effect of deleterious wild alleles for traits such as seed shattering, sterility, undesirable growth habit and small fruit that could otherwise interfere with later measurements of yield and other agronomic traits (Tanksley and Nelson 1996). Because in the advanced backcross families analyzed the recurrent parent's alleles are at a much higher frequency, the probability is

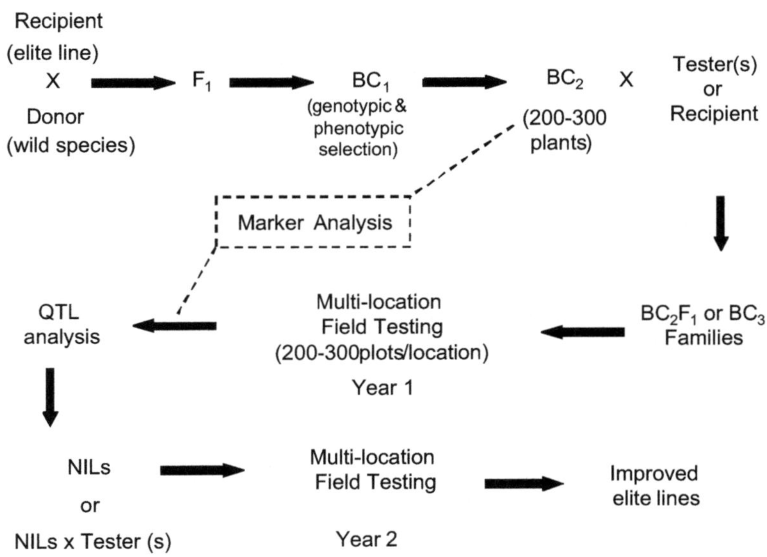

Figure 1. Scheme of the "Advanced Backcross QTL mapping strategy"

reduced for the detection of QTL requiring epistatic interactions among alleles from the donor parent. Instead, there should be a higher probability of detecting additive QTL which will more likely continue to function as predicted once they are transferred in the recurrent parent background. For crops where open pollinated varieties are the norm, field testing can be conducted on BC_2S_1 or BC_3S_1. On the other hand, for crops where commercial hybrids are more commonly used, the BC_2 or BC_3 plants are crossed with a tester variety to generate BC_2F_1 or BC_3F_1 families.

Another advantage of the AB-QTL method is that, once favorable QTL alleles are detected, only a few additional marker-assisted generations are required to generate near isogenic lines (NILs) or introgression lines (ILs) that can be field tested in order to confirm the QTL effect and subsequently used for variety development. Therefore, a cycle of AB-QTL analysis (i.e. QTL discovery, NIL/IL development and testing) represents a direct test of the underlying assumption of QTL breeding: that favorable alleles detected in segregating populations (i.e. BC_2 or BC_3 in the case of AB-QTL) will continue to exert their positive effects once they are placed in the genetic background of elite lines.

The AB-QTL strategy was initially developed and tested in tomato (Tanksley et al 1996). Since then, it has been adapted for use in other crops including rice (Xiao et al 1996, 1998; Moncada et al. 2001; Thomson et al. 2003; Septiningsih et al. 2003a,b), maize (Ho et al. 2002), wheat (Huang et al. 2003; Narasimhamoorthy et al. 2006), pepper (Rao et al. 2003), barley (Pillen et al. 2003, 2004; von Korff et al. 2005, 2006; Li et al. 2005a), bean (Blair et al. 2006).

2.1. Advanced Backcross QTL Analysis in Tomato

The tomato AB-QTL mapping project started in 1995 as a molecular marker-assisted breeding experiment applied in processing tomatoes. For this purpose a commercially acceptable publicly available open-pollinated processing variety (cv. E6203) was chosen as recurrent parent, and the only source of genetic variation permitted for improvement of E6203 was QTL alleles derived from wild tomato species. The main objectives of the project were: i) to test wild germplasm as a source of novel, useful QTL; ii) to test a new marker-assisted breeding scheme for maximizing chances of QTL discovery; iii) to develop new lines that would outperform elite commercial varieties by focusing on improving soluble solids content, while maintaining or improving other important traits for the processing industry, including yield, viscosity, firmness, color and fruit size.

The use of a common recurrent parent allows more direct cross-species comparisons of the genetic control of the analyzed traits. In order to increase the probability of identifying in each separate study a high proportion of useful, new QTL, exotic germplasm donors for use with the AB-QTL method were selected on the basis of genetic uniqueness, representing the broadest possible spectrum of wild species maintained in seed banks.

So far five AB-QTL studies have been conducted in tomato involving crosses with the five wild *Solanum* species: *S. pimpinellifolium* (acc. LA1589) (Grandillo and

Tanksley 1996; Tanksley et al. 1996), *S. peruvianum* (acc. LA1708) (Fulton et al. 1997), *S. habrochaites* (acc. LA1777) (Bernacchi et al. 1998a,b), *S. neorickii* (acc. LA2133) (Fulton et al. 2000), and *S. pennellii* (acc. LA1657) (Frary et al. 2004). All these wild species have been the sources of many major resistance genes. However, no effort has been made to take full advantage of the high level of genetic variation available for the improvement of quantitative traits.

For four studies marker analysis was conducted on BC_2 populations while BC_3 or BC_2F_1 families or both were used for phenotypic analysis (Table 1). The *S. peruvianum* AB-QTL study represents the only case in which genotypic analysis was postponed until the BC_3 generation, and phenotypes were evaluated on the derived BC_4 families. The estimated percentage of genome covered with molecular markers ranged from 55% for the *S. pennelli* study to up to 94% for the *S. pimpinellifolium* and *S. habrochaites* studies (Grandillo and Tanksley 2005).

All populations were field tested, in several locations worldwide, for numerous traits important for the tomato processing industry, ranging from a minimum of 19 traits measured in the *S. habrochaites* study up to 35 traits evaluated in the case of the *S. peruvianum* AB-population (Table 1). In all cases, total yield, red yield and main fruit quality characteristics, including soluble solids content or brix, fruit color, viscosity, firmness and fruit pH were measured. Due to the frequent negative relationship existing between brix and yield, the derived parameter brix x yield was considered as a more comprehensive biological and agricultural estimate for the productivity of processing tomatoes (Eshed and Zamir 1995, Tanksley et al. 1996). Moreover, the advanced backcross populations obtained with the four wild species, *S. pimpinellifolium*, *S. peruvianum*, *S. habrochaites* and *S. neorickii* have also been used to identify QTL influencing flavor as assessed by a taste panel, and QTL for biochemical properties that may contribute to the flavor of processed tomatoes, such as sugars and organic acids (Fulton et al. 2002).

In all five interspecific AB-QTL populations analyzed so far, favorable wild QTL alleles have been detected for more than 45% of the evaluated traits (Grandillo and Tanksley 2005). For example, in the *S. habrochaites* AB population, of the 101 QTL identified for 17 traits for which allelic effects could be deemed as favorable or unfavorable, 17 (17%) QTL corresponding to 8 traits (47%), had trait-improving alleles derived from *S. habrochaites* (Bernacchi et. al 1998a; Grandillo and Tanksley 2005). Approximately the same percentage of traits with favorable wild-alleles were obtained with *S. pennelli* (48%, Frary et al. 2004), while even higher percentages were observed for *S. pimpinellifolium* (88%, Tanksley et al. 1996), *S. peruvianum* (73%, Fulton et al. 1997) and *S. neorickii* (69%, Fulton et al. 2000).

For 12 traits it was possible to compare the percentages of positive QTL alleles detected in each wild tomato accession (Grandillo and Tanksley 2005). Favorable wild alleles were identified not only for traits for which the unadapted species showed a superior phenotype (e.g., soluble solids content, puffiness and cover) but also for those traits for which the wild phenotype was, in most cases, agronomically inferior (e.g., total yield, fruit weight and fruit color). The average percentage of favorable wild QTL alleles estimated across the five wild species ranged between a

Table 1. AB-QTL mapping studies for yield-related and quality traits in crops

Crop	Donor parent	Mapping population[a]	Number of traits evaluated[b]	Number of QTL identified	Average No. of QTL/trait	References
Tomato	*Solanum pimpinellifolium*	BC_2/BC_2F_1 & BC_3	31	120	3.8	Tanksley et al. 1996; Fulton et al. 2002
	Solanum peruvianum	BC_3/BC_4	47	269	5.7	Fulton et al. 1997; Fulton et al. 2002
	Solanum habrochaites	BC_2/BC_3	34	155	4.6	Bernacchi et al. 1998a; Fulton et al. 2002
	Solanum neorickii	BC_2/BC_3	42	251	6.0	Fulton et al. 2000; Fulton et al. 2002
	Solanum pennellii	BC_2/BC_2F_1	25	84	3.4	Frary et al. 2004
Rice	*Oryza rufipogon*	BC_2/BC_2F_1	12*	68	6.0	Xiao et al. 1996, 1998
	Oryza rufipogon	BC_2/BC_2F_2	8	25	3.1	Moncada et al. 2001
	Oryza rufipogon	BC_2/BC_2F_1 & BC_2F_2	13	76	5.8	Thomson et al. 2003
	Oryza rufipogon	BC_2/BC_2F_2	12	42	3.5	Septiningsih et al. 2003a
	Oryza rufipogon	BC_2/BC_2F_2	14*	23	1.6	Septiningsih et al. 2003b
	Oryza glaberrima	BC_3/BC_3F_1	16*	11	0.7	Li et al. 2004
Barley	*H. vulgare ssp. spontaneum*	$BC_2F_2/BC_2F_{2:5}$ & $BC_2F_{2:6}$	13	86	6.6	Pillen et al. 2003
	H. vulgare ssp. spontaneum	$BC_2F_2/BC_2F_{2:5}$ & $BC_2F_{2:6}$	13	108	8.3	Pillen et al. 2004
	H. vulgare ssp. spontaneum	BC_3DH	11	25	2.3	Li et al. 2005a
	H. vulgare ssp. spontaneum	BC_2DH	9	86	9.6	von Korff et al. 2006
Bean	wild common bean (G24404)	$BC_2F_{3:5}$	8	41	5.1	Blair et al. 2006
Maize	Inbred line Iodent (RD3013)	BC_2/BC_2TC	3	13	4.3	Ho et al. 2002
Pepper	*Capsicum frutescens*	BC_2/BC_2 & BC_2S_1	10	58	5.8	Rao et al. 2003
Wheat	synthetic wheat (W-7984)	BC_2F_2/BC_2F_3	5	40	8.0	Huang et al. 2003
	synthetic wheat (TA4152-4)	$BC_2F_3/BC_2F_{2:4}$	11	10	0.9	Narasimhamoorthy et al. 2006

[a] typically between 100-300 families were analyzed per AB-QTL population
[b] "*"indicates populations phenotyped in a single environment

minimum of 3% for total red yield to a maximum of 88% for soluble solids content. Over ten traits common to all five studies, the highest percentage of positive QTL was identified in the *S. pimpinellifolium* study (44%), followed by the *S. peruvianum* (41%), *S. neorickii* (28%), *S. pennellii* (27%) and *S. habrochaites* (15%) studies (Grandillo and Tanksley 2005).

Overall, these results have shown that in tomato, on average, for approximately 30% of QTL for any given trait, the wild species allele is expected to be superior (from an agricultural viewpoint) to the cultivated parent allele. Furthermore, after having sampled several wild species genomes the rate of discovery of "new" QTL alleles is still approximately 50% (Fulton et al. 2000; Frary et al. 2004). These results suggest that continued sampling of exotic germplasm should guarantee the discovery of new and useful QTL alleles.

2.2. Advanced Backcross QTL Analysis in Other Species

The AB QTL method has so far been tested also in rice, wheat, maize, barley, pepper and bean (Table 1).

In rice four parallel AB-QTL studies have been conducted for yield and yield components using the same wild accession of *Oryza rufipogon* (IRGC 105491) as donor parent and four different elite varieties as recurrent parent: the high-yielding Chinese hybrid V20/Ce64 (Xiao et al. 1996, 1998), the upland *Oryza sativa* subsp. *japonica* rice variety Caiapo from Brazil (Moncada et al. 2001), the U.S. long-grain tropical *japonica* cultivar Jefferson (Thomson et al. 2003), and the elite tropical cultivar IR64 (Septiningsih et al. 2003a). The use of the same *O. rufipogon* accession as donor parent offers the advantage of being able to compare the effects of wild QTL alleles in different genetic backgrounds and environments, and to identify QTL that are likely to be most stable when transferred to a new genetic background and/or evaluated under different environmental conditions.

In the first study conducted by Xiao et al. (1996, 1998) an interspecific BC_2 testcross population of 300 families was evaluated for 12 agronomically important traits under high-input conditions. Although the *O. rufipogon* accession was phenotypically inferior for all traits analyzed, transgressive segregants that outperformed the elite hybrid variety, V20A/Ce64, were observed for all 12 traits. A total of 68 significant QTL were identified, 35 (51%) of which had beneficial alleles deriving from the donor wild parent. Interestingly, 19 (54%) of these beneficial QTL alleles were free of deleterious effects on other traits. This was the case, for example, for the two QTL on chromosomes 1 and 2, for which the *O. rufipogon* alleles were associated with an 18% and 17% increase in grain yield per plant, respectively, without increasing plant height or delaying maturity (Xiao et al. 1996, 1998).

In the study conducted by Moncada et al. (2001) 274 BC_2F_2 families were evaluated for eight agronomic traits under the low-input conditions of the drought-prone acid soils to which Caiapo was adapted. Although *O. rufipogon* was phenotypically inferior for seven of the eight traits analyzed, 56% of trait-enhancing QTL identified were derived from this wild donor parent. These results showed that the AB-QTL method offers a useful strategy for the genetic improvement also of cultivars adapted to stress-prone environments.

A similar high percentage (53%) of favorable *O. rufipogon* alleles were detected for yield and yield component QTL in a study conducted using the cultivar Jefferson as recurrent parent (Thomson et al. 2003). On the other hand, a lower percentage (33%) of favorable *O. rufipogon* alleles were identified for the same yield-related traits using the cultivar IR64 as recurrent parent (Septiningsih et al. 2003a). The same population was also analyzed for 14 seed quality traits (Septiningsih et al. 2003b). Although a low proportion of *O. rufipogon* favorable alleles were identified for the quality QTL, it is worth noting that all but one of the positive *O. rufipogon*-derived yield and yield component QTL reported in Septiningsih et al. (2003a) were not linked to the negative grain quality QTL detected. This suggests that there is not likely to be a large amount of linkage drag associated with grain quality if markers are used to selectively introgress positive yield QTL from *O. rufipogon* into an IR64 background. Overall, these results indicate that one of the closest wild relatives of cultivated rice, *O. rufipogon,* despite its overall inferior appearance, contains QTL alleles that are likely to substantially improve the performance of elite rice germplasm for agronomically important traits, including yield.

The use of advanced backcross generations for the identification of useful QTL has also been applied to an interspecific population of rice, derived from a cross between the two cultivated species *O. sativa* (cv. V20A, a popular male-sterile line used in Chinese rice hybrids) and *O. glaberrima* (acc. IRGC#103544 from Mali) (Li et al. 2004). Approximately 300 BC_3F_1 hybrid families were used to identify QTL associated with grain quality and grain morphology. Eleven significant QTL were identified for seven of the 16 grain-related traits analyzed, with favorable alleles coming from *O. glaberrima* at eight (73%) loci.

In barley five AB-QTL studies have been conducted using four different interspecific crosses (Pillen et al. 2003, 2004; von Korff et al. 2005, 2006; Li et al. 2005a). In the first two studies, Pillen et al. (2003, 2004) conducted two separated AB-QTL analyses on BC_2F_2 populations derived from crosses between the wild barley accession ISR101-23 (*Hordeum vulgare* ssp. *spontaneum*) and the two German spring barley varieties 'Apex' (*H. vulgare* ssp. *vulgare*) (A x 101) and 'Harry' (H x 101), respectively. Both populations were evaluated for 13 agronomic quantitative traits measured in a maximum of six environments. In the A x 101 population a relatively high proportion (34%) of the total significant 86 QTL identified had favorable effects derived from the exotic parent, for seven of the 13 traits investigated. In one case the exotic parent allele was associated with a yield increase of 7.7% averaged across the six environments tested. In the H x 101 study an even higher percentage (48%) of favorable wild QTL alleles was detected out of the total 108 putative QTL identified. A comparison of the two AB-QTL studies showed that, in all, 26% of the putative QTL could be detected in both AB populations, suggesting a high degree of epistatic genetic interactions between the detected QTL and the genomic background.

Given the favorable results obtained on yield with the wild barley ISR101-23, a different exotic barley accession was tested using a modified AB-QTL scheme, where a BC_2-double haploid (DH) population derived from a cross between the spring barley cultivar 'Scarlett' and the wild barley accession ISR42-8 (*H. vulgare* ssp. *spontaneum*) was evaluated for nine agronomic traits in up to eight environments (von Korff et al. 2005, 2006). A total of 86 putative QTL were detected for nine agronomic traits (von Korff et al. 2006) and for 31 (36%) of them the wild alleles had a favorable effect. The same BC_2DH population has also been used to detect resistance genes against powdery mildew (*Blumeria graminis* f.sp. *hordei* L.), leaf rust (*Puccinia hordei* L.) and scald [*Rhynchosporium secalis* (Oud.) J. Davis]. For the majority of resistance QTL (61%) the wild parent contributed the favorable allele (von Korff et al. 2005).

A modified AB-QTL scheme was applied to spring barley also by Li et al. (2005a). In this study a BC_3-doubled haploid (DH) population derived from the cross between the German spring barley cultivar 'Brenda' (*H. vulgare* ssp. *vulgare*) and the wild species line 'HS213' (*H. vulgare* ssp. *spontaneum*) used as donor, was evaluated for yield and its components as well as malting quality traits. A total of 25 significant QTL were identified, and positive wild QTL alleles were found for 5 (20%) QTL. Due to the low percentage (6.25%) of donor-parent genome, the BC_3-DH lines could be directly used for the development of near-isogenic lines.

Wild barley germplasm (*H. vulgare* ssp. *spontaneum* acc. HOR11508) has also proven to be a good source of QTL alleles with favorable effects on yield and other agronomically important traits under conditions of water deficit in Mediterranean countries (Talamé et al. 2004). Of the total 80 significant QTL identified by Talamé et al. (2004), 42 (52%) had beneficial alleles derived from the donor wild parent *H. spontaneum*.

In wheat the first report on AB-QTL analysis is the study conducted by Huang et al. (2003). A BC_2F_2 population derived from a cross between the German winter wheat variety 'Prinz' and the synthetic hexaploid wheat line W-7984 developed by CIMMYT and derived from *Triticum taushii* for the D genome, was used to identify QTL for yield and yield component traits (Huang et al. 2003). Of the total 40 significant QTL identified for the five traits analyzed, 24 (60%) of them, had favorable alleles derived from the synthetic wheat W-7984, despite the fact that synthetic wheat was overall inferior with respect to agronomic appearance and performance. For four of the seven QTL identified for yield the wild allele had an effect that increased total yield, and the increases associated with the wild allele ranged from 5% to 15%.

Another AB-QTL study was conducted in hard winter wheat (*Triticum aestivum* L.) by Narasimhamoorthy et al. (2006). In this case, a population of 190 $BC_2F_{2:4}$ lines derived from a cross between the hard red winter wheat variety 'Karl 92' and the synthetic wheat line TA 4152-4, was evaluated in two environments and analyzed for 11 yield-related traits as well as for resistance to wheat soilborne mosaic virus (WSBMV). Of the ten putative QTL identified the favorable allele was contributed by the synthetic parent at three (30%) QTL, namely for grain hardness, kernels per spike, and tiller number.

In pepper AB-QTL analysis has been used in an interspecific BC_2 population derived by crossing the bell-type *Capsicum annuum* cv. Maor to the small oval-fruited wild *C. frutescens* BG 2816 accession (Rao et al. 2003). The BC_2 and the BC_2S_1 families were evaluated for ten yield-related traits, and a total of 58 QTL were identified. For six (10%) QTL, alleles with opposite effects to those expected from the phenotype were detected in the wild species. The relatively low percentage of transgressive and favorable QTL alleles originating from the wild donor could in part be due to the choice of the wild parent which is quite closely related to *C. annuum*. Therefore, in order to get a more comprehensive picture of the potential of marker utilization of exotic germplasm in pepper improvement, additional crosses with more distantly related *Capsicum* species need to be analyzed.

Recently, Blair et al. (2006) used the AB-QTL analysis approach to identify QTL for agronomic performance in a population of $BC_2F_{3:5}$ introgression lines generated from the cross of a Colombian large red-seeded commercial cultivar, ICA Cerinza, and a wild common bean accession, G24404. A total of 41 significant QTL were identified for the eight traits measured, 14 (34%) of which showed positive alleles derived from the wild parent.

The AB-QTL method has been tested also in maize (Ho et al. 2002), an allogamous crop species which, unlike tomato and rice, has retained abundant genetic variation (Eyre-Walker et al. 1998). Ho et al. (2002) showed that the AB-QTL method can be extended to BC_2TC progeny derived from two elite heterotic inbreds for the identification and transfer of agronomically useful QTL as well as for the maintenance and selection of favorable epistatic interactions.

3. INTROGRESSION LINES AND 'EXOTIC LIBRARIES'

Favorable wild QTL alleles become a useful resource for breeding programs after they have been fixed in isogenic lines and after the superior performance of the isogenic line is confirmed in comparison to the cultivated recurrent parent in replicated field experiments. Isogenic lines can be generated by systematic backcrossing and introgressing of marker-defined donor segments in the recurrent parent background. In plants, also depending on the strategy used for their development, these lines have been referred to as near isogenic lines (NILs) or QTL-NILs (Eshed and Zamir 1996; Tanksley and Nelson 1996; Monforte and Tanksley 2000a; Monforte et al. 2001; Lecomte et al. 2004; Yates et al. 2004; Frary et al. 2003; Eduardo et al. 2005; Chaïb et al. 2006; Thomson et al. 2006), introgression lines (ILs) (Eshed and Zamir 1995, 1996; von Korff et al. 2004; Canady et al. 2005; Li et al. 2005b; Xu et al. 2005; Zygier et al. 2005; Liu et al. 2006; Tian et al. 2006; Petsova et al. 2001, 2006), backcross inbred lines (BILs) (Jeuken and Lindhout 2004), backcross recombinant inbred lines (BCRIL) (Monforte and Tanksley 2000a), recombinant chromosome substitution lines (RCSLs) (Matus et al. 2005), chromosome segment substitution lines (CSSLs) (Wan et al. 2004) and

'Stepped Aligned Inbred Recombinant Strains' (STAIRS) (Koumproglou et al. 2002). Introgression lines (ILs) represent near-isogenic lines (NILs) with relatively large average introgression length, while BILs and BCRILs are backcross populations generally containing multiple donor introgressions per line. Similar genetic structures have been developed and used also in mice and rat genetics and they are referred to as chromosome substitution strains (CSSs), and more specifically as consomic and congenic strains when either the entire chromosome or part of a chromosome in an inbred strain has been substituted from a different inbred, respectively (Nadeau et al. 2000; Singer et al. 2004). For simplicity, hereafter we will use the term introgression lines (ILs) for plant lines containing a single marker-defined homozygous donor segment, pre-ILs for lines which still contain multiple homozygous and/or heterozygous donor segments.

ILs allow either the screening for QTL of entire genomes (Eshed and Zamir 1995), or focusing on a specific region of interest for fine mapping QTL (Paterson et al. 1990). Given the properties of introgression lines and the potential of exotic germplasm as a source of genetic variation which has overcome the pressure of natural selection during evolution, Zamir (2001) proposed to invest in the development of a genetic infrastructure of "exotic libraries" in order to enhance the rate of progress of introgression breeding. An exotic library consists of a set of introgression lines (ILs), each of which carries a single, possibly homozygous, marker-defined chromosomal segment that originates from a donor exotic parent, in an otherwise homogeneous elite genetic background; the entire donor genome would be represented in a set of introgression lines. Since populations of introgression lines have been developed also by using adapted germplasm as donor parents, in more general terms these series of introgression lines can be referred to as libraries of introgression lines or IL libraries.

While the production of such a congenic and permanent resource was not a trivial task in the early days, when molecular markers were still being developed (Ramsay et al. 1996), the availability of numerous marker-screening technologies has now made the development of such libraries a more efficient process that can be completed after ten generations of crossing and marker analysis (Young 1999).

3.1. Introgression Lines for the Analysis of Complex Traits

Several features of these libraries of introgression lines contribute to their efficiency in detecting and mapping QTL underlying traits of agronomic importance: 1) the lines in the library differ from the recurrent parent by only a single, defined chromosomal segment derived from the donor parent; therefore, their phenotypes generally resemble that of the recipient parent, which, in the case of crosses between cultivated and exotic germplasm, reduces the sterility problems that occur in other breeding-population structures characterized by a higher frequency of the exotic parent genome, and also allows the lines to be evaluated for yield-associated traits; 2) the ability to statistically identify small phenotypic effects is increased because all the phenotypic variation between a line in the library, or the hybrid of the recurrent parent with an IL (ILH), and the nearly isogenic recurrent parent is associated with the introgressed segment; 3) the statistical procedure to detect QTL is simplified as it relies on the comparison of each IL with the background recurrent line for the trait of interest, and is therefore less affected by the need for experiment-wise error. A significant difference for any one comparison indicates the presence of one or more QTL on the differential chromosome segment defining the introgression line; 4) the epistatic effects that are mediated by other regions of the donor genome, with the exception of the loci contained in the same introgression line, are eliminated; 5) IL libraries provide a permanent resource with a characterized genotype, and therefore the phenotypic value of each introgression can be tested on multiple replicates, reducing the effect of the environment and increasing the power of QTL detection. Moreover, replicated trials of the same line can be analyzed in different years and/or environments, which allows us to determine more precisely the effect of each QTL in different environments and estimate the extent of QTL by environment interactions (Monforte et al. 2001; Liu et al. 2003; Gur and Zamir 2004). The permanent nature of these lines not only facilitates more accurate estimates of the mean phenotypic values but it also allows several laboratories to collect data for different traits on the same lines, thereby creating a comprehensive phenotypic database for general access (Zamir 2001).

The map resolution of a population of ILs is defined by the overlap between contiguous segments (bins) to which genes or QTL can be assigned by comparing lines (Pan et al. 2000; Liu et al. 2003). Bin lengths vary across the genome, depending on the number, length, and overlap of adjacent segments. Although the initial ILs contributing to an IL library generally provide a relatively low level of map resolution, they represent the starting point by which the phenotypic effects of QTL can be fine-mapped to smaller intervals (Paterson et al. 1990). High resolution mapping of QTL not only allows us to assess whether the effect on the phenotype is due to a single QTL or to several tightly linked QTL affecting the same trait (Fridman et al. 2002; Monna et al. 2002; Thomson et al. 2003), but also to verify whether possible undesirable effects are caused by linkage drag of other genes or by pleiotropic effects of the selected QTL (Eshed and Zamir 1996; Monforte and Tanksley 2000b; Monforte et al. 2001; Frary et al. 2003; Yates et al. 2004). Besides reducing linkage drag, the development of lines with smaller introgressions (sub-ILs) allows molecular markers to be found which are more tightly linked to the QTL of interest that can be used for marker-assisted breeding (MAB).

Once introgressed chromosome segments have been sub-divided and targeted, and QTL-containing lines have been created, crosses between the lines can be used to study the phenotypic effects of QTL interactions, to better understand the nature of epistasis (Eshed and Zamir 1996; Lin et al. 2000; Yamamoto et al. 2000). ILs can also be used to obtain more precise estimates of the magnitude of QTL x genetic background interaction (Eshed and Zamir 1995, 1996; Monforte et al. 2001; Lecomte et al. 2004; Gur and Zamir 2004; Chaïb et al. 2006).

Introgression lines are also a powerful tool to study the genetic basis of heterosis, since homozygous lines in a library can be crossed to different tester lines, allowing the effects of heterozygosity on the phenotype to be investigated (Semel et al. 2006). Finally, ILs have also proven to be very efficient tools for the positional cloning of key genes underlying quantitative traits (Frary et al. 2000; Fridman et al. 2000, 2004; Yano et al. 2000; Takahashi et al. 2001; El-Din-El-Assal et al. 2001; Kojima et al. 2002).

One of the first examples of the development of this type of library was that by Kuspira and Unrau (1957), who used whole-chromosome substitution lines to analyze polygenic traits in common wheat. In tomato, in order to gain an insight into the underlying genetic factors that govern differences between the cultivated tomato and its wild relatives, Zamir and colleagues used RFLP (restriction fragment length polymorphism) markers to develop a full-coverage exotic library in the form of 50 introgression lines from a cross between the wild green-fruited species *S. pennellii* (acc LA716) and the cultivated tomato *Solanum lycopersicum* (cv. M82) (Eshed and Zamir 1995). This population allowed the identification of yield-associated QTL, and to examine their epistatic and environmental interactions (Eshed and Zamir 1995, 1996). These studies also highlighted the higher efficiency with which IL populations can detect QTL compared with conventional segregating populations such as F_2, BC_1 or recombinant inbreds (RIs) (Zamir and Eshed 1998). For example, while Eshed and Zamir (1995) detected a minimum of 18 and 23 QTL for fruit size and brix, respectively, only a maximum of 7 and 4 QTL were detected for the same traits when using standard mapping populations. To increase the mapping resolution of this 'exotic library' additional 26 sub-ILs have been added and the resulting 76 lines partition the entire genetic map into 107 bins, which are defined by singular or overlapping segments (Pan et al. 2000; Liu et al. 2003). Over the past 10 years the 76 ILs and their hybrids have been assayed for 20 different yield-associated, fruit morphology and biochemical traits. The resulting data are presented, *in silico*, in a search engine 'Real Time QTL' that displays a range of statistical and graphical outputs that describe in a user-friendly way the components of the genetic variation (http://zamir.sgn.cornell.edu/; Gur et al. 2004). This exotic library has also been used to identify the QTL controlling leaf dissection (Holtan and Hake 2003), fruit nutritional and antioxidant contents (Rousseaux et al. 2005), and tomato aroma (Tadmor et al. 2002). The latter study, identified *malodorous*, a wild species allele negatively affecting tomato aroma that was selected against during domestication, thus providing a genetic explanation of one of the aroma changes that occurred during the domestication of tomato. Recently, Semel et al. (2006) have used the 76 *S. pennellii* ILs to assess the contribution of overdominant (ODO) effects to heterosis in the absence of epistasis. Thirty-five different traits for yield and fitness were measured in the field on homozygous and heterozygous plants, and a total of 841 QTL were identified. ODO QTL were detected only for the reproductive traits, which suggested that the true ODO model involving a single functional Mendelian locus is a more likely explanation for the heterosis observed in the ILs. than the pseudoODO model.

3.2. Library Resources

Since the pioneer studies conducted by Kuspira and Unrau (1957) and by Eshed and Zamir (1995, 1996) and the theoretical landmark laid by Tanksley and Nelson (1996), sets of introgression lines representing different fractions of the exotic (wild species and/or landrace varieties) parent genome have been developed for various crops including tomato (Bernacchi et al. 1998b; Monforte and Tanksley 2000a; Chetelat and Meglic 2000; Canady et al. 2005), rice (Li et al. 2005b; Tian et al. 2006), lettuce (Jeuken and Lindhout 2004), wheat (Petsova et al. 2001, 2006; Liu et al. 2006) and barley (von Korff et al. 2004) (Table 2). In other cases, such as in rice (Lin et al. 1998; Li et al. 2005b; Wan et al. 2004; Xu et al. 2005; Mei et al. 2006), and in melon (Eduardo et al. 2005), libraries of introgression lines have been developed starting from crosses between cultivated parents. Sets of introgression lines have also been developed for the model species *Arabidopsis thaliana* using the three accessions Columbia, Landsberg and Niederzenz (Koumprouglou et al. 2002).

Several strategies have been used to develop ILs and pre-ILs; in several cases marker-assisted selection has been applied since the first generations of backcrossing (Eshed and Zamir 1995; Fulton et al. 1997; Bernacchi et al. 1998a,b; Chetelat and Meglic 2000; Monforte and Tanksley 2000a; Eduardo et al. 2005), while in other studies molecular characterization and selection of pre-ILs and ILs was postponed after several cycles of random backcrossing (Jeuken and Lindhout 2004; Matus et al. 2003; Liu et al. 2006; Tian et al. 2006). In wheat, sets of single chromosome substitution lines were used as starting material to develop ILs (Petsova et al. 2001, 2006).

From the tomato AB-QTL populations MAS has been used to develop ILs and pre-ILs that contain specific QTL alleles derived from the wild donors *S. habrochaites*, *S. pimpinellifolium*, and *S. peruvianum*, and that are able to significantly improve the performance of the elite variety (Tanksley et al. 1996; Bernacchi et al. 1998b, Monforte and Tanksley 2000a,b; Monforte et al. 2001; Yates et al. 2004). Evaluation of the agronomic performance, in five locations worldwide, of 23 ILs and pre-ILs containing either *S. habrochaites* or *S. pimpinellifolium* introgressions, revealed that a high percentage (88%) of quantitative factors exhibited, in at least one location, the phenotypic effect as had been detected in the previous QTL analysis of the BC_2/BC_3 populations (Bernacchi et al. 1998b). However, the significance at which QTL/factors were detected in the BC_3 families as well as the degree of conservation of QTL across locations seemed to be modest predictors for those realized in the derived ILs and pre-ILs.

From the *S. habrochaites* AB-QTL population (Bernacchi et al. 1998a,b), Monforte and Tanksley (2000a) developed a set of 99 ILs and pre-ILs which provided a coverage of approximately 85% of the wild donor genome, and therefore represent a useful tool for the identification of valuable QTL deriving from the wild donor parent. The lines differ in many traits including yield, leaf morphology, trichome density, and fruit traits such as shape, size and color; favorable wild QTL alleles were detected for several of these traits (Monforte and Tanksley 2000b; Monforte et al. 2001; Yates et al. 2004). The lines also differ in biochemical

Table 2. Libraries of introgression lines derived from interspecific crosses

Plant	Donor parent	No. of ILs developed and/or tested[a]	Estimated genome coverage (%)	Traits analyzed	References
Tomato					
	Solanum pennellii	50	100	yield-related leaf dissection	Eshed and Zamir 1995; Holtan and Hake 2003
	Solanum pennellii	75	100	fruit color fruit size and composition metabolic profiling	Pan et al. 2000; http://www.sgn.cornell.edu/ Causse et al. 2004 Schauer et al. 2006
	Solanum habrochaites	99	~ 85	NA	Monforte and Tanksley 2000b
	Solanum lycopersicoides	90	~ 96	NA	Chetelat and Meglic 2000; Canady et al. 2005
	Solanum chmielewskii	NA	NA	NA	Peleman and van der Voort 2003; Peleman JD unpublished data
Rice					
	Oryza glumaepatula	59*	~ 100	NA	Sobrizal et al. 1996
	Oryza glaberrima	39*	~ 100	NA	Doi et al. 1997
	Oryza meridionalis	61*	~ 100	NA	Kurakazu et al. 2001
	Oryza rufipogon	159	67.5	yield-related	Tian et al. 2006
	Hordeum vulgare ssp. spontaneoum	49*	~ 98	days until heading	von Korff et al. 2004
	Hordeum vulgare ssp. spontaneoum	43*	~ 93	days until heading	von Korff et al. 2004
Bread wheat					
	Sear's "Synthetic 6x" (derived from tetraploid emmer x wild grass *Aegilops tauschii*)	84	~ 100	yield-related	Petsova et al. 2006

[a] "*"indicates sub-set of candidate pre-ILs

NA = data not available

composition including sesquiterpenes (Van der Hoeven et al. 2000), soluble solids content of the fruits (Monforte and Tanksley 2000b; Monforte et al. 2001), and anthocyanin content (Oyanedel 1999). Currently, a new set of *S. habrochaites* ILs, each containing single homozygous wild introgressions, is being developed in order to ensure whole genome coverage and increase the mapping resolution of the population (Grandillo and Tanksley, personal communication).

Starting from the *S. pimpinellifolium* (acc. LA1589) BC_2 population developed as part of the AB-QTL strategy, Doganlar et al. (2002) derived a set of 196 inbred backcross lines (IBLs). The 196 pre-ILs were evaluated for 22 quantitative traits and a total of 71 significant QTL were identified. For 48% of these QTL the wild allele was associated with improved agronomic performance. To facilitate the use of this population, a subset of 100 of the 196 lines was selected which ensures the most uniform genome coverage and map resolution.

In tomato, another exotic library has been developed using the wild tomato-like nightshade *Solanum lycopersicoides* (accession LA2951) as donor parent, in the genetic background of cultivated tomato (*Solanum lycopersicum* cv. VF36) (Chetelat and Meglic 2000; Canady et al. 2005). The population consists of a primary subset of 56 lines which ensure maximum coverage of the *S. lycopersicoides* genome (approximately 96% of the total map units), homozygosity whenever possible, and a minimum number of introgressed segments per line; and a secondary subset of 34 lines which provides increased map resolution for specific regions. For this population, homozygotes were not recovered for certain introgressed segments, and therefore several lines have to be maintained at the heterozygote level.

From the two barley AB-QTL populations generated by introgressing the wild accession (ISR42-8, from Israel) of *Hordeum vulgare* spp. *spontaneum* into two different spring barley cultivars, Scarlett (S) and Thuringia (T), two sets with 49 (S42) and 43 (T42) pre-ILs, respectively, were developed using marker-assisted selection (von Korff et al. 2004). The two sets of pre-ILs cover approximately 98% (S42) and 93% (T42) of the exotic genome, and contain on average 2 (S42) and 1.5 (T42) additional non-target introgressions. Pure ILs are currently being generated by marker-assisted backcrossing of the pre-ILs , and an additional set of pre-ILs is also being developed using a winter barley cultivar as the recurrent parent and a different exotic accession as the donor (von Korff et al. 2004).

In rice, over the last decade several libraries of introgression lines have been developed, which are in some cases termed 'chromosome segment substitution lines, or CSSLs' (Sobrizal et al. 1996; Kurakazu et al. 2001; Ahn et al. 2002; Kubo et al. 2002; Doi et al. 1997; Wan et al. 2004; Ebitani et al. 2005; Li et al. 2005; Xu et al. 2005; Mei et al. 2006; Tian et al. 2006). Of particular relevance are the findings of Li et al. (2005b) who report the results of a large backcross (BC) breeding program – part of the International Rice Molecular Breeding Program – conducted at the International Rice Research Institute (IRRI) to introgress the genetic diversity of the primary gene pool of rice into three elite genetic backgrounds: two high-yielding varieties, IR64 (indica) and Teqing (indica), and a new plant type breeding line (NPT, tropical japonica). A total of 195 accessions, including commercially grown cultivars and landraces, were used as donor parents in the backcross program, and over 20,000 ILs and pre-ILs were developed in the three elite rice genetic backgrounds, which contain allelic diversity for a wide range of quantitative traits. This large set of ILs and pre-ILs was developed as genetic stocks of wide genome coverage that could complement genome-wide insertional and deletion mutants genetic stock for large-scale functional genomic research in rice (Li et al. 2005b).

The Chinese common wild rice *(Oryza rufinpogon* Griff.) has been used as donor parent for the development of a population of 159 (BC_4F_4) ILs and pre-ILs in the background of Indica cultivar (*O. sativa* L.) Guichao 2 (Tian et al. 2006). The 159 lines represented 67.5% of the wild parent genome. The mean number of homozygous and heterozygous donor segments were 2 (ranging 0-8) and 1 (ranging 0–7), respectively, and the majority of the introgressions have sizes smaller than 10 cM. The 159 lines were evaluated in two locations for seven yield-related traits, and favorable wild QTL alleles were found for the three traits panicles per plant, grains per panicle and filled grains per plant.

Exotic libraries have been developed also in wheat. Petsova et al. (2001, 2006) reported the development of 84 ILs generated from a set of *Triticum aestivum* cv. Chinese Spring/'Synthetic 6x' single chromosome substitution lines for the D-genome, where individual chromosomes of the wild grass *Aegilops tauschii* replaced the homologous chromosomes of the cv. Chinese Spring. The genome of the exotic parent is fully represented in these lines, with the exception of three telomeric regions and a region of less than 24 cM on the chromosome arm 3DL. A subset of 52 ILs were evaluated for six quantitative traits including flowering time, plant height, ear length, spikelet number, fertility and grain weight per ear. Favorable wild QTL alleles were detected for nine (53%) of the 17 significant QTL identified.

Another synthetic hexaploid wheat genotype, Am3, obtained by crossing *Triticum carthlicum* with *Aegilops taushii*, was used as exotic parent for the development of 97 BC_4F_3 lines (16 ILs and 66 pre-ILs) in the genetic background of the common wheat Chinese cultivar Laizhou953 (Liu et al. 2006). The 97 lines cover 37.7% of the donor parent genome. The lines were evaluated for nine yield-related traits in field trials conducted in three consecutive years. For every trait there were lines showing a better performance than the recurrent parent, indicating that favorable QTL alleles for the traits of agronomic importance have been transferred from the exotic parent to the elite wheat variety.

In lettuce, Jeuken and Lindhout (2004) developed a set of 28 backcross inbred lines (BILs) or pre-ILs of the wild species *Lactuca saligna* (CGN 5271) in the cultivated background of *L. sativa* (cv. Olof), covering at least 96% of the wild genome. Most of the lines (20 out of the total 28) contained a single homozygous wild introgression. At least 77% of the *L. saligna* genome is represented in 24 lines that are completely homozygous (BILs and doubleBILs). The lines were used to map 12 simple morphological traits.

In melon, interspecific crosses between *Cucumis melo* and wild *Cucumis* species are not viable; therefore, in order to develop a collection of ILs for this species, Eduardo et al. (2005) used an intraspecific cross between two distantly related cultivars: a Spanish cultivar "Piel de Sapo" (PS), belonging to the horticultural group *inodorous*, which was used as recipient parent, and the exotic Korean cultivar "Songwhan Charmi" (accession PI161375) (SC), included in the horticultural group *conomon*, as the donor genotype. The genetic distance between the two cultivars is one of the highest distances observed between melon cultivars (Monforte et al. 2003). A collection of 57 ILs was obtained, with each line containing a single independent introgression from the SC parent in the PS genetic background, and covering overall at least 85% of the SC genome. Three ILs have already been used to verify the QTL influencing fruit shape, external color and flesh color (Monforte et al. 2004).

With the aim of increasing the genetic diversity of upland cotton (*Gossypium hirsutum* L., 2n = 52), Saha et al. (2006) reported the development of 14 BC_5S_1 chromosome substitution lines carrying specific chromosomes or chromosome arms from *G. barbadense* L. substituted into *G. hirsutum*. *G. barbadense* is the only 52-chromosome relative of Upland cotton that is cultivated. It represents a good source of genes for improving fiber length and quality, whereas Upland cotton is more valued for its high yield. The lines, together with the derived F_2 families, have been evaluated for eight agronomic and fiber traits. The results showed that fiber quality of *G. hirsutum* can be improved by introgressing specific genomic regions of *G. barbadense* without genetic drag effect of poor agronomic qualities.

In order to provide a direct approach to QTL mapping and improve the power of resolution Koumproglou et al. (2002) developed a resource of lines for *Arabidopsis thaliana* that facilitate QTL localization first to a particular chromosome, then to successively smaller regions within a chromosome ($\leq$ 0.5 cM) by means of simple comparisons among a few lines. This resource consists of the five single whole Chromosome Substitution Strains (CSS1-5) plus a large number of homozygous lines derived from each CSS and that are referred to as 'Stepped Aligned Inbred Recombinant Strains' (STAIRS) to reflect their structural relationship. By using both resources a QTL can be located, in three steps, first to the chromosome by comparing the 5 CCSs, then to 5-10 cMs and finally to a $\leq$ 1 cM region. At every step only a limited number of lines are required, which allows high replication. The final step provides two lines that are identical except for the short differential region. These pairs of isogenic lines are very valuable for the analysis of QTL, for identifying candidate genes and for gene expression studies.

3.3. Multiple-Introgression Lines for Breeding

The results obtained from the tomato QTL mapping studies together with those obtained for other crops indicate that it is unlikely that the introgression of a single QTL will result in a substantial improvement in yield-associated phenotypes as well as for other agriculturally important traits. On the other hand, MAS pyramiding of newly-discovered favorable wild QTL alleles from the same or from different wild donor species to obtain multi-QTL ILs could be the strategy to greatly improve crop performance. A similar scenario was also observed, for example, for the acyl-sugar-mediated insect resistance that characterizes *S. pennelli,* as transferring this resistance from the wild species into the cultivated tomato requires at least 5 QTL, without which no acyl-sugars are accumulated (Lawson et al. 1997).

From the tomato AB-QTL project, four favorable wild QTL were pyramided that together produce an effect on brix and brix x yield that goes far beyond any other commercial cultivar and nearly doubles the brix x yield over the original starting material (cv. E6203). By means of MAS, Gur and Zamir (2004) developed a multiple-introgression line (IL789) by pyramiding three independent yield-promoting genomic regions derived from the drought-tolerant green-fruited wild species *S. pennellii* into the genetic background of the cultivated recipient genotype M82. In order to assess the potential of the wild QTL in the context of high-yield genetic backgrounds – those close to the "yield barrier"– the IL789 was crossed with four inbred tester lines, whose hybrids with M82 exhibit the highest brix x yield values (Gur and Zamir 2004). Yield of the hybrids between IL789 and the four testers was more than 50% higher than that of a leading commercial tomato hybrid (BOS3155) that was used as control, and this higher performance was observed under both wet and dry field conditions that received 10% of the irrigation water. Moreover, the effectiveness of the wild introgressions in diverse genetic backgrounds indicated that alleles similar to those of the wild species are not present in the cultivated tomato gene pool. These results underline the potential of exotic germplasm to improve yield stability in different environments, which has long been recognized as an important objective in plant breeding.

The validity of the QTL pyramiding approach for crop improvement has been demonstrated also in rice (Ashikari and Matsuoka 2006). For this crop high grain productivity and short plant height are both important traits. QTL analysis conducted on progeny of an intraspecific cross of rice identified QTL for grain number (*Gn1*) and for plant height (*Ph1*), and two NILs, NIL-*Gn1* and NIL-*Ph1*, were developed carrying QTL alleles that increased grain number and reduced plant height, respectively. To combine both positive phenotypes, the two lines were crossed and by MAS a bi-QTL NIL, NIL-*Gn1* + *Ph1*, was obtained that carried both favorable QTL alleles in the 'Koshihikari' genetic background. This pyramiding line showed increased grain production (+23%) and reduced plant height (–20%) compared with the recurrent control 'Koshihikari'.

Overall these results show that QTL pyramiding is a successful approach for producing new varieties. In addition, the tomato examples provide solid evidence that exotic germplasm represents a rich source of new valuable QTL. As more wild accessions will be screened by means of the AB-QTL method and by means of the "exotic libraries" approach, it will be these new combinations of QTL, from various accessions, that will really break the curve in plant improvement.

3.4. Introgression Breeding in the '-omics' Age

One of the most challenging tasks facing modern biology is unraveling the molecular basis of complex phenotypes, a knowledge that should positively impact on practical breeding programs.

The attributes of QTL mapping in plants have facilitated the cloning of QTL in the model plant *Arabidopsis* and in crops such as wheat, tomato, maize and rice (see the review of Paran and Zamir 2003; Salvi and Tuberosa 2005 and chapter 9 in this book; Varshney et al. 2006). These studies have shown that, similarly to the variation found for numerous genes that control quality traits, variation in QTL alleles in plants has been identified in both coding and regulatory regions. However, the number of plant QTL that have been molecularly identified so far is still low (about twenty), and most of these QTL have large phenotypic effects that allow them to be treated as major-effect genes during the cloning process (Paran and Zamir 2003; Salvi and Tuberosa 2005; Varshney et al. 2006). This prevents us from defining a general molecular model that underlies quantitative variation (Morgante and Salamini 2003).

Although an extremely powerful and unbiased approach, delimiting a QTL to a single gene using genetic approaches is still a time-consuming and technically demanding process (Fridman et al. 2000, 2004). In order to accelerate the rate of QTL discovery it is necessary to invest in genomic technologies and biological resources as well as in methodologies that will enable integration of genetic components of QTL variation in genomic databases.

Biological resources that seem promising to improve the efficiency of QTL cloning in plants include germplasm collections that allow fine QTL mapping via linkage disequilibrium (Rafalski 2002) and introgression lines. We have already shown how the use of ILs, which isolate a single QTL region, transformed the task of QTL cloning into one similar to that performed for simple Mendelian traits, with the exception that phenotyping requires more detailed measurements. However, as such, the approach is still far from being easy and rapid. Any additional information that could be associated to the observed traits in the introgression lines would therefore be useful in identifying the allele(s) responsible for a particular phenotype. In this regard, integrated strategies can be pursued to reduce the list of candidate genes for target QTL (Wayne and McIntyre 2002). Whenever available, data mining of DNA sequences and gene function in the public domain can help identify candidate genes. ESTs can be mapped to the target IL as they can also provide candidate genes should their map position fall within the QTL region. High-throughput expression technologies such as transcriptional analysis via microarray and gene chips, proteomic analyses and metabolic profiling can be applied to selected ILs in order to obtain the expressional candidates and related genes for target QTL. Transcriptomic analysis of a selected IL can identify not only genes with expression-level differences that map to the QTL and that are therefore candidate genes, but also differentially expressed genes that map outside the region and that therefore reveal downstream changes as a result of the introgression. Transcriptional profiling of ILs containing sufficiently short differential regions thus provide a valuable tool to identify the members of gene networks that may be regulated by QTL (Juenger et al. 2006). Once candidate genes for target QTL are identified, they can be verified either by conventional genetic complementation or by molecular and functional analyses of allelic diversity at candidate loci, by RNAi knockout/knockdown (Ahlquist 2002).

Given that the functional diversity of QTL alleles detected at the phenotypic level must reflect their diversity at the molecular level, an alternative way for QTL verification can be based on the analysis of multiple functional alleles at QTL identified in ILs derived from different donor parents (Fridman et al. 2004: Li et al. 2005b).

Along these lines, the *S. pennellii* IL population has been used to explore the potential of the 'candidate gene approach' to identify candidate genes for QTL influencing the intensity of tomato fruit color (Liu et al. 2003) as well as tomato fruit size and composition (Causse et al. 2004). In both cases a QTL mapping analysis was conducted for the quantitative traits of interest along with the mapping analysis of genes encoding, respectively, enzymes of the carotenoid biosynthesis pathway and enzymes involved in the fruit primary carbon metabolism. While in the first study the number of QTL that co-segregated with the same bins that contained the candidate gene was close to the number that is expected by chance alone, in the second study, Causse et al. (2004) found a number of obvious links between the presence of *S. pennellii* alleles of these genes and the observed trait. In order to provide additional definition of the biochemical traits that are altered in each line, metabolomic profiling of the *S. pennellii* ILs was pursued (Overy et al. 2005), and a comprehensive metabolic profiling and phenotyping of the *S. pennellii* IL population allowed identification of 889 quantitative fruit metabolic loci and 326 loci that modify yield-associated traits (Schauer et al. 2006). The analysis indicates that at least 50% of the metabolic loci are associated with QTL that influence whole-plant yield-associated traits. Finally, Baxter et al. (2005) conducted a study of transcriptomic changes in six non-overlapping *S. pennellii* introgression lines that share the common trait of increased ripe fruit soluble solids and increased accumulation of fruit carbohydrate. The analysis provided evidence of genome-wide transcriptional changes and revealed links to mapped QTL and described traits.

This way ILs provide a new paradigm to increase the efficiency in discovery, candidate gene identification and cloning of target QTL based on convergence of evidence deriving from QTL position, expression profiling data, functional and molecular diversity analyses of candidate genes (Li et al. 2005b). Given the large amount of data that will be generated by these integrative strategies, and the major role that extensive and precise phenotyping will play, another challenge we have to face is how to develop a framework for presenting, *in silico*, in a user-friendly bioinformatics management system, the range of statistical outputs that result from QTL studies; for example, homozygous, heterozygous, pleiotropic, epistatic and environmental effects. This framework, which can be based on the genetic or physical sequence map, will form the basis for further integration of QTL databases with genome information that includes gene content, expression and function.

4. CONCLUSIONS AND FUTURE PERSPECTIVES

Plant evolution under domestication led to increased productivity, but at the same time narrowed the genetic basis of crops. The challenges facing modern plant breeders are to develop higher yielding, nutritious and environmentally friendly varieties that will improve the quality of human life. This review demonstrates that wild ancestors of crop plants can be employed to enrich the genetic variation that was lost during domestication. This can be done through the advanced-backcross or introgression line genetics following the concept of breeding by design (Peleman and van der Voort 2003). For this purpose we should invest more in educating plant breeders about the value of exotic variation tailored to the discovery of useful QTL and develop statistical tools to detect and validate traits in a wide multitude of population structures.

ACKNOWLEDGEMENTS

Research in the laboratories of S Grandillo and D Zamir is supported in part by the European Union (EU) program EU-SOL (contract PL 016214-2 EU-SOL). This work was in part supported also by the Italian CNR Short-Term Mobility Program to S Grandillo. We thank colleagues who provided us with unpublished information and apologize to those authors whose work could not be discussed due to space limitations.

REFERENCES

Ahlquist P (2002) RNA-dependent RNA polymerases, viruses, and RNA silencing. Science 296:1270–1273

Ahn SN, Suh JP, Oh CS, Lee SJ, Suh HS (2002) Development of introgression lines of weedy rice in the background of Tongil-type rice. Rice Genet Newsl 19:14

Ashikari M, Matsuoka M (2006) Identification, isolation and pyramiding of quantitative trait loci for rice breeding. Trends Plant Sci 11(7):344–350

Baxter CJ, Sabar M, Quick WP, Sweetlove LJ (2005) Comparison of changes in fruit gene expression in tomato introgression lines provides evidence of genome-wide transcriptional changes and reveals links to mapped QTLs and described traits. J Exp Bot 56:1591–1604

Bernacchi D, Beck-Bunn T, Eshed Y, Lopez J, Petiard V, Uhlig J, Zamir D, Tanksley S (1998a) Advanced backcross QTL analysis in tomato. I. Identification of QTLs for traits of agronomic importance from *Lycopersicon hirsutum.* Theor Appl Genet 97:381–397

Bernacchi D, Beck-Bunn T, Emmatty D, Eshed Y, Inai S, Lopez J, Petiard V, Sayama H, Uhlig J, Zamir D, Tanksley S (1998b) Advanced backcross QTL analysis of tomato. II. Evaluation of near-isogenic lines carrying single-donor introgressions for desirable wild QTL-alleles derived from *Lycopersicon hirsutum* and *L. pimpinellifolium.* Theor Appl Genet 97:170–180 and 1191–1196

Bessey CE (1906) Crop improvement by utilizing wild species. Am Breed Assoc II:112–118

Blair MW, Iriarte G, Beebe S (2006) QTL analysis of yield traits in an advanced backcross population derived from a cultivated Andean × wild common bean (*Phaseolus vulgaris* L.) cross. Theor Appl Genet 112:1149–1163

Burbank L (1914) How plants are trained to work for man, Vol. 1. P. F. Collier and Son, New York, p 302

Canady MA, Meglic V, Chetelat RT (2005) A library of *Solanum lycopersicoides* introgression lines in cultivated tomato. Genome 48:685–697

Causse M, Duffe P, Gomez MC, Buret M, Damidaux R, Zamir D, Gur A, Chevalier C, Lemaire-Chamley M, Rothan C (2004) A genetic map of candidate genes and QTLs involved in tomato fruit size and composition. J Exp Bot 55:1671–1685

Chaïb J, Lecomte L, Buret M, Causse M (2006) Stability over genetic backgrounds, generations and years of quantitative trait locus (QTLs) for organoleptic quality in tomato. Theor Appl Genet 112:934–944

Chetelat RT, Meglic V (2000) Molecular mapping of chromosome segments introgressed from *Solanum lycopersicoides* into cultivated tomato (*Lycopersicon esculentum*). Theor Appl Genet 100: 232–241

de Vicente MC, Tanksley SD (1993) QTL analysis of transgressive segregation in an interspecific tomato cross. Genetics 134:585–596

Doganlar S, Frary A, Ku H-M, Tanksley SD (2002) Mapping quantitative trait loci in inbred backcross lines of *Lycopersicon pimpinellifolium* (LA1589). Genome 45:1189–1202

Doi K, Iwata N, Yoshimura A (1997) The construction of chromosome substitution introgression lines of African rice (*Oryza glaberrima* Steud.) in the background of japonica (*O. sativa* L.). Rice Genet Newsl 14:39–41

Ebitani T, Takeuchi Y, Nonoue Y, Yamamoto T, Takeuchi K, Yano M (2005) Construction and evaluation of chromosome segment substitution lines carrying overlapping chromosome segments of indica rice cultivar 'Kasalath' in a genetic background of japonica elite cultivar 'Koshihikari'. Breed Sci 55:65–73

Eduardo I, Arús P, Monforte AJ (2005) Development of a genomic library of near isogenic lines (NILs) in melon (*Cucumis melo* L.) from the exotic accession PI161375. Theor Appl Genet 112:139–148

El-Din El-Assal S, Alonso-Blanco C, Peeters AJM, Koornneef M (2001) A QTL for flowering time in *Arabidopsis* reveals a novel allele of CRY2. Nat Genet 29:435–440

Eshed Y, Zamir D (1995) An introgression line population of *Lycopersicon pennellii* in the cultivated tomato enables the identification and fine mapping of yield-associated QTL. Genetics 141:1147–1162

Eshed Y, Zamir D (1996) Less-than-additive epistatic interactions of quantitative trait loci in tomato. Genetics 143:1807–1817

Eyre-Walker A, Gaut RL, Hilton H, Feldman DL, Gaut BS (1998) Investigation of the bottleneck leading to the domestication of maize. Proc Natl Acad Sci USA 95:4441–4446

Frary A, Nesbitt TC, Frary A, Grandillo S, Van der Knaap E, Cong B, Liu J, Meller J, Elber R, Alpert KB, Tanksley SD (2000) *fw-2.2*: a quantitative trait locus key to the evolution of tomato fruit size. Science 289:85–88

Frary A, Doganlar S, Frampton A, Fulton T, Uhlig J, Yates H, Tanksley S (2003) Fine mapping of quantitative trait loci for improved fruit characteristics from *Lycopersicon chmielewskii* chromosome 1. Genome 46:235–243

Frary A, Fulton TM, Zamir D, Tanksley SD (2004) Advanced backcross QTL analysis of a *Lycopersicon esculentum* × *L. pennellii* cross and identification of possible orthologs in the Solanaceae. Theor Appl Genet 108:485–496

Fridman E, Pleban T, Zamir D (2000) A recombination hotspot delimits a wild-species quantitative trait locus for tomato sugar content to 484 bp within an invertase gene. Proc Natl Acad Sci USA 97:4718–4723

Fridman E, Liu YS, Carmel-Goren L, Gur A, Shoresh M, Pleban T, Eshed Y, Zamir D (2002) Two tightly linked QTLs modify tomato sugar content via different physiological pathways. Mol Genet Genomics 266:821–826

Fridman E, Carrari F, Liu YS, Fernie AR, Zamir D (2004) Zooming in on a quantitative trait for tomato yield using interspecific introgressions. Science 305:1786–1789

Fulton TM, Beck-Bunn T, Emmatty D, Eshed Y, Lopez J, Petiard V, Uhlig J, Zamir D, Tanksley SD (1997) QTL analysis of an advanced backcross of *Lycopersicon peruvianum* to the cultivated tomato and comparisons with QTLs found in other wild species. Theor Appl Genet 95:881–894

Fulton TM, Grandillo S, Beck-Bunn T, Fridman E, Frampton A, Lopez J, Petiard V, Uhlig J, Zamir D, Tanksley SD (2000) Advanced backcross QTL analysis of a *Lycopersicon esculentum* x *L. parviflorum* cross. Theor Appl Genet 100:1025–1042

Fulton TM, Bucheli P, Voirol E, López J, Pétiard V, Tanksley SD (2002) Quantitative trait loci (QTL) affecting sugars, organic acids and other biochemical properties possibly contributing to flavor, identified in four advanced backcross populations of tomato. Euphytica 127:163–177

Grandillo S, Tanksley SD (1996) QTL analysis of horticultural traits differentiating the cultivated tomato from the closely related species *Lycopersicon pimpinellifolium*. Theor Appl Genet 92:935–951

Grandillo S, Tanksley SD (2005) Advanced backcross QTL analysis: results and perspectives. In: Tuberosa R, Phillips RL, Gale M (eds) The wake of the double helix: from the green revolution to the gene revolution. Edizioni Avenue Media, Bologna, pp 115–132

Gur A, Zamir D (2004) Unused natural variation can lift yield barriers in plant breeding. PLoS Biol 2:e245

Gur A, Semel Y, Cahaner A, Zamir D (2004) Real time QTL of complex phenotypes in tomato interspecific introgression lines. Trends Plant Sci 9:107–109

Ho JC, McCouch SR, Smith ME (2002) Improvement of hybrid yield by advanced backcross QTL analysis in elite maize. Theor Appl Genet 105:440–448

Holtan HE, Hake S (2003) Quantitative trait locus analysis of leaf dissection in tomato using *Lycopersicon pennellii* segmental introgression lines. Genetics 165:1541–50

Huang XQ, Cöster H, Ganal MW, Röder MS (2003) Advanced backcross QTL analysis for the identification of quantitative trait loci alleles from wild relatives of wheat (*Triticum aestivum* L.). Theor Appl Genet 106:1379–1389

Juenger TE, Wayne T, Boles S, Vaughan SV, McKay J, Coughlan SJ (2006) Natural genetic variation in whole-genome expression in *Arabidopsis thaliana*: the impact of physiological QTL introgression. Mol Ecol 15:1351–1365

Jeunken MJW, Lindhout P (2004) The development of lettuce backcross inbred lines (BILs) for exploitation of the *Lactuca saligna* (wild lettuce) germplasm. Theor Appl Genet 109:394–401

von Korff M, Wang H, Léon J, Pillen K (2004) Development of candidate introgression lines using an exotic barley accession (*Hordeum vulgare* ssp. *spontaneum*) as donor. Theor Appl Genet 109:1736–1745

von Korff M, Wang H, Léon J, Pillen K (2005) AB-QTL analysis in spring barley. I. Detection of resistance genes against powdery mildew, leaf rust and scald introgressed from wild barley. Theor Appl Genet 111:583–590

von Korff M, Wang H, Leon J, Pillen K (2006) AB-QTL analysis in spring barley: II. Detection of favourable exotic alleles for agronomic traits introgressed from wild barley (*H. vulgare* ssp. *spontaneum*). Theor Appl Genet 112:1221–1231

Kojima S, Takahashi Y, Kobayashi Y, Monna L, Sasaki T, Araki T, Yano M (2002) *Hd3a*, a rice ortholog of the *Arabidopsis* FT gene, promotes transition to flowering downstream of *Hd1* under short-day conditions. Plant Cell Physiol 43:1096–1105

Koumproglou R, Wilkes TM, Towson P, Wang XY, Beyon J, Pooni HS, Newbury HJ, Kearsey MJ (2002). Plant J 31:355–364

Kubo T, Aida Y, Nakamura K, Tsunematsu H, Doi K, Yoshimura A (2002) Reciprocal chromosome segment substitution series derived from Japonica and Indica cross of rice (*Oryza sativa* L.). Breed Sci 52:319–325

Kurakazu T, Sobrizal K, Ikeda K, Sanchez PL, Doi K, Angeles ER, Khush GS, Yoshimura A (2001) *Oryza meridionalis* chromosomal segment introgression lines in cultivated rice, *O. sativa* L. Rice. Genet Newsl 18:81–82

Kuspira J, Unrau J (1957) Genetic analysis of certain characters in common wheat using all chromosome substitution lines. Can J Plant Sci 37:300–326

Ladizinsky G (1998) Plant evolution under domestication. Kluwer Academic Press, Dordrecht, p 262

Lawson DM, Lunde CF, Mutschler MA (1997) Marker-assisted transfer of acylsugar-mediated pest resistance from the wild tomato, *Lycopersicon pennellii*, to the cultivated tomato, *Lycopersicon esculentum*. Mol Breed 3:307–317

Lecomte L, Saliba-Colombani V, Gautier A, Gomez-Jimenez MC, Duffé P, Buret M, Causse M (2004) Fine mapping of QTLs of chromosome 2 affecting the fruit architecture and composition of tomato. Mol Breed 13:1–14

Li J, Xiao J, Grandillo S, Jiang L, Wan Y, Deng Q, Yuan L, McCouch SR (2004) QTL detection for rice grain quality traits using an interspecific backcross population derived from cultivated Asian (*O. sativa* L.) and African (*O. glaberrima* S.) rice. Genome 47:697–704

Li JZ, Huang XQ, Heinrichs F, Ganal MW, Röder MS (2005a) Analysis of QTLs for yield, yield components, and malting quality in a BC_3-DH population of spring barley. Theor Appl Genet 110:356–363

Li Z-K, Fu B-Y, Gao Y-M, Xu J-L, Ali J, Lafitte HR, Jiang Y-Z, Rey JD, Vijayakumar CHM, Maghirang R, Zheng T-Q, Zhu L-H (2005b) Genome-wide introgression lines and their use in genetic and molecular dissection of complex phenotypes in rice (*Oryza sativa* L.). Plant Mol Biol 59:33–52

Lin SY, Sasaki T, Yano M (1998) Mapping quantitative trait loci controlling seed dormancy and heading date in rice, *Oryza sativa* L., using backcross inbred lines. Theor Appl Genet 96:997–1003

Lin HX, Yamamoto T, Sasaki T, Yano M (2000) Characterization and detection of epistatic interactions of 3 QTLs, *Hd-1*, *Hd-2* and *Hd-3*, controlling heading date of rice using nearly isogenic lines. Theor Appl Genet 101:1021–1028

Liu Y-S, Gur A, Ronen G, Causse M, Damidaux R, Buret M, Hirschberg J, Zamir D (2003) There is more to tomato fruit colour than candidate carotenoid genes. Plant Biotech J 1:195–207

Liu S, Zhou R, Dong Y, Li P, Jia J (2006) Development, utilization of introgression lines using a synthetic wheat as donor. Theor Appl Genet 112:1360–1373

McCouch S (2004) Diversifying selection in plant breeding. PLoS Biol 2:e347

Matus I, Corey A, Filchkin T, Hayes PM, Vales MI, Kling J, Riera-Lizarazu O, Sato K, Powell W, Waugh R (2003) Development and characterization of recombinant chromosome substitution lines (RCSLs) using *Hordeum vulgare* subsp. *spontaneum* as a source of donor alleles in a *Hordeum vulgare* subsp. *vulgare* background. Genome 46:1010–1023

Mei HW, Xu JL, Li ZK, Yu XQ, Guo LB, Wang YP, Ying CS, Luo LJ (2006) QTLs influencing panicle size detected in two reciprocal introgressive line (IL) populations in rice (*Oryza sativa* L.). Theor Appl Genet 112:648–656

Miller JC, Tanksley SD (1990) RFLP analysis of phylogenetic relationships and genetic variation in the genus *Lycopersicon*. Theor Appl Genet 80:437–448

Moncada P, Martínez CP, Borrero J, Chatel M, Gauch Jr H, Guimaraes E, Tohme J, McCouch SR (2001) Quantitative trait loci for yield and yield components in an *Oryza sativa* × *Oryza rufipogon* BC_2F_2 population evaluated in an upland environment. Theor Appl Genet 102:41–52

Monforte AJ, Tanksley SD (2000a) Development of a set of near isogenic and backcross recombinant inbred lines containing most of the *Lycopersicon hirsutum* genome in a *L. esculentum* genetic background: a tool for gene mapping and gene discovery. Genome 43:803–813

Monforte AJ, Tanksley SD (2000b) Fine mapping of a quantitative trait locus (QTL) from *Lycopersicon hirsutum* chromosome 1 affecting fruit characteristics and agronomic traits: breaking linkage among QTLs affecting different traits and dissection of heterosis for yield. Theor Appl Genet 100:471–479

Monforte AJ, Friedman E, Zamir D, Tanksley SD (2001) Comparison of a set of allelic QTL-NILs for chromosome 4 of tomato: deductions about natural variation and implications for germplasm utilization. Theor Appl Genet 102:572–590

Monforte AJ, Garcia-Mas J, Arús P (2003) Genetic variability in melon based on microsatellite variation. Plant Breed 122:1–6

Monforte AJ, Oliver M, Gonzalo MJ, Alvarez JM, Dolçet-Sanjuan R, Arús P (2004) Identification of quantitative trait loci involved in fruit quality traits in melon. Theor Appl Genet 108:750–758

Monna L, Lin, HX, Kojima S, Sasaki T, Yano M (2002) Genetic dissection of a genomic region for quantitative trait locus, *Hd3*, into two loci, *Hd3a* and *Hd3b*, controlling heading date in rice. Theor Appl Genet 104:772–778

Morgante M, Salamini F (2003) From plant genomics to breeding practice. Curr Opin Biotech 14:214–219

Nadeau JH, Singer JB, Matin A, Lander ES (2000) Analysing complex genetic traits with chromosome substitution strains. Nat Genet 24:221–225

Narasimhamoorthy B, Gill BS, Fritz AK, Nelson JC, Brown-Guedira GL (2006) Advanced backcross QTL analysis of a hard winter wheat × synthetic wheat population. Theor Appl Genet 112:787–796

Overy, SA, Walker HJ, Malone S, Howard TP, Baxter CJ, Sweetlove LJ, Hill SA, Quick WP (2005) Application of metabolite profiling to the identification of traits in a population of tomato introgression lines. J Exp Bot 56:287–296

Oyanedel EA (1999) Quantitative trait loci analysis of chilling tolerance in tomato. PhD Dissertation, Cornell University, Ithaca, NY

Pan Q, Liu YS, Budai-Hadrian O, Sela M, Carmel-Goren L, Zamir D, Fluhr R (2000) Comparative genetics of nucleotide binding site-leucine rich repeat resistance gene homologues in the genomes of two dicotyledons: tomato and *Arabidopsis.* Genetics 155:309–322

Paran I, Zamir D (2003) Quantitative traits in plants: beyond the QTL. Trends Genet 19:303–306

Paterson AH, DeVerna JW, Lanini B, Tanksley SD (1990) Fine mapping of quantitative trait loci using selected overlapping recombinant chromosomes, in an interspecies cross of tomato. Genetics 124:735–742

Peleman JD, van der Voort JR (2003) Breeding by design. Trends Plant Sci 8:330–334

Pestsova EG, Börner A, Röder MS (2001) Development of a set of *Triticum aestivum-Aegilops tauschii* introgression lines. Hereditas 135:139–43

Pestsova EG, Börner A, Röder MS (2006) Development and QTL assessment of *Triticum aestivum-Aegilops tauschii* introgression lines. Theor Appl Genet 112:634–647

Pillen K, Zacharias A, Léon J (2003) Advanced backcross QTL analysis in barley (*Hordeum vulgare* L.). Theor Appl Genet 107:340–352

Pillen K, Zacharias A, Léon J (2004) Comparative AB-QTL analysis in barley using a single exotic donor of *Hordeum vulgare* ssp. *spontaneum.* Theor Appl Genet 108:1591–1601

Plunknett DL, Smith NJH, Williams JT, Murthi-Anishetti N (1987) Gene banks and the world's food Princeton University Press, Princeton, N.J

Rafalski A (2002) Applications of single nucleotide polymorphisms in crop genetics. Curr Opin Plant Biol 5:94–100

Ramsay LD, Jennings DE, Bohuon EJR, Arthur AE, Lydiate DJ, Kearsey MJ, Marshal DF (1996) The construction of a substitution library of recombinant backcross lines in *Brassica oleracea* for the precision mapping of quantitative trait loci. Genome 39:558–567

Rao GU, Ben Chaim A, Borovsky Y, Paran I (2003) Mapping of yield-related QTLs in pepper in an interspecific cross of *Capsicum annuum* and *C. frutescens.* Theor Appl Genet 106:1457–1466

Rousseaux MC, Jones CM, Adams D, Chetelat R, Bennett A, Powel A (2005) QTL analysis of fruit antioxidants in tomato using *Lycopersicon pennellii* introgression lines. Theor Appl Genet 111:1396–1408

Saha S, Jenkins JN, Wu J, McCarty JC, Gutiérrez OA, Percy RG, Cantrell RG, Stelly DM (2006) Effects of chromosome-specific introgression in upland cotton on fiber and agronomic traits. Genetics 172:1927–1938

Salvi S, Tuberosa R (2005) To clone or not to clone plant QTLs: present and future challenges. Trends Plant Sci 10:297–304

Schauer N, Semel Y, Roessner U, Gur A, Balbo I, Carrari F, Pleban T, Perez-Melis A, Bruedigam C, Kopka J, Willmitzer L, Zamir D, Fernie AR (2006) Comprehensive metabolic profiling and phenotyping of interspecific introgression lines for tomato improvement. Nat Biotechnol 24:447–454

Semel Y, Nissenbaum J, Menda N, Zinder M, Krieger U, Issman N, Pleban T, Lippman Z, Gur A, Zamir D (2006) Overdominant quantitative trait loci for yield and fitness in tomato. Proc Natl Acad Sci USA 103:12981–12986

Septiningsih EM, Pratsetiyono J, Lubis E, Tai TH, Tjubaryat T, Moeljopawiro S, McCouch SR (2003a) Identification of quantitative trait loci for yield and yield components in an advanced backcross population derived from *Oryza sativa* variety IR64 and the wild relative *O. rufipogon.* Theor Appl Genet 107:1419–1432

Septiningsih EM, Trijatmiko KR, Moeljopawiro S, McCouch SR (2003b) Identification of quantitative trait loci for quality in an advanced backcross population derived from *Oryza sativa* variety IR64 and the wild relative *O. rufipogon.* Theor Appl Genet 107:1433–1441

Simmonds NW (1976) Evolution of crop plants. Longman, London, New York

Singer JB, Hill AE, Burrage LC, Olszens KR, Song J, Justice M, O'Brien WE, Conti DV, Witte JS, Lander ES, Nadeau JH (2004) Genetic dissection of complex traits with chromosome substitution strains of mice. Science 304:445–448

Sobrizal K, Ikeda K, Sanchez PL, Doi K, Angeles ER, Kush GS, Yoshimua A (1996) Development of *Oryza glumaepatula* introgression lines in rice, *O. sativa* L. Rice Genet Newsl 16:107

Tadmor Y, Fridman E, Gur A, Larkov O, Lastochkin E, Ravid U, Zamir D, Lewinsohn E (2002) Identification of *malodorous*, a wild species allele affecting tomato aroma that was selected against during domestication. J Agric Food Chem 50:2005–2009

Takahashi Y, Shomura A, Sasaki T, Yano M (2001) *Hd6*, a rice quantitative trait locus involved in photoperiod sensitivity, encodes the alpha-subunit of protein kinase CK2. Proc Natl Acad Sci USA 98:7922–7927

Talamè V, Sanguineti MC, Chiapparino E, Bahri H, Ben Salem M, Forster BP, Ellis RP, Rhouma S, Zoumarou W, Waugh R, Tuberosa R (2004) Identification of *Hordeum spontaneum* QTL alleles improving field performance of barley grown under rainfed conditions. Ann Appl Biol 144:309–319

Tanksley SD (1993) Mapping polygenes. Annu Rev Genet 27:205–233

Tanksley SD, Nelson JC (1996) Advanced backcross QTL analysis: a method for the simultaneous discovery and transfer of valuable QTLs from unadapted germplasm into elite breeding lines. Theor Appl Genet 92:191–203

Tanksley SD, McCouch SR (1997) Seed banks and molecular maps: unlocking genetic potential from the wild. Science 277:1063–1066

Tanksley SD, Grandillo S, Fulton TM, Zamir D, Eshed Y, Petiard V, Lopez J, Beck-Bunn T (1996) Advanced backcross QTL analysis in a cross between an elite processing line of tomato and its wild relative *L. pimpinellifolium*. Theor Appl Genet 92:213–224

Thomson MJ, Tai TH, McClung AM, Lai X-H, Hinga ME, Lobos KB, Xu Y, Martinez CP, McCouch SR (2003) Mapping quantitative trait loci for yield, yield components and morphological traits in an advanced backcross population between *Oryza rufipogon* and the *Oryza sativa* cultivar Jefferson. Theor Appl Genet 107:479–493

Thomson MJ, Edwards JD, Septiningsih EM, Harrington SE, McCouch SR (2006) Substitution mapping of *dth1.1*, a flowering-time quantitative trait locus (QTL) associated with transgressive variation in rice, reveals multiple sub-QTL. Genetics 172:2501–2514

Tian F, Li de J, Fu Q, Zhu ZF, Fu YC, Wang XK, Sun CQ (2006) Construction of introgression lines carrying wild rice (*Oryza rufipogon* Griff.) segments in cultivated rice (*Oryza sativa* L.) background and characterization of introgressed segments associated with yield-related traits. Theor Appl Genet 112:570–580

Van der Hoeven RS, Monforte AJ, Breeden D, Tanksley SD, Steffens JC (2000) Genetic control and evolution of sesquiterpene biosynthesis in *Lycopersicon esculentum* and *L. hirsutum*. Plant Cell 12:2283–2294

Varshney RK, Hoisington DA, Tyagi AK (2006) Advances in cereal genomics and applications in crop breeding. Trends Biotech 24:490–499

Wan XY, Wan JM, Su CC, Wang CM, Shen WB, Li JM, Wang HL, Jiang L, Liu SJ, Chen LM, Yasui H, Yoshimura A (2004) QTL detection for eating quality of cooked rice in a population of chromosome segment substitution lines. Theor Appl Genet 110:71–79

Wang ZY, Second G, Tanksley SD (1992) Polymorphism and phylogenetic relationships among species in the genus *Oryzae* as determined by analysis of nuclear RFLPs. Theor Appl Genet 83: 565–581

Wayne ML, McIntyre LM (2002) Combining mapping and arraying: an approach to candidate gene identification. Proc Natl Acad Sci USA 99:14903–14906

Xiao J, Li J, Grandillo S, Ahn SN, Yuan L, McCouch SR, Tanksley SD (1996) Genes from wild rice improve yield. Scientific correspondence, Nature 384:223–224

Xiao J, Li J, Grandillo S, Ahn SN, Yuan L, Tanksley SD, McCouch SR (1998) Identification of trait-improving quantitative trait loci alleles from a wild rice relative, *Oryza rufipogon*. Genetics 150:899–909

Xu JL, Lafitte HR, Gao YM, Fu BY, Torres R, Li ZK (2005) QTLs for drought escape and tolerance identified in a set of random introgression lines of rice. Theor Appl Genet 111:1642–1650

Yamamoto T, Lin HX, Sasaki T, Yano M (2000) Identification of heading date quantitative trait locus *Hd-6* and characterization of its epistatic interactions with *Hd-2* in rice using advanced backcross progeny. Genetics 154:885–891

Yano M, Katayose Y, Ashikari M, Yamanouchi U, Monna L, Fuse T, Baba T, Yamamoto K, Umehara Y, Nagamura Y, Sasaki T (2000) *Hdl*, a major photoperiod sensitivity quantitative trait locus in rice, is closely related to the *Arabidopsis* flowering time gene CONSTANS. Plant Cell 12:2473–2484

Yates HE, Frary A, Doganlar S, Frampton A, Eannetta NT, Uhlig J, Tanksley SD (2004) Comparative fine mapping of fruit quality QTLs on chromosome 4 introgressions derived from two wild tomato species. Euphytica 135:283–296

Young ND (1999) A cautiously optimistic vision for marker-assisted breeding. Mol Breed 5:505–510

Zamir D (2001) Improving plant breeding with exotic genetic libraries. Nat Rev Genet 2:983–989

Zamir D, Eshed Y (1998) Tomato genetics and breeding using nearly isogenic introgression lines derived from wild species. In: Paterson AH (ed) Molecular dissection of complex traits, CRC Press Inc. Boca Raton FL, pp 207–217

Zygier S, Chaim AB, Efrati A, Kaluzky G, Borovsky Y, Paran I (2005) QTLs mapping for fruit size and shape in chromosomes 2 and 4 in pepper and a comparison of the pepper QTL map with that of tomato. Theor Appl Genet 111:437–445

CHAPTER 7

GENOMELESS GENOMICS IN CROP IMPROVEMENT

KEAN JIN LIM, SINI JUNTTILA, VIDAL FEY AND STEPHEN RUDD*

Centre for Biotechnology, Tykistökatu 6, FIN-20521 Turku, Finland

Abstract: The DNA sequencing of the entire nuclear genomes from *Arabidopsis*, rice and poplar has facilitated the broad-adoption of contemporary research techniques that extend far beyond the study of individual genes. New post-genomic technologies such as microarray based genome-scale gene expression profiling and proteome analysis are absolutely dependent on deep sequence coverage of the gene-space, but have an immense potential to drive the research community in exciting new directions. Facets of many crop genomes currently preclude them from complete sequencing, but the broadest adoption of post-genomic technologies is essential to make in-roads in plant biotechnology and crop improvement. A variety of technologies are available that can be used to establish a genomics foothold in even the most recalcitrant of crop plant species. In this review we address the fundamental technologies that are being widely adopted within the crop-plant research community to gain such a foothold. By considering primarily the expressed sequence tag resources, we have explored how even moderately simple genomic resources may be exploited in molecular marker development, candidate gene selection and microarray-based gene expression profiling.

1. INTRODUCTION

The prospect of genomics without a genome sequence might appear an oxymoron. Genomic and post-genomic technologies such as DNA-microarray based expression profiling and large-scale proteomics clearly require access to a well-assembled and comprehensively annotated genome sequence. This basic requirement is, however, largely inaccessible to many species of crop plant. Such crops may have fantastically large and repetitive genomes that preclude them from complete genome sequencing due to the underlying technical and practical limitations of today's DNA sequencing capabilities. Even when a species has an accessible genome size,

*Corresponding Author: stephen.rudd@btk.fi

R.K. Varshney and R. Tuberosa (eds.), Genomics-Assisted Crop Improvement: Vol. 1: Genomics Approaches and Platforms, 151–176.

the size and wealth of the research community, and the anthropocentric value of the species within agriculture may preclude adequate coverage genome sampling. Regardless of the need for the genome sequence itself, emerging techniques and technologies coming from such fields as functional genomics and system biology have the power to dramatically change the repertoire of tools available to the contemporary molecular biologist, plant breeder and geneticist. Not adopting these technologies would enhance the ever-increasing divide that separates our model plant species (e.g. *Arabidopsis thaliana*) and agriculturally and scientifically focused crop species (e.g. *Oryza sativa*, *Lycopersicon esculentum* or *Glycine max*) from the more 'humble' crop species (arguably such species as *Beta vulgaris*, *Daucus carota* or *Musa acuminata*)

The plant kingdom is a rich source of biodiversity. An estimated 270,000 extant species of vascular plant (May 1990) represent the half aeon of molecular evolution and genome adaptation since plants first colonised land. This landing on the Gondwana super-continent, and it's timing during the mid-Ordovician period of the Palaeozoic era is supported by tetrahedral spore observations from the fossil record (Friedman and Cook 2000; Wellman and Gray 2000; Wellman et al. 2003). The resulting diversification of the land plants now supports the bulk of the terrestrial food chain, but not surprisingly, relatively few domesticated species of plant account for the bulk of human nutrition. The United Nations Food and Agriculture Organisation (FAO) estimates the volumes of different domesticated crop plants that are farmed globally, and the five species of greatest relevance in terms of yield would be sugarcane, maize, wheat, rice and potato (http://en.wikipedia.org/wiki/Agriculture). The anthropocentric value of these 'crop' plants is therefore immense, and the understanding of the basic and applied aspects of their biology has the potential to greatly influence agriculture. This agricultural focus within the crop sciences clearly structures and influences both research philosophies and direction. The new methods, paradigms and resources currently available to the model species are also being widely adopted within crop research. The topic of this review aims to highlight and illustrate how the crop plant community is actively 'catching-up' with the genomic technologies that are now well refined in the plant species that have the luxury of a complete genome sequence.

It is reassuring that the arsenal of tools available to the crop research community has not remained static since the publication of the *Arabidopsis* and rice genome sequences. Several technologies that predate the first plant genome sequences have come to the forefront as pioneer technologies for accessing the wealth of information that is often cryptically encoded within the genome. These technologies have been joined by other cutting edge methods that again enrich the landscape of resources available to today's biologist. In this chapter we address how some of the alternative strategies to complete genome sequencing are being applied within the context of crop research to at least partially fill the genomic void. We also address how the resulting resources are being utilised and exploited within contemporary biology, and reciprocally how contemporary technologies are increasingly utilising the available genomic substrate.

2. AN INTRODUCTION TO PLANT GENOMES

Sequencing any crop-genome is, in essence, something that should be completely analogous to sequencing either the *Arabidopsis* or rice genome. The strategies that may be employed to access a plant genome have been reviewed recently (Paterson 2006). A genome contains a number of chromosomes and, in turn, each of the chromosomes contains a rather large number of nucleotides. The decoding and representation of this chromosomal information as a textual pseudo-molecule is the ultimate goal of a genome assembly. Once a crop genome has been selected for sequencing, a decision needs to be made as to whether the whole genome would be sampled using a clone-by-clone sequencing approach or if it will be sequenced using a whole genome shotgun sequencing approach. This philosophical decision will then direct the specific needs for DNA libraries and the experiments that need to be performed to establish these reference genomic foundations. A genome-sequencing centre that is well equipped with the needed robotics, fluidics stations and automated DNA sequencing machines can then start the process of sequencing the genome.

At first sight this seems like quite a manageable endeavour. While there are certainly some logistical hurdles in managing the tens of millions of sequencing reactions, the DNA sequencing of whole genomes seems straightforward. In reality, while complicated by the needs for robotics, massive data handling capabilities and an extremely well managed logistics infrastructure, the DNA sequencing of the volumes of data encoded within the genome is not especially problematic. The problems faced by the genome sequencing projects lie within the data generated, not with the data generation itself. The most intractable of the problems faced by the genome projects it that of converting the millions of individual DNA sequences into the pseudo-molecule scaffolds that represent the individual chromosomes. This process of building the genome scaffold is a process called 'assembly'. To understand the problems encountered within assembling a plant genome require that first we consider what is already known about plant genomes, their sizes and the underlying populations of structures and repeat elements, and the way that plant genomes evolve.

2.1. Variability of Plant Genome Sizes

Plant genomes tend to be large. Whereas the *Arabidopsis* genome is cited as having a genome size of approximately 125 Mbp (AGI 2000), and thus being equivalent to reference invertebrate genomes, *Caenorhabditis elegans* (97 Mbp, C. elegans Sequencing Consortium, 1998) and *Drosophila melanogaster* (~130 Mbp (Adams et al. 2000)), *Arabidopsis* is not at all representative of the plant kingdom. One of the reasons that *Arabidopsis* was selected for genome sequencing was that it's genome is amongst the smallest characterised plant genomes. The C-value paradox (Vendrely and Vendrely 1948) is especially true within plant species.

Within the angiosperms, *Fragaria viridis* represents the species currently with the smallest measured haploid genome size (98 Mbp) (Antonius and Ahokas 1996).

The lily, *Fritillaria assyriaca* (124,852 Mbp) may represent the largest angiosperm genome listed in the plant C-values database (Bennett and Leitch 2003), but *Trillium rhombifolium* (109,270 Mbp) has the largest published genome (Grif et al. 1980). With further sampling of plant species, it is reasonable to assume that plant genome sizes will continue to span over four orders of magnitude in size. Within the crop plants there is again much diversity in genome size. Rice, one of the smaller genome sized grass species has a haploid genome size of approximately 490 Mbp. *Lycopersicon esculentum* (1005 Mbp) and *Hordeum vulgare* (5439 Mbp) represent the next size order and species such as *Triticum aestivum* (16979 Mbp), again, have even larger genomes. In (Paterson 2006), Andrew Paterson has argued that performing a meaningful depth (8x coverage) sequencing of the genomes from the 200 most critical crop plant species would require the sampling 3.4×10^{12} nucleotides of sequence. This is 72 times the amount of all DNA sequence currently available in the GenBank database.

Size in itself is not a critical limitation to the sequencing of a genome although the expense of sequencing 10^{12} nucleotides is certainly insurmountable. Technically though, genome centres with sufficient funding could adequately sample the genomes. The underlying issues lie with the observation that larger genomes contain more repetitive sequence content.

2.2. Jumping Genes and Repetitive Elements

We have established that plant genomes can be very large, and that crop species have genome sizes that are considerably larger than those of either *Arabidopsis*, populus or rice (2,352 Mbp on average (Paterson 2006)). The problem, however, is not the size of the genome, but the underlying repetitivity (see (Peterson et al. 2002b) for a review). The repetitive nature of a sequence is important, 'complex' sequences have a single or few repeats whereas low-complexity, or repetitive, sequences may be found hundreds or thousands of times throughout the genome. The essence of this repetitive nature problem is that genomic sequence complexity is not uniform. Within any given genome sequence, the total genomic DNA may be considered as belonging to one of four general classes. These four classes include (a) single-copy or low complexity DNA, (b) moderately-repetitive DNA (c) highly-repetitive DNA and (d) 'foldback' DNA.

This DNA complexity may be studied using kinetics during re-association experiments (e.g. (Peterson et al. 2002a; Peterson et al. 2002b)). The kinetic experiments reveal that the fraction of the genome that represents non-redundant sequence is lower in larger genomes. Therefore as genome size increases, the absolute amount of low complexity DNA remains relatively static but there is a disproportionate increase in the amount of moderately and highly repetitive DNA content. The further study of re-association kinetics suggests that the moderately or highly repetitive molecules are likely present in hundreds to many-thousands of copies per genome (Peterson et al. 2002a). These abundant molecules are therefore unlikely to represent the protein-coding genes that genomic research is currently interested in.

The massive effort to sequence a large genome will therefore involve the sequencing of relatively few 'type' regions that are present thousands of times, while the needed protein coding sequences would remain relatively poorly sequenced.

A more detailed investigation of the repetitive content of plant genomes reveals that the sequences are biased to specific classes of mobile DNA, the long-terminal repeat (LTR)-retrotransposons (Kumar and Bennetzen 2000). It has been suggested that these LTR-retrotransposons may account for more than half of the larger nuclear genomes (Lagudah et al. 1997; Meyers et al. 2001; Morgante 2006; SanMiguel et al. 1998). Specific classes of LTR-retroelements (mainly *copia*-like and *gypsy*-like elements), in maize have proliferated within the last five million years leading to a doubling in the genome size (SanMiguel et al. 1998). This recent proliferation means that the bulk of repeats are highly similar, having not had enough time to diverge greatly at the molecular level (SanMiguel et al. 1998). This again compounds the issues of sequence disambiguation; e.g. it cannot be established as to whether a sequence from a copia-like element stems from element number 1 or element number 1001!

2.3. Polyploidy

Another factor that has precluded some species from further genomic investigation is that many crop plants are polyploid (Adams and Wendel 2005). Polyploidy is a significant force in plant genome evolution, and investigation of large sequence collections reveals evidence that polyploidisation events are recurrent for most species of plant (Blanc and Wolfe 2004a, b). While genome doubling events are followed by massive differential gene loss, some species still contain the hallmarks of recent genome duplication events and exist as polyploids with multiple copies of related genomes. Such polyploids may exist as autopolyploids e.g. sugarcane (Ming et al. 2001) (where chromosomes from the ancestral genome can pair with each other) or as allopolyploids e.g. cotton (Nekrutenko and Baker 2003) (where clearly defined chromosome pairs have been re-established). A crop species showing recent polyploidy will again share many of the genome sampling issues as faced by species containing significant repetitive sequence; how can one establish which copy of a particular genomic segment is being assembled? In reality, unless existing as a recent autopolyploid event, this would only likely pose a serious problem within the functional and sequence conserved regions of the genome.

2.4. Genome Context Summary

It is not the fact that genomes are highly biased to relatively few repetitive elements that precludes a species from genome sequencing. The critical problem lies in both the repetitive elements themselves and the limitations of today's dideoxy sequencing technologies. If we consider first the sequencing reaction, only as much as a thousand nucleotides of DNA may be read reliably within a single reaction.

For the expedited reading of the whole genome, shotgun sequencing of either the whole genome or individual sub-cloned elements is performed. This has the result that the assembler algorithm is faced with a large number of sequences that are neither ordered nor oriented when compared to the reference starting sequence. When reading through regions of high complexity with low repeat content, long identical substrings within different reads may be used to join sequences with a high degree of confidence (unambiguously). Once there is a degree of repetitivity within the sequence collection then it becomes increasingly difficult to compute if the repeat is indeed a repeat, and how it relates to other non-repeat and repeat sequence. It must be assumed that any given sequence read may overlap with any other read, and a significant background of erroneously assembled but unrelated sequences must be expected. (Jaffe et al. 2003; Myers 2005; Myers et al. 2000; Myers et al. 2002) While approaches have been optimised to allow for assembly with significant repetitive content, the process is not trivial. As a result shotgun sequencing is typically performed on a case-by-case basis and paired-sequences from libraries with different insert sizes are sequenced such that blocks of repetitive sequence can be tolerated and meaningful genome scaffolds may be constructed. Even with a sensible sequencing strategy, a significant number of contigs will be sequenced that cannot be structured or oriented into higher-order super-contigs without additional manual curation and directed chromosome walking to close the most intractable of gaps.

While these observations are best documented following the human and drosophila genome assemblies, this problem is likely to be more acute within plant genomes. Study of the maize genome have shown that while it has a rather average 'plant' genome in both terms of genome size and re-association kinetics, the bulk of sequenced retrotransposons have been inserted within the last five million years (SanMiguel et al. 1998). Plant genomes thus having more repetitive DNA than animals and having fewer distinguishing mutations cannot be easier to assemble! (Haberer et al. 2005; Paterson 2006) The general consensus within the plant genome sequencing community is that alternative approaches to whole genome sequencing are needed. We will address the contemporary methods that are being utilised to establish genomic resources within various species of crop plant in the next sections.

3. EXPRESSED SEQUENCE TAGS AND THE TRANSCRIPTOME

The application of expressed sequence tag (EST) data within plant biology has been well reviewed elsewhere within the scope of managing the divide between the model and non-model plant species (Mayer and Mewes 2002) and as reviews of the available resources (e.g. (Rudd 2003, 2005)). ESTs continue to represent a basic commodity within the analysis of genomes and their genes. Whereas the complete sequencing of a genome may utilise either a clone-by-clone approach or a whole genome shotgun approach to acquiring adequate coverage to assemble a meaningful scaffold, EST sequencing is directed at the quick, cheap and simple sequencing of partial gene transcripts.

ESTs are typically sequenced from cDNA libraries. mRNA is purified from a given tissue, organ or whole organism that may be described according to an ontology representing development, biotic stress, abiotic stress and environmental conditions. Within this mRNA population the ratio of transcripts faithfully represents the expression levels of all underlying transcribed genes and the stability of the individual transcripts. The resulting sequence collection is therefore rather heterogeneous with most transcripts coming from relatively few genes and many more transcripts having much lower representation levels. Following the mRNA isolation the transcripts may be reverse transcribed thus creating a pool of cDNA sequences that can in turn be cloned into a plasmid vector. The cDNA sequence collection may also be normalised prior to cloning. Such a normalisation step is aimed at reducing the bias within the sequence collection by selectively removing most of the most abundant transcripts. While the logic for cDNA normalisation is clear, many researchers continue to sequence from non-normalised libraries because of the wealth of pseudo-expression information that can be gleaned from the comparison of gene representation between sequence libraries. Regardless of normalisation, the resulting plasmid constructs are typically cultured in 96 well microtitre plates and the EST is obtained by sequencing into either the 5' or 3' end of the cloned cDNA insert. The decision as to whether the 5', 3' or if both ends should be sequenced again depends on the rationale underlying the individual project.

Since the first description of an EST collection (Adams et al. 1991), and subsequently the first systematically acquired *Arabidopsis* ESTs (Delseny et al. 1997), there are now 9.3 million plant ESTs in the public domain (EMBL Release 86, daily updates included until 15th April 2006), which in turn have been prepared from over 400 plant species. Table 1 lists the 40 plant species with most available ESTs. From these 'top' species, 20 are crop plants and most of the others are commercially grown within horticulture, viticulture or forestry. Of the 442 plant species represented within the EST databases, 254 species have at least 500 ESTs and 91 species have in excess of 10,000 ESTs. Plotting the richness and depth of plant EST sequence collections chronologically (data sampled from the dbEST database, but are not shown) reveals that sequence content continues to grow unabated and that in addition to the addition of increasingly large amounts of sequence data the number of species reflected within the dataset also continues to grow.

Whilst at first glance, the heterogeneity of the plant EST collections paints a rosy picture of transcriptomic sequence availability, a closer inspection reveals that the sequences are rather biased towards certain species. Figure 1 illustrates the taxonomic grouping of large plant EST collections (an arbitrary filter of collections with more than 10k sequences has been imposed). From this approximate cladogram of the plant kingdom, it can be seen that the available sequences are highly biased towards the agriculturally relevant angiosperms. This is certainly reassuring to the crop plant community, the direct beneficiaries of these massive sequence resources, but this is at the same time a little disappointing when we consider a more holistic

Table 1. A summary of the 40 plant species with most publicly available EST sequences. Sequences were downloaded from the EST division of the EMBL database (release 86, 28th Feb 2006). In addition to the species name, common names where available are shown in addition to the number of EST sequence reads. These species represent 80% of the total 9.3 million plant ESTs and show a reasonable range of taxonomy and agricultural or academic relevance

Species name	Common name	Number of ESTs
Oryza sativa	rice	1186294
Zea mays	maize	692030
Arabidopsis thaliana	Thale cress	622791
Triticum aestivum	wheat	601402
Hordeum vulgare	barley	437321
Glycine max	soybean	356805
Pinus taeda	Loblolly pine	329469
Saccharum officinarum	sugarcane	246301
Medicago truncatula	barrel medic	221123
Solanum tuberosum	potato	219765
Sorghum bicolor	sorghum	208466
Lycopersicon esculentum	tomato	199875
Malus x domestica	apple	199036
Vitis vinifera	European grapevine	196013
Chlamydomonas reinhardtii	Chlamydomonas	167641
Picea glauca	White spruce	132624
Physcomitrella patens subsp. patens	Physcomitrella	120702
Lotus japonicus	Trefoil	111623
Gossypium hirsutum	cotton	108424
Citrus sinensis	orange	92521
Populus trichocarpa	black cottonwood	89943
Aquilegia formosa x Aquilegia pubescens	Columbine hybrid	85039
Populus tremula x Populus tremuloides	aspen hybrid	82239
Picea sitchensis	Sitka spruce	80789
Brassica napus	oilseed rape	72362
Helianthus annuus	sunflower	66098
Gossypium raimondii	another cotton	63577
Ipomoea nil	Japanese morning glory	62282
Lactuca serriola	prickly lettuce	55490
Lactuca sativa	lettuce	54822
Populus trichocarpa x Populus deltoides	black cottonwood hybrid	53116
Euphorbia esula	leafy spurge	47543
Coffea canephora	Robusta coffee	46907
Festuca arundinacea	tall fescue	44342
Prunus persica	peach	41541
Gossypium arboreum	cotton	39223
Nicotiana tabacum	tobacco	38857
Trifolium pratense	red clover	38109
Zingiber officinale	ginger	38083
Populus tremula	aspen	31234
	Total	7581822

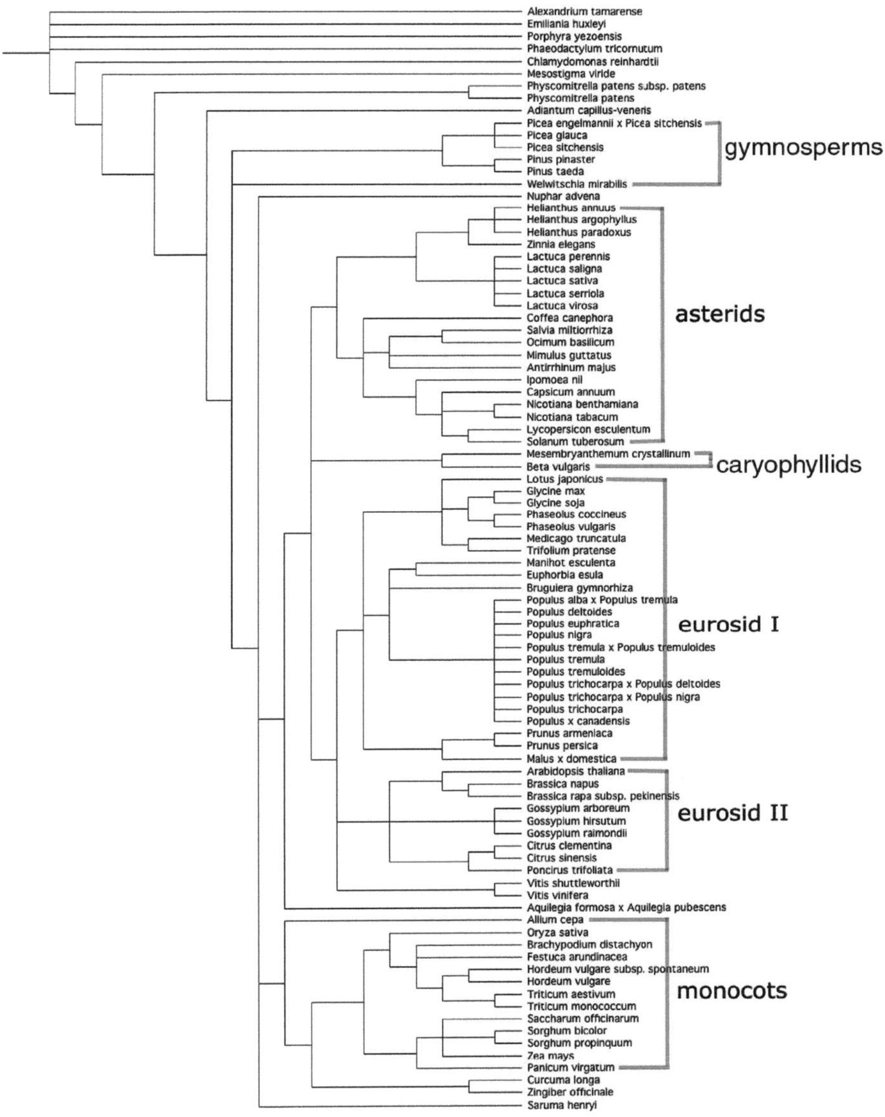

Figure 1. An approximate cladogram showing the taxonomic relationship between plant species with more than 10,000 ESTs. The cladogram structure has been derived from the NCBI taxonomy database. The EST collections have been extracted from within the EST division of the EMBL database (release 86). Super-imposed on the figure are basic legends that identify groups of related species that are of relevance to agriculture and forestry. These clades include the gymosperms and the angiosperm clade that contains the rosids (eurosid I and eurosid II are highlighted), the asterids, the caryophyllids and the monocots. These 'rough' descriptive clades describe 80% of the species with more than 10,000 ESTs.

view of the plant kingdom. There is a wealth of comparative and functional data within the poorly sampled clades that will undoubtedly enlighten the research community as to the clusters of genes that are manifested throughout plant life and that account for the developmental, metabolic and ecological adaptations of the constituent plants (Pryer et al. 2002). This rather general criticism of the sequence content of the EST databases is however minor, and a number of research groups (ours included) is actively working towards filling in some of these taxonomic gaps (e.g. sequencing ESTs from untouched clades such as the hornworts) and in increasing the comparative genomic resolution within the poorly sampled plant lineages.

It is unclear as to whether EST collection can ever substitute a 'completed genome' quality genome assembly with associated annotation and analysis. EST sequencing has been adopted on such a large scale because ESTs are simple and cheap to produce when the genome sequence itself is impractical to sequence. ESTs are not actively sequenced as an alternative to the genome, but rather as an adequate interim solution. Substantial EST resources are an extremely valuable resource during the gene modelling and annotation of a complete genome sequence, so this 'interim sequence solution' has several valuable roles. It may therefore be concluded that ESTs are of interest in the broader studies of crop genomics. They may be sequenced to establish a glimpse of the sequence diversity reflected within the genome, or may be produced to facilitate the construction of cDNA microarrays (this will be discussed in more detail later in this chapter).

4. REDUCED REPRESENTATION SEQUENCING

The robust sequencing of significant numbers of ESTs is a solid and accepted route to sampling a collection of genes that are actively expressed within a given tissue representing the developmental, homeostatic, biotic and abiotic environment. This snap-shot is however biased towards the biological system. While this is a primary objective of many EST sequencing strategies, the bias also means that many gene sequences that have either low expression levels or are expressed in only a few cells following precise signals will remain unsampled. For many genomic approaches, this bias makes no sense. There are solutions to this conundrum that may fall within the realm of sub-genomic sampling. The technologies that we will address here include draft or partial genome sequencing, BAC end sequencing, methylation-filtration of the gene space and high-C_0T selection of low complexity DNA. Each

Figure 1. Looking at the sampled species reveals two new facets of EST biology. Compared to earlier reviews that have considered EST collections and the taxonomic content of the collections (e.g. (Rudd 2003, 2005)), there is a more meaningful sample of species that fall outside of the angiosperm crop species (e.g. *Adiantum*, *Nuphar* and *Saruma*). Also of considerable note is that for several genus multiple species have been sampled (e.g. *Populus*, *Lactuca*, *Pinus* and *Gossypium*) illustrating that the power of comparative genomics is being applied to 'genomeless' organisms of these approaches is directed at sampling a fraction of the genome in a gene expression independent manner.

4.1. Draft Genome Sequencing

The objectives of a typical genome project are to sequence the bulk of accessible euchromatic DNA, and to assemble the resulting sequence reads into a minimal set of "contigs" representing large continguous stretches of genomic DNA. The process of sequencing requires a sufficient amount of sequence redundancy such that the total pool of underlying sequences can be unequivocally assembled into super-scaffolds whilst allowing for the random sample effects. The *Arabidopsis* genome has been sequenced to approximately seven genome equivalents (AGI 2000) meaning that the average DNA residue represents a consensus sequence of at least 7 read nucleotides. A 1x genome sequencing strategy would therefore imply that a single residue will represent the consensus of a single sequence read, so, through chance alone some bases will be sequenced more frequently whilst others may remain un-sequenced. The strategy for partial genome shotgun sequencing where a 0.5x genome coverage is therefore reasonable. Whilst using 10-times fewer reagents, and thus costing 10-times less, than a 5x genome, such a strategy does not preclude the species from further completion in the future. This technique has been applied successfully within the crop species *Brassica oleracea* (Ayele et al. 2005; Katari et al. 2005). The results of this strategy revealed the potential of genome survey sequencing for comparative genomic analysis and as a tool for gene identification. *Brassica oleracea*, however, has a compact genome of 650 Mbp with the result that 35% of brassica sequence corresponded to protein coding gene sequence in the close relative species, *Arabidopsis thaliana*. When we consider large species such as wheat, such an approach would seem impractical. A draft genome sequencing approach might therefore make sense for the smaller crop plant genomes, and could be extremely valuable in surveying closely related species, but this is not a reliable route for the larger plant genomes!

4.2. BAC End Sequencing

Bacterial artificial chromosomes (BACs) are large genomic inserts that have been cloned into bacterial constructs. BAC-by-BAC sequencing arguably provides a more efficient strategy for complete genome sequencing than the shotgun sequencing approach. A BAC library may be constructed that contains a certain number of genome equivalents. By sequencing both ends from a large enough number of BACs, a sequence resource can be generated that provides some insight into the content of the underlying genome. As with the genome survey approach in the previous section on draft genome sequencing, this approach is largely impotent for gene discovery within the larger genomes, but at least can be used to establish the types of retroelement that will be encountered within a more thorough genome sampling, and can provide a route for the discovery of molecular markers. Large

BAC libraries are available for a very large number of species and are widely used within map-based cloning and the discovery of candidate genes (e.g. (Ling and Chen 2005) and (Gaafar et al. 2005)). The sequencing of BAC ends within the context of genome survey has been described for at least ginseng (Hong et al. 2004), soybean (Marek et al. 2001) and maize (Gardiner et al. 2004). The deep sequencing of BAC ends only really becomes appropriate once a whole genome sequencing strategy has been adopted, but nevertheless there remain significant Genome Survey Sequence (GSS) resources in the public domain.

4.3. Methylation Filtration of the Gene Space

Current biology and post-genomic technologies are largely biased towards the understanding of the protein coding genes and their regulatory elements. Random genome sampling techniques appear unsuitable for large genomes so alternative approaches to sampling the parts of the genome enriched for the gene space in a largely unbiased manner have been sought. It is well known that the bulk of a larger plant genome is repetitive and that much of this repetitive content consists of retroelements. It has also been shown that much of this repetitive DNA is hyper-methylated in comparison to the hypo-methylated 'gene-space.' These observations have been applied to the development of technologies that can be used to create libraries that are enriched for the hypo-methylated genome fraction. This is achieved by the isolation and shearing of genomic DNA. The DNA, which is of mixed methylation states, is cloned and propagated in an *E. coli* strain containing a 5-methyl cytosine restriction system. This has the result that methylated DNA is cut, and unmethylated fragments will be successfully cloned and propagated. The construction of such libraries has been coined "methylation filtration" (Rabinowicz et al. 2003; Rabinowicz et al. 1999). The technologies have been demonstrated in both maize (Palmer et al. 2003) and in sorghum (Bedell et al. 2005).

4.4. High-C_0T Sequencing

High-C_0T sequencing is based upon renaturation of sheared genome fragments, and has been elegantly demonstrated, again, using the sorghum (Peterson et al. 2002a) and maize genomes (Yuan et al. 2003b). Genomic DNA is isolated and sheared into fragments that are sufficiently small that 'gene' sequence can likely be dissociated from any adjacent retroelement or other repeat; the experimental fragment size being 1.8 kbp on average for maize (Yuan et al. 2003b). The resulting fragments are then melted, and renaturation is performed in a controlled and gradual manner. The study of the resulting population of DNA fragments reveals the C_0t-values, where C_0 is the nucleotide concentration and t is the reassociation time (see (Paterson 2006) for illustrated review). With the establishment of a C_0T curve, particular fractions may be identified that contain high complexity DNA fragments, and which should therefore also contain significantly less retroelement or other repeat. Shotgun sequencing of clones isolated from high-C_0T fractions has indeed revealed that there indeed is a clear enrichment for gene sequence with a concomitant reduction in the amount of repeat or retroelement sequence. This technology has been at least demonstrated in sorghum (Peterson et al. 2002a), the maize genome (Yuan et al. 2003b) and wheat (Lamoureux et al. 2005). The selection of high- C_0t libraries thus demonstrates another route into a preferential sampling of the gene-space.

4.5. Reduced Representation Summary

There are a number of extremely powerful techniques that may be used to access the content of the protein coding space within a genome. While EST sequencing has been very widely adopted throughout the research community, the other approaches to the sampling of the gene space have been demonstrated as effective and powerful approaches, but have been adopted only within the context of very large pilot projects within the scope of further complete genome sequencing. These true genome sampling technologies in addition to providing routes to the unbiased collection of protein-coding genes, gives access to transcribed but non-polyadenylated features, to random slices of the genome, and to hypo-methylated genome fragments, or to any continuum of features within. With DNA as the starting material rather than a poly-adenylated RNA there seems to be a much greater versatility for genome sampling and genomeless genomics, especially since these technologies will provide access to the regulatory regions upstream of the genes themselves and can provide access to both intronic and exonic sequence.

These methodological advances in conquering the plant genome have been focused by applying the traditional di-deoxy sequencing methods to sequencing from DNA libraries that have been constructed using complex techniques. The sequencing method has remained largely unchanged throughout, albeit with greater automation. Common sense would demand that in order to solve the insurmountable issues of plant genome sequencing we would need access to technologies capable of sequencing significantly longer DNA regions. It is therefore counter intuitive, in that one emerging technology excels in the production of shorter sequence reads (Margulies et al. 2005). The process of genome sequencing using 454 sequencing yields as many as 500,000 sequences in parallel from a single run. Each read is significantly shorter than a typical 'Sanger read' in that the average length may be only 110 nucleotides. While this technology does not solve the issues of plant genome sequencing, the ability to rapidly sequence vast amounts of short sequence from the genome, the transcriptome or any other reduced representation library does open up some rather fantastic opportunities to the plant research community.

In future we will undoubtedly see a much wider adoption of these non-EST technologies, but it seems that for the moment at least, the crop plant community has the best access to EST resources. ESTs can certainly provide an answer to many questions posed, and for the remainder of this review I will focus solely on the EST sequences and their applications.

5. COMPUTERS, DATABASES AND THE REPRESENTATION OF CROP EST SEQUENCE DATA

The volumes of publicly available EST data have created a formidable resource for the research community as exemplified the dbEST database (Boguski et al. 1993). This resource however has not been designed for the needs of the biologist. The dbEST database (or it's siblings such as the EST division of the EMBL database at the EBI (Cochrane et al. 2006)) has been designed as a sequence repository where researchers are both free to contribute their sequences, and as a repository they are expected to deposit their sequences upon publication. The sequence repository is therefore humble in it's offerings and the sequences and their critical annotations and associations are maintained as a flat and textual representation of the data, rather than as a structured, maintained or curated collection of sequences. This does not mean that these sequences alone are without use. Public BLAST (Altschul et al. 1990) servers such as the NCBI BLAST server use the available EST resources as a substrate against which user sequences may be compared. The data from within these primary databases is also freely available to download in full. This free data accessibility also provides the gateway into secondary sequence databases that exploit the fuller potential of the contained information.

The plant research community is no stranger to databases and web-based methods for the presentation and dissemination of biological knowledge. The *Arabidopsis* (Garcia-Hernandez et al. 2002; Hubbard et al. 2005; Schoof et al. 2004) and rice (Karlowski et al. 2003; Schoof et al. 2005; Yuan et al. 2003a) genome databases have paved the way for the exploitation of genome data in research, and further 'generic' database infrastructures for the description and exploitation of plant genomic data have been discussed (e.g. (Hubbard et al. 2005; Lawrence et al. 2005; Schoof et al. 2005)) so that the crop plant research community will be a direct beneficiary of the forthcoming crop-plant genome initiatives. The critical aspect of a meaningful database is that data should be stored and maintained in a structure such that biologically meaningful queries can be addressed to the data collection and meaningful results can be retrieved quickly and simply. The *Arabidopsis*, rice and maize genome databases act as a repository for the raw genome sequences – while important this sequence is of little direct relevance to the community. Onto this data substrate additional annotation and analyses of varying dimensionality are layered (Reed et al. 2006). The information added typically includes gene-models, map positions, similarity and identity to known genes and descriptions that relate to function, structure, ontology or domain content.

When we consider the primitive data-types that are associated with sequences in a genome database the content is not completely dissimilar to the information content that could (or should) be associated with EST sequences. As argued earlier, an EST collection is fragmentary at best with a significant background of sequencing error (empirically shown as approximately 1.5% mismatch error per nucleotide using *Arabidopsis* EST sequence, data not shown) and massive sequence redundancy. To make sense of the sequence data, it is therefore imperative that the EST sequences be cleaned, clustered and assembled to produce a minimal 'unigene-set'. This

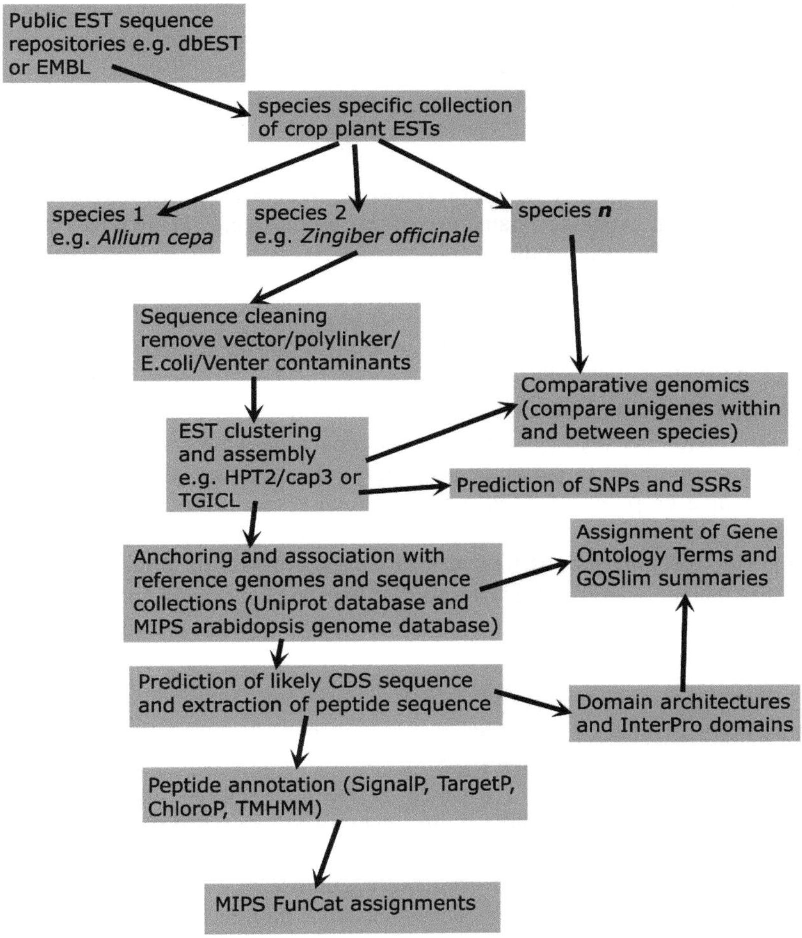

Figure 2. A diagram illustrating the flow of information within the openSputnik EST sequence and annotation database (Rudd 2005). Collections of EST sequence are downloaded from public sequence databases such as the dbest and are used to build species centric databases. The EST sequences are aggressively trimmed of known and probable sources of contamination such as *E.coli* sequence, cloning vector etc. The 'filtered' sequences are then clustered and assembled using specialist software resulting in a 'unigene set' of reduced redundancy and increased complexity. These unigene sequences are used to predict probable molecular markers and in conjunction with other assembled sequence collections are used to establish comparative genomics resources. The comparison of the unigene set to the fully sequenced genomes and to reference protein databases such as UniProt allos for the tentative assignment of role and function and facilitates the unigene assignment to the Gene Ontology. Peptide sequence is predicted on the basis of codon usage and maximum likelihood assessment of the sequence and the resulting peptides are annotated for protein domains, signal peptides, transmembrane domains and sub-cellular localisation. The fullest repertoire of annotations ensures that a simple EST collection may be exhaustively searched for biological context that may be used for the selection of candidate genes for crop improvement

unigene sequence collection can then be annotated and analysed to build a sequence resource that may be used for comparative and functional genomics and that may act as a plentiful resource of candidate genes for further analysis. Figure 2 shows the analytical graph that is used within the openSputnik EST sequence database (Rudd 2005) to assign meaning to ESTs and their parent unigenes.

The openSputnik database (Rudd 2005) is not alone in providing a repository of processed EST data for a collection of crop species. Other databases may contain significant volumes of processed data for several species (e.g. plantGDB (Dong et al. 2004), the TIGR gene indices (Lee et al. 2005) or the NCBI UniGene resource (Wheeler et al. 2006)) or may contain focussed and deep annotation and analysis for a more restricted collection of species (e.g. parasitic plants (Torres et al. 2005), flowering plants (Albert et al. 2005), peach (Lazzari et al. 2005), pineapple (Moyle et al. 2005), or many others).

It is reasonable to summarise that EST databases are an essential resource for migrating information between genomes, for understanding what an EST might do, and where else it might be found and as a general resource for the understanding of the content of a crop species, and a fundamental resource in the selection of candidate genes during the process of crop improvement. The assumption that a large crop EST collection may be treated as a genome-project in miniature, while correct, is also naïve. There is much more that can be done using the large and already existing sequence resources, or with sequence collections that may be created for a specific need.

6. GENOME SEQUENCE, SEQUENCE HETEROGENEITY AND MOLECULAR MARKERS

It seems likely that for the foreseeable future many of the crop species that we are currently reliant upon will remain unsequenced. EST sequencing has demonstrated a technology that may be applied to establish a glimpse of the underlying gene content and databases have been established that work around the limitations and caveats of EST sequence to provide the maximal available context to interested researchers. Meanwhile, some of the caveats of EST sequence data (namely the vast redundancy within sequence collections) have been turned to the advantage of the community. Mining EST collections for molecular markers is routinely performed and robust methods have been developed for this.

While genomics is a relatively new approach within the crop plant community, genetics is a traditional approach whereby researchers and breeders attempt to identify the chromosomal intervals in which traits that enhance performance, value or the scientific merit of the plant are delimited. To make sense of breeding populations breeders have been constructing both genetic and physical maps from many genomes. The genetic maps are reliant upon molecular markers that can be of a few different types. The most popular markers include simple sequence repeat markers (SSR), single nucleotide polymorphism (SNP) markers or more recently conserved ortholog set (COS) markers. While SSR and SNP markers were traditionally identified manually, the vast volumes of data in the EST databases have led to the development of automated processed for candidate marker selection.

SSRs, also known as microsatellite markers, consist of a variable number of typically di-nucleotide or tri-nucleotide repeats. The variability in number of repeat elements will segregate between breeding populations and again may be used in the construction of a genetic map. ESTs have been well used within the generation of SSR markers within several grass species (and in other crops) e.g. (Barkley et al. 2005; Graham et al. 2004; Gupta et al. 2003; Mian et al. 2005; Saha et al. 2004; Thiel et al. 2003). The aim of these experiments was to identify potential microsatellites and to investigate if length variability could be identified within other populations. SSR markers are cheap to test and develop, and will remain a favourite of the crop research community.

SNPs are an enticing form of molecular marker; they are simple in that they are composed of a nucleotide difference at a single position within the much larger chromosomal context. This single difference reliably differentiates between given varieties, cultivars or ecotypes. There has been much discussion recently on computational methods that may be used to select SNPs from within the redundancy of large sequence collection (Huntley et al. 2006; Kota et al. 2003; Marth 2003; Matukumalli et al. 2006; Weil et al. 2004). Regardless of the underlying methodologies, the selection of candidate SNPs has provided an expedited route to the discovery and validation of novel markers.

Not only can plant breeders utilise large sequence collections that reside in the public domain for the selection of candidate markers, but there are also suggestions (in mammalian systems at least) that SNP markers developed in one system may be applicable to other systems (Grapes et al. 2006). The selection of likely candidate SNPs from pig protein coding sequences and their comparison to known human SNPs has revealed that there is a reasonable correlation in gene-to-gene variability across species opening the prospects for site-directed mining of SNPs between species.

7. CROP PLANTS AND THE DNA MICROARRAY

While ESTs have often been sequenced to access both gene sequence from a given genome and to derive an approximate expression level for the sequenced transcripts on the basis of the underlying sequence redundancy, new technologies have come to the forefront that allow for the parallel investigation of both relative and comparative expression levels within series of experiments.

The northern blot (Alwine et al. 1977) has become an indispensable method for the quantification of RNA molecules that have been resolved on a gel and immobilised onto a solid substrate. A labelled DNA 'probe' is hybridised to these immobilised RNAs and the resulting quantification of label provides a view of transcript abundance. A reverse procedure was demonstrated using 45 *Arabidopsis* genes. A probe to each gene was immobilised on a glass slide and free-labelled RNA extracts were hybridised to the first arrays. Quantification of label from each gene allowed for the parallel investigation of expression across the whole set (Schena et al. 1995). This simple demonstration of an array technology has naturally evolved and now 48,000 probes may be comfortably fitted on a single glass slide, hundreds-of-thousands of features may be placed on other more proprietary platforms such as those from companies including Illumina, Agilent, Affymetrix or Nimblegen.

7.1. cDNA Arrays

An EST sequence stems from a cDNA clone. The cDNA insert that has been sequenced may be mechanically applied to a treated glass-slide to create a cDNA array. The cDNA may be spotted prior to sequencing such that a large number of candidate genes that are differentially expressed following a particular treatment may be identified and then sequenced e.g. (Lim 2005). Alternatively the cDNAs may be arrayed after the sequencing step so that in addition to the selection of new candidate genes, already known sequences may be investigated for quantitative differences within an experiment. The application of cDNA arrays in contemporary crop biology has become extremely widespread and brief literature review identifies publications relating to *Arabidopsis* (Kim and von Arnim 2006) (Oono et al. 2006), cotton (Shi et al. 2006), medicago (Tesfaye et al. 2006), sorghum (Buchanan et al. 2005), wild rice (Kim et al. 2005), potato (Rensink et al. 2005) (Schmidt et al. 2005), the genus Senecio (Hegarty et al. 2005), poplar (Taylor et al. 2005), cassava (Lopez et al. 2005), citrus (Forment et al. 2005), gerbera (Laitinen et al. 2005), eucalyptus (Duplessis et al. 2005), *Brassica oleracaea* (Soeda et al. 2005), strawberry (Aharoni et al. 2004), tobacco (Matsuoka et al. 2004) and pine (Egertsdotter et al. 2004).

The continued popularity of cDNA microarrays is in part driven by the relative inexpensiveness of physically arraying small aliquots of DNA solution onto a glass slide. Since no *a priori* knowledge as to the content and structure of the genes expressed within a tissue is needed, cDNA arrays are inexpensive to set-up and are amenable to customisation (groups of target genes may be easily added to the array). The array construction process can be further simplified by arraying the DNA solution onto nylon filters yielding 'macro-arrays'. Macroarrays have a much lower feature density and a typical filter may contain only a few thousand features at most. Such arrays continue to be used within genomics research e.g. (Beldade et al. 2006; Derory et al. 2006; Jia et al. 2006; Nakano et al. 2006; Puthoff and Smigocki 2007), but the availability of many academic and commercial service providers have driven the popularity of both cDNA and oligonucleotide microarrays. Another critical consideration of the macroarray technology is the physical size of the array (tens of square centimetres) and the necessary volumes of hybridisation solutions that are required. Macroarrays are therefore unsuited to some of the more contemporary and sensitive techniques within gene expression profiling. It might be argued that macroarrays are best suited to pilot projects, small numbers of candidate genes, or to preselected clusters of pre-classified genes.

7.2. Oligonucleotide Arrays

Oligonucleotide arrays instead of being reliant upon a cloned and amplified cDNA molecule use instead the sequence to select for long DNA oligonucleotide sequences that may be between 25 and 80 nucleotides long. These oligonucleotides may be synthesised and mechanically arrayed onto a glass slide, they may be synthesised on micro-beads and arrayed or may be synthesised directly on an array as exemplified by the Affymetrix photolithography process. Commerical oligonucleotide arrays such as those provided by Affymetrix have been widely adopted by the research community since they may provide greater reproducibility and sensitivity than cDNA arrays.

Since the manufacture of oligonucleotide arrays requires access to deep quality sequence information this has recently been restricted to the model organisms. However, the demands of the crop research community has been such that oligonucleotide arrays are commercially available on the Affymetrix platform for *Arabidopsis thaliana*, *Hordeum vulgare*, *Zea mays*, *Medicago truncatula*, *Oryza sativa*, *Populus trichocarpa*, *Glycine max*, *Saccharum officinarium*, *Lycopersicon esculentum*, *Triticum aestivum* and *Vitis vinifera*. Other companies such as Illumina and Nimblegen have methods for probe design and optimisation and a ready to prepare oligonucleotide arrays to suit the needs of the crop research community. It would seem that with today's broad and deep EST collections that meaningful oligonucleotide arrays could be synthesised to address many questions and to identify candidate genes involved in many biological processes.

The popularity of microarrays as a fundamental technology to view differential gene expression and as a bridge-technology into the field of system biology or functional genomics is clear (Allison et al. 2006). It has been argued that if significant EST, or other genomic resource, exist then eventually a microarray will be produced (Richmond and Somerville 2000). The varied crop plants for which cDNA and oligonucleotide arrays are already available show that this argument is indeed completely true; furthermore, there are cases where ESTs have undoubtedly been sequenced as a step in the construction of a microarray.

The roles of microarrays within plant genomics continue to diversify. The development of techniques for array-based single nucleotide polymorphism (SNP) classification and the concomitant genome-scale genotyping and haplotyping strategies are opening exciting new developments for the plant breeders (Borevitz 2006). Array-based SNP technologies have already been demonstrated in at least potato (Rickert et al. 2005) and rice (Shirasawa et al. 2006) and will undoubtedly be described in many more species. This exciting new direction is perhaps limited by our ability to define the starting SNPs rather than in their subsequent detection (Chevreux et al. 2004).

8. CONCLUSIONS AND PROSPECTS

The crop species are in a position where they are already benefiting from the rewards of the genomics era, or where they will soon join the species that already benefit. It is rather simple to produce and normalise cDNA libraries that represent the genes actively expressed within any plant, it's constituent tissues and their development, and a wide continuum of biotic and abiotic stresses. These plant EST resources will contain over 10 million ESTs by the end of 2006 and the data contained within the sequence collections will be routinely used by molecular biologists working outside the model species with their complete genomes. These EST data drive research and deeper mapping projects accompanied by new technologies such as microarray and proteomics-based approaches to biology will justify the resources.

We suspect however that the face of plant genomics will fundamentally change in the near future. While EST sequencing has demonstrated fantastic results for hundreds of species, alternative reduced representation approaches to the genomic sampling are perhaps more complete and valuable.

New technologies such as Illumina beadarrays™ for SNP genotyping and expression analysis, or Nimblegen arrays™ for gene expression are opening up fantastic new research avenues for the crop plant community. As the price of DNA sequencing continues to drop, and with the development of further new post-genomic technologies it may be that soon ESTs will have had their day. While the focus of the crop plant research community has been towards the proteins and the genetic mapping of traits, the new empowering post-genomic technologies will provide a plethora of novel and increasingly valuable tools with which in addition to surveying genes and their proteins we can additionally address gene expression, gene regulation and functional networks.

The repetitive and complex structure of many plant genomes may preclude the broadest definitions of plant genomics, but methylation-filtration and high-C_0t filtering of the gene space are undoubtedly a better solution than the patchy transcriptomic representation found within EST collections. There are certainly caveats to the transition, and not all crop species will be surveyed at first, but everything that can be done with EST can also be done with methylation-filtered or high-C_0t sequence.

REFERENCES

Adams KL, Wendel JF (2005) Polyploidy and genome evolution in plants. Curr Opin Plant Biol 8:135–141

Adams MD, Celniker SE, Holt RA, Evans CA, Gocayne JD, Amanatides PG, Scherer SE, Li PW, Hoskins RA, Galle RF, George RA, Lewis SE, Richards S, Ashburner M, Henderson SN, Sutton GG, Wortman JR, Yandell MD, Zhang Q, Chen LX, Brandon RC, Rogers YH, Blazej RG, Champe M, Pfeiffer BD, Wan KH, Doyle C, Baxter EG, Helt G, Nelson CR, Gabor GL, Abril JF, Agbayani A, An HJ, Andrews-Pfannkoch C, Baldwin D, Ballew RM, Basu A, Baxendale J, Bayraktaroglu L, Beasley EM, Beeson KY, Benos PV, Berman BP, Bhandari D, Bolshakov S, Borkova D, Botchan MR, Bouck J, Brokstein P, Brottier P, Burtis KC, Busam DA, Butler H, Cadieu E, Center A, Chandra I, Cherry JM, Cawley S, Dahlke C, Davenport LB, Davies P, de Pablos B, Delcher A, Deng Z, Mays AD, Dew I, Dietz SM, Dodson K, Doup LE, Downes M, Dugan-Rocha S, Dunkov BC, Dunn P, Durbin KJ, Evangelista CC, Ferraz C, Ferriera S, Fleischmann W, Fosler C, Gabrielian AE, Garg NS, Gelbart WM, Glasser K, Glodek A, Gong F, Gorrell JH, Gu Z, Guan P, Harris M, Harris NL, Harvey D, Heiman TJ, Hernandez JR, Houck J, Hostin D, Houston KA, Howland TJ, Wei MH, Ibegwam C, Jalali M, Kalush F, Karpen GH, Ke Z, Kennison JA, Ketchum KA, Kimmel BE, Kodira CD, Kraft C, Kravitz S, Kulp D, Lai Z, Lasko P, Lei Y, Levitsky AA, Li J, Li Z, Liang Y,

Lin X, Liu X, Mattei B, McIntosh TC, McLeod MP, McPherson D, Merkulov G, Milshina NV, Mobarry C, Morris J, Moshrefi A, Mount SM, Moy M, Murphy B, Murphy L, Muzny DM, Nelson DL, Nelson DR, Nelson KA, Nixon K, Nusskern DR, Pacleb JM, Palazzolo M, Pittman GS, Pan S, Pollard J, Puri V, Reese MG, Reinert K, Remington K, Saunders RD, Scheeler F, Shen H, Shue BC, Siden-Kiamos I, Simpson M, Skupski MP, Smith T, Spier E, Spradling AC, Stapleton M, Strong R, Sun E, Svirskas R, Tector C, Turner R, Venter E, Wang AH, Wang X, Wang ZY, Wassarman DA, Weinstock GM, Weissenbach J, Williams SM, Woodage T, Worley KC, Wu D, Yang S, Yao QA, Ye J, Yeh RF, Zaveri JS, Zhan M, Zhang G, Zhao Q, Zheng L, Zheng XH, Zhong FN, Zhong W, Zhou X, Zhu S, Zhu X, Smith HO, Gibbs RA, Myers EW, Rubin GM, Venter JC (2000) The genome sequence of *Drosophila melanogaster.* Science 287:2185–2195

Adams MD, Kelley JM, Gocayne JD, Dubnick M, Polymeropoulos MH, Xiao H, Merril CR, Wu A, Olde B, Moreno RF et al (1991) Complementary DNA sequencing: expressed sequence tags and human genome project. Science 252:1651–1656

AGI T (2000) Analysis of the genome sequence of the flowering plant *Arabidopsis thaliana.* Nature 408:796–815

Aharoni A, Giri AP, Verstappen FW, Bertea CM, Sevenier R, Sun Z, Jongsma MA, Schwab W, Bouwmeester HJ (2004) Gain and loss of fruit flavor compounds produced by wild and cultivated strawberry species. Plant Cell 16:3110–3131

Albert VA, Soltis DE, Carlson JE, Farmerie WG, Wall PK, Ilut DC, Solow TM, Mueller LA, Landherr LL, Hu Y, Buzgo M, Kim S, Yoo MJ, Frohlich MW, Perl-Treves R, Schlarbaum SE, Bliss BJ, Zhang X, Tanksley SD, Oppenheimer DG, Soltis PS, Ma H, DePamphilis CW, Leebens-Mack JH (2005) Floral gene resources from basal angiosperms for comparative genomics research. BMC Plant Biol 5:5

Allison DB, Cui X, Page GP, Sabripour M (2006) Microarray data analysis: from disarray to consolidation and consensus. Nat Rev Genet 7:55–65

Altschul SF, Gish W, Miller W, Myers EW, Lipman DJ (1990) Basic local alignment search tool. J Mol Biol 215:403–410

Alwine JC, Kemp DJ, Stark GR (1977) Method for detection of specific RNAs in agarose gels by transfer to diazobenzyloxymethyl-paper and hybridization with DNA probes. Proc Natl Acad Sci USA 74:5350–5354

Antonius K, Ahokas H (1996) Flow cytometric determination of polyploidy level in spontaneous clones of strawberries. Hereditas 124:285

Ayele M, Haas BJ, Kumar N, Wu H, Xiao Y, Van Aken S, Utterback TR, Wortman JR, White OR, Town CD (2005) Whole genome shotgun sequencing of *Brassica oleracea* and its application to gene discovery and annotation in *Arabidopsis*. Genome Res 15:487–495

Barkley NA, Newman ML, Wang ML, Hotchkiss MW, Pederson GA (2005) Assessment of the genetic diversity and phylogenetic relationships of a temperate bamboo collection by using transferred EST-SSR markers. Genome 48:731–737

Bedell JA, Budiman MA, Nunberg A, Citek RW, Robbins D, Jones J, Flick E, Rholfing T, Fries J, Bradford K, McMenamy J, Smith M, Holeman H, Roe BA, Wiley G, Korf IF, Rabinowicz PD, Lakey N, McCombie WR, Jeddeloh JA, Martienssen RA (2005) Sorghum genome sequencing by methylation filtration. PLoS Biol 3:e13

Beldade P, Rudd S, Gruber JD, Long AD (2006) A wing expressed sequence tag resource for *Bicyclus anynana* butterflies, an evo-devo model. BMC Genomics 7:130

Bennett M, Leitch I (2003) Plant DNA C-values database. release 2.0, January 2003 edn

Blanc G, Wolfe KH (2004a) Functional divergence of duplicated genes formed by polyploidy during *Arabidopsis* evolution. Plant Cell 16:1679–1691

Blanc G, Wolfe KH (2004b) Widespread paleopolyploidy in model plant species inferred from age distributions of duplicate genes. Plant Cell 16:1667–1678

Boguski MS, Lowe TM, Tolstoshev CM (1993) dbEST – database for “expressed sequence tags”. Nat Genet 4:332–333

Borevitz J (2006) Genotyping and mapping with high-density oligonucleotide arrays. Methods Mol Biol 323:137–145

Buchanan CD, Lim S, Salzman RA, Kagiampakis I, Morishige DT, Weers BD, Klein RR, Pratt LH, Cordonnier-Pratt MM, Klein PE, Mullet JE (2005) Sorghum bicolor's transcriptome response to dehydration, high salinity and ABA. Plant Mol Biol 58:699–720

C. elegans Sequencing Consortium (1998) Genome sequence of the nematode C. elegans: a platform for investigating biology. Science 282:2012–2018

Chevreux B, Pfisterer T, Drescher B, Driesel AJ, Muller WE, Wetter T, Suhai S (2004) Using the miraEST assembler for reliable and automated mRNA transcript assembly and SNP detection in sequenced ESTs. Genome Res 14:1147–1159

Cochrane G, Aldebert P, Althorpe N, Andersson M, Baker W, Baldwin A, Bates K, Bhattacharyya S, Browne P, van den Broek A, Castro M, Duggan K, Eberhardt R, Faruque N, Gamble J, Kanz C, Kulikova T, Lee C, Leinonen R, Lin Q, Lombard V, Lopez R, McHale M, McWilliam H, Mukherjee G, Nardone F, Pastor MP, Sobhany S, Stoehr P, Tzouvara K, Vaughan R, Wu D, Zhu W, Apweiler R (2006) EMBL nucleotide sequence database: developments in 2005. Nucleic Acids Res 34:D10–15

Delseny M, Cooke R, Raynal M, Grellet F (1997) The *Arabidopsis thaliana* cDNA sequencing projects. FEBS Lett 403:221–224

Derory J, Leger P, Garcia V, Schaeffer J, Hauser MT, Salin F, Luschnig C, Plomion C, Glossl J, Kremer A (2006) Transcriptome analysis of bud burst in sessile oak (*Quercus petraea*). New Phytol 170:723–738

Dong Q, Schlueter SD, Brendel V (2004) PlantGDB, plant genome database and analysis tools. Nucleic Acids Res 32:D354–359

Duplessis S, Courty PE, Tagu D, Martin F (2005) Transcript patterns associated with ectomycorrhiza development in Eucalyptus globulus and Pisolithus microcarpus. New Phytol 165:599–611

Egertsdotter U, van Zyl LM, MacKay J, Peter G, Kirst M, Clark C, Whetten R, Sederoff R (2004) Gene expression during formation of earlywood and latewood in loblolly pine: expression profiles of 350 genes. Plant Biol (Stuttg) 6:654–663

Forment J, Gadea J, Huerta L, Abizanda L, Agusti J, Alamar S, Alos E, Andres F, Arribas R, Beltran JP, Berbel A, Blazquez MA, Brumos J, Canas LA, Cercos M, Colmenero-Flores JM, Conesa A, Estables B, Gandia M, Garcia-Martinez JL, Gimeno J, Gisbert A, Gomez G, Gonzalez-Candelas L, Granell A, Guerri J, Lafuente MT, Madueno F, Marcos JF, Marques MC, Martinez F, Martinez-Godoy MA, Miralles S, Moreno P, Navarro L, Pallas V, Perez-Amador MA, Perez-Valle J, Pons C, Rodrigo I, Rodriguez PL, Royo C, Serrano R, Soler G, Tadeo F, Talon M, Terol J, Trenor M, Vaello L, Vicente O, Vidal C, Zacarias L, Conejero V (2005) Development of a citrus genome-wide EST collection and cDNA microarray as resources for genomic studies. Plant Mol Biol 57:375–391

Friedman WE, Cook ME (2000) The origin and early evolution of tracheids in vascular plants: integration of palaeobotanical and neobotanical data. Philos Trans R Soc Lond B Biol Sci 355:857–868

Gaafar RM, Hohmann U, Jung C (2005) Bacterial artificial chromosome-derived molecular markers for early bolting in sugar beet. Theor Appl Genet 110:1027–1037

Garcia-Hernandez M, Berardini TZ, Chen G, Crist D, Doyle A, Huala E, Knee E, Lambrecht M, Miller N, Mueller LA, Mundodi S, Reiser L, Rhee SY, Scholl R, Tacklind J, Weems DC, Wu Y, Xu I, Yoo D, Yoon J, Zhang P (2002) TAIR: a resource for integrated *Arabidopsis* data. Funct Integr Genomics 2:239–253

Gardiner J, Schroeder S, Polacco ML, Sanchez-Villeda H, Fang Z, Morgante M, Landewe T, Fengler K, Useche F, Hanafey M, Tingey S, Chou H, Wing R, Soderlund C, Coe Jr. EH (2004) Anchoring 9,371 maize expressed sequence tagged unigenes to the bacterial artificial chromosome contig map by two-dimensional overgo hybridization. Plant Physiol 134:1317–1326

Graham J, Smith K, MacKenzie K, Jorgenson L, Hackett C, Powell W (2004) The construction of a genetic linkage map of red raspberry (*Rubus idaeus* subsp. *idaeus*) based on AFLPs, genomic-SSR and EST-SSR markers. Theor Appl Genet 109:740–749

Grapes L, Rudd S, Fernando RL, Megy K, Rocha D, Rothschild MF (2006) Prospecting for pig single nucleotide polymorphisms in the human genome: have we struck gold? J Anim Breed Genet 123:145–151

Grif VG, Valovich EM, Levbedeva NV (1980) Parameters of the mitotic cycle in two species of *Trillium* L. Tsitologiya 22:1331–1338

Gupta PK, Rustgi S, Sharma S, Singh R, Kumar N, Balyan HS (2003) Transferable EST-SSR markers for the study of polymorphism and genetic diversity in bread wheat. Mol Genet Genomics 270:315–323

Haberer G, Young S, Bharti AK, Gundlach H, Raymond C, Fuks G, Butler E, Wing RA, Rounsley S, Birren B, Nusbaum C, Mayer KF, Messing J (2005) Structure and architecture of the maize genome. Plant Physiol 139:1612–1624

Hegarty MJ, Jones JM, Wilson ID, Barker GL, Coghill JA, Sanchez-Baracaldo P, Liu G, Buggs RJ, Abbott RJ, Edwards KJ, Hiscock SJ (2005) Development of anonymous cDNA microarrays to study changes to the *Senecio* floral transcriptome during hybrid speciation. Mol Ecol 14:2493–2510

Hong CP, Lee SJ, Park JY, Plaha P, Park YS, Lee YK, Choi JE, Kim KY, Lee JH, Lee J, Jin H, Choi SR, Lim YP (2004) Construction of a BAC library of Korean ginseng and initial analysis of BAC-end sequences. Mol Genet Genomics 271:709–716

Hubbard T, Andrews D, Caccamo M, Cameron G, Chen Y, Clamp M, Clarke L, Coates G, Cox T, Cunningham F, Curwen V, Cutts T, Down T, Durbin R, Fernandez-Suarez XM, Gilbert J, Hammond M, Herrero J, Hotz H, Howe K, Iyer V, Jekosch K, Kahari A, Kasprzyk A, Keefe D, Keenan S, Kokocinsci F, London D, Longden I, McVicker G, Melsopp C, Meidl P, Potter S, Proctor G, Rae M, Rios D, Schuster M, Searle S, Severin J, Slater G, Smedley D, Smith J, Spooner W, Stabenau A, Stalker J, Storey R, Trevanion S, Ureta–Vidal A, Vogel J, White S, Woodwark C, Birney E (2005) Ensembl 2005. Nucleic Acids Res 33:D447–453

Huntley D, Baldo A, Johri S, Sergot M (2006) SEAN: SNP prediction and display program utilizing EST sequence clusters. Bioinformatics 22:495–496

Jaffe DB, Butler J, Gnerre S, Mauceli E, Lindblad-Toh K, Mesirov JP, Zody MC, Lander ES (2003) Whole-genome sequence assembly for mammalian genomes: arachne 2. Genome Res 13:91–96

Jia Y, Anderson JV, Horvath DP, Gu YQ, Lym RG, Chao WS (2006) Subtractive cDNA libraries identify differentially expressed genes in dormant and growing buds of leafy spurge (*Euphorbia esula*). Plant Mol Biol 61:329–344

Karlowski WM, Schoof H, Janakiraman V, Stuempflen V, Mayer KF (2003) MOsDB: an integrated information resource for rice genomics. Nucleic Acids Res 31:190–192

Katari MS, Balija V, Wilson RK, Martienssen RA, McCombie WR (2005) Comparing low coverage random shotgun sequence data from *Brassica oleracea* and *Oryza sativa* genome sequence for their ability to add to the annotation of *Arabidopsis thaliana*. Genome Res 15:496–504

Kim BH, von Arnim AG (2006) The early dark-response in *Arabidopsis thaliana* revealed by cDNA microarray analysis. Plant Mol Biol 60:321–342

Kim KM, Cho SK, Shin SH, Kim GT, Lee JH, Oh BJ, Kang KH, Hong JC, Choi JY, Shin JS, Chung YS (2005) Analysis of differentially expressed transcripts of fungal elicitor- and wound-treated wild rice (*Oryza grandiglumis*). J Plant Res 118:347–354

Kota R, Rudd S, Facius A, Kolesov G, Thiel T, Zhang H, Stein N, Mayer K, Graner A (2003) Snipping polymorphisms from large EST collections in barley (*Hordeum vulgare* L.). Mol Genet Genomics 270:24–33

Kumar A, Bennetzen JL (2000) Retrotransposons: central players in the structure, evolution and function of plant genomes. Trends Plant Sci 5:509–510

Lagudah ES, Moullet O, Appels R (1997) Map-based cloning of a gene sequence encoding a nucleotide-binding domain and a leucine-rich region at the Cre3 nematode resistance locus of wheat. Genome 40:659–665

Laitinen RA, Immanen J, Auvinen P, Rudd S, Alatalo E, Paulin L, Ainasoja M, Kotilainen M, Koskela S, Teeri TH, Elomaa P (2005) Analysis of the floral transcriptome uncovers new regulators of organ determination and gene families related to flower organ differentiation in *Gerbera hybrida* (Asteraceae). Genome Res 15:475–486

Lamoureux D, Peterson DG, Li W, Fellers JP, Gill BS (2005) The efficacy of Cot-based gene enrichment in wheat (*Triticum aestivum* L.). Genome 48:1120–1126

Lawrence CJ, Seigfried TE, Brendel V (2005) The maize genetics and genomics database. The community resource for access to diverse maize data. Plant Physiol 138:55–58

Lazzari B, Caprera A, Vecchietti A, Stella A, Milanesi L, Pozzi C (2005) ESTree db: a tool for peach functional genomics. BMC Bioinform 6:S16

Lee Y, Tsai J, Sunkara S, Karamycheva S, Pertea G, Sultana R, Antonescu V, Chan A, Cheung F, Quackenbush J (2005) The TIGR gene indices: clustering and assembling EST and known genes and integration with eukaryotic genomes. Nucleic Acids Res 33:D71–74

Lim KJ (2005) Utilization of microarray technology for identification of disease response genes in banana (*Musa* spp.). Universiti Putra Malaysia

Ling P, Chen XM (2005) Construction of a hexaploid wheat (*Triticum aestivum* L.) bacterial artificial chromosome library for cloning genes for stripe rust resistance. Genome 48: 1028–1036

Lopez C, Soto M, Restrepo S, Piegu B, Cooke R, Delseny M, Tohme J, Verdier V (2005) Gene expression profile in response to *Xanthomonas axonopodis* pv. manihotis infection in cassava using a cDNA microarray. Plant Mol Biol 57:393–410

Marek LF, Mudge J, Darnielle L, Grant D, Hanson N, Paz M, Huihuang Y, Denny R, Larson K, Foster-Hartnett D, Cooper A, Danesh D, Larsen D, Schmidt T, Staggs R, Crow JA, Retzel E, Young ND, Shoemaker RC (2001) Soybean genomic survey: BAC-end sequences near RFLP and SSR markers. Genome 44:572–581

Margulies M, Egholm M, Altman WE, Attiya S, Bader JS, Bemben LA, Berka J, Braverman MS, Chen YJ, Chen Z, Dewell SB, Du L, Fierro JM, Gomes XV, Godwin BC, He W, Helgesen S, Ho CH, Irzyk GP, Jando SC, Alenquer ML, Jarvie TP, Jirage KB, Kim JB, Knight JR, Lanza JR, Leamon JH, Lefkowitz SM, Lei M, Li J, Lohman KL, Lu H, Makhijani VB, McDade KE, McKenna MP, Myers EW, Nickerson E, Nobile JR, Plant R, Puc BP, Ronan MT, Roth GT, Sarkis GJ, Simons JF, Simpson JW, Srinivasan M, Tartaro KR, Tomasz A, Vogt KA, Volkmer GA, Wang SH, Wang Y, Weiner MP, Yu P, Begley RF, Rothberg JM (2005) Genome sequencing in microfabricated high-density picolitre reactors. Nature 437:376–380

Marth GT (2003) Computational SNP discovery in DNA sequence data. Methods Mol Biol 212:85–110

Matsuoka K, Demura T, Galis I, Horiguchi T, Sasaki M, Tashiro G, Fukuda H (2004) A comprehensive gene expression analysis toward the understanding of growth and differentiation of tobacco BY-2 cells. Plant Cell Physiol 45:1280–1289

Matukumalli LK, Grefenstette JJ, Hyten DL, Choi IY, Cregan PB, Van Tassell CP (2006) Application of machine learning in SNP discovery. BMC Bioinform 7:4

May RM (1990) How many species? Phil Trans Roy Soc B330:292–304

Mayer K, Mewes HW (2002) How can we deliver the large plant genomes? Strategies and perspectives. Curr Opin Plant Biol 5:173–177

Meyers BC, Tingey SV, Morgante M (2001) Abundance, distribution, and transcriptional activity of repetitive elements in the maize genome. Genome Res 11:1660–1676

Mian MA, Saha MC, Hopkins AA, Wang ZY (2005) Use of tall fescue EST-SSR markers in phylogenetic analysis of cool-season forage grasses. Genome 48:637–647

Ming R, Liu SC, Moore PH, Irvine JE, Paterson AH (2001) QTL analysis in a complex autopolyploid: genetic control of sugar content in sugarcane. Genome Res 11:2075–2084

Morgante M (2006) Plant genome organisation and diversity: the year of the junk! Curr Opin Biotechnol 17:168–173

Moyle RL, Crowe ML, Ripi-Koia J, Fairbairn DJ, Botella JR (2005) PineappleDB: an online pineapple bioinformatics resource. BMC Plant Biol 5:21

Myers EW (2005) The fragment assembly string graph. Bioinformatics 21:ii79–ii85

Myers EW, Sutton GG, Delcher AL, Dew IM, Fasulo DP, Flanigan MJ, Kravitz SA, Mobarry CM, Reinert KH, Remington KA, Anson EL, Bolanos RA, Chou HH, Jordan CM, Halpern AL, Lonardi S, Beasley EM, Brandon RC, Chen L, Dunn PJ, Lai Z, Liang Y, Nusskern DR, Zhan M, Zhang Q, Zheng X, Rubin GM, Adams MD, Venter JC (2000) A whole-genome assembly of *Drosophila.* Science 287:2196–2204

Myers EW, Sutton GG, Smith HO, Adams MD, Venter JC (2002) On the sequencing and assembly of the human genome. Proc Natl Acad Sci USA 99:4145–4146

Nakano T, Suzuki K, Ohtsuki N, Tsujimoto Y, Fujimura T, Shinshi H (2006) Identification of genes of the plant-specific transcription-factor families cooperatively regulated by ethylene and jasmonate in *Arabidopsis thaliana.* J Plant Res 119:407–413

Nekrutenko A, Baker RJ (2003) Subgenome-specific markers in allopolyploid cotton *Gossypium hirsutum*: implications for evolutionary analysis of polyploids. Gene 306:99–103

Oono Y, Seki M, Satou M, Iida K, Akiyama K, Sakurai T, Fujita M, Yamaguchi-Shinozaki K, Shinozaki K (2006) Monitoring expression profiles of *Arabidopsis* genes during cold acclimation and deacclimation using DNA microarrays. Funct Integr Genomics:1–23

Palmer LE, Rabinowicz PD, O'Shaughnessy AL, Balija VS, Nascimento LU, Dike S, de la Bastide M, Martienssen RA, McCombie WR (2003) Maize genome sequencing by methylation filtration. Science 302:2115–2117

Paterson AH (2006) Leafing through the genomes of our major crop plants: strategies for capturing unique information. Nat Rev Genet 7:174–184

Peterson DG, Schulze SR, Sciara EB, Lee SA, Bowers JE, Nagel A, Jiang N, Tibbitts DC, Wessler SR, Paterson AH (2002a) Integration of Cot analysis, DNA cloning, and high-throughput sequencing facilitates genome characterization and gene discovery. Genome Res 12:795–807

Peterson DG, Wessler SR, Paterson AH (2002b) Efficient capture of unique sequences from eukaryotic genomes. Trends Genet 18:547–550

Pryer KM, Schneider H, Zimmer EA, Banks JA (2002) Deciding among green plants for whole genome studies. Trends Plant Sci 7:550–554

Puthoff DP, Smigocki AC (2007) Insect feeding-induced differential expression of *Beta vulgaris* root genes and their regulation by defense-associated signals. Plant Cell Rep 26:71–84

Rabinowicz PD, McCombie WR, Martienssen RA (2003) Gene enrichment in plant genomic shotgun libraries. Curr Opin Plant Biol 6:150–156

Rabinowicz PD, Schutz K, Dedhia N, Yordan C, Parnell LD, Stein L, McCombie WR, Martienssen RA (1999) Differential methylation of genes and retrotransposons facilitates shotgun sequencing of the maize genome. Nat Genet 23:305–308

Reed JL, Famili I, Thiele I, Palsson BO (2006) Towards multidimensional genome annotation. Nat Rev Genet 7:130–141

Rensink WA, Iobst S, Hart A, Stegalkina S, Liu J, Buell CR (2005) Gene expression profiling of potato responses to cold, heat, and salt stress. Funct Integr Genomics 5:201–207

Richmond T, Somerville S (2000) Chasing the dream: plant EST microarrays. Curr Opin Plant Biol 3:108–116

Rickert AM, Ballvora A, Matzner U, Klemm M, Gebhardt C (2005) Quantitative genotyping of single-nucleotide polymorphisms by allele-specific oligonucleotide hybridization on DNA microarrays. Biotechnol Appl Biochem 42:93–96

Rudd S (2003) Expressed sequence tags: alternative or complement to whole genome sequences? Trends Plant Sci 8:321–329

Rudd S (2005) openSputnik – a database to ESTablish comparative plant genomics using unsaturated sequence collections. Nucleic Acids Res 33 Database Issue:D622–627

Saha MC, Mian MA, Eujayl I, Zwonitzer JC, Wang L, May GD (2004) Tall fescue EST-SSR markers with transferability across several grass species. Theor Appl Genet 109:783–791

SanMiguel P, Gaut BS, Tikhonov A, Nakajima Y, Bennetzen JL (1998) The paleontology of intergene retrotransposons of maize. Nat Genet 20:43–45

Schena M, Shalon D, Davis RW, Brown PO (1995) Quantitative monitoring of gene expression patterns with a complementary DNA microarray. Science 270:467–470

Schmidt DD, Voelckel C, Hartl M, Schmidt S, Baldwin IT (2005) Specificity in ecological interactions: attack from the same lepidopteran herbivore results in species-specific transcriptional responses in two solanaceous host plants. Plant Physiol 138:1763–1773

Schoof H, Ernst R, Nazarov V, Pfeifer L, Mewes HW, Mayer KF (2004) MIPS *Arabidopsis thaliana* Database (MAtDB): an integrated biological knowledge resource for plant genomics. Nucleic Acids Res 32:D373–376

Schoof H, Spannagl M, Yang L, Ernst R, Gundlach H, Haase D, Haberer G, Mayer KF (2005) Munich information center for protein sequences plant genome resources: a framework for integrative and comparative analyses 1(W). Plant Physiol 138:1301–1309

Shi YH, Zhu SW, Mao XZ, Feng JX, Qin YM, Zhang L, Cheng J, Wei LP, Wang ZY, Zhu YX (2006) Transcriptome profiling, molecular biological, and physiological studies reveal a major role for ethylene in cotton fiber cell elongation. Plant Cell 18:651–664

Shirasawa K, Shiokai S, Yamaguchi M, Kishitani S, Nishio T (2006) Dot-blot-SNP analysis for practical plant breeding and cultivar identification in rice. Theor Appl Genet 113:147–155

Soeda Y, Konings MC, Vorst O, van Houwelingen AM, Stoopen GM, Maliepaard CA, Kodde J, Bino RJ, Groot SP, van der Geest AH (2005) Gene expression programs during *Brassica oleracea* seed maturation, osmopriming, and germination are indicators of progression of the germination process and the stress tolerance level. Plant Physiol 137:354–368

Taylor G, Street NR, Tricker PJ, Sjodin A, Graham L, Skogstrom O, Calfapietra C, Scarascia-Mugnozza G, Jansson S (2005) The transcriptome of populus in elevated CO2. New Phytol 167:143–154

Tesfaye M, Samac DA, Vance CP (2006) Insights into symbiotic nitrogen fixation in *Medicago truncatula.* Mol Plant Microbe Interact 19:330–341

Thiel T, Michalek W, Varshney RK, Graner A (2003) Exploiting EST databases for the development and characterization of gene-derived SSR-markers in barley (*Hordeum vulgare* L.). Theor Appl Genet 106:411–422

Torres MJ, Tomilov AA, Tomilova N, Reagan RL, Yoder JI (2005) Pscroph, a parasitic plant EST database enriched for parasite associated transcripts. BMC Plant Biol 5:24

Vendrely R, Vendrely C (1948) La teneur du noyau cellulaire en acide désoxyribonucléique à travers les organes, les individus et les espèces animales: techniques et premiers résultats. Experientia 4:434–436

Weil MM, Pershad R, Wang R, Zhao S (2004) Use of BAC end sequences for SNP discovery. Methods Mol Biol 256:1–6

Wellman CH, Gray J (2000) The microfossil record of early land plants. Philos Trans R Soc Lond B Biol Sci 355:717–731; discussion 731–712

Wellman CH, Osterloff PL, Mohiuddin U (2003) Fragments of the earliest land plants. Nature 425:282–285

Wheeler DL, Barrett T, Benson DA, Bryant SH, Canese K, Chetvernin V, Church DM, DiCuccio M, Edgar R, Federhen S, Geer LY, Helmberg W, Kapustin Y, Kenton DL, Khovayko O, Lipman DJ, Madden TL, Maglott DR, Ostell J, Pruitt KD, Schuler GD, Schriml LM, Sequeira E, Sherry ST, Sirotkin K, Souvorov A, Starchenko G, Suzek TO, Tatusov R, Tatusova TA, Wagner L, Yaschenko E (2006) Database resources of the National Center for Biotechnology Information. Nucleic Acids Res 34:D173–180

Yuan Q, Ouyang S, Liu J, Suh B, Cheung F, Sultana R, Lee D, Quackenbush J, Buell CR (2003a) The TIGR rice genome annotation resource: annotating the rice genome and creating resources for plant biologists. Nucleic Acids Res 31:229–233

Yuan Y, SanMiguel PJ, Bennetzen JL (2003b) High-Cot sequence analysis of the maize genome. Plant J 34:249–255

CHAPTER 8

COMPARATIVE GENOMICS OF CEREALS

JÉRÔME SALSE AND CATHERINE FEUILLET*

UMR INRA-UBP 1095, Amélioration et Santé des Plantes, Domaine de Crouelle 234, Avenue du Brézet, 63100 Clermont-Ferrand, France

Abstract: Cereals such as wheat, barley, maize, sorghum, millet and rice belong to the grass family and comprise some of the most important crops for human and animal nutrition. Comparative genomic studies in cereals have been pioneering the field of plant comparative genomics in the past decade. The first comparative studies were performed at the genetic map level. They have revealed a very good conservation of the order (colinearity) of molecular markers and of QTL for agronomic traits along the chromosomes thereby establishing evolutionary relationships between the cereal genomes. For this reason and because of its small size, rice was promoted as a model and was chosen to be the first cereal genome sequenced. Further, the development of large EST collections and the first inter- and intra-specific comparative studies of BAC sequences from maize, sorghum, rice, wheat and barley have increased the resolution of comparative analyses and have shown that a number of rearrangements disrupting microcolinearity have occurred during the evolution of the cereal genomes in the past 50–70 million years.

This chapter reviews comparative studies that have been performed at the macro- and micro- levels in cereals and discusses what was learned about the mechanisms underlying genome evolution in these important crop species. It describes how this knowledge can be applied to support gene discovery and cereal crop improvement and presents the opportunities that will be available within the next few years as the sequencing of several cereal genomes in addition to rice will be completed.

1. INTRODUCTION

Cereals (grass species that are cultivated for their edible seeds) such as wheat, rice, maize, barley, oat, sorghum or millets constitute over 50% of the total crop production worldwide (http://www.fao.org/). Since the beginning of agriculture, their grains have represented one the most important renewable resources for human food and domestic animal feed. Moreover, cereal seeds and straw represent

*Corresponding Author: catherine.feuillet@clermont.inra.fr

R.K. Varshney and R. Tuberosa (eds.), Genomics-Assisted Crop Improvement: Vol. 1: Genomics Approaches and Platforms, 177–205.

an increasingly important feedstock for non-food products and for bioenergy production to supplement or replace fossil fuel based products and energy. All cereal crop species are members of the grass (*Poaceae)* family that is the fourth largest family of flowering plants. With about 10‘000 species growing under nearly all climates and latitudes, grasses exceed all other families in ecological dominance and economic importance. In terms of genome organisation they represent a very diverse family with basic chromosome numbers ranging from 4 to 266 and genome sizes ranging from 400 Mb to 17 Gb (Feuillet and Keller 2002). Within the grass family, the cereals are represented in four of the five main sub-families (Figure 1): Sorghum, maize, pearl millet and foxtail millet are members of the *Panicoideae*; finger millet belongs to the *Chloridoideae*; rice to the *Ehrhartoideae;* and wheat, barley, oats, and rye are *Pooideae* representatives. Fossil data and phylogenetic studies have estimated that the grasses have diverged from a common ancestor 50 to 70 MYA (for reviews see Kellogg 2001 and Gaut 2002). Archaeological records suggest that farming started concomitantly in a least three widely separated regions 10‘000-5‘000 years ago during the late Neolithic period. The three most important cereals were domesticated independently in three centres: wheat in southwestern Asia in the Fertile Crescent region, maize in Mexico and rice in both southeast Asia and west Africa (Harlan 1992; Zohary and Hopf 2000; Piperno and Flannery 2001). Due to their economic

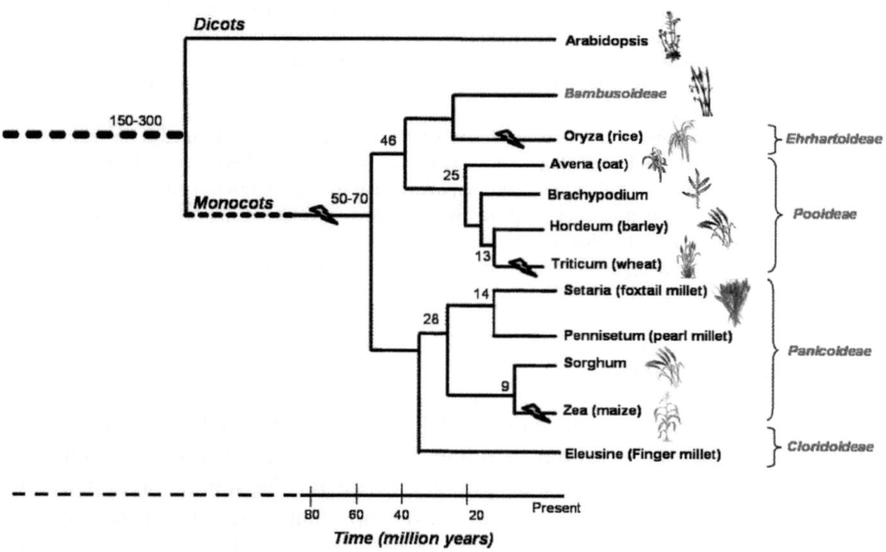

Figure 1. Phylogenic relationships between cereal genomes
Divergence times from a common ancestor between the different species are indicated on the branches of the phylogenic tree (in million years). Duplication or polyploidization events are shown with an arrow. The five main grass subfamilies are indicated in grey and italic

importance, broad diversity, and relatively recent evolutionary history, grasses have been subjected to extensive research. Early comparative studies were performed using isozymes but it is only with the development of DNA markers and the "genomic revolution" in the early 1980's that a "common language" was found to compare genomes.

Comparative genomics studies the relationships between genomes of different species. It enables the identification of the portion of genomes that are conserved and those that are unique, thereby allowing one to relate specific changes in genome structure and content to differences in the biology of the different species. It gives insight into the mechanisms of genome evolution and speciation as well as provides tools for a variety of studies and applications that range from the densification of DNA markers on genetic maps to the identification of conserved genes and regulatory sequences. Comparative genomics can be performed at different levels (genetic map, partial or whole genome sequence) depending on the genomic information available in the species that are compared. It allows one to utilize genomic resources from model species to accelerate gene discovery in species for which genomic tools or sequences are not yet available. Comparative genomics in families with a relatively recent history, such as the cereal crops, has great potential because it provides access to and understanding of the basis of diversity and adaptation that, in turn, allows for increased exploitation of the genetic resources for crop improvement.

Comparative genomics between grasses and mostly cereals such as barley, wheat, maize, rice, and sorghum has been the focus of intense research during the past 8 years. Early results indicated a good level of conservation of marker order at the genetic map level (macrocolinearity) and thus with its small genome size and well studied genetics, rice was promoted as a reference genome for grasses. The release of the first rice genome sequence drafts in 2002 and the development of a number of genomic resources (EST collections, BAC libraries) from different cereal species then enabled inter- and intra-specific comparisons at the sequence level (microcolinearity). This shed new light onto the level of conservation between the cereal genomes and provided the first insights into the mechanisms that have shaped these genomes during 50–70 million years of evolution.

2. MACROCOLINEARITY: COMPARATIVE GENOMICS AT THE GENETIC MAP LEVEL

Early comparative genetic mapping studies have indicated that despite large differences in ploidy level, chromosome number, and haploid DNA content, the linear order (colinearity) of markers remained largely conserved between grass species over several million years of evolution (reviewed in Devos and Gale 2000, Feuillet and Keller 2002). The estimated level of colinearity has evolved at the same time that the level of resolution of the analysis has increased with the saturation of genetic maps with new markers and the availability of the various sequences of the rice genome after 2002.

2.1. Building the "Crop Circles" Model

Initial comparisons between the genomes of all important grass species were performed with restriction fragment length polymorphisms (RFLP) markers. This provided compelling evidence that, except for a few large rearrangements, the linear order of markers on the chromosomes was conserved (macrocolinearity) despite 50-70 million years of divergent evolution. These data were brought together into the famous "crop circles" (Moore et al. 1995; Devos and Gale 1997) that provided a representation of the relationships between orthologous chromosomes in eight species belonging to three grass subfamilies: rice (*Ehrhartoideae*), foxtail millet, sugar cane, sorghum, pearl millet, maize (*Panicoideae*), and the triticeae and oats (*Pooideae*). The degree of macrocolinearity led to the consideration of grasses as a single genetic system built from 30 rice linkage blocks that possibly represented linkage blocks of the ancestral grass genome (Moore et al. 1995). These results, however, were obtained from low resolution genetic maps with an average of one marker every 10 cM that allowed the detection of only dramatic rearrangements. Moreover, the maps were constructed with low copy RFLP markers that were selected for their ability to provide a signal in cross hybridizations, thereby limiting the detection of whole or partial genome duplication events and making it difficult to assess orthologous and paralogous relationships of gene families. Finally, as comparisons based on the genetic maps overemphasize polymorphic regions, the overall genomes were not evenly represented and this was especially true for the centromeric regions.

A reassessment of the colinearity among the grass genomes was performed by Gaut in 2002 utilising collated data from different comparative studies to estimate the probability for one marker that is found in the vicinity of another to be in a colinear region. The data indicated that the average probability for moving from one marker into a colinear region was not very high (about 50% on average) even between closely related species such as maize and sorghum. These results suggested extensive rearrangements between the grass genomes and questioned the concept of using small grass genomes (rice or sorghum) as a proxy for more complex genomes (maize or wheat) (Gaut 2002).

2.2. Sequence-based Macrocolinearity Studies

In the past 5 years, the release of the rice genome sequences (Feng et al. 2002, Goff et al. 2002, Sasaki et al. 2002, Yu et al. 2002, The Rice Chromosome 10 Sequencing Consortium 2003, International Rice Genome Sequencing Project 2005; The Rice Chromosome 3 Sequencing Consortium 2005) and the development of large EST collections from other cereals have provided new insights into the level of colinearity between the cereal genomes. For wheat and barley, the International Triticeae EST Cooperative (ITEC; http://wheat.pw.usda.gov/genome/) resulted in increasing the number of ESTs present in the public databases from 6 in 1998 to more than one million (879,909 for wheat; 461,471 for barley) by mid-2006. For maize and sorghum, 753,411 and 237,054 ESTs have been released into the public domain, respectively (http://www.ncbi.nlm.nih.gov/dbEST/dbEST_summary.html; 09/06/06 release). Large programs aiming at globally assigning (in genetic or deletion bins) or mapping precisely ESTs have resulted in the mapping of 7,107 EST singletons (16,099 loci) into a chromosome bin map using sets of euploid and aneuploid lines in wheat (Qi et al. 2004) and the addition of 1,454 new candidate genes to the 4,821 existing loci in maize (http://www.maizemap.org/iMapDB/iMap.html) (Falque et al. 2005).

With these data, it became possible to compare *in silico* the sequences of EST markers mapped in different cereal species to each other and to the rice genome sequence and thus, to study macrocolinearity at a higher resolution. Because these comparisons are based on sequence alignments and because in most of the cases it is difficult to infer orthologous and paralogous relationships, statistical analysis needs to be performed to evaluate objectively whether the association between two or more genes in the same order on two chromosomal segments occurs by chance or if it reveals significant colinearity. Several software programs such as LineUP (Hampson et al. 2003), ADHoRE (Automatic Detection of Homologous Regions, Vandepoele et al. 2002) and FISH (Fast Identification of Segmental Homology, Calabrese et al. 2003) have been developed recently for this purpose. A number of programs e.g. Cmap (Fang et al. 2003) and websites such as Gramene (for the last update see Jaiswal et al. 2006) have also been established to visualize the sequence-based colinearity data obtained between the grass genomes. Outputs of the sequenced-based macrocolinearity between the rice genome sequence and 9,332 wheat genetic marker sequences (http://www.tigr.org/tdb/synteny/wheat/description.shtml), 1,569 maize markers (http://www.tigr.org/tdb/synteny/maize_IBMn/description.shtml), and 447 sorghum markers (http://www.tigr.org/tdb/synteny/sorghum/description.shtml) can be found at TIGR.

These studies greatly enhanced the resolution of comparative mapping and revealed additional features of the conservation between cereal genomes. In maize, more than 2,600 mapped sequence markers identified 656 (46%) putative orthologous genes in the rice genome (Salse et al. 2004). The high resolution provided by this sequence-based approach identified six new colinear regions between maize chromosomes 1, 4, 5, and 6 and rice chromosomes 9-12, 6-8, 6, and 1, respectively. It also provided evidence for duplications events within the rice genome that had not been found before (Figure 2). In wheat, similar studies with 4,485 ESTs increased the resolution of comparative mapping with rice by 25-30 fold (Sorells et al. 2003; Sorrells 2004) and have allowed the specification of the degree of conservation between orthologous chromosomes. Thus, we know now that chromosome 3 is the most conserved and chromosome 5 is the least conserved of all the wheat chromosomes when compared to rice (La Rota and Sorrels 2004). In an extensive colinearity study using RFLP, candidate genes, and EST sequences from the short arm of wheat chromosome 1A and rice chromosome 5S, Guyot et al. (2004) found frequent disruptions of the marker order resulting in a mosaic conservation of genes in this region. Studies focussing on single chromosome groups or regions have been performed recently as well for rice chromosome 3 compared to the wheat and maize (Buell et al. 2005, The Rice Chromosome 3 Sequencing Consortium 2005)

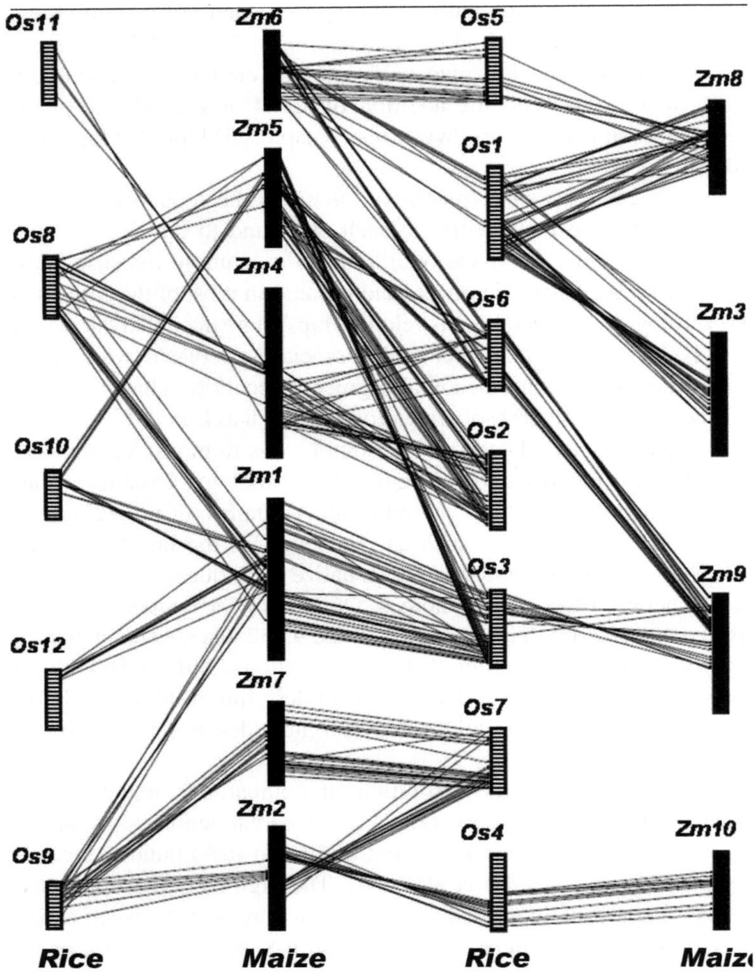

Figure 2. Macrocolinearity relationships between the rice and maize genomes (updated from Salse et al. 2004)

Schematic representation of the macrocolinearity identified through EST comparisons between the 10 maize chromosomes (Zm1–10) and the 12 rice chromosomes (Os1–12). Chromosomes are represented by thick vertical bars (black for maize, dashed for rice) and orthologous genes are linked by thin bars. Regions of maize chromosomes 1, 5, 6 (or 8) and 9 define duplicated regions on rice chromosomes 8-12, 2-6 and 6-10, 1-5 and 3-6, respectively

and for rice chromosome 11 compared to wheat (Singh et al. 2004). In addition to ESTs, low-pass BAC sequences also can be used for sequence based comparative studies. Klein et al. (2003) have used them successfully in sorghum to align chromosome 3 BAC sequences against the rice chromosome 1 sequence and to identify a previously undetected inversion between the two chromosomes.

Thus, increasing the resolution of map-based comparative studies as well as applying statistical tests to the sequence based comparisons has revealed additional chromosomal rearrangements between rice and the grass genomes than those reported previously that were based on RFLP analysis and, thus, led to the revision of the "concentric crop circles" model (Devos 2005). Further, it provided a more complex picture of the orthologous relationships between these genomes and raised significant questions about using rice as a model for the direct transfer of information to the other grass species.

3. MICROCOLINEARITY: COMPARATIVE GENOMICS AT THE SEQUENCE LEVEL

Increasing evidence that rearrangements limit the extent of colinearity between the grass genomes has led cereal geneticists working with non-model genomes to develop large insert BAC libraries from their own species to perform map-based cloning and study genome structure and evolution. Technological improvements have allowed the construction of BAC libraries with a sufficient number of clones to provide reasonable coverage even from large and complex genomes such as those of wheat and barley (Chalhoub et al. 2004). In addition, for polyploid species such as wheat, recent advances in flow sorting techniques (Kubalakova et al. 2002) have allowed the isolation of DNA in sufficient amounts and quality to construct libraries from single chromosomes or chromosome arms (Safar et al. 2004; Janda et al. 2004; 2006). To date a number of BAC libraries are available from different cereal species, subspecies, and even from different varieties (Table 1), permitting microcolinearity studies at different levels.

3.1. Interspecific Comparative Studies: Looking at 50–70 MY of Speciation

One of the first microcolinearity studies was performed at the *Shrunken 2/ Anthocyaninless1* (*sh2/a1*) orthologous regions in maize, sorghum, and rice (Chen et al. 1997; 1998). Despite large differences in the length of the intergenic regions in maize compared to rice and sorghum and a tandem duplication of one gene (A1) in sorghum, the linear order of the four genes (*Sh2*, *X1*, *X2* an*d A1*) present at this locus was remarkably conserved between the three species. In contrast, in the Triticeae, the colinearity was limited to the conservation of the *Sh2* and *X1* genes on chromosome 1L; whereas, the two other genes, *X2* and *A,1* were found on a non orthologous chromosome (3L). This indicated that numerous rearrangements including genes translocation have occurred at this locus since the divergence between the Triticeae and the other grasses (Li and Gill 2002). Similarly, substantial rearrangements were observed at the *adh1* gene in maize, rice, and sorghum. Nine genes were found in a colinear order between maize and sorghum but three additional genes were present in sorghum in an interval that was more than 3 fold larger in maize (Tikhonov et al. 1999). Additional comparisons with

Table 1. BAC libraries of cereal genomes. Nb clones = Number of clones. References are given for BAC libraries that represent more than 3 fold genome coverage. The web address of the Clemson University Genomics Institute (CUGI) is given below the table

Plant species	Nb Clones	coverage	insert size (kb)	Reference
Zea mays B73	247680	14	137	CUGI*
Zea mays BSSS53	70000	3	100	Song et al. 2001
Zea mays B73	331776	20	130	Yim et al. 2002
Sorghum bicolor	13440	2.8	157	Woo et al. 1994
Sorghum bicolor	110592	17	120	CUGI*
Sorghum propinquum	73728	13	132	CUGI*
Oryza sativa cv IRBB21	11000	3.5	125	Wang et al. 1995
Oryza sativa cv Teqing	14208	4.4	130	Zhang et al. 1996
Oryza sativa cv IR64	18.432	3.3	107	Yang et al, 1997
Oryza sativa diverse cultivars	8 libraries	2.5–10	107–150	CUGI*
Oryza wild species	12 libraries	10.8–19.3	123–161	Ammiraju et al. 2006
Triticum monococcum cv DV92	276480	5.6	115	Lijavetzski et al. 1999
Triticum urartu	163200	3.7	110	Akhunov et al. 2005
Aegilops tauschii	144000	3.7	119	Moullet et al. 1999
Aegilops tauschii	181248	4.1	115	Akhunov et al. 2005
Aegilops speltoides	237312	5.4	115	Akhunov et al. 2005
Triticum durum cv Landgon	516096	5.1	131	Cenci et al. 2003
Triticum aestivum cv Glenlea	650,000	3.1	79	Nilmalgoda et al. 2003
Triticum aestivum cv Chinese Spring	1000320	7	140	Allouis et al. 2003
Triticum aestivum cv Chinese Spring	395136	3.4	157	Shen et al. 2005
Triticum aestivum 3B (Chinese Spring)	67968	6.2	103	Safar et al. 2004
Triticum aestivum 1D, 4D, 6D (Chinese Spring)	87168	3.4	85	Janda et al. 2004
Triticum aestivum 1BS (Chinese Spring)	65,280	14.5	82	Janda et al. 2006
Hordeum vulgare	313344	6.3	106	Yu et al. 2000
Secale cereale	373632	6	125-150	Shi and Gustafson (Pers.Com.)

* **http://www.genome.clemson.edu/groups/bac/**

rice revealed a very complex history of rearrangements at this locus involving differential gene translocations, insertions, and deletions. It also indicated that the rice genome is highly stable whereas maize has undergone a high frequency of gene deletions during its evolution (Tarchini et al. 2000; Bennetzen and Ramakrishna 2002; Ilic et al. 2003). Since these first studies, several other microcolinearity studies have been performed in cereals at different orthologous loci that harbor genes involved in resistance (e.g. *Lrk*, *Rp1*, *Rph7*), development (e.g. *Vrn1*, *lg2/lrs1*, *PhdH1*), and quality (e.g. *Zein*, *Ha*, *r/b*, *Glutenin*) (Table 2). All of these studies confirmed that many small-scale genic rearrangements, such as single or multiple gene insertions and/or deletions, tandem duplications, inversions, and translocations that were previously overlooked by comparative mapping, occurred during the evolution of the cereal genomes (for reviews see Feuillet and Keller 2002; Bennetzen and Ramakrishna 2002; Devos 2005). Depending on the chromosomal location and type of locus, the extent of conservation can vary from single gene differences to complete disruption of colinearity due to translocations. The comparisons have been very helpful in identifying some of the mechanisms involved in the rearrangements that have shaped the grass genomes during their evolution. It clearly showed that retroelements have played a major role in the expansion of the large genomes of maize, barley, and wheat through nested insertions and that numerous small deletions caused by unequal homologous recombination and illegitimate recombination have counteracted this expansion (for a recent review see Bennetzen et al. 2005).

Table 2. Inter and intra specific microcolinearity studies in cereals:
List of the various loci that have been compared at the sequence level through BAC sequencing between different cereal species. The asterix indicates comparisons that have also been performed at the intraspecific level

Locus	Compared plant species	Reference
Lrk	Wheat, barley, maize, rice	Feuillet and Keller 1999
Rph7	Barley*, rice	Brunner et al. 2003
		Scherrer et al. 2005
adh1/adh2	Maize, sorghum, rice	Tikhonov et al. 1999
		Ilic et al. 2003
		Tarchini et al. 2000
Vrn1	Wheat, barley, sorghum, rice	Ramakrishna et al. 2002a
lg2/lrs1	Maize, rice	Langham et al. 2004
sh2/a1	Maize, sorghum, rice, wheat	Chen et al. 1997
		Chen et al. 1998
		Li and Gill 2002
		Bennetzen and Ma 2003
Zein	Maize*, sorghum, rice	Song et al. 2002
		Song and Messing 2003
Ha	Barley, rice, wheat*	Caldwell et al. 2004
		Chantret et al. 2005
r1/b1	Maize, sorghum, rice	Swigonova et al. 2005
Orp1/Orp2	rice sorghum	Ma et al. 2005
Rp1	Maize, sorghum	Ramakrishna et al. 2002b
Phd-H1	Barley , rice	Dunford et al. 2002
Glutenin	Rice, wheat*	Wicker et al. 2003
		Gu et al. 2004
Bz	Maize*, rice	Fu and Dooner 2002
		Lai et al. 2005

3.2. Intraspecific Comparisons: Looking at Less Than a Few MY of Speciation

With the development of BAC libraries from different subspecies in rice and from wheat species at different ploidy levels, microcolinearity studies have been performed within species with divergence times of less than 5 million years. Several studies compared BAC sequences of the homoeologous A, B, and D genomes of wheat that are estimated to have diverged from a common ancestor between 2.5 and 4.5 million years ago (Huang et al. 2002). The first two studies compared orthologous glutenin gene loci in the A and B genomes of *T. durum* and the D genome of *Ae. tauschii* (Gu et al. 2004), as well as the homoeologous A genomes of *T. durum* and *T. monococcum* (Wicker et al. 2003). In both cases, conservation between the different genomes was restricted mostly to the gene space; whereas, sequence rearrangements in the intergenic regions were due mainly to the insertions of retrotransposons and illegitimate recombination events. BAC sequences that originate from two different haplotypes identified at the disease resistance locus *Lr10* were compared in diploid (*T. monococcum*), tetraploid (*T. durum*), and hexaploid wheat (*T. aestivum*) (Isidore et al. 2005). Insertions as well as deletions and unequal crossing over between transposable elements have reduced the overall percentage of sequence conservation between the three orthologous regions to 33% and very few elements were conserved in the intergenic regions even within the same haplotype. A good degree of conservation of the gene content and order was found between the diploid and tetraploid sequences that belong to the same haplotype; however, a large rearrangement involving a deletion followed by a large inversion was observed in the second haplotype in hexaploid wheat. Finally, this work enabled the determination of the estimated divergence time between the A genomes of wheat at 2 MY (Isidore et al. 2005). Comparative sequencing was also performed at the *Ha* locus that controls grain hardness in wheat. Orthologous BACs were compared in *Triticum aestivum*, *Triticum durum* and the diploid relatives *Triticum monococcum* and *Aegilops tauschii* (Chantret et al. 2005). Rearrangements, such as transposable element insertions, sequence deletions, duplications, and inversion involving illegitimate recombination, were shown to be responsible for the major differences observed between the same genomes at different ploidy levels (Figure 3). These comparisons provided an explanation for the previously reported loss of the *Pina* and *Pinb* genes in tetraploid wheat, through large deletions that occurred independently in the A and B genomes following polyploidization (Figure 3). Together, these data allowed the identification of major mechanisms involved in both expansion and reduction of the wheat genomes. They suggest that TEs have been very active since the divergence of the A, B, and D genomes as well as after polyploidization and that illegitimate DNA recombination, leading to various genomic rearrangements, is one of the major evolutionary mechanisms in these genomes.

Draft or complete sequences from the two major rice subspecies *Oryza sativa ssp. Japonica* and *Oryza sativa ssp. indica* were released between 2002 and 2005 (Goff et al. 2002; Yu et al. 2002; IRGP, 2005) permitting comparative studies between genomes that diverged less than 0.5 MYA (Ma and Bennetzen 2004). Before the release of the genome drafts, small BACs isolated from *japonica* and *indica* subspecies at the maize zein storage protein z1C-1 orthologous locus were compared. This indicated a nearly complete conservation between the sequences of the two rice subspecies with only few differences in intergenic regions (Song et al. 2002). Feng et al. (2002) aligned 2.3 Mb of orthologous chromosome 4 sequences from *indica* and *japonica*. Again, extensive conservation was observed but the larger size of the alignment identified deviation from colinearity even within genic regions. More than 9,000 Single Nucleotide Polymporphisms (SNPs) as well as 63 and 138 Insertion/Deletions (Indels) were observed for the *indica* and *japonica* sequences, respectively. Together, these data suggest that the rice genome has been very stable since the divergence between the *japonica* and *indica* subspecies. In the near future, additional rice genomes will be investigated though the Oryza Mapping Alignment project (OMAP; http://www.omap.org/index.html) and extensive sequence comparison between the 10 different rice genome types will provide very significant insights on the evolution of the genomes within a species.

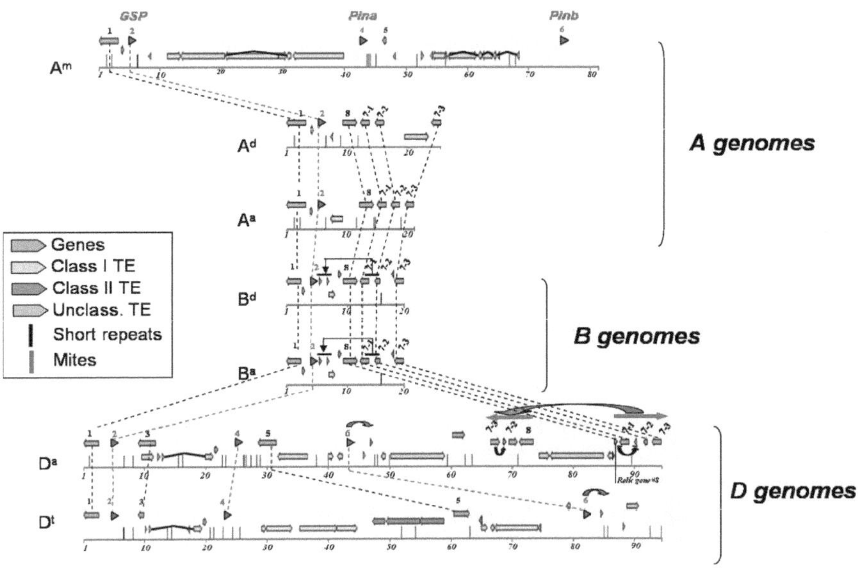

Figure 3. Microcolinearity studies at the Hardness locus in wheat (adapted from Chantret et al. 2005) Schematic representation of BAC sequence comparisons at the wheat *Ha* locus from the A (A^m : *T. monococcum;* A^a : *T. aestivum*; A^d : *T. durum*) , B (B^a : *T. aestivum* ; B^d : *T. durum*) and D (D^a : *T. aestivum*; D^t : *Ae. tauschii*) genomes in different polyploidy context. Genes (CDS) (light blue), class I TEs (yellow), class II TEs (green), unclassified elements (gray), MITEs (red), and short repeats (black) are indicated. Orthologous CDS between the different genomes are linked by dashed bars whereas CDS duplications and deletion events are indicated by arrows. The *GSP*, *Pina* and *Pinb* genes that were lost in tetraploid wheat following polyploidisation are highlighted in red and are numbered respectively as gene 2, 4, 6

Preliminary data using BAC end sequences from the existing 12 BAC libraries revealed a linear relationship between the genome size and the LTR retrotransposon content thereby indicating the predominant role of this class of repeats in the evolution of the rice genomes (Ammiraju et al. 2006). Recently, Monna et al. (2006) identified 7,805 polymorphic loci (SNP, Indels) within 1,117 predicted intergenic sequences that were obtained from eight rice cultivars and a wild *O.rufipogon* accession and demonstrated the potential of intraspecific comparisons for association studies in rice.

With the release of the sequence drafts, whole genome sequence comparisons have been performed between the *japonica* and *indica* subspecies (Feltus et al. 2004; Shen et al. 2004). Depending on the stringency of the analysis, 400,000 to more than one million SNPs and indels have been identified and are available now in public databases (http://www.plantgenome.uga.edu/snp; http://shenghuan.shnu.edu.cn/ricemarker). Beyond the information that these comparisons provide on the divergence between the subspecies, they represent an extremely useful source of markers for genetic mapping and map-based cloning in populations derived from crosses between the two subspecies since the Indels are conserved beyond the sequenced cultivars within each subspecies (Shen et al. 2004).

Beside BAC sequence or whole genome sequence comparisons, large scale comparisons can be performed as well using BAC end sequences (BESs). Recently, BAC ends and physical maps have been used to compare additional *indica* and *japonica* cultivars to the Nipponbare sequence. For example, the comparison of 12,170 BESs from the *indica* variety Kasalath with the 12 Nipponbare pseudomolecules identified an average SNP rate of 0.71% on chromosome 1, 4 and 10 (Katagiri et al. 2004). The Nipponbare chromosome 3 sequence also was compared with a reconstructed chromosome 3 from the wild relative *Oryza nivara.* No major rearrangements were observed but the results of the alignments of paired BESs suggested that the *japonica* chromosome is 21 % larger than *nivara* chromosome 3 and that there is high variability in the intergenic regions. In wheat, 11 Mb of BAC end sequences have been very obtained recently from chromosome 3B of hexaploid wheat (Paux et al. 2006). Comparison with 2.9 Mb of random sequences from *Aegilops tauschii*, the D- genome donor of bread wheat (Li et al. 2004), suggested that the larger size of the B-genome compared to the D-genome of wheat is due to a higher content in repetitive elements and provided insight into which families of TE are responsible mostly for differential expansion of the homoeologous wheat genomes during evolution (Paux et al. 2006).

3.3. Intervarietal Comparisons: Looking at Less than 10,000 Years of Speciation

The observed absence of colinearity at the intraspecific level between recently diverged species raised the question of sequence rearrangements within different lines or varieties. A first study, comparing BAC sequences from the bronze (*bz1*) locus in two maize inbred lines (McC and B73), surprisingly revealed as dramatic of differences between the two lines as those observed between orthologous loci in two different species (Fu and Dooner 2002). Violation of microcolinearity did not concern only the length and composition of the intergenic regions, it affected also the gene density and content. Four out of ten genes present in McC were absent in B73. Such a dramatic variation in gene copy number and insertion of different transposable elements was found also at the orthologous loci containing the zein storage protein gene cluster *z1C-1* between the maize inbred lines B73 and BSSS53 (Song and Messing 2003). In addition, expression analysis of the genes present at the *z1C-1* locus demonstrated that even though approximately the same number of genes is expressed in the two inbred lines, only three of the genes exist in both lines. Recently, Brunner et al. (2005) extended these studies by comparing DNA sequences from four allelic chromosomal regions in the Mo17 and B73 maize inbreds. Almost 50% of the total sequence analyzed was not shared between the two inbreds. Most of it consisted of LTR-retrotransposons and other mobile elements but there were also considerable differences in the genic sequences. In total, 23 out of 68 putative genes (34%) were present only in either Mo17 or B73. In contrast to the *z1C-1* locus where half of the non-shared sequence originated from extensive local duplications, the non-shared sequences corresponded to clusters of genes fragments. Interestingly, in contrast to the shared genes, the non-shared genes were not present at colinear positions in rice suggesting that they likely originate from insertions rather than deletions. Only very recently, Morgante et al. (2005) identified the non shared pseudogene clusters as part of non-autonomous Helitrons, a new type of eukaryotic transposable elements. These transposons appear to have copied and incorporated genic segments from different genomic locations of the host, clustered them together and duplicated these arrangements via a copy-past transposition mechanism to non-allelic loci across the maize genome (for references see Morgante et al. 2005).

More than 300 kb of sequence spanning the *Rph7* leaf rust disease resistance gene have been compared recently between two barley cultivars (Scherrer et al. 2005). Colinearity was restricted to five genic and two intergenic regions representing less than 35% of the two sequences. In each interval separating the conserved regions, the number and type of repetitive elements were completely different and a single gene that was identified later as an helitron (C. Feuillet, personal communication) was absent in one cultivar. In both cultivars, the non-conserved regions consisted of ~53% repetitive sequences mainly represented by long-terminal repeat retrotransposons that were inserted less than 1 million years ago. PCR-based analysis of intergenic regions at the *Rph7* locus and at three other independent loci in 41 *H. vulgare* lines indicated rapid and recent divergence at homologous loci in the cultivated barley genome (Scherrer et al. 2005). The rearrangements observed in barley were less dramatic than those found between maize inbreds as well as those observed between rice subspecies suggesting that maize has a highly unstable genome compared to the other grasses.

No comparative analysis has been performed yet between different wheat varieties of the same ploidy level. However, BAC libraries from three hexaploid *T. aestivum* cultivars are available (Table 1) and comparative studies are underway (http://www.intl-pag.org/14/abstracts/PAG14_W30.html). It will be interesting to compare the rate and mechanisms of evolution at a similar time scale in barley and wheat, two species that are closely related but that have very different population histories. Future intraspecific comparisons will provide greater understanding of the evolutionary differences between the cereal genomes.

4. GRASS GENOME DUPLICATION

Polyploidy, *i.e.* the presence of multiple sets of chromosomes in the same nucleus, is an important evolutionary mechanism in angiosperms. The analysis of EST collections, macro- and microcolinearity studies, as well as whole genome sequence comparisons clearly indicate that genome duplications have been a significant driving force in the evolution of plant genomes. The identification of a common duplication event between monocots and dicots suggests that all angiosperms are actually ancient polyploids (Bowers et al. 2003). Polyploid species are found also in cereals such as wild rice (tetraploid) and wheat (tetraploid, hexaploid), while maize has been recognized as an ancient tetraploid (Gaut 2001).

Cytological studies had suggested long ago that diploid cereal genomes are ancient polyploids (McClintock 1930; Ting 1966). In the 1990‘s, the use of molecular markers revealed the presence of duplicated loci on the genetic maps in different cereals suggesting ancestral genome duplications and polyploidization events in the history of species that are now identified as diploids. In 1993, Ahn and Tanksley found 72% of the genetic markers at two loci on their reference genetic map in maize, while in rice, pairing of RFLP mapping suggested that chromosome 1 and 5 (Kishimoto et al. 1994) as well as chromosome 11 and 12 (Nagamura et al. 1995) were ancient duplicates. With the release of the rice genome sequence drafts in 2002, whole genome duplication analysis was undertaken. Analysis of the 370 Mbp of *Oryza sativa* ssp *indica* sequence revealed 10 paralogous blocks involving 47% of the annotated genes. Ancient duplication events were estimated to have occurred 70 MYA, the more recent one involving chromosomes 11 and 12, 5 MYA (Yu et al. 2002; Yu et al. 2005). Duplications involving a large number of genes along the length of chromosome representing 65% of the sequenced genome with 18 pairs of duplicated segments were identified in the *japonica* sequence (Paterson et al. 2004; IRGSP 2005). It confirmed that most of the duplicated segments result from an ancient whole genome duplication event that occurred before the radiation of the cereal genomes and that the largest duplicated fragment involving rice chromosomes 11 and 12 is independent and more recent. Identification of duplicated blocks in the rice genome sequence has been updated recently by TIGR based on 42,662 non-transposable element related rice protein sequences and using specific alignment criteria (http://www.tigr.org/tdb/e2k1/osa1/segmental_dup/index.shtml). The origin of the duplications in rice are still a subject of controversy as it has been proposed that rice is either an ancient aneuploid (Vandepoele et al. 2003) or an ancient paleopolyploid (Paterson et al. 2003). These two interpretations are based on distinct considerations about gene tandem duplications within blocks of paralogous genes that bias the dating procedure of the duplication events. In other species such as wheat, maize, and several dicots, evidence for aneuploidy or paleoploidy has been suggested by comparative analysis of EST databases (Blanc and Wolfe 2004). In many genomes, genome duplication is followed generally by diploidization that involves gene loss; one copy may be retained at one locus in the first genome but is lost in the other genome while the second copy is retained. Consequently, diploidization often results in disruption of microcolinearity and the observation that genes that are not found at orthologous positions are nevertheless present elsewhere in the genomes.

Even if genetic mapping can indicate ancient duplications, there is nothing more powerful than a genome sequence to identify the origin and the mechanisms of ancient duplications. There is no doubt that the projects that are underway for sequencing the maize and sorghum genomes (see chapter 6) will provide additional information and shed new light into the duplication events that have affected differentially the grass genomes during their evolution.

5. COMPARATIVE GENOMICS AS A TOOL FOR GENE DISCOVERY AND MARKER DEVELOPMENT

Comparative genomic studies have increased knowledge about the level of conservation between the cereal genomes and led to the generation of genomic tools that can be used to define efficient strategies for genetic studies and gene isolation in these genomes.

5.1. Colinearity-Based Gene Cloning in Cereals

In some cases, the conservation of sequence at orthologous positions between the genomes can reflect the conservation of a gene with a similar function between species. Early comparative genetic studies using RFLP identified that a number of genes and quantitative trait loci (QTL) for developmental and domestication traits, such as shattering, plant height, vernalisation, flowering time, row number, and kernels per row, were at orthologous positions in cereal genomes (Lin et al. 1995; Paterson et al. 1995; Bailey et al. 1999). The concomitant discovery of colinearity between rice and the other cereals opened up opportunities to use the rice genome data to support positional cloning of genes from the other genomes in a so called "cross genome map-based cloning" approach even before the rice genome sequence was completed (Killian et al. 1997). The best example of colinearity in gene type and function and in the efficient use of rice for direct gene cloning in other cereals is the isolation of the "green revolution" dwarfing genes *Sd1* (Monna et al. 2002), *Rht-1* in wheat, *D8* in maize (Peng et al.1999). In the last years, the isolation of genes by map-based cloning in barley, wheat, and maize has revealed as well examples of conservation between genes at orthologous positions in cereals. Thus, the wheat vernalisation gene, *Vrn1* (Yan et al. 2003) and the barley photoperiod *PPD-H1*gene have orthologous genes in rice (Turner et al. 2005). In other cases, gene conservation has been suggested based on the conservation of genetic locations for similar phenotypes e.g. the maize *barren stalk1* mutation was mapped in a region colinear with the rice *lax panicle* gene (Gallavotti et al. 2004). In regions where microcolinearity is high, candidate genes can be identified directly from the rice sequence even if the target trait has not been mapped at a colinear position in rice. This has been used successfully to support the isolation of the powdery mildew resistance gene *Ror2* (Collins et al. 2003) and the *sw3* dwarfism gene in barley (Gottwald et al. 2004). In other cases, similar functions do not seem to be associated with similar genes. For example, in a study comparing QTL for heading time in rice and barley, Griffiths et al. (2003) have shown that in rice a number of QTL belong to the CONSTANS gene family but that in barley none of the homologous CONSTANS genes are associated with any of the known QTL for flowering time. Thus, generally genes and QTL involved in developmental processes and that have been selected during domestication show good conservation between cereal genomes and rice genes are good candidates for direct gene isolation.

In contrast, other types of genes do not show colinearity between the cereal genomes. Indeed, there is no example of colinearity for disease resistance (R) genes in grasses and, so far, map-based cloning of R genes in cereals was not profiting significantly from the rice genome information. The non-syntenic location of these genes between cereals has been already identified through comparative genetic analysis (Leister et al. 1998) and, in many cases, the attempts to use colinearity with rice for isolating R genes have revealed the limits of colinearity between the cereal genomes. The first example that questioned the extent of the utility of using rice for map-based cloning of disease R genes was the work with the barley stem rust resistance gene *Rpg1*. Despite a certain degree of some colinearity retained at the orthologous locus in rice (Killian et al. 1997), no orthologous gene is present in the rice genome and map-based cloning of *Rpg1* has been achieved in barley (Brueggeman et al. 2002). In some cases, such as with the leaf rust *Lr10* and the powdery mildew *Pm3* fungal disease R genes, the rice genome contains genes homologous to the wheat genes but at non-orthologous positions, indicating massive genome rearrangements (Guyot et al. 2004). Both of these genes were cloned using alternative strategies (see below). The only known exception to this lack of colinearity between R genes has been reported recently by Chen et al. (2005) who showed that a QTL conferring resistance to the blast fungus *Magnaporthe grisea* is conserved in rice and barley at the same homologous location and with the same race specificity.

Even if the gene is not present at its orthologous position in rice, the flanking genes are often conserved enough to provide a collection of markers than can be used to saturate the target region in the other cereal genomes. For example, rice ESTs were used to reduce the genetic interval around the disease R loci *Rpg1* and *Rph7* in barley to a density that allowed initiation of chromosome walking in barley

(Brueggeman et al. 2002, Brunner et al. 2003). Colinearity between rice, sorghum, and sugarcane was used also to generate markers and saturate the genetic region for the map-based cloning of *Bru1,* a major leaf brown rust resistance gene from sugarcane (Asnaghi et al. 2004). There are now many additional examples of the use of rice EST derived markers to saturate genetic regions in other cereals and this approach is routinely used now in laboratories that are involved in cereal gene cloning worldwide (Collins et al. 2003; Yan et al. 2003; Yan et al. 2004; Gallavotti et al. 2004; Bortiri et al. 2006). Recently, a new model species, *Brachypodium*, has been proposed (Draper et al. 2001; Vogel et al. 2006) for temperate cereals such as wheat and barley and it was used successfully in combination with rice to isolate *Ph1*, one of the key gene controlling pairing in polyploid wheat (Griffith et al. 2006).

In cases where colinearity is too low, alternative strategies, such as transposon-tagging, the use of more closely related species, or direct map-based cloning in the species of interest have to be applied. Such an example is given by the maize *Ramosa1* gene, that controls the architecture of the tassel. This gene is specific for the *Andropogoneae* tribe and is lacking in rice. It was isolated recently using a transposon-tagging strategy (Vollbrecht et al. 2005). In wheat, so called "subgenome map-based cloning" (Stein et al. 2000) has been used to isolate the *Lr10* and *Pm3* disease R genes both of which are located on the short arm of chromosome 1A in a non colinear region with rice (Guyot et al. 2004). In this strategy, genetic mapping was performed in hexaploid wheat and physical chromosome walking was done with BACs from the A genome diploid relative *T. monococcum* (Feuillet et al. 2003; Yahiaoui et al. 2004).

5.2. Gene Annotation and Marker Development

Complete genome sequences provide the basis for understanding the gene structure and function within species. As genes are the most conserved features between genomes, the availability of a genome sequence can help greatly to predict genes in other genomes. Even between distantly related genomes such as the one of rice and *A. thaliana* the ancestors of which diverged 200 million years ago and do not show extensive macrocolinearity, a large number of genes have been conserved (Salse et al. 2002). Thus, the rice genome sequence represents a unique tool to support gene annotation in other cereals, a critical issue in view of the sequencing of additional cereal genomes expected to be completed in the next decade (see chapter 6). Conversely, the alignment of ESTs from other cereal species with the rice genome sequence can help to predict new genes from rice. The generation of a large set of full-length cDNAs in rice (The Rice Full-Length cDNA Consortium 2003) is particularly useful for gene annotation of other cereal genomic sequences as it can help to validate intron/exon boundaries and can be used to train gene predictors. The identification of intron/exon boundaries is helpful as well for the development of new markers. Indeed, SNP frequencies are higher in introns than in exons and the possibility to design PCR primers that amplify intronic sequences improves polymorphism detection in species, such as wheat, that chronically suffer from a lack of polymorphism. This concept has been applied recently in pearl millet by Bertin et al. (2005) where millet ESTs were aligned against the rice gene sequences to predict the location of introns and to amplify products across the sequences followed by the detection by Single Strand Conformational Polymorphism (SSCP). The SSCP-SNP marker technique has great potential for the development of COS (Conserved Orthologous Set) markers for comparative mapping in cereals as comparisons can be performed between sequences from many different species and used to define perfect match primers.

5.3. Functional Comparative Genomics in Cereals

The development of genomic resources, in particular EST collections for several crops, has enabled genome-wide studies of gene expression based on various types of DNA chips. DNA chips are now available for rice, wheat, barley and maize. In the beginning, the lack of standardization and the use of home-made chips made it nearly impossible to compare different experiments. However, this situation is changing rapidly as a result of improvements in the technology, the commercialisation of high quality DNA chips, and strict requirements by most journals for standardisation of data presentation (Brazma and Vilo 2001). Although there are still some limitations (e.g., poor annotation, incomplete representation of the genome for most crops, different kinetics of development and phenotypic stages, variable experimental conditions), it has become possible to compare gene expression profiles in similar physiological and biological situations in different cereals. Since the various crops often show different adaptive responses, these comparisons should be particularly helpful to unravel key regulatory genes and investigate control of their expression. A number of transcriptomic studies of stress responses using various types of DNA chips have been reported in various cereals, but they have not yet been extensively compared. Recently, a comparative micro-array analysis between winter and spring wheat with a set of 974 unigenes led to the identification of 65 candidate genes differentially expressed under cold treatment (Gulick et al. 2005). Expression profiling experiments in cereals will accumulate and be stored in databases (http://barleybase.org/; http://www.ricearray.org/; http://www.maizearray.org/) enabling thereby the rapid development of meta-analysis of expression patterns across cereal species.

6. CEREAL GENOMES SEQUENCING

The recognition of rice as a model for cereal crops positioned it for genome sequencing in 1997 through the International Rice Genome Sequencing Project (IRGSP) that aimed at sequencing the *japonica* cultivar Nipponbare through a clone-by-clone shotgun (CBC) approach. In addition to the first IRGSP sequences (Sasaki et al. 2002), drafts were obtained from whole genome shotgun (WGS) sequencing from the same cultivar by Syngenta (Goff et al. 2002) and from the *indica* cultivar

'93-11' by the Beijing Genomics Institute (BGI) (Yu et al. 2002). Recently, the complete and accurate sequencing of Nipponbare has been achieved (International Rice Genome Sequencing Programme, 2005). Already, these resources have boosted cereal genomics and rice breeding and have demonstrated that sequencing of large genomes is feasible. They also have paved the way for further cereal genomes to be sequenced by demonstrating the relative advantages and limits of the CBC and WGS strategies (Yu and Wing 2004; Paterson 2006). Although the drafts using Whole Genome Shotgun sequencing approaches were released before the BAC-by-BAC sequenced genome, the later approach has provided a much more accurate and useful sequence. On the other hand, sequences that are refractory to cloning are represented in WGS and therefore a hybrid approach using a combination of both methods appears to be the best strategy to follow (Green 2001).

As in animals, sequencing additional cereal genomes will accelerate and enhance the impact of comparative genomics in the identification of key regulators that underlie genetic variation and adaptation of these species to their environment and support their improvement. Recently, an additional small sized genome, *Brachypodium distachyon* (diploid, 10 pairs of chromosomes, a genome size of 335 Mbp) that belongs to the Pooideae family and is therefore phylogenetically closer to wheat than rice, has been proposed as new model for temperate grasses (Draper et al. 2001), in particular, for the study of basic developmental processes specific to monocots, such as cell wall synthesis (Vogel et al. 2006). BAC libraries have been constructed from two *Brachypodium distachyon* ecotypes (ABR1 and ABR2) and the colinearity between Brachypodium, rice, and other Poaceae sequences has been investigated through PCR screening and fluorescent *in situ* hybridization (Hasterok et al. 2006). In addition, the US DOE/JGI has announced its intention to provide a 8X WGS sequencing of this genome. Although the extent of colinearity needs to be studied further, this new resource should facilitate physical mapping and sequencing of the Triticeae genomes.

After rice, the next cereal crop with a relatively small genome (738 Mb) that has been chosen for sequencing is sorghum (diploid, 10 chromosome pairs), a species of major economic importance that can also serve as a reference genome for tropical cereals. Physical maps of *Sorghum bicolour* and *Sorghum propinquum* have been genetically anchored (Bowers et al. 2005) and methyl filtration sequences have been produced for this species (Bedell et al. 2005). These resources are being used in combination with a 8X WGS that is being done by the US DOE/JGI to produce genetically oriented pseudomolecules (http://www.jgi.doe.gov/sequencing/why/CSP2006/sorghum.html).

While there have been significant reductions in sequencing costs over the past five years, the complete sequencing of larger and more repetitive cereal genomes such as those of maize and wheat remains relatively expensive given the limited financial resources available for plant genomics. Thus, for these genomes, currently the strategy is to first focus on sequencing the gene space while waiting for a revolution in sequencing technologies that will significantly reduce sequencing costs (Service, 2006) and can still handle repeated sequences. Gene enrichment (GE) methods such as methyl and Cot filtration have been used to successfully increase the representation of the genic regions in maize and wheat leading to enrichment factors up to 13.7 fold (Whitelaw et al. 2003; Yuan et al. 2003; Springer et al. 2004; Lamoureux et al. 2005; Rabinowicz et al. 2005). Combining GE (5X) with low redundancy BAC (3X) and whole genome shotgun (2X) sequences is therefore a promising approach to ensure a sufficient coverage (˜10X) of the genic regions for these genomes (Rabinowicz and Bennetzen 2006).

For maize (diploid, 2.4 Gb, 10 pairs of chromosomes), genetic maps and physical maps have been established already (Coe et al. 2002) and a "gold standard" for genome sequencing has been defined by the maize community. This "gold standard" is the complete sequence and structure of all maize genes with their locations identified on both the genetic and physical maps of maize using B73 as the reference. A project is now under way (http: //www.nsf.gov/awardsearch/showAward.do?AwardNumber=0527192) and sequencing will be carried out using a combination of WGS, BAC sequencing, and GE sequences that will be completed by MSLL (Methylation Spanning Linker Libraries) and HMPR (HypoMethylated Partial Restriction libraries) sequencing to ensure that most of the gene space is captured (Rabinowicz and Bennetzen 2006).

With 17 Gb (hexaploid, 21 chromosome pairs), bread wheat, is 40 times larger than rice and 6 times larger than maize. Its recent polyploidization represents an additional difficulty in physical mapping and sequence assembly but also makes it a very suitable model to study the effect of polyploidy on genome evolution and for determining the best strategies for sequencing polyploid genomes. In 2004, a workshop was held to identify the foundation needed for sequencing wheat (Gill et al. 2004) and an International Wheat Genome Sequencing Consortium (IWGSC) was created in 2005 with the goal of establishing a physical map of the 21 chromosomes of hexaploid wheat and sequencing the wheat gene space in the first place (http://wheatgenome.org). A number of pilot projects are underway currently to determine the best strategy for the construction of the physical map: whole genome fingerprinting and/or chromosome specific strategies. In addition, a project to construct a physical map of *Ae. tauschii*, the wild diploid D genome donor species of hexaploid wheat, is underway (http://wheat.pw.usda.gov/PhysicalMapping/) and will serve as a good framework for assembling the hexaploid D genome chromosomes. Like wheat and rye, barley (diploid, 5 Gb, 7 pairs of chromosomes) is a member of the Triticeae, and therefore also represents a potential target for genome sequencing in this major tribe. A physical mapping project is in preparation (http://pgrc.ipk-gatersleben.de/etgi/publications/whitepaper_barley_physmap_and_sequence.pdf) and a project to identify, fingerprint and contig BACs containing expressed genes (ESTs) was launched in 2003 (http://phymap.ucdavis.edu:8080/barley/).

A key question at this time is the extent to which the rice genome sequence can facilitate the construction of the physical maps of other cereal genomes. BAC end sequencing followed by *in silico* mapping on the rice genome has been used efficiently to order BACs from *Oryza sativa ssp Kassalath* on the japonica sequence as part of the Oryza Map Alignment Project (Ammiraju et al 2006) and there is no doubt that this approach will be very effective for genomes that are closely related. However, the level of rearrangements between rice and the other cereal genomes and the high amount of repeated sequences that are not conserved between these genomes will likely hamper any useful alignment of BES from wheat, maize, sorghum, or barley to the rice sequence. Thus, the utility of the rice genome sequence in constructing physical maps in other cereals will be limited more than likely to providing a source of additional markers to anchor the physical maps to the genetic maps. For this reason, it is important to obtain the sequence from genomes in different branches of the phylogenetic tree (Paterson, 2006). In this regard, sorghum will likely be very useful for maize, and it remains to be determined the extent to which extend Brachypodium can serve as a reference for assembling the physical maps of wheat and barley.

7. SUMMARY AND OUTLOOK

The past decade has witnessed the first breakthrough in cereal genomics and has illustrated the power of comparative studies in these economically important species. Comparative studies in cereals led to improved genetic maps, the development of accurate markers for breeding, and the map-based isolation of the first genes of agronomic interest. It also provided insight into the evolution of the cereal genomes, unravelling some of the major mechanisms that have shaped their evolution during the past 50-70 million years, and highlighted the differences in their stability. However, with only one species sequenced, the power of comparative genomics has been limited mostly to the identification of structural differences between the cereal genomes. Over the next ten years, we can expect significant breakthroughs as the sequencing of additional cereal genomes will allow the identification of elements that have been conserved during evolution and that have a functional significance as has resulted from animal comparative genomics. In mammals, comparisons of human, mouse and rat genomes indicated that about 3% of the genome corresponding to non protein coding sequences are ultra conserved across genomes and have been under purifying selection (Bejerano et al. 2004). This has led to the idea that while waiting for high-quality genomic sequences from many other mammals it should be possible to detect highly conserved functional elements by comparing low-redundancy sequence data (about 2 fold redundancy) that would be obtained from species chosen to maximize the representation of the mammalian genomes (Margulies et al. 2005). Paterson (2006) has suggested recently that such a "phylogenetic shadowing" concept could be applied to the angiosperms and that in addition to those genomes currently underway, low redundancy sequencing of 16 additional genomes chosen across 28 taxa might provide some clues about conserved functional elements in plants. In addition, the author suggested that sequencing wild and domesticated representatives would provide information about the genomic features that underlie domestication. Conserved non coding regions (CNS), i.e. conserved sequences located in the non coding regions of genes (introns or upstream regulatory sequences), have been surveyed in cereals (maize vs rice) and mammals (human vs mouse) by Freeling and collaborators (Kaplinsky et al. 2002; Inada et al. 2003). They showed that CNSs are more abundant in regulatory genes such as transcription factors and that despite similar divergence times from their common ancestors, grass genes have dramatically fewer (5- to 20-fold) and smaller CNSs than mammalian genes. One possible explanation is that in contrast to vertebrate genomes, plant genomes have been subjected to several complete rounds of whole genome and/or segmental duplications and polyploidization events that have affected profoundly their organisation with the subfonctionalisation of duplicated genes leading to a greater loss of CNS per gene (Lockton and Gaut 2005). Future comparative genome sequencing in cereals will help to confirm these features and provide clues to the relationship between CNS, regulation, and phenotypes. In addition, CNSs also represent great targets for PCR primer binding sites that can be used to design a new generation of COS markers for high density mapping in cereal genomes.

For the past decade, comparative genomics in cereals has pioneered in many ways the field of plant comparative genomics. There is no doubt that with the ongoing efforts, comparative studies in cereals will continue to provide invaluable information for a better understanding of the adaptation of plants to their environment and open new areas for breeding strategies, plant protection, and conservation of biodiversity.

REFERENCES

Ahn S, Tanksley SD (1993) Comparative linkage maps of the rice and maize genomes. Proc Natl Acad Sci USA 90:7980–4

Akhunov ED, Akhunova AR, Dvorak J (2005) BAC libraries of *Triticum urartu*, *Aegilops speltoides* and *Ae. tauschii*, the diploid ancestors of polyploid wheat. Theor Appl Genet 111:1617–22

Allouis S, Moore G, Bellec A, Sharp R, Faivre Rampant P, Mortimer K, Pateyron S, Foote TN, Griffiths S, Caboche M, Chalhoub B (2003) Construction and characterisation of a hexaploid wheat (*Triticum aestivum* L.) BAC library from the reference germplasm 'Chinese Spring'. Cereal Res Comm 31:331–338

Ammiraju JSS, Luo M, Goicoechea JL, Wang W et al (2006) The *Oryza* bacterial artificial chromosome library resource: construction and analysis of 12 deep-coverage large-insert BAC libraries that represent the 10 genome types of the genus *Oryza*. Genome Res 16:140–147

Asnaghi C, Roques D, Ruffel S, Kaye C, Hoarau JY, Telismart H, Girard JC, Raboin LM, Risterucci AM, Grivet L, D'Hont A (2004) Targeted mapping of a sugarcane rust resistance gene (Bru1) using bulked segregant analysis and AFLP markers. Theor Appl Genet 108:759–64

Bailey PC, McKibbin RS, Lenton JR, Holdsworth MJ, Flintham JE, Gale MD (1999) Genetic map locations for orthologous Vp1 genes in wheat and rice. Theor Appl Genet 98:281–284

Bedell J, Budiman MA, Nunberg A, Citek RW et al (2005) *Sorghum* genome sequencing by methylation filtration. PLoS Biol 3:103–115

Bejerano G, Pheasant M, Makunin I, Stephen S, Kent WJ, Mattick JS, Haussler D (2004) Ultraconserved elements in the human genome. Science 304:1321–1325

Bennetzen JL, Ramakrishna W (2002) Numerous small rearrangements of gene content, order and orientation differentiate grass genomes. Plant Mol Biol 48:821–827

Bennetzen JL, Ma J (2003) The genetic colinearity of rice and other cereals on the basis of genomic sequence analysis. Curr Opin Plant Biol 6:128–33

Bennetzen JL, Ma J, Devos KM (2005) Mechanisms of recent genome size variation in flowering plants. Ann Bot 95:127–32

Bertin I, Zhu JH, Gale MD (2005) SSCP-SNP in pearl millet – a new marker system for comparative genetics. Theor Appl Genet 110:1467–72

Blanc G, Wolfe KH (2004) Widespread paleopolyploidy in model plant species inferred from age distributions of duplicate genes. Plant Cell 16:1667–78

Bortiri E, Jackson D, Hake S (2006) Advances in maize genomics: the emergence of positional cloning. Curr Opin Plant Biol 9:164–71

Bowers JE, Chapman BA, Rong JK, Paterson AH (2003) Unravelling angiosperm genome evolution by phylogenetic analysis of chromosomal duplication events. Nature 422:433–438

Bowers JE, Arias MA, Asher R, Avise JA et al (2005) Comparative physical mapping links conservation of microsynteny to chromosome structure and recombination in grasses. Proc Natl Acad Sci USA 102:13206–13211

Brazma A, Vilo J (2001) Gene expression data analysis. Microbes Infect 3:823–9

Brueggeman R, Rostoks N, Kudrna D, Kilian A, Han F, Chen J, Druka A, Steffenson B, Kleinhofs A (2002) The barley stem rust-resistance gene Rpg1 is a novel disease – resistance gene with homology to receptor kinases. Proc Natl Acad Sci USA 99:9328–9333

Brunner S, Keller B, Feuillet C (2003) A large rearrangement involving genes and low copy DNA interrupts the microcolinearity between rice and barley at the *Rph7* locus. Genetics 164:673–683

Brunner S, Fengler K, Morgante M, Tingey S, Rafalski A (2005) Evolution of DNA sequence nonhomologies among maize inbreds. Plant Cell 17:343–360

Buell CR, Yuan Q, Ouyang S, Liu J et al (2005) Sequence, annotation, and analysis of synteny between rice chromosome 3 and diverged grass species. Genome Res 15:1284–91

Calabrese PP, Chakravarty S, Vision TJ (2003) Fast identification and statistical evaluation of segmental homologies in comparative maps. Bioinformatics 19:i74–80

Caldwell KS, Langridge P, Powell W (2004) Comparative sequence analysis of the region harboring the hardness locus in barley and its colinear region in rice. Plant Physiol 136:3177–3190

Cenci A, Chantret N, Xy K, Gu Y, Anderson OD, Fahima T, Distelfeld A, Dubcovsky J (2003) A half million clones bacterial artificial chromosome (BAC) library of durum wheat. Theor Appl Genet 107:931–939

Chalhoub B, Belcram H, Caboche M (2004) Efficient cloning of plant genomes into bacterial artificial chromosome (BAC) libraries with larger and more uniform insert size. Plant Biotech J 2: 181–188

Chantret N, Salse J, Sabot F, Rahman S et al (2005) Molecular basis of evolutionary events that shaped the hardness locus in diploid and polyploid wheat species (*Triticum* and *Aegilops*). Plant Cell 17:1033–1045

Chen M, SanMiguel P, de Oliveira AC, Woo SS, Zhang H, Wing RA, Bennetzen JL (1997) Microcolinearity in sh2-homologous regions of the maize, rice, and *sorghum* genomes. Proc Natl Acad Sci USA 94:3431–3435

Chen M, SanMiguel P, Bennetzen JL (1998) Sequence organization and conservation in sh2/a1-homologous regions of *sorghum* and rice. Genetics 148:435–443

Chen H, Wang S, Xing Y, Xu C, Hayes PM, Zhang Q (2005) Comparative analyses of genomic locations and race specificities of loci for quantitative resistance to *Pyricularia grisea* in rice and barley. Proc Natl Acad Sci USA 100:2544–2549

Coe E, Cone K, McMullen M, Chen SS et al (2002) Access to the maize genome: an integrated physical and genetic map. Plant Physiol 128:9–12

Collins NC, Thordal-Christensen H, Lipka V, Bau et al (2003) SNARE-protein-mediated disease resistance at the plant cell wall. Nature 425:973–977

Devos KM, Gale MD (1997) Comparative genetics in the grasses. Plant Mol Biol 35:3–15

Devos KM, Gale MD (2000) Genome relationships: the grass model in current research. Plant Cell 12:637–646

Devos KM (2005) Updating the 'crop circle'. Curr Opin Plant Biol 8:155–162

Draper J, Mur LJ, Jenkins G, Ghosh-Biswas C, Bablak P, Hasterok R, Routledge APM (2001) Brachypodium distachyon. A new model system for functional genomics in grasses. Plant Physiol 127: 1539–1555

Dunford RP, Yano M, Kurata N, Sasaki T, Huestis G, Rocheford T, Laurie DA (2002) Comparative mapping of the barley Ppd-H1 photoperiod response gene region, Which lies close to a junction between two rice linkage segments. Genetics 161:825–834

Falque M, Decousset L, Dervins D, Jacob AM, Joets J, Martinant JP, Raffoux X, Ribiere N, Ridel C, Samson D, Charcosset A, Murigneux A (2005) Linkage mapping of 1454 new maize candidate gene loci. Genetics 170:1957–66

Fang Z, Polacco M, Chen S, Schroeder S, Hancock D, Sanchez H, Coe E (2003) cMap: the comparative genetic map viewer. Bioinformatics 19:416–7

Feltus FA, Wan J, Schulze SR, Estill JC, Jiang N, Paterson AH (2004) An SNP resource for rice genetics and breeding based on subspecies Indica and Japonica genome alignments. Genome Res 14:1812–1819

Feng Q, Zhang Y, Hao P, Wang S et al (2002) Sequence and analysis of rice chromosome 4. Nature 420:316–320

Feuillet C, Keller B (1999) High gene density is conserved at syntenic loci of small and large grass genomes. Proc Natl Acad Sci USA 96:8265–8270

Feuillet C, Keller B (2002) Comparative genomics in the grass family: molecular characterization of grass genome structure and evolution. Ann Bota 89:3–10

Feuillet C, Travella S, Stein N, Albar L, Nublat A, Keller B (2003) Map-based isolation of the leaf rust disease resistance gene Lr10 from the hexaploid wheat (*Triticum aestivum* L.) genome. Proc Natl Acad Sci USA 100:15253–15258

Fu H, Dooner HK (2002) Intraspecific violation of genetic colinearity and its implications in maize. Proc Natl Acad Sci USA 99:9573–9578

Gallavotti A, Zhao Q, Kyozuka J, Meeley RB, Ritter MK, Doebley JF, Pe ME Schmidt RJ (2004) The role of barren stalk1 in the architecture of maize. Nature 432:630–635

Gaut BS (2001) Patterns of chromosomal duplication in maize and their implications for comparative maps of the grasses. Genome Res 11:55–66

Gaut BS (2002) Evolutionary dynamics of grass genomes. New Phytol 154:15–28

Gill BS, Appels R, Botha-Oberholster A-M, Buell CR, Bennetzen JL, Chalhoub B, Chumley F, Dvorak J, Iwanaga M, Keller B, Li W, McCombie WR, Ogihara Y, Quetier F, Sasaki T (2004) A workshop report on wheat genome sequencing: international genome research on wheat consortium. Genetics 168:1087–1096

Goff SA, Ricke D, Lan TH, Presting G et al (2002) A draft sequence of the rice genome (*Oryza sativa* L. ssp. japonica). Science 296:92–100

Gottwald S, Stein N, Borner A, Sasaki T, Graner A (2004) The gibberellic-acid insensitive dwarfing gene sdw3 of barley is located on chromosome 2HS in a region that shows high colinearity with rice chromosome 7L. Mol Genet Genomics 271:426–436

Green ED (2001) Strategies for the systematic sequencing of complex genomes. Nat Rev Genet 2:573–583

Griffiths S, Dunford RP, Coupland G, Laurie DA (2003) The evolution of CONSTANS-like gene families in barley, rice, and *Arabidopsis*. Plant Physiol 131:1855–1867

Griffiths S, Sharp R, Foote TN, Bertin I, Wanous M, Reader S, Colas I, Moore G (2006) Molecular characterization of Ph1 as a major chromosome pairing locus in polyploid wheat. Nature 439:749–752

Gu YQ, Coleman-Derr D, Kong X, Anderson OD (2004) Rapid genome evolution revealed by comparative sequence analysis of orthologous regions from four *triticeae* genomes. Plant Physiol 135: 459–470

Gulick PJ, Drouin S, Yu Z, Danyluk J, Poisson G, Monroy AF, Sarhan F (2005) Transcriptome comparison of winter and spring wheat responding to low temperature. Genome 48:913–923

Guyot R, Yahiaoui N, Feuillet C, Keller B (2004) In silico comparative analysis reveals a mosaic conservation of genes within a novel colinear region in wheat chromosome 1AS and rice chromosome 5S. Funct Integr Genomics 4:47–58

Hampson S, McLysaght A, Gaut B, Baldi P (2003) LineUp: statistical detection of chromosomal homology with application to plant comparative genomics. Genome Res 13:1–12

Harlan JR (1992) Origins and processes of domestication. In Chapman GP (ed) Grass evolution and domestication, Cambridge University Press, Cambridge, pp 159–175

Hasterok R, Marasek A, Donnison IS, Armstead I, Thomas A, King IP, Wolny E, Idziak D, Draper J, Jenkins G (2006) Alignment of the genomes of brachypodium distachyon and temperate cereals and grasses using bacterial artificial chromosome landing with fluorescence in situ hybridization. Genetics 173:349–62

Huang SX, Sirikhachornkit A, Faris JD, Su XJ, Gill BS, Haselkorn R, Gornicki P (2002) Phylogenetic analysis of the acetyl-CoA carboxylase and 3- phosphoglycerate kinase loci in wheat and other grasses. Plant Mol Biol 48:805–820

Ilic K, SanMiguel PJ, Bennetzen JL (2003) A complex history of rearrangement in an orthologous region of the maize, *sorghum*, and rice genomes. Proc Natl Acad Sci USA 100:12265–12270

Inada DC, Bashir A, Lee C, Thomas BC, Ko C, Goff SA, Freeling M (2003) Conserved noncoding sequences in the grasses. Genome Res 13:2030–2041

International Rice Genome Sequencing Project (2005) The map-based sequence of the rice genome. Nature 436:793–800

Isidore E, Scherrer B, Chaloub B, Feuillet C, Keller B (2005) Ancient haplotypes resulting from extensive molecular rearrangements in the wheat A genome have been maintained in species of three different ploidy levels. Genome Res 15:526–536

Jaiswal P, Ni J, Yap I, Ware D et al (2006) Gramene: a bird's eye view of cereal genomes. Nucleic Acids Res 34:D717–23

Janda J, Bartos J, Safar J, Kubalakova M et al (2004) Construction of a subgenomic BAC library specific for chromosomes 1D, 4D and 6D of hexaploid wheat. Theor Appl Genet 109:1337–1345

Janda J, Šafář J, Kubaláková M, Bartoš J et al (2006) Advanced resources for plant genomics: BAC library specific for the short arm of chromosome 1B. Plant J, 47:977–986

Kaplinsky NJ, Braun DM, Penterman J, Gof SA, Freeling M (2002) Utility and distribution of conserved noncoding sequences in the grasses. Proc Natl Acad Sci 99:6147–6151

Katagiri S, Wu J, Ito Y, Karasawa W, Shibata M, Kanamori H, KatayoseY, Namiki N, Matsumoto T, Sasaki T (2004) End sequencing and chromosomal in silico mapping of BAC clone derived from an indica rice cultivar, Kasalath. Breed Sci 54:273–279

Kellogg EA (2001) Evolutionary history of the grasses. Plant Phys 125:1198–1205

Kilian A, Chen J, Han F, Steffenson B, Kleinhofs A (1997) Towards map-based cloning of the barley stem rust resistance genes rpg1 and rpg4 using rice as an intergenomic cloning vehicle. Plant Mol Biol 35:187–195

Kishimoto N, Higo H, Abe K, Arai S, Saito A, Higo K (1994) Identification of the duplicated segments in rice chromosomes 1 and 5 by linkage analysis of cDNA markers of known functions. Theor Appl Genet 88:722–726

Klein PE, Klein RR, Vrebalov J, Mullet JE (2003) Sequence-based alignment of sorghum chromosome 3 and rice chromosome 1 reveals extensive conservation of gene order and one major chromosomal rearrangement. Plant J 34:605–621

Kubalakova M, Vrana J, Cihalikova J, Simkova H, Dolezel J (2002) Flow karyotyping and chromosome sorting in bread wheat (*Triticum aestivum* L.). Theor Appl Genet 104:1362–1372

Lai J, Li Y, Messing J, Dooner HK (2005) Gene movement by helitron transposons contributes to the haplotype variability of maize. Proc Natl Acad Sci USA 102:9068–9073

Langham RJ, Walsh J, Dunn M, Ko C, Goff SA et al (2004) Genomic duplication, fractionalization and the origin of regulatory novelty. Genetics 166:935–945

La Rota M, Sorrells ME (2004) Comparative DNA sequence analysis of mapped wheat ESTs reveals the complexity of genome relationships between rice and wheat. Funct Integr Genomics 4:34–46

Lamoureux D, Peterson DG, Li W, Fellers JP, Gill BS (2005) The efficacy of Cot-based gene enrichment in wheat (*Triticum aestivum* L.). Genome 48:1120–1126

Leister D, Kurth J, Laurie DA, Yano M, Sasaki T, Devos K, Graner A, Schulze-Lefert P (1998) Rapid reorganization of resistance gene homologues in cereal genomes. Proc Natl Acad Sci USA 95:370–375

Li W, Gill BS (2002) The colinearity of the Sh2/A1 orthologous region in rice sorghum and maize is interrupted and accompanied by genome expansion in the Triticeae. Genetics 160:1153–1162

Li W, Zhang P, Fellers JP, Friebe B, Gill BS (2004) Sequence composition, organization, and evolution of the core Triticeae genome. Plant J 40:500–511

Lijavetzky D, Muzzi G, Wicker T, Keller B, Wing R, Dubcovsky J (1999) Construction and characterization of a bacterial artificial chromosome (BAC) library for the A genome of wheat. Genome 42:1176–1182

Lin YR, Schertz KF, Paterson AH (1995) Comparative analysis of QTLs affecting plant height and maturity across the poaceae, in reference to an interspecific *sorghum* population. Genetics 141: 391–411

Lockton S, Gaut BS (2005) Plant conserved non-coding sequences and paralogue evolution. Trends Genet. 21:60–65

McClintock B (1930) A cytological demonstration of the location of an interchange between two non-homologous chromosomes of *Zea mays*. Proc Natl Acad Sci USA 16:791–796

Ma J, Bennetzen JL (2004) Rapid recent growth and divergence of rice nuclear genomes. Proc Natl Acad Sci USA 101:12404–12410

Ma J, SanMiguel P, Lai J, Messing J, Bennetzen JL (2005) DNA rearrangement in orthologous orp regions of the maize, rice and *sorghum* genomes. Genetics 170:1209–1220

Margulies EH, Vinson JP, Miller W, Jaffe DB, Lindblad-Toh K, Chang JL, Green ED, Lander ES, Mullikin JC, Clamp M (2005) An initial strategy for the systematic identification of functional elements in the human genome by low-redundancy comparative sequencing. Proc Natl Acad Sci USA 102:4795–4800

Monna L, Kitazawa N, Yoshino R, Suzuki J, Masuda H, Maehara Y, Tanji M, Sato M, Nasu S, Minobe Y (2002) Positional cloning of rice semidwarfing gene, sd-1: rice 'green revolution gene' encodes a mutant enzyme involved in gibberellin synthesis. DNA Res 9:11–17

Monna L, Ohta R, Masuda H, Koike A, Minobe Y (2006) Genome-wide searching of single-nucleotide polymorphisms among eight distantly and closely related rice cultivars (*Oryza sativa* L.) and a wild accession (*Oryza rufipogon* Griff.). DNA Res 13:43–51

Moore G, Devos KM, Wang Z, Gale MD (1995) Cereal genome evolution: grasses, line up and form a circle. Curr Biol 5:737–739

Morgante M, Brunner S, Pea G, Fengler K, Zuccolo A, Rafalski A (2005) Gene duplication and exon shuffling by helitron-like transposons generate intraspecies diversity in maize. Nat Genet 37: 997–1002

Moullet O, Zhang HB, Lagudah ES (1999) Construction and characterisation of a large DNA insert library from the D genome of wheat. Theor Appl Genet 99:305–313

Nagamura Y, Inoue T, Antonio B, Shimano T et al (1995) Conservation of duplicated segments between rice chromosomes 11 and 12. Breed Sci 45:373–376

Nilmalgoda SD, Cloutier S, Walichnowski AZ (2003) Construction and characterization of a bacterial artificial chromosome (BAC) library of hexaploid wheat (*Triticum aestivum* L.) and validation of genome coverage using locus-specific primers. Genome 46:870–878

Paterson AH, Lin YR, Li ZK, Schertz KF, Doebley JF, Pinson SRM, Liu SC, Stansel JW, Irvine JE (1995) Convergent domestication of cereal crops by independent mutations at corresponding genetic loci. Science 269: 1714–1718.

Paterson AH, Bowers JE, Peterson DG., Estill JC, and Chapman BA (2003) Structure and evolution of cereal genomes. Curr Opin Genet Devel 13:644–650

Paterson AH, Bowers JE, Chapman BA (2004) Ancient polyploidization predating divergence of the cereals, and its consequences for comparative genomics. Proc Natl Acad Sci USA 101:9903–9908

Paterson AH (2006) Leafing through the genomes of our major crop plants: strategies for capturing unique information. Nat Rev Genet 7:174–184

Paux E, Roger D, Badaeva E, Gay G, Bernard M, Sourdille P, Feuillet C (2006). Characterizing the composition and evolution of homoeologous genomes in hexaploid wheat through BAC-end sequencing on chromosome 3B. Plant J 48:463–474

Peng JR, Richards DE, Hartley NM, Murphy GP et al (1999) 'Green revolution' genes encode mutant gibberellin response modulators. Nature 400:256–261

Piperno DR, Flannery KV (2001) The earliest archaeological maize (*Zea mays* L.) from highland Mexico: new accelerator mass spectrometry dates and their implications. Proc Natl Acad Sci USA 98:2101–2103

Qi LL, Echalier B, Chao S, Lazo GR et al (2004) A chromosome bin map of 16,000 expressed sequence tag loci and distribution of genes among the three genomes of polyploid wheat. Genetics 168:701–712

Rabinowicz PD, Citek R, Budiman MA, Nunberg A, Bedell JA, Lakey N, O'Shaughnessy AL, Nascimento LU, McCombie WR, Martienssen RA (2005) Differential methylation of genes and repeats in land plants. Genome Res 15:1431–40

Rabinowicz PD, Bennetzen JL (2006) The maize genome as a model for efficient sequence analysis of large plant genomes. Curr Opin Plant Biol 9:149–156

Ramakrishna W, Dubcovsky Y, Park YJ, Busso CS, Emberton J, SanMiguel P, Bennetzen JL (2002a) Different types and rates of genome evolution detected by comparative sequence analysis of orthologous segments from four cereal genomes. Genetics 162:1389–1400

Ramakrishna W, Emberton J, Ogden M, SanMiguel P, Bennetzen JL (2002b) Structural analysis of the maize Rp1 complex reveals numerous sites and unexpected mechanisms of local rearrangement. Plant Cell 14:3213–3223

Safar J, Bartos J, Janda J, Bellec A et al (2004) Dissecting large and complex genomes: flow sorting and BAC cloning of individual chromosomes from bread wheat. Plant J 39:960–968

Salse J, Piegu B, Cooke R, Delseny M (2002) Synteny between *Arabidopsis thaliana* and rice at the genome level: a tool to identify conservation in the ongoing rice genome sequencing project. Nucleic Acids Res 30:2316–2328

Salse J, Piegu B, Cooke R, Delseny M (2004) New in silico insight into the synteny between rice (*Oryza sativa* L.) and maize (*Zea mays* L.) highlights reshuffling and identifies new duplications in the rice genome. Plant J 38:396–409

Sasaki T, Matsumoto T, Yamamoto K, Sakata K et al (2002) The genome sequence and structure of rice chromosome 1. Nature 420:312–316

Shen B, Wang DM, McIntyre CL, Liu CJ (2005) A 'Chinese Spring' wheat (*Triticum aestivum* L.) bacterial artificial chromosome library and its use in the isolation of SSR markers for targeted genome regions. Theor Appl Genet 111:1489–1494

Scherrer B, Isidore E, Klein P, Kim JS, Bellec A, Chalhoub B, Keller B, Feuillet C (2005) Large intraspecific haplotype variability at the Rph7 locus results from rapid and recent divergence in the barley genome. Plant Cell 17:361–374

Service RF (2006) Gene sequencing. The race for the $1000 genome. Science 311:1544–1546

Shen YJ, Jiang H, Jin JP, Zhang ZB et al (2004) Development of genome-wide DNA polymorphism database for map-based cloning of rice genes. Plant Physiol 135:1198–1205

Singh NK, Raghuvanshi S, Srivastava SK, Gaur A et al (2004) Sequence analysis of the long arm of rice chromosome 11 for rice-wheat synteny. Funct Integr Genomics 4:102–117

Song R, Llaca V, Linton E, Messing J (2001) Sequence, regulation, and evolution of the maize 22-kD zein gene family. Genome Res 11:1817–1825

Song R, Llaca V, Messing J (2002) Mosaic organization of orthologous sequences in grass genomes. Genome Res 12:1549–1555

Song R and Messing J (2003) Gene expression of a gene family in maize based on noncollinear haplotypes. Proc Natl Acad Sci 100:9055–9060

Sorrells ME, La Rota M, Bermudez-Kandianis CE, Greene RA et al (2003) Comparative DNA sequence analysis of wheat and rice genomes. Genome Res 13:1818–1827

Sorrells M (2004) Cereal genomics research in the post-genomic era. In: Gupta PK, Varshney RK (eds) Cereal genomics. Kluwer Academic Publishers, Dordrecht, pp 559–584

Springer NM, Xu X, Barbazuk WB (2004) Utility of different gene enrichment approaches toward identifying and sequencing the maize gene space. Plant Physiol 136:3023–3033

Stein N, Feuillet C, Wicker T, Schlagenhauf E, Keller B (2000) Subgenome chromosome walking in wheat: a 450-kb physical contig in *Triticum monococcum* L. spans the *Lr10* resistance locus in hexaploid wheat (*Triticum aestivum* L.). Proc Natl Acad Sci USA 97:13436–13441

Swigonova Z, Bennetzen JL, Messing J (2005) Structure and evolution of the r/b chromosomal regions in rice maize and *sorghum*. Genetics 169:891–906

Tarchini R, Biddle P, Wineland R, Tingey S, Rafalski A (2000) The complete sequence of 340 kb of DNA around the rice adh1-adh2 region reveals interrupted colinearity with maize chromosome 4. Plant Cell 12:381–391

The Rice Chromosome 10 Sequencing Consortium (2003) In-depth view of structure, activity, and evolution of rice chromosome 10. Science 300:1566–1569

The Rice Chromosome 3 Sequencing Consortium (2005) Sequence, annotation, and analysis of synteny between rice chromosome 3 and diverged grass species. Genome Res 15:1284–1291

The Rice Full-Length cDNA Consortium (2003) Collection, mapping, and annotation of over 28,000 cDNA clones from japonica rice. Science 301:376–379

Tikhonov A, SanMiguel P, Nakajima Y, Gorenstein N, Bennetzen J et al (1999) Colinearity and its exceptions in orthologous adh regions of maize and *sorghum*. Proc Natl Acad Sci USA 96:7409–7414

Ting YC (1966) Duplications and meiotic behavior of the chromosomes in haploid maize (*Zea mays* L.). Cytologia 31:324–329

Turner A, Beales J, Faure S, Dunford RP, Laurie DA (2005) The pseudo-response regulator Ppd-H1 provides adaptation to photoperiod in barley. Science 310:1031–1034

Vandepoele K, Saeys Y, Simillion C, Raes J, Van de Peer Y (2002) The automatic detection of homologous regions (ADHoRe) and its application to microcolinearity between *Arabidopsis* and rice. Genome Res 12:1792–1801

Vandepoele K, Simillion C, Van de Peer Y (2003) Evidence that rice and other cereals are ancient aneuploids. Plant Cell 15:2192–2202

Vogel JP, Gu YQ, Twigg P, Lazo GR, Laudencia-Chingcuanco D, Hayden DM, Donze TJ, Vivian LA, Stamova B, Coleman-Derr D (2006) EST sequencing and phylogenetic analysis of the model grass *Brachypodium distachyon*. Theor Appl Genet DOI:10.1007/s00122-006-0285-3

Vollbrecht E, Springer PS, Goh L, Buckler ES, Martienssen R (2005) Architecture of floral branch systems in maize and related grasses. Nature 436:1119–1126

Wang GL, Holsten TE, Song WY, Wang HP, Ronald PC (1995) Construction of a rice bacterial artificial chromosome library and identification of clones linked to the Xa-21 disease resistance locus. Plant J 7:525–533

Whitelaw CA, Barbazuk WB, Pertea G, Chan AP et al (2003) Enrichment of gene-coding sequences in maize by genome filtration. Science 302:2118–2120

Wicker T, Yahiaoui N, Guyot R, Schlagenhauf E, Liu ZD, Dubcowski J, Keller B (2003) Rapid genome divergence at orthologous low molecular weight glutenin loci of the A and Am genomes of wheat. Plant Cell 15:1186–1197

Woo SS, Jiang JM, Gill BS, Paterson AH, Wing RA (1994) Construction and characterization of a bacterial artificial chromosome library of *Sorghum* bicolor. Nucleic Acids Res 23:4922–4931

Yahiaoui N, Srichumpa P, Dudler R, Keller B (2004) Genome analysis at different ploidy levels allows cloning of the powdery mildew resistance gene *Pm3b* from hexaploid wheat. Plant J 37:528–538

Yan L, Loukoiannov A, Tranquilli G, Helguera M, Fahima T, Dubcovsky J (2003) Positional cloning of the wheat vernalization gene VRN1. Proc Natl Acad Sci USA 100:6263–6268

Yan L, Loukoiannov A, Blechl A, Tranquilli G, Ramakrishna W, SanMiguel P, Bennetzen JL, Echenique V, Dubcovsky J (2004) The wheat VRN2 gene is a flowering repressor down-regulated by vernalization. Science 303:1640–1644

Yang D, Parco A, Nandi S, Subudhi P, Zhu Y, Wang G, Huang N (1997) Construction of a bacterial artificial chromosome (BAC) library and identification of overlapping BAC clones with chromosome 4-specific RFLP markers in rice. Theor Appl Genet 95:1147–1154

Yim YS, Davis GL, Duru NA, Musket TA, Linton EW, Messing JW, McMullen MD, Soderlund CA, Polacco ML, Gardiner JM, Coe Jr. EH (2002) Characterization of three maize bacterial artificial chromosome libraries toward anchoring of the physical map to the genetic map using high-density bacterial artificial chromosome filter hybridization. Plant Physiol 130:1686–1696

Yu Y, Tomkins JP, Waugh R, Frisch DA, Kudrna D, Kleinhofs A, Brueggeman RS, Muehlbauer GJ, Wise RP, Wing RA (2000) A bacterial artificial chromosome library for barley (*Hordeum vulgare* L.) and the identification of clones containing putative resistance genes. Theor Appl Genet 101: 1093–1099

Yu J, Hu S, Wang J, Wong GKS et al (2002) A draft sequence of the rice genome (*Oryza sativa* L. ssp. indica). Science 296:79–92

Yu Y, Wing RA (2004) Whole genome sequencing: methodology and progress in cereals. In: Gupta PK, Varshney RK (eds) Cereal genomics. Kluwer Academic Publishers, Dordrecht, pp 385–423

Yu J, Wang J, Lin W, Li S et al (2005) The genomes of *Oryza sativa*: a history of duplications. PLoS Biol 3:266–281

Yuan YN, SanMiguel PJ, Bennetzen JL (2003) High-Cot sequence analysis of the maize genome. Plant J 34:249–255

Zhang HB, Choi S, Woo SS, Li Z, Wing RA (1996) Construction and characterization of two rice bacterial artificial chromosome libraries from the parents of a permanent recombinant inbred mapping population. Mol Breed 2:11–24

Zohary D, Hopf M (2000) Domestication of plants in the old world. 3rd edn, Oxford University Press, Oxford

CHAPTER 9

CLONING QTLS IN PLANTS

SILVIO SALVI* AND ROBERTO TUBEROSA

Department of Agroenvironmental Sciences and Technology, University of Bologna, Viale Fanin 44, 40127 Bologna, Italy

Abstract: The utilization of a number of genomics platforms and analytical methods allows us to fine map and clone major quantitative trait loci (QTLs) responsible for the genetic control of quantitatively inherited traits. To date, most plant QTLs that were successfully cloned have been dissected by means of a positional cloning approach within a biparental cross. In some cases, an association between allelic variation at a candidate gene and a phenotype has been established through the analysis of existing genetic accessions. The effectiveness of these strategies can be enhanced by using appropriate genetic materials (e.g. introgression libraries, panels of unrelated accessions, etc.) and the latest developments in forward- and reverse-genetic platforms. Under this respect, the 'omics' platforms provide a new paradigm to identifiy candidate genes and clues for their function. Completion of genome sequences and improved bioinformatics will facilitate *in silico* cross-matching of candidate sequences with QTLs in programmes of positional cloning or association mapping. Several QTLs have been associated to candidate genes solely based on map information and further circumstantial observation, and without completing a formal cloning procedure. Although QTL mapping and cloning have so far been almost synonymous with the dissection of the genetic control of naturally available phenotypic differences, genes involved in controlling quantitative traits could be identified also by combining quantitative genetics with insertional mutagenesis. Although QTL analysis and cloning addressing naturally occurring genetic variation will continue to shed light on mechanisms of plant adaptation, a greater emphasis on approaches relying on mutagenesis and candidate gene validation is likely to accelerate the discovery of the genes underlying QTLs.

1. INTRODUCTION

For most phenotypic traits, variation among individuals within one species cannot be accounted for by allelic differences at one single locus. Instead, the action of multiple loci, their interactions and random environmental effects are involved

*Corresponding Author: silvio.salvi@unibo.it

R.K. Varshney and R. Tuberosa (eds.), Genomics-Assisted Crop Improvement: Vol. 1: Genomics Approaches and Platforms, 207–225.

in determining phenotypes. Early work indicated that loci with major effects on quantitative traits could be identified and mapped on chromosomes by evaluating the correlation between trait values and the allelic state at genetic markers (Sax 1923; Thoday 1961). This led to the definition of quantitative trait locus (QTL; Geldermann 1975) as a genetic locus where functionally different alleles segregate and cause significant effects on a quantitative trait. With the advent of molecular marker technology, QTL mapping on chromosome linkage maps has become a standard procedure in quantitative genetics (Paterson et al. 1988; Tanksley 1993; Lynch and Walsh 1998; Hackett 2002). By coupling marker technology with genomics resources such as bacterial artificial chromosome (BAC) libraries and physical maps, and by exploiting appropriately developed plant materials, it is now possible to clone single QTLs and identify the DNA polymorphisms responsible for a target QTL (Paran and Zamir 2003; Salvi and Tuberosa 2005). The impact of QTL mapping and cloning on our understanding of plant biology is remarkable: for the first time, we have the opportunity to unravel and describe the genetic complexity (i.e. the number and the type of action of genes) behind quantitatively inherited processes/traits such as adaptation to photoperiod conditions, extreme environments, domestication and many others, including yield and its stability. Such description is at the core of evolutionary genetics and plant breeding.

This chapter highlights the major methodological trends toward QTL cloning and some preliminary indications on the molecular nature of quantitative variation. Clearly, genetic adaptation also involves selection for mutations with a strong effect on the phenotype which are usually classified as Mendelian genes rather than QTLs. Examples of such loci are the major genes (*FLC* and *Frigida*) involved in the vernalization response and flowering time of *Arabidopsis* (Michaels and Amasino 1999; Johanson et al. 2000), the vernalization (*Vrn1-3*) loci in wheat and barley (Yan et al. 2003, 2004, 2006) and the photoperiod response *Ppd-H1* locus in barley (Turner et al. 2005). Because the cloning and the characterization of such loci did not require the QTL mapping and cloning toolbox, the relevant results have not been considered for this review.

2. AVENUES TOWARD QTL CLONING

QTL analysis for a given trait in plants usually begins with a primary (or coarse) QTL mapping step which localizes all major loci responsible for the trait variation observed in a given biparental population. Subsequently, a QTL is mapped within a chromosome supporting interval of ca. 10–30 cM (Lynch and Walsh 1998; Doerge 2002) which can include several hundred genes. The challenge is then to enhance the genetic resolution so that the QTL is confined to a chromosome segment ideally including only one gene. Positional cloning and association mapping are the two main approaches that have been deployed for cloning QTLs. Both approaches exploit linkage disequilibrium (LD; i.e. the level of non-random assortment of alleles at different loci) in order to verify the correlation between the shortest chromosome region tagged by molecular markers and the trait value.

In positional cloning, the increase in mapping resolution is obtained by producing a new, large mapping population (ca. 2000 or more F_2 plants) derived from the cross of two nearly-isogenic parental lines (see below) carrying functionally different alleles at the target QTL. In association mapping (see Gupta et al. 2005), the phenotype/marker correlation is carried out across a set of unrelated individuals (e.g. cultivars, germplasm accessions, etc). Because entries are seperated by many generations, hence meiotic events, the genetic resolution is expected to be higher than the one usually obtained by positional cloning (Flint-Garcia et al. 2003). The genes found to co-segregate with the target QTL are then functionally tested in order to identify the actual candidate gene and to gain further independent evidence about its involvement in controlling trait expression.

3. POSITIONAL CLONING OF QTLS

The key to success in the positional cloning of a QTL is the preparation of appropriate genetic material. With few exceptions, all QTLs cloned so far have required the production of a population from the cross of nearly isogenic lines (NILs) differing only for the allele composition at the target QTL region. Such parental lines have often been indicated as QTL-NILs (Salvi and Tuberosa 2005). In such a population, due to the reduction or absence of other segregating QTLs, the target QTL becomes the main genetic source of variation and, according to the heritability of the trait, a major source of the total phenotypic variation, thus enabling the detection of significant differences between phenotypic means of the QTL genotypic classes (+/+, −/− and, when present, −/+). The level of replication and/or progeny testing is generally based upon the heritability of the trait considered. Under appropriate experimental conditions, the QTL is considered Mendelized (Alonso-Blanco and Koornneef 2000) and genetic distances between a QTL and the nearby molecular markers can be more precisely estimated. In only two cases in *Arabidopsis* (BRX and TE1/ERECTA; Table 1), the large proportion of phenotypic variance explained (0.80 and 0.21–0.64, respectively) by the QTLs allowed their fine mapping for positional cloning directly into the primary mapping populations.

The NILs suitable for positional cloning of QTLs can be produced by a number of designs (Tuinstra et al. 1997; Alonso-Blanco and Koornneef 2000): (i) selfing BC_2 or BC_3 progenies that, based on marker analysis, have recovered most of the genome of the recurrent parent and remained heterozygous at the QTL; (ii) crossing a parental line with a NIL differing only at the target QTL region and obtained after several cycles of backcross and selfing; (iii) selfing a residual heterozygous individual within a highly homozygous family (Tuinstra et al. 1997).

QTL-NILs can also be efficiently identified within introgression libraries (ILs), i.e. collections of lines where each line is isogenic to a background elite parental line with the exception of a single short chromosome segment introgressed from a donor (Zamir 2001), frequently a wild or unadapted accession. Remarkably, the same IL of the wild tomato *Lycopersicon pennellii* within the cultivated tomato

Table 1. Summary of the main characteristics of the QTLs cloned in plants

Species	Trait	QTL/gene	Function	Molecular identification	Candidate gene [a]	R^2 (%)[b]	Plants (no.)[c]	ORF (no.)[e]	Resolution (kb)[d]	Identification of QTN	Functional proof	References
Arabidopsis	Dormancy	DOG1	Unknown	Pos. cloning & cand. gene	No	12	NA	22	86	No, possibly regulatory	Complementation	Bentsink et al. 2006
	Flowering time	ED1/CRY2	Cryptochrome	Pos. cloning	Yes (L)	28-56	1,822	15	45	Amino acid substitution	Transformation	El-Din El-Assal et al. 2001
	Flowering time	FLW/FLM	Transcription factor	Pos. cloning	Yes (E)	27	NA	38	138	Deletion of whole gene	Transformation	Werner et al. 2005
	Glucosinolates content	ESM1	Myrosinase associated protein	Pos. cloning	No	NA	1,344	36	100	Possibly regulatory	Knockout and transformation	Zhang et al. 2006
	Glucosinolates structure	GS-elong/MAM	MAM synthase	Pos. cloning	Yes (E)	NA	4,600	NA	NA	Nucleotide and gene indels	NA	Kroymann et al. 2003
	Root morphology	BRX	Transcription factor	Pos. cloning	No	80	860	10	45	Premature stop codon	Transformation	Mouchel et al. 2004
	Transpiration efficiency	TE1/Erecta	leucine-rich repeat receptor-like kinase	Candidate gene	Yes (L)	21-64	NA	37	NA	Amino acid substitution	Transformation	Masle et al. 2005
Maize	Flowering time	Vgt1	Transcription factor	Pos. cloning	No	15	4,526	1	2	No, regulatory	Transformation and association	Salvi et al. 2007
	Glume architecture	Tga1	Transcription factor	Pos. cloning	No	NA	3,106	1	1	Amino acid substitution	Recovery of mutant	Wang et al. 2005
	Plant architecture	Tb1	Transcription factor	Candidate gene	Yes (E)	17-31	NA	NA	NA	No, possibly regulatory	Complementation	Doebley et al. 1995, 1997
Rice	Heading time	Hd1/Se1	Transcription factor	Pos. cloning	Yes (L)	67	1,505	2	12	No	Transformation	Yano et al. 2000
	Heading time	Hd6/αCK2	Protein kinase	Pos. cloning	No	NA	2,807	1	26	Premature stop codon	Transformation	Takahashi et al. 2001
	Heading time	Hd3a	Unknown	Pos. cloning	Yes (L)	NA	2,207	4	20	No	Transformation	Kojima et al. 2002

	Heading time	Ehd1	B-type response regulator	Pos. cloning	No	NA	>2,500	3	16	Amino acid substitution	Transformation	Doi et al. 2004
	Grain size and length	GS3	VWFC membrane protein	Pos. cloning	No	NA	1,384	1	7.9	Premature stop codon	No	Fan et al. 2006
	Grain number	Gn1/CKX2	Cytokinin oxidase/ dehydrogenase	Pos. Cloning	No	44	13,000	1	6.3	Several	Transformation	Ashikari et al. 2005
	Regenerability	PSR1	Nitrite reductase	Pos. cloning	No	NA	3,800	4	51	Possibly amino acid substitution	Transformation	Nishimura et al. 2005
	Seed shattering	qSH-1/RPL	BEL1-homeobox	Pos. cloning	No	69	10,388	1	0.6	Regulatory	Complementation	Konishi et al. 2006
	Seed shattering	sh4	Transcription factor	Pos. cloning	No	69	12,000	1	1.7	Amino acid substitution	Transformation	Li et al. 2006
	Salt tolerance	SKC1	HKT transporter	Pos. cloning	No	40	2,973	1	7.4	Amino acid substitution	Transformation	Ren et al. 2005
	Submergence tolerance	Sub1	Transcription factor	Pos. cloning	No	70	4,022	13	182	No	Transformation	Xu et al. 2006
	UV resistance	qUVR-10	CPD photlyase	Pos. cloning	Yes (L)	37–41	1,850	6	27	Amino acid substitution	Transformation	Ueda et al. 2005
Tomato	Fruit shape	Ovate	Unknown	Pos. cloning	No	40–67	3,000	8	55	Premature stop codon	Transformation	Liu et al. 2002
	Fruit sugar content	Brix9–2–5/Lin5	Invertase	Pos. cloning	Yes (L)	NA	7,000	1	0.5	Amino acid substitution	Complementation	Fridman et al. 2000, 2004
	Fruit weight	fw2.2	Unknown	Pos. cloning	No	30	3,472	4	92	Unknown regulatory variant	Transformation	Frary et al. 2000; Cong et al. 2002

[a] Evidence for candidate gene. (E) indicates early evidence, after primary QTL analysis; (L) indicates late evidence, after physical mapping and/or sequencing.
[b] Proportion of phenotypic variance explained by the QTL in the primary cross.
[c] Dimension of the population utilized for fine mapping.
[d] DNA physical interval completely linked with the QTL.
NA: not applicable or not available.

genetic background (Eshed and Zamir 1994) provided the source of the QTL-NILs utilized for the cloning of three tomato QTLs. NILs suitable for positional cloning are also produced by the advanced backcross QTL analysis (ABQA) method, which combines backcrossing chromosome segments from a wild accession within an elite line with some level of phenotypic selection against extreme phenotypes with undesirable characteristics (Tanksley and Nelson 1996).

An important innovation for QTL analysis and mapping is the concept of multi-parental intercrossed population as proposed by Mott et al. (2000). This type of population is generated by crossing a panel of parental lines chosen in order to capture a considerable portion of the genetic variation of the species, followed by performing several cycles of intermating to enhance genetic resolution. This approach promises to increase the efficiency of QTL mapping both in terms of detection (segregation is expected at many loci) and genetic resolution due to the repeated cycles of intermating). It should be noted that a substantial increase in genetic resolution can also be obtained by repeated intercrossing of F_2 plants of standard biparental populations (Lee et al. 2002).

During the fine mapping step, the resolution of the target QTL in two or more linked loci can bring positional cloning projects to an end when the proportion of phenotypic variability explained by each QTL is too small to be revealed with a realistically manageable number of replications. QTL clusters have indeed been observed in plants (Khavkin and Coe 1997; Tuberosa et al. 2002, 2003; Chen and Tanksley 2004). On the other hand, cloning was accomplished when one of the linked QTLs retained most of the effect (Fridman et al. 2002; Kojima et al. 2002).

The recruitment of polymorphic markers required for the fine mapping of a QTL is a rather simple procedure for species where the genome has been sequenced or information is available in terms of ESTs (expressed sequenced tags). However, in species for which detailed sequence information is not available or cannot be deduced from syntenic, related species, a large number of molecular markers (e.g. AFLPs) need to be screened in genotypes contrasted at the target region (e.g. pair of QTL-NILs).

Arguably, a major improvement in the positional cloning of QTLs will be indirectly provided by the implementation of marker technologies (e.g. single feature polymorphisms on array platforms; Borevitz et al. 2003) enabling the genotyping of a large population in a fraction of the current time and cost, therefore boosting the development of nearly isogenic materials and the use of very large mapping populations for fine mapping.

3.1. Physical Mapping and Candidate Sequences

When the genetic resolution approaches the cM level, the markers closest to the target QTL are used for anchoring the genetic map to the physical map, i.e. the genomic sequence or, when sequence information is unavailable, a BAC (bacterial artificial chromosome) contig covering the QTL region. An early transfer of the information to the physical map allows for the efficient generation of new single-copy markers useful for refining the genetic mapping and the search for candidate genes. Even if only a BAC contig is available, sequenced BAC ends can often be transformed in genetic markers and low-pass, shot-gun sequencing can provide a glimpse of local gene content. At this stage, exploitation of synteny and microcolinearity is particularly useful in species where a contigued library or the genome sequence is not available (see chapter in this book by Salse and Feuillet). For example, the relatively high microcolinearity of wheat with rice and *Brachypodium* helped the cloning of the *Ph1* locus (Griffiths et al. 2006) and is now being explored to clone an important QTL for resistance to *Fusarium* head blight (Cuthbert et al. 2006; Liu et al. 2006).

It is worth reporting that a recent observation of Price (2006) based on the results of several QTL cloning studies in plants indicates that the actual position of the polymorphism responsible for a given QTL was always very close (from a few to less than 1 cM) to the position of the QTL peak originally mapped in the primary population. Nonetheless, this hypothesis of high accuracy of coarse QTL mapping should be critically considered in view of the fact that it is based solely on major QTLs that were successfully cloned.

Among the studies herein considered, six (*Tga1* in maize; *Gn1*, *GS3*, *Hd6*, *sh4* and *SKC1* in rice; Table 1) managed to obtain a genetic resolution sufficiently high to reduce the number of genes co-segregating with the target QTL to one, and in some cases to a portion of the target gene (*Brix-9-2-5* in tomato: Fridman et al. 2004) or the regulatory sequence (*qSH1* in rice: Konishi et al. 2006; *Vgt1* in maize: Salvi et al. 2007) containing one or very few allelic sequence polymorphisms between parental alleles. Nonetheless, QTLs have been cloned even when the physical region identified after fine mapping spanned a large number of genes (up to 38; Werner et al. 2005). In this case, the general approach has been to select candidates for further testing via function prediction (e.g. *Cry2*, *FLM*, *Hd1* and *Hd3a*; Table 1). When multiple coding sequences with no obvious candidate gene are identified, two possible options are to further increase the mapping resolution and/or to functionally test each open reading frame (ORF). It is interesting to note that while QTL cloning was accomplished in rice and maize by exploiting positional cloning at its best (i.e. delimiting the target chromosome region to a portion containing one or very few genes), in *Arabidopsis* this was never accomplished (Table 1). Conversely, researchers engaged in QTL cloning in *Arabidopsis* instead of further refining the mapping resolution at the target region have preferred to test directly the function of a rather large number of genes. This approach is possible only for those species where well-annotated genome sequence and reverse-genetics techniques/platforms suitable for high-throughput, functional gene testing are available.

3.2. Validation of a Candidate Sequence

The functional testing of a candidate gene/s can be performed by over-expressing or down-regulating the target gene through genetic engineering or RNAi (Waterhouse and Helliwell 2003), by genetic complementation of a known mutant (Doebley et al. 1997) or by rescuing and phenotypically and molecularly characterizing mutants at the candidate gene (Wang et al. 2005). If available within the species under investigation, reverse genetics tools such as T-DNA or transposon-tagged populations (Maes et al. 1999) and/or TILLING (Targeting Induced Local Lesions in Genomes; McCallum et al. 2000; see chapter in this book by Till et al.) can also be exploited. As compared to transposon tagging, TILLING and RNAi are appealing alternatives for their almost universal applicability and for providing subtle changes of gene functionality comparable to those observed naturally. Gene replacement, still in its infancy but already reported in rice (Iida and Terada 2004) can be considered the ultimate tool for validating candidate genes.

The validation of a QTL mapping in non-coding regions remains one of the current major challenges faced by those engaging in QTL cloning. Regulatory regions close to (e.g. promoters) or far from (e.g. enhancers/silencers) from the target gene have been shown to host sequence polymorphisms causing variation in quantitative phenotypes. In one case, a single nucleotide substitution located ca. 12 kb upstream of a transcription factor was responsible for its regulation contributing to the non-shattering phenotype typical of cultivated rice (Konishi et al. 2006). However, other mutations within the coding sequence were also required to fully explain the non-shattering phenotype (Konishi et al. 2006). In other cases, the functional polymorphisms were mapped to promoters (Cong et al. 2002; Bentsink et al. 2006) and/or enhancers (Clark et al. 2006; Salvi et al. 2007), but because of insufficient map resolution and/or large number of allelic differences within the DNA region co-segregating with the phenotype, it was not possible to pinpoint a single polymorphism, nor to identify any molecular mechanisms (e.g. methylation, chromatin folding, DNA-protein interaction, etc.) responsible for the actual effect on transcription.

Notably, in all these cases the *cis*-regulatory effects of QTL regions on the downstream gene were tested by means of allele-specific gene expression assays (Pastinen and Hudson 2004; Wittkop et al. 2004; Salvi et al. 2007). In such cases, the level of the two allelic mRNAs can be independently quantified by means of SNPs detection techniques based on quantitative PCR on cDNA from F_1 plants obtained from the cross between QTL-NILs. In these conditions, the detection of differences in the expression level of the two alleles can be ascribed to differences in *cis*-regulatory regions since *trans*-acting regulators should act homogeneously in the nucleus.

After considering all of the above-mentioned aspects, it is clear that positional cloning of QTLs in plants remains a rather demanding and daunting undertaking. Additionally, positional cloning has so far been limited exclusively to major QTLs, since all the cloned QTLs showed, in the primary genetic analysis, an R^2 value higher than 15% (Table 1). It should be noted that (i) R^2 values based on primary mapping can be grossly overestimated (Beavis 1994) due to statistical artefacts and (ii) epistasis can modify the genetic effect of the target QTL when the genetic background changes (Doebley et al. 1995), for instance during QTL-NIL preparation. Therefore, an independent evaluation of the QTL effect (e.g. by developing and testing QTL-NILs; Landi et al. 2005) is recommended before embarking on QTL cloning.

4. CLONING QTLS BY ASSOCIATION MAPPING

As an alternative to positional cloning, QTLs can be molecularly dissected through association mapping by searching for a statistical association between allelic variants at marker or candidate loci and the mean of the analyzed trait within a set of unrelated genotypes characterized by low LD (Cardon and Bell 2001). The analysis evaluates the trait mean change caused by the substitution of one allele with another. For QTL cloning in plants, the interest lies in (i) the possibility of finding chromosome regions important for controlling quantitative traits without the costly and time-consuming production of large experimental populations (Morgante and Salamini 2003), (ii) the potentially high genetic resolution provided by the many meiotic events which occurred during past generations, and (iii) the possibility of surveying a large number of functionally diverse alleles per locus.

The major factor to be considered in association mapping is the level of LD among the tested accessions. In plants, the extensive LD analyses conducted in *Arabidopsis* and maize have indicated that while LD persists over hundreds of kb in *Arabidopsis*, in maize LD decays after a few kb, although it can extend significantly farther in collections of elite germplasm (Flint-Garcia et al. 2003, 2005, see chapter in this book by Ersoz et al.). With high LD values (i.e. in the ca. 1- to 5-cM range) marker-trait association can theoretically be revealed with a manageable number of molecular markers (Maccaferri et al. 2005). In this case, the expected mapping resolution will only be sufficient for the discovery and coarse mapping of the QTL. On the other hand, with germplasm panels with low LD (i.e. < 0.1-0.5 cM) the diagnostic power of a single marker will only extend for a short distance, thus requiring a prohibitively high number of markers for a whole-genome scan. In this case, association mapping can still be used to fine map the QTL at the gene level after the QTL is positioned using standard mapping procedures. Based on this, it is conceivable that different sets of genotypes, characterized by high or low LD, can be assembled and used for QTL discovery or candidate gene validation, respectively, as suggested for human genetics (Reich et al. 2001). Notably, the presence of population structure, i.e. the possible presence of hidden subgroups (e.g. due to relatedness, selection, etc.) with an unequal distribution of alleles may influence the efficacy of this approach by causing spurious trait-marker associations (Pritchard et al. 2000).

A powerful approach for identifying different haplotypes (i.e. combinations of allelic variants) at target loci and making them available for association mapping is provided by EcoTILLING (Comai et al. 2004), a technique which allows for the identification of virtually all single nucleotide polymorphisms (SNPs) and small insertion/deletions within a ca. 1-kb window in a set of genotypes at a fraction of the sequencing cost. This notwithstanding, the necessity to screen also regulatory regions often quite distant from the effector genes indicates that the selection of candidate sequences to be tested for association mapping is not a trivial task if the genomic scan aims to be comprehensive. Examples of the identification of association between haplotype variation at a candidate gene and a quantitative trait were reported in *Arabidopsis* (Olsen et al. 2004), *Brassica* (Osterberg et al. 2002; Gupta et al. 2004), potato (Simko et al. 2004) and in maize (Thornsberry et al. 2001; Whitt et al. 2002; Guillet-Claude et al. 2004; Palaisa et al. 2004; Wilson et al. 2004; Szalma et al. 2005). The identification of a statistically significant association between haplotype variation at a candidate gene or sequence and a quantitative phenotype should be followed by validation experiments similar to those described within the positional cloning approach.

A partially different approach for identifying genes involved in processes such as domestication and adaptation to modern cropping system was proposed by Yamasaki et al. (2005). The underlining hypothesis is that genes which underwent strong selection through domestication and during early farming practices should show a sizeable reduction in molecular diversity when comparing wild germplasm, old landraces and modern cultivars. Therefore, an "historical" loss of molecular diversity should highlight genes important for adaptation to modern cultivation practices and to determine crop productivity.

5. FUNCTIONAL GENOMICS AND QTL CLONING

Functional genomics is contributing to many aspects of QTL analysis and cloning. Transcriptional profiling between contrasting QTL genotypes can quickly provide a list of genes differentially expressed; subsequently, those genes functionally related to the target trait and mapping at the QTL region can be selected as candidates (Wayne and McIntyre 2002; Giuliani et al. 2005). Unfortunately, the number of QTLs cloned so far in plants is too small to test the validity of this approach. Indeed, when the QTL caused a difference in gene expression level between alleles, those differences were either too low (ca. two-fold; Doebley et al. 1997) or showed too strong of a spatial and/or temporal pattern (Cong et al. 2002) to allow for their identification with a standard microarray-based transcriptome analysis. Other profiling platforms, such as MPSS (massively parallel signature sequencing; Brenner et al. 2000) and SAGE (serial analysis of gene expression; Gowda et al. 2004; see chapter in this book by Sharma et al.) are better suited to detect subtle differences in gene expression. Transcript profiling can reach the sub-tissue level of resolution if carried out in combination with laser-capture microscopy (Schnable et al. 2004).

The expression profiling of a mapping population at the mRNA or protein level allows us to treat the level of expression of a single gene as a quantitative trait and to dissect its genetic control by QTL analysis (Jansen and Nap 2001; Brem and Kruglyak 2005). The loci controlling the level of gene expression have alternatively been named transcript quantity loci (TQLs), expression QTLs (eQTLs; Schadt et al. 2003; see Figure 1) or protein quantity loci (PQLs; Damerval et al. 1994;). Correspondences between eQTLs and/or PQLs for candidate genes with QTLs for

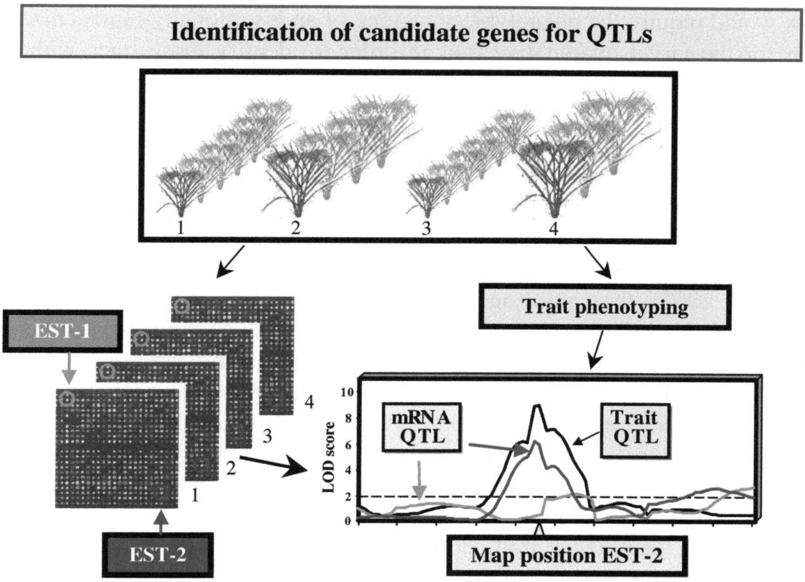

Figure 1. Expression profiling of a mapping population at the mRNA level via microarray analysis to identify expression QTLs (eQTLs) for specific cDNA and therefore genes. Correspondence between an eQTL peak for a specific cDNA (e.g. cDNA-2) and a QTL peak for a trait causally linked to the function of the protein encoded by the cDNA provides circumstantial evidence supporting the role of the cDNA as a candidate gene for the target trait *(see plate 7)*

morpho-physiological traits have already been observed in small- or medium-scale experiments (de Vienne et al. 1999; Francia et al. 2004; Guillaumie et al. 2004). Microarray-based studies have mapped eQTLs both at the same location of the gene whose expression was measured, thus indicating a role for *cis*-regulatory allelic variation, and also at distant chromosome positions (Brem et al. 2002; Schadt et al. 2003; West et al. 2007). The same studies highlighted the presence of eQTL "hot spots", i.e. chromosome regions apparently responsible for controlling the simultaneous expression of many genes (see the chapter in this book by Kirst and Yu).

6. QTL TAGGING

QTL mapping and cloning have been so far almost synonymous for the dissection of the genetic control of naturally available phenotypic differences. However, genes involved in controlling subtle and environmentally affected traits can be identified also by combining quantitative genetics with mutagenesis. Indeed, it has been argued that mutagenesis could be more efficient for dissecting the genetic basis of quantitative traits than QTL analysis, which only provides "accidents of history" allelic variants as stated by Nadeau and Frankel (2000). One way proposed is to utilize a tagging (insertional) approach (Robertson 1985; Soller and Beckmann 1987). Such framework would require the phenotypic screening of an insertionally-mutagenized population for the target quantitative trait in order to identify those lines with a phenotypic mean value outside a predicted range due to environmental effects. The complete screening experiment would involve a manageable number of plants (e.g. from a few thousand up to tens of thousands) if multiple insertion systems are employed and several quantitative traits are concurrently evaluated (Soller and Beckmann 1987). The gene functionally modified or inactivated by the insertional event can be rescued using standard molecular procedures. Following a similar approach, QTL tagging has already been successfully accomplished in *Magnaporthe* (Fujimoto et al. 2002), the causal agent of rice blast, and in *Drosophila* (Norga et al. 2003). In plants, QTL tagging could be carried out with a number of different approaches, based on T-DNA as well as DNA-transposons and retrotransposons. However, systems relying on callus cultures (e.g. activation of rice TOS-17 retrotransposon; Hirochika et al. 1996) should be considered with caution due to the occurrence of somaclonal variation, i.e. the *de novo* variation observed in plants regenerated from tissue culture and caused by changes in DNA-methylation, transposon activity and others molecular events (Kaeppler and Phillips 1993) that can potentially alter any quantitative trait and therefore hinder the identification of the tagged QTL. Instead, interesting resources are the *Ac-Ds*-based insertional populations developed in rice (Jeon and An 2001): in these cases, following the introduction of the heterologous transposons, the majority of the mutational events were created by new transposition activity. In maize, a *Mu*-based insertional population has been developed in a non-segregating genetic background (McCarty et al. 2005), where most of the quantitative variability can be attributed to the segregation of the tagged QTLs.

7. THE ROLE OF CANDIDATE GENES

Classically, a link between a gene and a quantitative trait can be hypothesized based on linkage information (all genes co-segregating with a QTL are positional candidates) or communality between the quantitative trait physiology and the biochemical function of the gene (functional candidate gene; Pflieger et al. 2001) or both. For example, completion of genome sequences and improved bioinformatics will facilitate *in silico* cross-matching of candidate sequences with QTLs in programmes of positional cloning or association mapping. Additionally, a deeper understanding of the mechanisms governing gene expression will extend the concept of candidate gene to include *cis*-acting regulatory sequences as well. Several QTLs have been associated to candidate genes solely based on map information and further circumstantial observation, and without completing a formal cloning procedure. Examples of this type are: a CBF-gene cluster associated to cold tolerance QTLs in barley and wheat (Francia et al. 2004; Tondelli et al. 2006), the phytoene-synthase gene associated to a major QTL for endosperm (and semolina) colour in durum wheat (Pozniak et al. 2007), glutamine-synthetase genes associated to grain yield QTLs in maize (Hirel et al. 2001, Martin et al. 2006), cellulose synthase-like genes associated to a QTL for accumulation of cell-wall glucans in barley (Burton et al. 2006) and a gene involved in inflorescence development (*ra1*) associated with a QTL for tassel branching in maize (Upadyayula et al. 2006). Based on these premises, it is conceivable that in the future, QTL cloning will increasingly rely on candidate gene information.

8. CONCLUDING REMARKS

Robertson (1985) suggested that qualitative mutant alleles and wild-type alleles at loci affecting quantitative traits are the extremes of a possible range of effects, with QTLs resulting from the segregation of naturally-available alleles with milder effects. To date, the cloning of plant QTLs has essentially confirmed Robertson's hypothesis since the type of mutations, genes and cellular and physiological pathways determining quantitative traits are not distinct from those underlining Mendelian traits and in some cases involve the same genes for which a strong mutation was already known. Among the QTLs cloned so far, the apparent abundance of regulatory genes or transcription factors, which potentially act on many downstream functions, was to a certain extent expected due to the complexity of the traits that are usually investigated in QTL analysis.

Almost invariably, the QTLs cloned had shown the largest phenotypic effect in the original experimental populations, often produced by wide crosses between subspecies. Additionally, targeting major QTLs simplifies the cloning process especially when it is based on positional cloning. However, the so-called minor QTLs (i.e. those showing a smaller effect on the trait in the original population) should be targeted with the same emphasis since they can represent potentially important genes and their identification as "minor" could simply be due to the segregation of alleles of rather similar effect in the experimental cross. For this reason, a vast number of minor but equally important QTLs (and genes) are expected to govern yield and agronomic traits in crosses involving elite germplasm, where plant breeding has already eliminated most undesired alleles. It is likely that the constant improvement of the molecular platforms, new types of genetic materials, progress in bioinformatics, high-throughput phenotyping and the increasing availability of tools for functionally testing candidate genes will offer the opportunity of targeting QTLs other than those with a major effect (Varshney et al. 2006).

Having identified an allele with a strong genetic effect in one genetic background by no means warrants a successful plant breeding intervention using marker-assisted selection or genetic engineering. It is probably naive and too optimistic to assume that complex traits resulting from the interaction of hundreds of cellular and developmental functions can be easily altered by acting on one single genetic component, if even the outcomes resulting from engineering interventions at the cellular level are difficult to predict (Stephanopoulos et al. 2004). Clearly, a model providing a simplified and manageable representation of the interacting physiological and developmental components is needed in order to identify the most promising entry steps (Hammer et al. 2006), that is, what are the genes/QTLs to preferentially act upon. Particularly challenging sources of complexities are gene interactions (which QTL analysis still handles rather poorly; Erickson et al. 2004) at transcriptional and post-transcriptional levels, metabolic fluxes, key developmental steps and their integrated responses to environmental cues.

From a more practical and applicative standpoint, we wish to underline the importance of an accurate and equally relevant phenotyping for the success of any QTL cloning effort; due to the elusive nature of most quantitatively inherited traits, precise phenotyping under appropriate conditions probably remains the most critical factor limiting our capacity to dissect QTLs. Although it is not possible to predict to what extent QTL cloning will impact molecular breeding in the next decades, we remain confident that progress toward a more targeted and effective tailoring of improved cultivars will be accelerated by the systematic dissection of QTLs governing the naturally occurring variation relevant for improving and sustaining crops' yield.

REFERENCES

Alonso-Blanco C, Koornneef M (2000) Naturally occurring variation in *Arabidopsis* : an underexploited resource for plant genetics. Trends Plant Sci 5:22–29

Ashikari M, Sakakibara H, Lin S, Yamamoto T, Takashi T, Nishimura A, Angeles ER, Qian Q, Kitano H, Matsuoka M (2005) Cytokinin oxidase regulates rice grain production. Science 309:741–745

Beavis WD (1994) The power and deceit of QTL experiments: lessons from comparative QTL studies. In: Proceedings of the 49th Ann Corn and Sorghum research conference american seed trait association. American Seed Trait Association, Washington D.C, USA pp 250–266

Bentsink L, Jowett J, Hanhart CJ, Koornneef M (2006) Cloning of DOG1, a quantitative trait locus controlling seed dormancy in Arabidopsis. Proc Natl Acad Sci USA 103:17042–17047

Borevitz JO, Liang D, Plouffe D, Chang HS, Zhu T, Weigel D, Berry CC, Winzeler E, Chory J (2003) Large-scale identification of single-feature polymorphisms in complex genomes. Genome Res 13:513–523

Brem RB, Kruglyak L (2005) The landscape of genetic complexity across 5,700 gene expression traits in yeast. Proc Natl Acad Sci USA 102:1572–1577

Brem RB, Yvert G, Clinton R, Kruglyak L (2002) Genetic dissection of transcriptional regulation in budding yeast. Science 296:752–755

Brenner S, Johnson M, Bridgham J, Golda G, Lloyd DH, Johnson D, Luo S, McCurdy S, Foy M, Ewan M, Roth R, George D, Eletr S, Albrecht G, Vermaas E, Williams SR, Moon K, Burcham T, Pallas M, DuBridge RB, Kirchner J, Fearon K, Mao J, Corcoran K (2000) Gene expression analysis by massively parallel signature sequencing (MPSS) on microbead arrays. Nat Biotechnol 18:630–634

Burton RA, Wilson SM, Hrmova M, Harvey AJ, Shirley NJ, Medhurst A, Stone BA, Newbigin EJ, Bacic A, Fincher GB (2006) Cellulose synthase-like *CslF* genes mediate the synthesis of cell wall (1,3;1,4)-beta-D-glucans. Science 311:1940–1942

Cardon LR, Bell JI (2001) Association study designs for complex diseases. Nat Rev Genet 2:91–99

Chen KY, Tanksley SD (2004) High-resolution mapping and functional analysis of *se2.1*: a major stigma exsertion quantitative trait locus associated with the evolution from allogamy to autogamy in the genus *Lycopersicum*. Genetics 168:1563–1573

Clark RM, Wagler TN, Quijada P, Doebley J (2006) A distant upstream enhancer at the maize domestication gene *tb1* has pleiotropic effects on plant and inflorescent architecture. Nat Genet 38:594–597

Comai L, Young K, Till BJ, Reynolds SH, Greene EA, Codomo CA, Enns LC, Johnson JE, Burtner C, Odden AR, Henikoff S (2004) Efficient discovery of DNA polymorphisms in natural populations by Ecotilling. Plant J 37:778–786

Cong B, Liu J, Tanksley SD (2002) Natural alleles at a tomato fruit size quantitative trait locus differ by heterochronic regulatory mutations. Proc Natl Acad Sci USA 99:13606–13611

Cuthbert PA, Somers DJ, Thomas J, Cloutier S, Brule-Babel A (2006) Fine mapping *Fhb1*, a major gene controlling fusarium head blight resistance in bread wheat (*Triticum aestivum* L.). Theor Appl Genet 112:1465–1472

Damerval C, Maurice A, Josse JM, de Vienne D (1994) Quantitative trait loci underlying gene product variation: a novel perspective for analyzing regulation of genome expression. Genetics 137:289–301

de Vienne D, Leonardi A, Damerval C, Zivy M (1999) Genetics of proteome variation for QTL characterization: application to drought stress responses in maize. J Exp Bot 50:303–309

Doebley J, Stec A, Gustus C (1995) *Teosinte branched1* and the origin of maize: evidence for epistasis and the evolution of dominance. Genetics 141:333–346

Doebley J, Stec A, Hubbard L (1997) The evolution of apical dominance in maize. Nature 386:485–488

Doerge RW (2002) Mapping and analysis of quantitative trait loci in experimental populations. Nat Rev Genet 3:43–52

Doi K, Izawa T, Fuse T, Yamanouchi U, Kubo T, Shimatani Z, Yano M, Yoshimura A (2004) *Ehd1*, a B-type response regulator in rice, confers short-day promotion of flowering and controls *FT*-like gene expression independently of *Hd1*. Genes Dev 18:926–936

El-Din El-Assal S, Alonso-Blanco C, Peeters AJ, Raz V, Koornneef M (2001) A QTL for flowering time in *Arabidopsis* reveals a novel allele of *CRY2*. Nat Genet 29:435–440

Erickson DL, Fenster CB, Stenoien HK, Price D (2004) Quantitative trait locus analyses and the study of evolutionary process. Mol Ecol 13:2505–2522

Eshed Y, Zamir D (1994) A genomic library of *Lycopersicon pennellii* in L. *esculentum*: a tool for fine-mapping of genes. Euphytica 79:175–179

Fan C, Xing Y, Mao H, Lu T, Han B, Xu C, Li X, Zhang Q (2006) GS3, a major QTL for grain length and weight and minor QTL for grain width and thickness in rice, encodes a putative transmembrane protein. Theor Appl Genet 112:1164–1171

Flint-Garcia SA, Thornsberry JM, Buckler ES IV (2003) Structure of linkage disequilibrium in plants. Annu Rev Plant Biol 54:357–374

Flint-Garcia SA, Thuillet A, Yu J, Pressoir G, Romero SM, Mitchell SE, Doebley JF, Kresovich S, Goodman MM, Buckler ES (2005) Maize association population: a high resolution platform for QTL dissection. Plant J 44:1054–1064

Francia E, Rizza F, Cattivelli L, Stanca AM, Galiba G, Toth B, Hayes PM, Skinner JS, Pecchioni N (2004) Two loci on chromosome 5H determine low-temperature tolerance in a ‘Nure’ (winter) ×‘Tremois’ (spring) barley map. Theor Appl Genet 108:670–680

Frary A, Nesbitt TC, Grandillo S, Knaap E, Cong B, Liu J, Meller J, Elber R, Alpert KB, Tanksley SD (2000) *fw2.2*: a quantitative trait locus key to the evolution of tomato fruit size. Science 289:85–88

Fridman E, Carrari F, Liu Y-S, Fernie AR, Zamir D (2004) Zooming in on a quantitative trait for tomato yield using interspecific introgressions. Science 305:1786–1789

Fridman E, Liu YS, Carmel-Goren L, Gur A, Shoresh M, Pleban T, Eshed Y, Zamir D (2002) Two tightly linked QTLs modify tomato sugar content via different physiological pathways. Mol Genet Genomics 266:821–826

Fridman E, Pleban T, Zamir D (2000) A recombination hotspot delimits a wild-species quantitative trait locus for tomato sugar content to 484 bp within an invertase gene. Proc Natl Acad Sci USA 97:4718–4723

Fujimoto D, Shi Y, Christian D, Mantanguihan JB, Leung H (2002) Tagging quantitative loci controlling pathogenicity in *Magnaporthe grisea* by insertional mutagenesis. Physiol Mol Plant Path P 61:77–88

Geldermann H (1975) Investigation on inheritance of quantitative characters in animals by gene markers. I. Methods. Theor Appl Genet 46:319–330

Giuliani S, Clarke J, Kreps J, Sanguineti MC, Salvi S, Landi P, Zhu T, Tuberosa R (2005) Microarray analysis of backcrossed-derived lines differing for *root-ABA1*, a major QTL controlling root characteristics and ABA concentration in maize. In: Tuberosa R, Phillips RL, Gale M (eds) In the wake of double helix – from the green revolution to the gene revolution. Edizioni Avenue Media, Bologna, pp 463–489

Gowda M, Jantasuriyarat C, Dean RA, Wang GL (2004) Robust-LongSAGE (RL-SAGE): a substantially improved LongSAGE method for gene discovery and transcriptome analysis. Plant Physiol 134:890–897

Griffiths S, Sharp R, Foote TN, Bertin I, Wanous M, Reader S, Colas I, Moore G (2006) Molecular characterization of *Ph1* as a major chromosome pairing locus in polyploid wheat. Nature 439:749–752

Guillaumie S, Charmet G, Linossier L, Torney V, Robert N, Ravel C (2004) Colocation between a gene encoding the bZip factor *SPA* and an eQTL for a high-molecular-weight glutenin subunit in wheat (*Triticum aestivum*). Genome 47:705–713

Guillet-Claude C, Birolleau-Touchard C, Manicacci D, Rogowsky PM, Rigau J, Murigneux A, Martinant JP, Barriere Y (2004) Nucleotide diversity of the *ZmPox3* maize peroxidase gene: relationships between a MITE insertion in exon 2 and variation in forage maize digestibility. BMC Genet 5:19

Gupta V, Mukhopadhyay A, Arumugam N, Sodhi YS, Pental D, Pradhan AK (2004) Molecular tagging of erucic acid trait in oilseed mustard (*Brassica juncea*) by QTL mapping and single nucleotide polymorphisms in *FAE1* gene. Theor Appl Genet 108:743–749

Gupta PK, Rustgi S, Kulwal PL (2005) Linkage disequilibrium and association studies in higher plants: present status and future prospects. Plant Mol Biol 57:461–485

Hackett CA (2002) Statistical methods for QTL mapping in cereals. Plant Mol Biol 48:585–599

Hammer G, Cooper M, Tardieu F, Welch S, Walsh B, van Eeuwijk F, Chapman S, Podlich D (2006) Models for navigating biological complexity in breeding improved crop plants. Trends Plant Sci 11:587–593

Hirel B, Bertin P, Quillere I, Bourdoncle W, Attagnant C, Dellay C, Gouy A, Cadiou S, Retailliau C, Falque M, Gallais A (2001) Towards a better understanding of the genetic and physiological basis for nitrogen use efficiency in maize. Plant Physiol 125:1258–1270

Hirochika H, Sugimoto K, Otsuki Y, Tsugawa H, Kanda M (1996) Retrotransposons of rice involved in mutations induced by tissue culture. Proc Natl Acad Sci USA 93:7783–7788

Iida S, Terada R (2004) A tale of two integrations, transgene and T-DNA: gene targeting by homologous recombination in rice. Curr Opin Biotechnol 15:132–138

Jansen RC, Nap JP (2001) Genetical genomics: the added value from segregation. Trends Genet 17:388–391

Jeon JS, An G (2001) Gene tagging in rice: a high throughput system for functional genomics. Plant Sci 161:211–219

Johanson U, West J, Lister C, Michaels S, Amasino R, Dean C (2000) Molecular analysis of *FRIGIDA*, a major determinant of natural variation in *Arabidopsis* flowering time. Science 290:344–347

Kaeppler SM, Phillips RL (1993) Tissue culture-induced DNA methylation variation in maize. Proc Natl Acad Sci USA 90:8773–8776

Khavkin E, Coe E (1997) Mapped genomic locations for developmental functions and QTLs reflect concerted groups in maize (*Zea mays* L.). Theor Appl Genet 95:343–352

Kojima S, Takahashi Y, Kobayashi Y, Monna L, Sasaki T, Araki T, Yano M (2002) *Hd3a*, a rice ortholog of the *Arabidopsis FT* gene, promotes transition to flowering downstream of *Hd1* under short-day conditions. Plant Cell Physiol 43:1096–1105

Konishi S, Izawa T, Lin SY, Ebana K, Fukuta Y, Sasaki T, Yano M (2006) An SNP caused loss of seed shattering during rice domestication. Science 312:1392–1396

Kroymann J, Donnerhacke S, Schnabelrauch D, Mitchell-Olds T (2003) Evolutionary dynamics of an *Arabidopsis* insect resistance quantitative trait locus. Proc Natl Acad Sci USA 100:14587–14592

Landi P, Sanguineti MC, Salvi S, Giuliani S, Bellotti M, Maccaferri M, Conti S, Tuberosa R (2005) Validation and characterization of a major QTL affecting leaf ABA concentration in maize. Mol Breed 15:291–303

Lee M, Sharopova N, Beavis WD, Grant D, Katt M, Blair D, Hallauer A (2002) Expanding the genetic map of maize with the intermated B73 × Mo17 (IBM) population. Plant Mol Biol 48:453–461

Li C, Zhou A, Sang T (2006) Rice domestication by reducing shattering. Science 311:1936–1939

Liu J, Van Eck J, Cong B, Tanksley SD (2002) A new class of regulatory genes underlying the cause of pear-shaped tomato fruit. Proc Natl Acad Sci USA 99:13302–13306

Liu S, Zhang X, Pumphrey MO, Stack RW, Gill BS, Anderson JA (2006) Complex microcolinearity among wheat, rice, and barley revealed by fine mapping of the genomic region harboring a major QTL for resistance to *Fusarium* head blight in wheat. Funct Integr Genomics 6:83–89

Lynch M, Walsh B (eds) (1998) Genetics and analysis of quantitative traits, Sinauaer Associates, Sunderland, Massachusetts, USA

Maccaferri M, Sanguineti MC, Noli E, Tuberosa R (2005) Population structure and long-range linkage disequilibrium in a durum wheat elite collection. Mol Breed 15:271–290

McCallum CM, Comai L, Greene EA, Henikoff S (2000) Targeting Induced Local Lesions IN Genomes (TILLING) for plant functional genomics. Plant Physiol 123:439–442

McCarty DR, Settles AM, Suzuki M, Tan BC, Latshaw S, Porch T, Robin K, Baier J, Avigne W, Lai J, Messing J, Koch KE, Hannah LC (2005) Steady-state transposon mutagenesis in inbred maize. Plant J 44:52–61

Maes T, De Keukeleire P, Gerats T (1999) Plant tagnology. Trends Plant Sci 4:90–96

Martin A, Lee J, Kichey T, Gerentes D, Zivy M, Tatout C, Dubois F, Balliau T, Valot B, Davanture M, Terce-Laforgue T, Quillere I, Coque M, Gallais A, Gonzalez-Moro MB, Bethencourt L, Habash DZ, Lea PJ, Charcosset A, Perez P, Murigneux A, Sakakibara H, Edwards KJ, Hirel B (2006) Two cytosolic glutamine synthetase isoforms of maize are specifically involved in the control of grain production. Plant Cell 18:3252–3274

Masle J, Gilmore SR, Farquhar GD (2005) The *ERECTA* gene regulates plant transpiration efficiency in *Arabidopsis*. Nature 436:866–870

Michaels SD, Amasino RM (1999) *FLOWERING LOCUS C* encodes a novel MADS domain protein that acts as a repressor of flowering. Plant Cell 11:949–956

Morgante M, Salamini F (2003) From plant genomics to breeding practice. Curr Opin Biotechnol 14:214–219

Mott R, Talbot CJ, Turri MG, Collins AC, Flint J (2000) A method for fine mapping quantitative trait loci in outbred animal stocks. Proc Natl Acad Sci USA 97:12649–12654

Mouchel CF, Briggs GC, Hardtke CS (2004) Natural genetic variation in *Arabidopsis* identifies *BREVIS RADIX*, a novel regulator of cell proliferation and elongation in the root. Genes Dev 18:700–714

Nadeau JH, Frankel WN (2000) The roads from phenotypic variation to gene discovery: mutagenesis versus QTLs. Nat Genet 25:381–384

Nishimura A, Ashikari M, Lin S, Takashi T, Angeles ER, Yamamoto T, Matsuoka M (2005) Isolation of a rice regeneration quantitative trait loci gene and its application to transformation systems. Proc Natl Acad Sci USA 102:11940–11944

Norga KK, Gurganus MC, Dilda CL, Yamamoto A, Lyman RF, Patel PH, Rubin GM, Hoskins RA, Mackay TF, Bellen HJ (2003) Quantitative analysis of bristle number in *Drosophila* mutants identifies genes involved in neural development. Curr Biol 13:1388–1396

Olsen KM, Halldorsdottir SS, Stinchcombe JR, Weinig C, Schmitt J, Purugganan MD (2004) Linkage disequilibrium mapping of *Arabidopsis CRY2* flowering time alleles. Genetics 167:1361–1369

Osterberg MK, Shavorskaya O, Lascoux M, Lagercrantz U (2002) Naturally occurring indel variation in the *Brassica nigra COL1* gene is associated with variation in flowering time. Genetics 161:299–306

Palaisa K, Morgante M, Tingey S, Rafalski A (2004) Long-range patterns of diversity and linkage disequilibrium surrounding the maize *Y1* gene are indicative of an asymmetric selective sweep. Proc Natl Acad Sci USA 101:9885–9890

Paran I, Zamir D (2003) Quantitative traits in plants: beyond the QTL. Trends Genet 19:303–306

Pastinen T, Hudson TJ (2004) Cis-acting regulatory variation in the human genome. Science 306:647–650

Paterson AH, Lander ES, Hewitt JD, Peterson S, Lincoln SE, Tanksley SD (1988) Resolution of quantitative traits into Mendelian factors by using a complete linkage map of restriction fragment length polymorphisms. Nature 335:721–726

Pflieger S, Lefebvre V, Causse M (2001) The candidate gene approach in plant genetics: a review. Mol Breed 7:275–291

Pozniak CJ, Knox RE, Clarke FR, Clarke JM (2007) Identification of QTL and association of a phytoene synthase gene with endosperm colour in durum wheat. Theor Appl Genet 114:525–537

Price AH (2006) Believe it or not, QTLs are accurate! Trends Plant Sci 11:213–216

Pritchard JK, Stephens M, Rosenberg NA, Donnelly P (2000) Association mapping in structured populations. Am J Hum Genet 67:170–181

Reich DE, Cargill M, Bolk S, Ireland J, Sabeti PC, Richter DJ, Lavery T, Kouyoumjian R, Farhadian SF, Ward R, Lander ES (2001) Linkage disequilibrium in the human genome. Nature 411:199–204

Ren ZH, Gao JP, Li LG, Cai XL, Huang W, Chao DY, Zhu MZ, Wang ZY, Luan S, Lin HX (2005) A rice quantitative trait locus for salt tolerance encodes a sodium transporter. Nat Genet 37:1141–1146

Robertson DS (1985) A possible technique for isolating genic DNA for quantitative traits in plants. J Theor Biol 117:1–10

Salvi S, Sponza G, Morgante M, Tomes D, Xiaomu Niu X, Fengler KA, Meeley R, Ananiev EV, Svitashev S, Bruggemann E, Li B, Haney CF, Radovic S, Zaina G, Rafalski J-A, Tingey SV, Miao G-H, Phillips RL, Tuberosa R (2007) Conserved non-coding genomic sequences controlling flowering time differences in maize. Proc Natl Acad Sci USA, 104:11376–11381

Salvi S, Tuberosa R (2005) To clone or not to clone plant QTLs: present and future challenges. Trends Plant Sci 10:297–304

Sax K (1923) The association of size differences with seed-coat pattern and pigmentation in *Phaseolus vulgaris*. Genetics 8:552–560

Schadt EE, Monks SA, Drake TA, Lusis AJ, Che N, Colinayo V, Ruff TG, Milligan SB, Lamb JR, Cavet G, Linsley PS, Mao M, Stoughton RB, Friend SH (2003) Genetics of gene expression surveyed in maize, mouse and man. Nature 422:297–302

Schnable PS, Hochholdinger F, Nakazono M (2004) Global expression profiling applied to plant development. Curr Opin Plant Biol 7:50–56

Simko I, Costanzo S, Haynes KG, Christ BJ, Jones RW (2004) Linkage disequilibrium mapping of a *Verticillium dahliae* resistance quantitative trait locus in tetraploid potato (*Solanum tuberosum*) through a candidate gene approach. Theor Appl Genet 108:217–224

Soller M, Beckmann JS (1987) Cloning quantitative trait loci by insertional mutagenesis. Theor Appl Genet 74:369–378

Stephanopoulos G, Alper H, Moxley J (2004) Exploiting biological complexity for strain improvement through system biology. Nat Biotechnol 22:1261–1267

Szalma SJ, Buckler ES, Snook ME, McMullen MD (2005) Association analysis of candidate genes for maysin and chlorogenic acid accumulation in maize silks. Theor Appl Genet 110:1324–1333

Takahashi Y, Shomura A, Sasaki T, Yano M (2001) *Hd6*, a rice quantitative trait locus involved in photoperiod sensitivity, encodes the α-subunit of protein kinase CK2. Proc Natl Acad Sci USA 98:7922–7927

Tanksley SD (1993) Mapping polygenes. Annu Rev Genet 27:205–233

Tanksley SD, Nelson JC (1996) Advanced backcross QTL analysis: a method for the simultaneous discovery and transfer of valuable QTLs from unadapted germplasm into elite breeding lines. Theor Appl Genet 92:191–203

Thoday JM (1961) Location of polygenes. Nature 191:368–370

Thornsberry JM, Goodman MM, Doebley J, Kresovich S, Nielsen D, Buckler ES IV (2001) *Dwarf8* polymorphisms associate with variation in flowering. Nat Genet 28:286–289

Tondelli A, Francia E, Barabaschi D, Aprile A, Skinner JS, Stockinger EJ, Stanca AM, Pecchioni N (2006) Mapping regulatory genes as candidates for cold and drought stress tolerance in barley. Theor Appl Genet 112:445–454

Tuberosa R, Salvi S, Sanguineti MC, Landi P, Maccaferri M, Conti S (2002) Mapping QTLs regulating morpho-physiological traits and yield: case studies, shortcomings and perspectives in drought-stressed maize. Ann Bot 89:941–963

Tuberosa R, Salvi S, Sanguineti MC, Maccaferri M, Giuliani S, Landi P (2003) Searching for quantitative trait loci controlling root traits in maize: a critical appraisal. Plant Soil 255:35–54

Tuinstra MR, Ejeta G, Goldsbrough PB (1997) Heterogeneous inbred family (HIF) analysis: a method for developing near-isogenic lines that differ at quantitative trait loci. Theor Appl Genet 95:1005–1011

Turner A, Beales J, Faure S, Dunford RP, Laurie DA (2005) The pseudo-response regulator *Ppd-H1* provides adaptation to photoperiod in barley. Science 310:1031–1034

Ueda T, Sato T, Hidema J, Hirouchi T, Yamamoto K, Kumagai T, Yano M (2005) *qUVR-10*, a major quantitative trait locus for ultraviolet-B resistance in rice, encodes cyclobutane pyrimidine dimer photolyase. Genetics 171:1941–1950

Upadyayula N, da Silva HS, Bohn MO, Rocheford TR (2006) Genetic and QTL analysis of maize tassel and ear inflorescence architecture. Theor Appl Genet 112:592–606

Varshney RK, Hoisington DA, Tyagi AK (2006) Advances in cereal genomics and applications in crop breeding. Trends Biotechnol 24:490–499

Wang H, Nussbaum-Wagler T, Li B, Zhao Q, Vigourox Y, Faller M, Bomblies K, Lukens L, Doebley JF (2005) The origin of the naked grains of maize. Nature 436:714–719

Waterhouse PM, Helliwell CA (2003) Exploring plant genomes by RNA-induced gene silencing. Nat Rev Genet 4:29–38

Wayne ML, McIntyre LM (2002) Combining mapping and arraying: an approach to candidate gene identification. Proc Natl Acad Sci USA 99:14903–14906

Werner JD, Borevitz JO, Warthmann N, Trainer GT, Ecker JR, Chory J, Weigel D (2005) Quantitative trait locus mapping and DNA array hybridization identify an *FLM* deletion as a cause for natural flowering-time variation. Proc Natl Acad Sci USA 102:2460–2465

West MAL, Kim K, Kliebenstein DJ, van Leeuwen H, Michelmore RW, Doerge RW, St. Clair DA (2007) Global eQTL mapping reveals the complex genetic architecture of transcript-level variation in *Arabidopsis*. Genetics 175:1441–1450

Whitt SR, Wilson LM, Tenaillon MI, Gaut BS, Buckler ES (2002) Genetic diversity and selection in the maize starch pathway. Proc Natl Acad Sci USA 99:12959–12962

Wilson LM, Whitt SR, Ibanez AM, Rocheford TR, Goodman MM, Buckler ES (2004) Dissection of maize kernel composition and starch production by candidate gene association. Plant Cell 16:2719–2733

Wittkopp PJ, Haerum BK, Clark AG (2004) Evolutionary changes in *cis* and *trans* gene regulation. Nature 430:85–88

Xu K, Xu X, Fukao T, Canlas P, Maghirang-Rodriguez R, Heuer S, Ismail AM, Bailey-Serres J, Ronald PC, Mackill DJ (2006). *Sub1A* is an ethylene-response-factor-like gene that confers submergence tolerance to rice. Nature 442:705–708

Yamasaki M, Tenaillon MI, Bi IV, Schroeder SG, Sanchez-Villeda H, Doebley JF, Gaut BS, McMullen MD (2005) A large-scale screen for artificial selection in maize identifies candidate agronomic loci for domestication and crop improvement. Plant Cell 17:2859–2872

Yan L, Fu D, Li C, Blechl A, Tranquilli G, Bonafede M, Sanchez A, Valarik M, Yasuda S, Dubcovsky J (2006) The wheat and barley vernalization gene *VRN3* is an orthologue of *FT*. Proc Natl Acad Sci USA 103:19581–19586

Yan L, Loukoianov A, Blechl A, Tranquilli G, Ramakrishna W, SanMiguel P, Bennetzen JL, Echenique V, Dubcovsky J (2004) The wheat *VRN2* gene is a flowering repressor down-regulated by vernalization. Science 303:1640–1644

Yan L, Loukoianov A, Tranquilli G, Helguera M, Fahima T, Dubcovsky J (2003) Positional cloning of the wheat vernalization gene *VRN1*. Proc Natl Acad Sci USA 100:6263–6268

Yano M, Katayose Y, Ashikari M, Yamanouchi U, Monna L, Fuse T, Baba T, Yamamoto K, Umehara Y, Nagamura Y, Sasaki T (2000) *Hd1*, a major photoperiod sensitivity quantitative trait locus in rice, is closely related to the *Arabidopsis* flowering time gene *CONSTANS*. Plant Cell 12:2473–2484

Zamir D (2001) Improving plant breeding with exotic genetic libraries. Nat Rev Genet 2:983–989

Zhang Z, Ober JA, Kliebenstein DJ (2006) The gene controlling the quantitative trait locus *EPITHIOSPECIFIER MODIFIER1* alters glucosinolate hydrolysis and insect resistance in *Arabidopsis*. Plant Cell 18:1524–1536

CHAPTER 10

USE OF SERIAL ANALYSIS OF GENE EXPRESSION (SAGE) FOR TRANSCRIPT PROFILING IN PLANTS

PRAKASH C. SHARMA[1,*], HIDEO MATSUMURA[2] AND RYOHEI TERAUCHI[2]

[1] *University School of Biotechnology, Guru Gobind Singh Indraprastha University, Delhi 110006, India*

[2] *Iwate Biotechnology Research Center, 22-174-4, Narita, Kitakami, Iwate 024-0003, Japan*

Abstract: Serial Analysis of Gene Expression (SAGE) is a powerful technique for genome wide analysis of gene expression. The SAGE technique quantifies a 'tag' which represents the transcriptome product of a gene. A tag for the purpose of SAGE, is a nucleotide sequence of a defined length, directly adjacent to the 3'-most restriction site for a particular restriction enzyme. Thus, data product of SAGE is a list of tags with their count values, providing a digital representation of cellular gene expression. Several technical modifications have been made to the original SAGE protocol to improve its efficiency, reducing the amount of input RNA, increasing the length of SAGE tags, and allowing further use of SAGE results. This chapter deals with the methodology of SAGE, problems associated with SAGE, various SAGE modifications attempted, comparison with other contemporary high-throughput methods like microarrays, and current applications of SAGE in plant transcript profiling. In plants, SAGE has been used to analyse: (1) host-pathogen interactions, (2) plant responses to various environmental and nutritional stresses, (3) metabolism of toxic compounds, and (4) transcript profiling of a particular tissue/organ. Although majority of these studies have confined to model plants i.e. rice and *Arabidopsis thaliana*, a few recent efforts have extended the use of SAGE to other plant species also.

1. INTRODUCTION

The goals of crop sciences are to increase crop productivity, improve crop quality, and maintain the environment. The increase that is required in the world food supply will have to rely on increase in economically viable and sustainable food

*Corresponding Author: deansbt@yahoo.co.in, sharmapc_meerut@yahoo.com

R.K. Varshney and R. Tuberosa (eds.), Genomics-Assisted Crop Improvement: Vol. 1: Genomics Approaches and Platforms, 227–244.

production. This may be possible by reducing pre- and post harvest losses due to pests and pathogens and stabilizing yields in poor soils and changing environments. Analysis and manipulation of plant genomes using molecular tools provide practical approaches to enhance the efficiency of agriculture by improving in both the quantity and quality of production. In plants, genetic information is now being uncovered *en mass* such that plant that used to be looked at in terms of its individual genes can now be examined in terms of its genome organization, expression and interaction.

Initiatives on whole genome sequencing efforts for more than 450 eukaryotic species have provided a wealth of sequence information available in the public domain. The publication of the complete genome sequence of model plants *Arabidopsis thaliana* and rice has equipped the biologists with the opportunity to describe a plant's basic genetic determinants (Arabidopsis Genome Initiative, 2000; International Rice Genome Sequencing Project, 2005). Information on both the physical and functional annotation of the genome can be gained through transcript profiling (Hughes et al. 2001; Shoemaker et al. 2001). In recent years, transcript profiling has become synonymous with gene expression analysis, largely because of the technical difficulties and greater molecular complexities of proteomics and metabolomics (Smith 2000). The wide accessibility of transcript profiling in recent years has led to the establishment of various high-throughput methodologies of gene expression analysis. These methodologies differ in their convenience, expense, number of transcripts assayed and sensitivity (Kuhn 2001). However, as in case of genome sequencing projects, automation and efficient data management are essential factors in all comprehensive transcript profiling systems.

2. METHODS OF TRANSCRIPT PROFILING

Understanding of underlying principles governing gene expression across genomes has generated immense interest in this area of experimental biology. Ueda et al. (2004) studied genome-wide gene expression in many experimental conditions in six model organisms namely *E. coli*, *S. cerevisiae*, *D. melanogaster*, *A. thaliana*, *M. musculus* and *H. sapiens*. This comprehensive study demonstrated that the gene expression dynamics follows the same and surprisingly simple principle from *E. coli* to human. Their findings provide a universal principle in the regulation of gene expression and show how complex and dynamic organization can emerge from simple underlying dynamics. Thus, there is a great deal of universality and flexibility in gene expression across a wide range of taxa.

Transcript profiling has been going on in one form or another for over 25 years (reviewed by Goldberg 2001). This period has exploited techniques such as northern transfer hybridization, S1 nuclease analysis and *in situ* hybridization. While these methods are characterized by good, well-defined sensitivities, they are time-consuming, and therefore well suited for the in-depth analysis of a small number of genes. The first genomics technology practiced on a large scale, sequencing the 3' ends of cDNAs to produce expressed sequence tags (ESTs) provided a cost effective and rapid approach to describe the collection of genes expressing at a given time in a particular tissue during the life cycle of an organism. By comparison, current high-throughput transcript profiling technologies have relatively poorly defined sensitivities. Therefore, these early methods provide both a valuable means of confirming and extending the results obtained with the more global approaches.

The high-throughput approaches can be divided into two classes: (1) Direct analysis, including procedures involving nucleotide sequencing [EST sequencing (Adams et al. 1995); serial analysis of gene expression (SAGE) (Velculescu et al. 1995); massively parallel signature sequencing (MPSS) (Brenner et al. 2000)], and fragment sizing [Differential display (DD) (Liang and Pardee 1992), cDNA-amplified fragment length polymorphism analysis (cDNA-AFLP) (Bachem et al. 1996)]; and (2) Indirect analysis, involving nucleic acid hybridization of mRNA or cDNA fragments [oligo chips (Lockhart et al. 1996), cDNA microarrays (Schena et al. 1995)]. Details regarding principle, cost involved and relative advantages and disadvantages of different methods are available elsewhere (Donson et al. 2002; Bals and Jany 2001; Cekan 2004; Meyers et al. 2004).

3. SERIAL ANALYSIS OF GENE EXPRESSION (SAGE)

Serial analysis of gene expression, or SAGE, is a technique designed to take advantage of high-throughput sequencing technology to obtain a quantitative profile of gene expression. Essentially, the SAGE technique measures not the expression level of a gene, but quantifies a 'tag' which represents the transcriptome product of a gene. A tag for the purpose of SAGE, is a nucleotide sequence of a defined length, directly adjacent to the 3'-most restriction site for a particular restriction enzyme. The data product of the SAGE technique is a list of tags, with their corresponding count values, and thus provides a digital representation of cellular gene expression. Theoretically, a sequence stretch, as short as 9 bp can distinguish 4^9that is 2,62,144 transcripts, provided a random nucleotide distribution throughout the genome. This ability appears sufficient to discriminate all the transcripts in higher plants and humans as well.

3.1. Methodology of SAGE

SAGE procedure starts with the synthesis of double-stranded cDNA from mRNA using a biotinylated oligo(dT) primer. The cDNA is then cleaved with a restriction enzyme (called anchoring enzyme, AE). *Nla*III is the most frequently used enzyme, but substitution like *Sau*3A and *Rsa*1 are possible. The 3' –most region of the cleaved cDNA with a common *Nla*III cohesive end at its 5' –terminus is then recovered by binding to streptavidin-coated beads. After dividing the reaction mixture into two portions, two independent linkers are ligated using *Nla*III cohesive termini to each portion. These linkers are designed to contain type IIS enzyme (usually *Fok*1 or *Bsm*F1, designated as tagging enzyme, TE) site near (or partially overlapping) the 3' –*Nla*III sequence. The reaction mixtures are digested with type IIS enzyme and released portions are recovered. Resulting staggered ends of the products are blunt-ended by T4 DNA polymerase. Two portions are mixed again and ligated. Since the 5' -ends of the linkers are blocked by amino group, only the mRNA derived termini can ligate in a tail-to-tail orientation. The products (ditags) are PCR-amplified, cleaved by *Nla*III, and then separated by polyacrylamide gel electrophoresis (PAGE). Ditag fragments flanked both ends with *Nla*III cohesive terminus are isolated and ligated to obtain concatemers. Highly concatenated products are recovered by PAGE, cloned into a suitable plasmid vector to create a SAGE library, followed by sequencing of individual clones. Alternatively, ditags can immediately be sequenced by 454 platform to reduce cost. The resulting sequence data are analysed using the SAGE software (Johns Hopkins University, Baltimore, MD) which extracts the tags from the sequence and determines their abundance and identity. Figure 1 shows a flow sheet of different steps of SAGE procedure. Since a typical SAGE experiment involves the comparison of at least

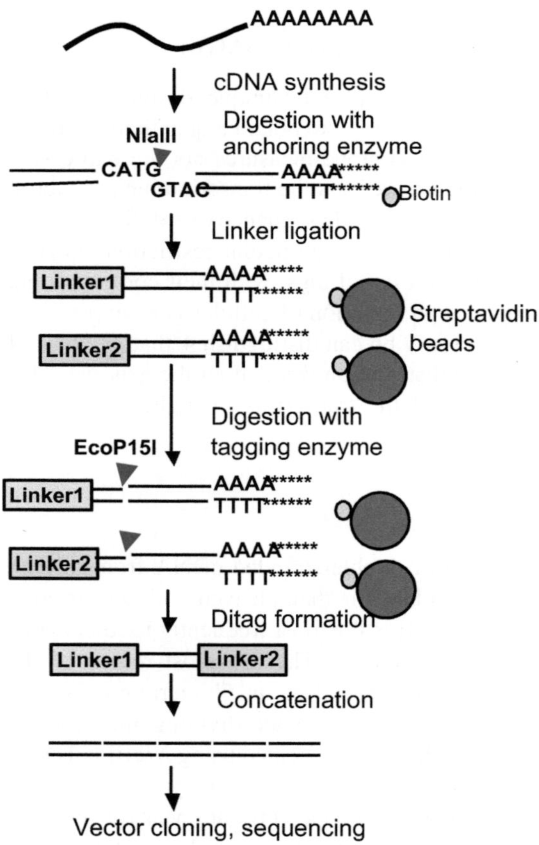

Figure 1. Schematic diagram of SAGE procedure (see text for details) *(see plate 8)*

two different libraries, the statistical significance of gene expression differences between libraries is calculated for each transcript detected. The various aspects of statistical analysis of SAGE data and the repertoire of bioinformatics tools for SAGE have been presented by Pylouster et al. (2005) and Tuteja and Tuteja (2004a).

3.2. Problems Associated with SAGE

Although SAGE produces a digital output, loss of fidelity may occur during conversion of an actual transcript and its expression level to a tag and its count value. Accuracy in both the assignment of tags to genes as well as the ability to quantify a gene's expression level are sacrificed in order to increase throughput, that in turn increase the speed and lower the cost of analysis.

So, two problems arise when dealing with SAGE data. The first deals with ensuring that the tags and their counts are a valid representation of transcripts and their level of expression, and the second, with making valid tag to gene assignments. In consideration of the first problem, the valid tag sequence data, sequencing error has the greatest effect. Assuming that there is an average 1% per base sequencing error rate, for ten bases, the chance of one or more errors occurring is roughly 10%. Therefore, such an error will lower or increase the correct tag count by one, or will establish a tag which really does not exist. Such a situation will not seriously affect the results for tags with relatively high counts but definitely create problem in cases where tag count is very low, particularly for those with a count of one. Therefore, one should either exclude the tags with such very low counts or should adopt other means of verification of the results.

The second problem, making valid tag to gene assignments, unspecific and ambiguous tag to gene assignments, as well as sequencing error, both play a role in creating confusion. In making tag to gene assignments, a certain degree of ambiguity is encountered. One solution to overcome these problems has been attempted by increasing the size of the SAGE tag. Although the original SAGE protocol yielded 9 base pair tag (Velculescu et al. 1995) further modifications in the technique increased the tag size to up to 14 bp, 21 bp in LongSAGE (Saha et al. 2002), and 26 bp in the SuperSAGE (Matsumura et al. 2003, 2005) by using the different tagging enzymes. The longer tag SAGE not only ensures more fidelity of results but the information so derived on gene expression profiling may be extended for further applications like SuperSAGE array (Matsumura et al. 2006). Another approach, known as GLGI (Generation of Longer cDNA fragments from SAGE tags for Gene Identification; Chen et al. 2000, 2002a, 2002b) is based on converting SAGE tags into their corresponding 3' ends using a PCR strategy. Lee et al. (2002) compared the two approaches and showed that significantly higher enhancement of tag specificity could be achieved using the GLGI approach compared to increasing the tag length from 10 to 35 bases. Since the conventional 14-bp SAGE tags may not always perform well in the GLGI approach due to length constraint, a combination of Long/SuperSAGE and GLGI should lead to more robust SAGE-based gene expression profiling circumventing the problem of decreased specificity in the conventional approach.

There will be instances in which multiple genes share the same tag as observed frequently in the gene families, and instances in which one gene has multiple tags as in the genes having alternate Poly (A) sites. A population polymorphism may also pose a similar problem. Moreover, messages without a poly (A) tail will also be excluded from SAGE analysis.

Some other technological problems also encounter the SAGE technology. The requirement of relatively high amount of starting material that is mRNA poses problems in constructing meaningful SAGE libraries. To counter this problem, various modifications of SAGE have been attempted (see next section) which allow up to 5000 fold reduction in the starting material.

Another technical problem often propping up in SAGE experiments is of contamination by linker-dimers. To minimize such contamination, Powell (1998) introduced the use of biotinylated primers. Kenzelmann and Muhlemann (1999) suggested a simple heating step during final ligation step that enhanced the length concatemers with an average of 67 tags. This eliminated the small sized concatemers and thereby increased the efficiency of tag collection by reducing the sequencing efforts.

A major problem of the SAGE approach is how to further analyze the unassigned tags. Since the original strategy (Velculescue et al. 1995) utilized the conventional oligonucleotide-based plaque lift method that may yield many false positives. Different workers have attempted to address the problem (RAST-PCR, van den Berg et al. 1999; Matsumura et al. 1999).

SAGE is generally believed to provide an unbiased and quantitative report of gene expression. However, it can under perform for a number of reasons. Certain transcripts may be missed due to the absence of a recognition site for the anchoring enzyme or GC content bias (Margulies et al. 2001). While the use of two anchoring enzymes may help bypass this problem, this would be financially unviable. Transcripts that produce multiple tags is a frequently reported problem in the literature (Welle et al. 1999; Neilson et al. 2000; Fizames et al. 2004). Alternate splicing and incomplete digestion with the anchoring enzyme during library construction have also been reported as source of multiple tags per transcript in humans (Welle et al. 1999). In barley, it appears that polyadenylation is more likely to be the source of such tags (Ibrahim et al. 2005). Unneberg et al. (2003) and Pleasance et al. (2003) have analyzed influence of important factors viz. tag length, restriction site and transcript database on transcript identification.

3.3. Modifications of SAGE

Several technical modifications have been made to the original SAGE protocol to improve its efficiency and for further use of SAGE results. These include:

Micro-SAGE (Datson et al. 1999) or *SAGE-lite* (Peters et al. 1999): The former technique uses a preliminary cDNA PCR step that could potentially introduce an amplification bias. The latter modification has formed the basis of performing the standard enzymatic steps on mRNA attached to a solid phase.

SAR-SAGE (small amplified RNA-SAGE) (Vilain et al. 2003) method involves T7 RNA polymerase dependent transcription of the mRNA segment between the tag and the poly (A) tail. This additional amplification step allows preparation of SAGE libraries with as low as 50 ng of total RNA. Moreover, since no PCR step is used to amplify the starting material, any PCR caused representation bias is overcome.

MiniSAGE (Ye et al. 2000) and *SADE (SAGE adaption for downsized extracts*; Virlon et al. 1999) developed to diminish the loss of material throughout the procedure permit lowering of input RNA.

LongSAGE : uses a different type IIS restriction endonuclease *Mme*I as the tagging enzyme that cuts 17-bp 3' from the anchoring site (Saha et al. 2002). The longer tags increase the power of identification of genes, while not diminishing the sensitivity of SAGE given by the use of PCR and concatenation.

RL-SAGE (Robust-LongSAGE) : An improved version of LongSAGE with better cloning efficiency and increased insert size (Gowda et al. 2004).

SuperSAGE : A method for the isolation of tag sequences >25 bp from defined positions of cDNA by using the type III restriction enzyme *Eco*P15I . Use of this type III endonuclease as the "tagging enzyme" dramatically improves the conventional SAGE protocol through a reliable identification of the corresponding genes and an accurate gene expression analysis (Matsumura et al. 2003, 2005).

Other modifications have improved cDNA synthesis by using more efficient polymerases, minimized contaminants that inhibit ditag formation, release, and concatenation by adding purification steps and improved the screening of concatemer inserts (Powell 1998; Virlon et al. 1999; Lee et al. 2001; Angelastro et al. 2000, 2002). A potentially confounding factor is the GC content of the freed ditags that may affect their stability hampering their ability to concatenate. This bias in favour of GC-rich ditags can be prevented by keeping them at a low temperature and by querying the GC content of the concatemer inserts. Munasinghe et al. (2001) have successfully applied SAGE with a few modifications to A-T rich genome of *Plasmodium falciparum.*

Because the transcripts are anchored to oligo(dT) beads, potential internal poly (A) priming has been addressed by Nam et al. (2002). Contaminating genomic DNA fragments containing poly(A) stretches can be eliminated by DNase pre-treatment but RNA species could create spurious tags. Whether oligo(dT) primers in solution or on solid phase magnetic beads have similar internal priming potential is not clear. In practice, this issue has not been a significant factor for the numerous SAGE projects performed.

3.4. Comparison of SAGE with Microarrays

SAGE and microarrays are the two high-throughput methods used to analyse complete transcriptome in different plant, animal and microbial systems. Microarrays allow large scale gene expression analysis in many samples at a time which is not feasible through SAGE. However, as a 'closed architecture' technology, microarrays allow detection of change of expression of spotted genes only and therefore does not allow the discovery of new genes. In contrast, SAGE is an 'open architecture' system that enables both the discovery of new expressed genes and the accurate quantification of the resulting transcripts (Boheler and Stern 2003; Wang 2003). Furthermore, tags of only 14 bp in the initial SAGE and 21 bp in the LongSAGE procedure are too short for unambiguous tag-to-gene annotations, hampering the application of SAGE to 'non-model' organisms for which genome sequence information is not available. SuperSAGE, an improved version of SAGE, allowing the isolation of 26 bp tags, greatly improves the efficiency of tag-to-gene annotation and has extended the application of SAGE to 'non-model' organisms (Matsumura et al. 2003). In a very recent experiment, Matsumura et al. (2006) have used these 26 bp long SuperSAGE tags as hybridization probes in microarray analysis called as SuperSAGE array. This platform combines the advantages of the highly quantitative SuperSAGE analysis with the high-throughput microarray technology. SuperSAGE array system thus represents a new paradigm for microarray construction, as no genomic or cDNA sequence data are required for its preparation.

A comparative study of gene expression pattern in developing barley caryopsis was performed using SAGE and Affymetrix GeneChip technology (Ibrahim et al. 2005). A good overall concurrence between the two methods was observed while reporting absolute expression levels and comparative analysis of gene expression profiles. However, the agreement between two methods was more acceptable for genes expressed at high levels than those expressed at low levels. In addition to this solitary report from plant system, many works involving other biological systems describe direct (Ishii et al. 2000; Kim 2003) and indirect comparisons between these two approaches (Evans et al. 2002; Iacobuzio-Donahue et al. 2003). They also report a good comparability between SAGE and microarray based systems. Nevertheless, some discrepancies between the two methods were evident which can result due to the complexity of the experimental protocols, and potential for associated measurement errors. In Arabidopsis, it was possible to detect transcription from 20243 different annotated genes from leaf tissue using combination of SAGE, MPSS and microarrays (Robinson et al. 2004), which represents approximately 70% of the predicted gene content. Although there may be some degree of error owing to differences in growth conditions and the developmental stage of the material used, SAGE was able to detect expression in leaf tissue from 64% of the annotated genes identified across the three profiling platforms, which is comparable to 66% for microarray analysis but is lower than the 78% detected using MPSS. Therefore, MPSS may have an advantage over SAGE in generating a complete representation of the whole transcriptome, but access to this technology is limited to specialized laboratories.

4. APPLICATIONS OF SAGE

SAGE was originally developed to study differential gene expression in cancerous and normal human tissues as an alternative to EST sequencing with a view to reduce sequencing labour and cost. Therefore, the early SAGE studies have focused mainly on human transcript profiling. Later, the use of technology was extended to other organisms including plants. Current holdings at SAGEdb maintained at NCBI include 625 SAGE libraries, of which only 18 belongs to plants. The total tag number has reached little over 42 million including about 1.7 million tags from plant sources (Table 1). Some other summary SAGE data is provided in Table 2. Important Web resources on SAGE are listed in Table 3.

SAGE has been used to identify tumor markers for a variety of cancers (Riggins 2001, Tuteja and Tuteja, 2004b and references therein). Simultaneously, this technology has been widely applied for other studies aiming to analyse the effect of drugs on tissues, to identify disease-related genes, and to provide insights into disease pathway (reviewed by Ye et al. 2002; Tuteja and Tuteja, 2004c). In all these studies, comparison of the normal and target tissues has led to the identification of a number of highly up-regulated and down-regulated genes. The future studies on these lines will certainly lead to the identification of candidate genes to be focused while aiming to develop effective disease control strategies.

In comparison to human and other animal systems, SAGE in plants has been attempted rather infrequently. Although, the first report of SAGE profiling in rice appeared seven years ago (Matsumura et al. 1999), some good use of the technology has been demonstrated only recently (see Dean and Lorenz 2004; Lorenz and Dean 2005; Meyers et al. 2004). In plants, SAGE has primarily been used to analyse: (1) host-pathogen interactions, (2) plant responses to various environmental and nutritional stresses (3) metabolism of toxic compounds, and (4) transcript profiling of a particular tissue/organ. Majority of these studies have confined to model plants i.e. rice and *Arabidopsis thaliana*. However, a few recent efforts have extended the use of SAGE to other plant and fungal species also.

Table 1. Summary of current holdings in SAGEdb (http:/www.ncbi.nlm.nih.gov/SAGE)

Organism	Number of libraries	Total number of tags	Number of unique tags
Homo sapiens	327	19300584	1296360
Mus musculus	213	16549657	1552119
Caenorhabditis elegans	17	1928482	211328
Drosophila melanogaster	5	489140	41225
Arabidopsis thaliana	7	248659	37798
Oryza sativa	4	805823	126663
Medicago truncatula	3	131599	8770
Pinus taeda	2	150885	42641
Zea mays	2	368519	59720
Musa acuminata	1	10196	5274
Magnaporthe grisea	1	246967	51927
Plants (Total)	**18**	**1705485**	**275592**
All Organism (18 species)	**625**	**42215451**	**3810871**

Table 2. Data on SAGE libraries in SAGEdb

Characteristic Tag length (bp)	No. of libraries
10	467
14	3
17	154
22	1
Anchoring Enzyme	
*Nla*III	594
*Sau*3A	28
*Rsa*1	3

4.1. Host-Pathogen Interactions

Great emphasis has been laid time-to-time on understanding molecular mechanisms underlying host-pathogen interactions in different biological systems. Availability of detailed information on this subject is necessary to devise strategies to control crop diseases, which reduce crop production by 10% worldwide. One step in this direction focuses on gene expression profiling in both, the host and the pathogen.

Matsumura et al. (2003a) has developed and used SuperSAGE technology to monitor gene expression profiles of both rice and its blast fungus *Magnaporthe grisea*, simultaneously making use of fully sequenced genomes of both the organisms. The *hydrophobin* gene was found to be the most actively transcribed *M. grisea* gene in blast-infected rice leaves. Earlier the same group (Matsumura et al. 2003b) has exploited normal SAGE to elucidate genes involved in cell death and associated processes in rice suspension culture cells treated with cell wall extract of *M. grisea*. Among the down regulated genes in the elicitor treated cells, a *BI-1* gene coding for the Bax inhibitor was identified. The functional role of this gene identified through SAGE was confirmed by overexpressing the gene in

Table 3. Important SAGE resources on the internet

Web site URL Description	
SAGEmap www.ncbi.nlm.nih.gov/SAGE	Public database repository by NCBI at National Institute of Health conjuction with NIH's Cancer Genome Anatomy Project (CGAP)
SAGEnet www.sagenet.org	Public SAGE database maintained by Vogelstein/Kinzler Lab at Johns Hopkins University, Baltimore
Genzyme www.genzyme.com/SAGE	Properietary SAGE database; company also offers SAGE services for commercial users
Mouse SAGE site http://mouse.biomed.cas.cz/sage/	Public database maintained at the Institute of Molecular Genetics, Academy of Sciences of Czech Republic

transgenic rice. Another study (Irie et al. 2003), used SAGE to identify genes involved in appressorium formation in *M grisea* by analysing expression profile of fungal conidia in the presence and absence of cAMP. RT-PCR of 13 randomly selected genes confirmed the SAGE results, verifying the fidelity of the SAGE data. Thomas et al. (2002) used SAGE to characterize changes in transcript accumulation during early development of barley mildew pathogen *Blumeria graminis* on barley leaves.

SuperSAGE has also been used to study gene expression changes preceding hypersensitive response (HR) caused by INF1 elicitor in *Nicotiana benthamiana*, a non-model plant recently used for a number of experiments on functional genomics. *NbCD1* and *NbCD3* genes encoding proteins that induce cell death upon overexpression were isolated from *N. benthamiana* (Nasir et al. 2005). Several up- or down-regulated genes were identified following *NbCD1* overexpression in *N. benthamiana* by infiltrating leaves with *Agrobacterium tumefaciens* carrying plasmids harbouring *NbCD1* and GFP (control) in the glucocorticoid –inducible gene expression cassette GVG. SuperSAGE along with RT-PCR validation identified 58 differentially expressed transcripts between the two samples. Several of the differentially expressed genes were previously linked to pathogen defense and hypersensitive death. The results on differential gene expression in response to *NbCD1* and *NbCD3* genes obtained through SuperSAGE and SuperSAGE array (Matsumura et al. 2003) were significantly similar as 74% of all the expressed genes showed identical expression patterns. These authors have strongly demonstrated the use of SuperSAGE, SuperSAGE array and RACE (rapid amplification of cDNA ends) techniques in combination for systematic and comprehensive transcription analysis as well as rapid isolation of target genes for functional analysis in even non-model organisms where no prior sequence databases are available. These tools certainly will find promising applications in human studies such as for cancer diagnostics and prognostics (Argani et al. 2001; Yasui et al. 2004).

One of the efficient and currently followed reverse genetics approach for functional genomics involves RNAi. SuperSAGE tags are long enough to be directly used for synthesizing short-hairpin (sh)RNA, which is expressed in eukaryotic cells to induce RNA interference with the corresponding gene. Tag sequence may be cloned in a suitable expression vector and transferred directly into fungal cell or plant cells via *Agrobacterium.* After RNAi of the corresponding gene is established, virulence (fungus) and resistance (plant) can be tested (Matsumura et al. 2006). Thus, SAGE has great potential as an important tool in functional genomics (Figure 2).

Mysore et al. (2001) identified tomato genes that are up-regulated and down-regulated by *Pti4* (Pto interacting protein), a tomato transcription factor, through SAGE analysis. Subsequently, in a more detailed study (Chakravarthy et al. 2003), SAGE was used to compare the transcripts in wild type and Pti4-expressing *Arabidopsis* plants. Comparative profiling revealed 78 differentially abundant transcripts encoding defense-related proteins, protein kinases, ribosomal proteins, transporters, and two transcription factors. Many of the genes identified were

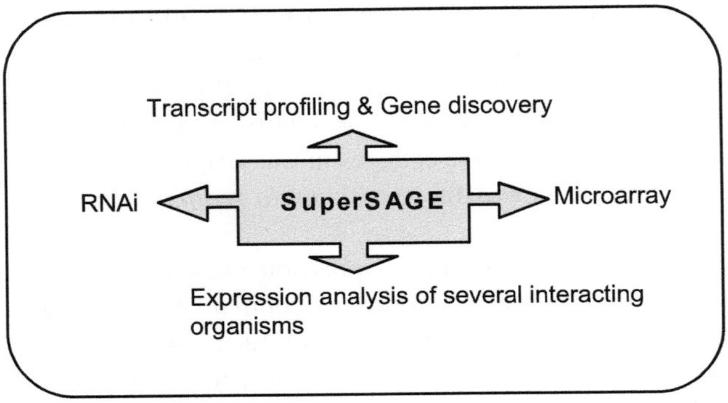

Figure 2. Applications of SuperSAGE in functional genomics

expressed differentially in wild-type *Arabidopsis* during infection by *Pseudomonas syringae* pv tomato, supporting a role for them in defense related processes. In cassava (*Manihot esculanta*), host-plant resistance to the cassava mosaic disease (CMD) was analysed using SAGE by comparing gene expression patterns in a bulk of 40 each of CMD resistant and susceptible genotypes drawn from a gene mapping population (Fregene et al. 2004). Annotation of more than 30 differentially expressed tags revealed several genes expressed during systemic acquired resistance (SAR) in plants and other genes involved in cell-to-cell and cytoplasm-to-nucleus virus trafficking. The genes identified through SAGE have provided a sound basis for further research towards identification of candidate genes for effective disease management.

4.2. Response to Environmental Stresses

Breeding crops adapted to extreme environments like cold, drought is one of the main objectives of modern plant breeding. Information regarding differential gene expression in these environments is therefore very important. While studying global gene expression in Arabidopsis leaves Jung et al. (2003) found that a majority of genes expressed in normal leaves were involved in energy metabolism processes, especially in photosynthesis. Further, cold treated leaves revealed 272 differentially expressed genes. Of these 82 genes were highly expressed in the normal leaves and 190 were highly expressed in the cold treated leaves. Cold stress, in general, induced genes involved in processes like cell rescue, defense, cell death and aging. These included various COR genes, lipid transfer protein genes, alcohol dehydrogenase, β amylase and many novel genes. During cold stress, genes involved in energy metabolism process like photosynthesis were down-regulated. This global gene expression pattern seen in normal and cold treated leaves is drastically different with that in normal and cold treated pollens in Arabidopsis (Lee and Lee 2003). These experiment provide insight into the mechanism of differential sensitivity of different plant organs to stress. Similarly, Arabidopsis root transcriptome (Fizames et al. 2004) differs markedly from other organs. During this study, 270 differentially expressed genes have been identified in roots in response to N sources i.e., different NO_3^- or NH_4NO_3. The maize root transcriptome analysis also documented that less than 5% of the most abundant transcript were shared between maize and Arabidopsis (Poroyko et al. 2005).

Ekman et al. (2005) used SAGE to assess transcriptome response to hexahydro-1,3,5-trinitro-1,3,5-triazine (RDX) in Arabidopsis roots. The RDX treatment induced genes already known to respond to a variety of general stresses. Among the highly induced genes, several encoded molecular chaperones and transcription factors as well as vacuolar proteins and peroxidases. Strongly repressed transcripts included ones encoding ribosomal proteins, a cyclophilin, a katanin and a peroxidase. The expression pattern in response to RDX was significantly different to that against 2,4,6-trinitrotoluene (TNT) earlier studied by Ekman et al. (2003). It gives an idea that plants employ different mechanism to cope with different chemicals. This type of information will certainly help in engineering crops better tolerant to explosives and similar compounds and to devise strategies for phytoremediation of degraded soils.

4.3. SAGE for Other Traits

SAGE technology has been extended to some non-model but otherwise important plant species also, for example, loblolly pine (*Pinus taeda*) (Lorenz and Dean 2002). Two SAGE libraries were generated based on lignifying xylem isolated from either the upper (crown) or lower (base) portions of the trunk in view to identifying genes specifically involved in wood formation and characterizing their roles in determining wood quality. Coemans et al. (2005) used SuperSAGE coupled with 3'RACE (3'-rapid amplification of cDNA ends; van den Berg et al. 1999) and TAIL-PCR (thermal asymmetric interlaced PCR; Liu et al. 1995) to characterize the global gene expression in another non-model plant, banana (*Musa acuminata*). As expected, and also reported for Arabidopsis (Jung et al. 2003), majority of the abundant transcripts are involved in energy production, mainly photosynthesis. The most abundant tag represented a type 3 metallothionein transcript which accounted for nearly 3% of the total transcript analyzed as in case of rice leaf also (Gibbings et al. 2003).

Asamizu et al. (2005) performed comprehensive transcript analysis pertaining to early stage of root nodulation by generating SAGE libraries from uninfected roots and nodulating roots of the model legume *Lotus japonicus*. They reported different levels of transcript induction among leghemoglobin gene paralogs indicating the effectiveness of SAGE in discriminating different gene family members. They found 11 antisense tags that increased during nodulation indicating that regulation of gene expression by antisense transcripts may occur in an organ dependent manner. In a project undergoing at Giessen University, Germany, attempts are being made to study gene expression in oilseed rape (*Brassica napus*) and to identify genes which are up- and down-regulated during seed development and are associated with synthesis of commercially valuable seed compounds, e.g. fatty acid metabolism (Obermeier and Snowdon 2006)

In a massive study from Beijing Genomics Institute, China, transcriptomes of three major tissues (panicles, leaves, and roots) of a super-hybrid rice (*Oryza sativa*) strain were compared with the two parents (Bao et al. 2005). A number of genes differentially expressed in the three target tissues in the hybrid. Most of the up-regulated genes were related to enhancing carbon- and nitrogen-assimilation, including photosynthesis in leaves, nitrogen uptake in roots, and rapid growth in both roots and panicles. An essential enzyme involved in photorespiration, alanine:glyoxylate aminotransferase 1 was the most conscipicuous among the down regulated genes in the hybrid. Such studies will provide additional data crucial for understanding of molecular mechanism of heterosis and gene regulation networks of the cultivated rice.

5. FUTURE PERSPECTIVES

Global transcript profiling provides a better insight into the transcribed genome of an organism, a cell or a tissue under a particular environmental condition. Amongst the various contemporary high-throughput technologies, SAGE is an 'Open Architecture' system as compared to DNA chip technology enabling accurate quantification and identification of newly expressed genes also. Recently developed SuperSAGE coupled with SuperSAGEArray allowing isolation and use of 26 tags has greatly overcome the limitation of short tags in original SAGE technology and has facilitated unambiguous gene annotations. GLGI (Generation of Longer fragments from SAGE tags For Gene Identification) has helped in further characterization of individual target genes. Moreover, a number of attempts have reduced drastically the amount of starting material for SAGE. The utility of SAGE in transcript profiling has already been demonstrated in various plant and animal systems for different traits and environmental conditions.

The advantages of SAGE including application to non-model organisms and in discovery of new genes will certainly extend the use of SAGE. Reduction in the cost of SAGE particularly for library sequencing and generation of longer SAGE tags will further enhance the use of SAGE in near future.

ACKNOWLEDGEMENTS

RT thanks "Program for Promotion of Basic Research Activities for Innovative Bioscience" and JSPS grant no. 18310136. HM thanks JSPS grant no. 18688001.

REFERENCES

Adams MD, Kerlavage AR, Fleischmann RD, Fuldner RA, Bult CJ, Lee NH, Kirkness EF, Weinstock KG, Gocayne JD, White O et al (1995) Initial assessment of human gene diversity and expression patterns based upon 83-million nucleotides of cDNA sequence. Nature 377:3–7

Angelastro JM, Klimaschewski LP, Vitolo OV (2000) Improved *NlaIII* digestion of PAGE-purified 102 bp ditags by addition of a single purification step in both the SAGE and microSAGE protocols. Nucleic Acids Res 28:E62

Angelastro JM, Ryu EJ, Torocsik B, Fiske BK, Greene LA (2002) Blue–white selection step enhance the yield of SAGE concatemers. Biotechniques 32:484–486

Arabidopsis Genome Initiative (2000) Analysis of the genome sequence of the flowering plant *Arabidopsis thaliana.* Nature 408:796–815

Argani P, Rosty C, Reiter RE, Wilentz RE, Murugesan SR, Leach SD, Ryu B, Skinner HG, Goggins M, Jaffee EM, Yeo CJ, Cameron JL, Kern SE, Hruban RH (2001) Discovery of new markers of cancer through serial analysis of gene expression: prostate stem cell antigen is overexpressed in pancreatic adenocarcinoma. Cancer Res 61:4320–4324

Asamizu E, Nakamura Y, Sato S, Tabata S (2005) Comparison of the transcript profiles from the root and nodulating root of the model legume *Lotus japonicus* by serial analysis of gene expression. Mol Plant Microbe Interact 18:487–498

Bachem CWB, van der Hoeven RS, de Bruijin SM, Vreugdenhil D, Zabeau M, Visser RGF (1996) Visualization of differential gene expression using a novel method of RNA fingerprinting based on AFLP: analysis of gene expression during potato tuber development. Plant J 9:745–753

Bals R, Jany B (2001) Identification of disease genes by expression profiling. Eur Respir J 18:882–889

Bao J, Lee S, Chen C, Zhang X, Zhang Y, Liu S, Clark T, Wang J, Cao M, Yang H, Wang SM, Yu J (2005) Serial analysis of gene expression study of a hybrid rice strain (LYP9) and its parental cultivars. Plant Physiol 138:1216–1231

Boheler KR, Stern MD (2003) The new role of SAGE in gene discovery. Trends Biotechnol 21:55–57

Brenner S, Johnson M, Bridgham J, Golda G, Lloyd DH, Johnson D, Luo S, McCurdy S, Foy M, Ewan M, Roth R, George D, Eletr S, Albrecht G, Vermaas E, Williams SR, Moon K, Burcham T, Pallas M, DuBridge RB, Kirchner J, Fearon K, Mao JI, Corcoran K (2000) Gene expression analysis by massively parallel signature sequencing (MPSS) on microbead array. Nature Biotechnol 18:630–634

Cekan SZ (2004) Methods to find out the expression of activated genes. Reprod Biol Endocrinol 2:68

Chakravarthy S, Tuori RP, D'Ascenzo MD, Fovert PR, Despres C, Martin GB (2003) The tomato transcription factor *Pti4* regulates defence-related gene expression via GCC box and non-GCC box cis elements. Plant Cell 15:3033–3050

Chen JJ, Lee SG, Zhou GL, Wang SM (2002b) High-throughput GLGI procedure for converting a large number of serial analysis of gene expression tag sequences into 3′ complementary DNAs. Genes Chromosomes Cancer 33:252–261

Chen JJ, Rowley JD, Wang SM (2000) Generation of longer cDNA fragments from serial analysis of gene expression tags for gene identification. Proc Natl Acad Sci USA 97:349–353

Chen JJ, Sun M, Lee SG, Zhou GL, Rowley JD, Wang S (2002a) Identifying novel transcripts and novel genes in the human genome by using novel SAGE tags. Proc Natl Acad Sci USA 99: 12257–12262

Coemans B, Matsumura H, Terauchi R, Remy S, Swennen R, Sagi L (2005) Super SAGE combined with PCR walking allows global gene expression profiling of banana (*Musa acuminata*), a non-model organism. Theor Appl Genet 111:1118–1126

Datson NA, van der perk-de Jong J, van den Berg MP, de Kloet ER, Vreugdenhil E (1999) MicroSAGE: a modified procedure for serial analysis of gene expression in limited amounts of tissue. Nucleic Acids Res 27:1300–1307

Dean JFD, Lorenz WW (2004) Transcriptional profiling in plants using serial analysis of gene expression (SAGE). AgBiotechNet 6:124–130

Donson J, Fang Y, Espiritu-Santo G, Xing W, Salazar A, Miyamoto S, Armendarez V, Volkmuth W (2002) Comprehensive gene expression analysis by transcript profiling. Plant Mol Biol 48: 75–97

Ekman DR, Lorenz WW, Przybyla AE, Wolfe NL, Dean JF (2003) SAGE analysis of transcriptome responses in *Arabidopsis* roots exposed to 2,4,6-trinitrotoulene. Plant Physiol 133:1397–406

Ekman DR, Wolfe NL, Dean JF (2005) Gene expression changes in *Arabidopsis thaliana* seedling roots exposed to the munition hexahydro-1,3,5-trinitro-1,3,5-triazine. Environ Sci Technol 39:6313–6320

Evans SJ, Datson NA, Kabbaj M, Thompson RC, Vreugdenhii E, De Kloet ER, Watson SJ, Akil H (2002) Evaluation of affymetrix gene chip sensitivity in rat hippocampal tissue using SAGE analysis. Eur J Neurosci 16:409–413

Fizames C, Muños S, Cazettes C, Nacry P, Boucherez J, Gaymard F, Piquemal D, Delorme V, Commes T, Doumas P, Cooke R, Marti J, Sentenac H, Gojon A (2004) The *Arabidopsis* root transcriptome by serial analysis of gene expression. Gene identification using the genome sequence. Plant Physiol 134:67–80

Fregene M, Matsumara H, Akano A, Dixon A, Terauchi R (2004) Serial analysis of gene expression (SAGE) of host-plant resistance to the cassava mosaic disease (CMD). Plant Mol Biol 56:563–571

Gibbings JG, Cook BP, Dufault MR, Madden SL, Khuri S, Turnbull CJ, Dunwell JM (2003) Global transcript analysis of rice leaf and seed using SAGE technology. Plant Biotechnol J 1:271–285

Goldberg RB (2001) From cot curves to genomics. How gene cloning established new concepts in plant biology. Plant Physiol 125:4–8

Gowda M, Jantasuriyarat C, Dean RA, Wang G-L (2004) Robust-LongSAGE (RL-SAGE): a substantially improved LongSAGE method for gene discovery and transcriptome analysis. Plant Physiol 134:890–897

Hughes TR, Mao M, Jones AR, Burchard J, Marton MJ, Shannon KW, Lefkowitz SM, Ziman M, Schelter JM,, Meyer MR, Kobayashi S, Cavet G, Walker WL, West A, Coffey E, Shoemaker DD, Stoughton R, Blanchard AP, Friend SH, Linsley PS (2001) Expression profiling using microarrays fabricated by an ink-jet oligonucleotide synthesizer. Nature Biotechnol 19:342–347

Iacobuzio-Donahue CA, Ashfaq R, Maitra A, Adsay NV, Shen-Ong GL, Berg K, Hollingsworth MA, Cameron JL, Yeo CJ, Kern SE et al (2003) Highly expressed genes in pancreatic ductal adenocarcinomas: a comprehensive characterization and comparison of the transcription profiles obtained from three technologies. Cancer Res 63:8614–8622

Ibrahim AFM, Hedley PE, Cardle L, Kruger W, Marshall DF, Muehlbauer GJ, Waugh R (2005) A comparative analysis of transcript abundance using SAGE and Affymetrix arrays. Funct Integr Genomics 5:163–174

International Rice Genome Sequencing Project (2005) The map-based sequence of the rice genome. Nature 436:793–800

Irie T, Matsumura H, Terauchi R, Saitoh H (2003) Serial analysis of gene expression (SAGE) of *Magnaporthe grisea*: genes involved in appressorium formation. Mol Genet Genomics 270: 181–189

Ishii M, Hashimoto S, Tsutsumi S, Wada Y, Matsushima K, Kodama T, Aburatani H (2000) Direct comparison of gene chip and SAGE on the quantitative accuracy in transcript profiling analysis. Genomics 68:136–143

Jung SH, Lee JY, Lee DH (2003) Use of SAGE technology to reveal changes in gene expression in *Arabidopsis* leaves undergoing cold stress. Plant Mol Biol 52:553–567

Kenzelmann M, Muhlemann K (1999) Substantially enhanced cloning efficiency of SAGE (serial analysis of gene expression) by adding a heating step to the original protocol. Nucleic Acids Res 27:917–918

Kim HL (2003) Comparison of oligonucleotide-microarray and serial analysis of gene expression (SAGE) in transcript profiling analysis of megakaryocytes derived from CD3 4(+) cells. Exp Mol Med 35:460–466

Kuhn E (2001) From library screening to microarray technology: strategies to determine gene expression profiles and to identify differentially regulated genes in plants. Ann Bot 87:139–155

Lee JY, Lee DH (2003) Use of serial analysis of gene expression technology to reveal changes in gene expression in arabidopsis pollen undergoing cold stress. Plant Physiol 132:517–529

Lee S, Chen J, Zhou G, Wang SM (2001) Generation of high-quality and quantity tag/ditag cDNAs for SAGE analysis. Biotechniques 31:348–354

Lee S, Clark T, Chen JJ, Zhou GL, Scott LR, Rowley JD, Wang SM (2002) Correct identification of genes from serial analysis of gene expression tag sequences. Genomics 79:598–602

Liang P, Pardee AB (1992) Differential display of eukaryotic messenger-RNA by means of the polymerase chain-reaction. Science 257:967–971

Liu Y, Mitsukawa N, Oosumi T, Whittier RF (1995) Efficient isolation and mapping *Arabidopsis thaliana* T-DNA insert junction by thermal asymmetric interlaced PCR. Plant J 8:457–463

Lockhart DJ, Dong HL, Byrne MC, Follettie MT, Gallo MV, Chee MS, Mittmann M, Wang CW, Kobayashi M, Horton H et al (1996) Expression monitoring by hybridization to high-density oligonucleotide arrays. Nature Biotechnol 14:1675–1680

Lorenz WW, Dean JFD (2002) SAGE profiling and demonstration of differential gene expression along the axial development gradient of lignifying xylem in loblolly pine (*Pinus taeda*). Tree Physiol 22:301–310

Lorenz WW, Dean JFD (2005) Studies of plant gene expression using SAGE. In: Wang SM, Wang M (eds) SAGE: current technologies and applications. Horizon Scientific Press, London, pp.189–206

Margulies EH, Kardia SLR, Innis JW (2001) Identification and prevention of a GC content bias in SAGE libraries. Nucleic Acids Res 29:E60

Matsumura H, Ito A, Saitoh H, Winter P, Kahl G, Reuter M, Krüger DH, Terauchi R (2005) SuperSAGE. Cell Microbiol 7:11–18

Matsumura H, Nasir KHB, Yoshida K, Ito A, Kahl G, Krüger DH, Terauchi R (2006) SuperSAGE array: the direct use of 26-base-pair transcript tags in oligonucleotide arrays. Nat Methods 3:469–474

Matsumura H, Nirasawa S, Kiba A, Urasaki N, Saitoh H, Ito M, Kawal-Yamada M, Uchimiya H, Terauchi R (2003b) Overexpression of Bax inhibitor supresses the fungal elicitor-induced cell death in rice (*Oryza sativa* L.) cells. Plant J 33:425–434

Matsumura H, Nirasawa S, Terauchi R (1999) Transcript profilling in rice (*Oryza sativa* L.) seedlings using serial analysis of gene expression (SAGE). Plant J 20:719–726

Matsumura H, Reich S, Ito A, Saitoh H, Kamoun S, Winter P, Kahl W, Reuter M, Krüger DH, Terauchi R (2003a) Gene expression analysis of plant host-pathogen interactions by SuperSAGE. Proc Natl Acad Sci USA 100:15718–15723

Meyers BC, Galbraith DW, Nelson T, Agarwal V (2004) Methods for transcription profiling in plants. Be fruitful and replicate. Plant Physiol 135:637–652

Munasinghe A, Patankar S, Cook BP, Madden SL, Martin RK, Kyle DE, Shoaibi A, Cummings LM, Wirth DF (2001) Serial analysis of gene expression (SAGE) in *Plasmodium falciparum* : application of the technique to A–T rich genomes. Mol Biochem Parasitol 113:23–34

Mysore KK, Tuori R, D'Ascenzo M, Debbie P, Gu Y, Crasta O, Folkerts O, Martin GB (2001) Expression profiling of genes that are induced or supressed during an incompatible plant–pathogen interaction in tomato. Plant and Animal Genome IX Conference, January 13–17, 2001, San Diego CA

Nam DK, Lee S, Zhou G, Cao X, Wang C, Clark T, Chen T, Rowley JD, Wang SM (2002) Oligo (dT) primer generates a high frequency of truncated cDNAs through internal Poly (A) priming during reverse transcription. Proc Nat Acad Sci USA 99:6152–6156

Nasir KHB, Takahashi Y, Ito A, Saitoh H, Matsumura H, Kanzaki H, Shimizu T, Ito M, Fujisawa S, Sharma PC, Ohme-Takagi M, Kamoun S, Terauchi R (2005) High-throughput *in planta* expression screening identifies a class II ethylene responsive element binding factor-like protein that regulates plant cell death and nonhost resistance. Plant J 43:491–505

Neilson L, Andalibi A, Kang D, Coutifaris C, Strauss JF, Stanton JAL, Green DPL (2000) Molecular phenotype of the human oocyte by PCR-SAGE. Genomics 63:13–24

Obermeier C, Snowdon R (2006) Serial analysis of gene expression (SAGE) in *Brassica napus* seed development for identification of genes and markers associated with fatty acid metabolism. Available at http : //www.uni-giessen.de/ ~ gh1262/pz/englisch/pro_chris.html

Peters DG, Kassa AB, Yonas H, O'Hare EH, Ferrell RE, Brufsky AM (1999) Comprehensive transcript analysis in small quantities of mRNAs by SAGE-Lite. Nucleic Acids Res 27:e39

Pleasance ED, Marra MA, Jones SJM (2003) Assessment of SAGE in transcription identification. Genome Res 13:1203–1215

Poroyko V, Hejlek LG, Spollen WG, Springer GK, Nguyen HT, Sharp RE, Bohnert HJ (2005) The maize transcriptome by serial analysis of gene expression. Plant Physiol 138:1700–1710

Powell J (1998) Enhanced concatemer cloning: a modification to the SAGE (serial analysis of gene expression) technique. Nucleic Acids Res 26:3445–3446

Pylouster J, Senamaud-Beaufort C, Saison-Behmoaras TE (2005) WEBSAGE: a web tool for visual analysis of differentially expressed human SAGE tags. Nucleic Acids Res 33:W693–W695

Riggins GJ (2001) Using serial analysis of gene expression to identify tumor markers and antigens. Dis Markers 17:41–48

Robinson SJ, Cram DJ, Lewis CT, Parkin IAP (2004) Maximizing the efficacy of SAGE analysis identifies novel transcripts in *Arabidopsis*. Plant Physiol 136:3223–3233

Saha S, Sparks AB, Rago C, Akamaev V, Wang CJ, Vogelstein B, Kinzler KW, Velculescu VE (2002) Using the transcriptome to annotate the genome. Nat Biotechnol 20:508–512

Schena M, Shalon D, Davis RW, Brown PO (1995) Quantitative monitoring of gene-expression patterns with a complementary-DNA microarrays. Science 270:467–470

Shoemaker DD, Schadt EE, Armour CD, He YD, Garrett-Engele P, McDonagh PD, Loerch PM, Leonardson A, Lum PY, Cavet G, Wu LF, Altschuler SJ, Edwards S, King J, Tsang JS, Schimmack G, Schelter JM, Koch J, Ziman M, Marton MJ, Li B, Cundiff P, Ward T, Castle J, Krowlewski M, Meyer MR, Mao M, Burchard J, Kidd MJ, Dai H, Phillips JW, Linsley PS, Stoughton R, Scherer S, Boguski MS (2001) Experimental annotation of the human genome using microarray technology. Nature 409:922–927

Smith RD (2000) Probing proteomics: seeing the whole picture? Nat Biotechnol 18:1041–1042

Thomas SW, Glaring MA, Rasmussen SW, Kinane JT, Oliver RP (2002) Transcript profiling in the barley mildew pathogen *Blumeria graminis* by serial analysis of gene expression (SAGE). Mol Plant Microbe Interact 15:847–856

Tuteja R, Tuteja N (2004a) Serial analysis of gene expression (SAGE): unravelling the bioinformatics tools. BioEssays 26:916–922

Tuteja R, Tuteja N (2004b) Serial analysis of gene expression (SAGE): application in cancer research. Med Sci Monit 10:RA132–RA140

Tuteja R, Tuteja N (2004c) Serial analysis of gene expression: applications in human studies. J Biomed Biotech 2:113–120

Ueda HR, Hayashi S, Matsuyama S, Yomo T, Hashimoto S, Kay SA, Hogenesch JB, Iino M (2004) Universality and flexibility in gene expression from bacteria to human. Proc Nat Acad Sci USA 101:3765–3769

Unneberg P, Wennborg A, Larsson M (2003) Transcript identification by analysis of short sequence tags-influence of tag length, restriction site and transcript database. Nucleic Acids Res 31:2217–2226

van den Berg A, van der Leij J, Poppema S (1999) Serial analysis of gene expression: rapid RT-PCR analysis of unknown SAGE tags. Nucleic Acids Res 27:E17(i–iii)

Velculescu VE, Zhang L, Vogelstein B, Kinzler KW (1995) Serial analysis of gene expression. Science 270:484–487

Vilain C, Libert F, Venet D, Costagliola S, Vassart G (2003) Small amplified RNA-SAGE: An alternative approach to study transcriptome from limiting amount of mRNA. Nucleic Acids Res 31:E24(1–7)

Virlon B, Cheval L, Bhule JM, Billon E, Doucet A, Elalouf JM (1999) Serial microanalysis of renal transcriptomes. Proc Natl Acad Sci USA 96:15286–15291

Wang SM (2003) Response: the new role of SAGE in gene discovery. Trends Biotechnol 21:57–58

Welle S, Bhatt K, Thornton CA (1999) Inventory of high-abundance mRNAs in skeletal muscle of normal men. Genome Res 9:506–513

Yasui W, Que N, Ito R, Kuraoka K, Nakayama H (2004) Search for new biomarkers of gastric cancer through serial analysis of gene expression and its clinical implications. Cancer Sci 95:385–392

Ye SQ, Usher DC, Zhang LQ (2002) Gene expression profiling of human diseases by serial analysis of gene expression. J Biomed Sci 9:384–394

Ye SQ, Zhang IQ, Zheng F, Virgil D, Kwiterovich PO (2000) MiniSAGE: gene expression profiling using serial analysis of gene expression from 1 μg total RNA. Anal Biochem 287:144–152

CHAPTER 11

GENETICAL GENOMICS: SUCCESSES AND PROSPECTS IN PLANTS

MATIAS KIRST* AND QIBIN YU

School of Forest Resources and Conservation, University of Florida, PO Box 110410, Gainesville, FL 32611, USA

Abstract: Sequencing of expressed genes from several plant species has revealed that there is a relatively high level of conservation in amino acid sequence among distantly related taxonomic groups, despite the tremendous phenotypic and developmental diversity in the plant kingdom. This diversity appears to be primarily created by polymorphisms that contribute to quantitative gene expression variation, rather than protein structure modification or creation of novel transcriptional units. A few studies have now demonstrated the heritability of gene expression and the dissection of its genetic control in plants. The approach – generally referred to as genetical genomics – relies on the transcript level and quantitative trait loci (QTL) analysis of the transcriptome in segregating populations. In this chapter we review the principles of genetical genomics, results of these studies in plants, and the use of this approach to dissect the genetic control of phenotypic traits of biological and agricultural interest. Although still in their infancy, pioneering genetical genomics studies have shown that this approach is valuable to unravel genetic networks implicated in transcription regulation, and for the identification of genes and pathways implicated in phenotypic variation. More important, they suggest that integrative genomic methods, that merge information from variation at the level of DNA, gene expression, protein and metabolites will be essential for understanding the complexity of plants.

1. INTRODUCTION

The diversity of eukaryotes arose primarily from the differential regulation of a similar group of genes, rather than by the creation of new ones (King and Wilson 1975; Baltimore 2001; Levine and Tjian 2003). Although plants and animals

*Corresponding Author: mkirst@ufl.edu

R.K. Varshney and R. Tuberosa (eds.), Genomics-Assisted Crop Improvement: Vol. 1: Genomics Approaches and Platforms, 245–265.

typically have a larger number of genes compared to simple, unicellular eukaryotes, this difference is generally due to duplication events among a common core set of genes, rather than by the appearance of novel units of transcription. The higher complexity of plants, much like in animals, became possible in part because of a larger variety of regulatory elements and the utilization of more sophisticated protein complexes to modulate and fine-tune gene expression (Levine and Tjian 2003).

Genetic variation in gene expression regulation may be the basis for the diversity of plant species, but is also key to intra-specific diversity and adaptation. Therefore, a genome-wide assessment of the genetic control of gene expression regulation could aid in explaining the morphological and developmental diversity of eukaryotes, including higher plants (Doebley and Lukens 1998; Purugganan 2000; Tautz 2000; Wray et al. 2003). The first studies in structured populations demonstrated that variation in gene expression is genetically controlled and heritable (Dumas et al. 2000; Karp et al. 2000; Brem et al. 2002; Wayne and McIntyre 2002; Schadt et al. 2003; Yvert et al. 2003), suggesting that transcript level variation could be genetically dissected as a typical quantitative trait. Understanding the genetic architecture of gene expression – i.e. defining when, how and why certain genes regulate and are regulated by other genes – could establish a hierarchy of gene action and define the genetic "rule book" for plant growth, development, environmental response and, ultimately, adaptation and evolution.

This chapter provides an overview of the strategies and discoveries made in studies of the genetic architecture of gene expression regulation through QTL analysis, in what is generally referred to as *genetical genomics*. The chapter begins with a description of the fundamental biological information that can be obtained in genetical genomics studies, followed by a discussion about the applications of this information to discover candidate genes that regulate complex traits. Next, genetical genomics studies carried out in plants are reviewed. In the last part, the pitfall and limitations, as well as future perspectives are discussed in the context of studies in plant species.

2. GENETICAL GENOMICS

The concept of genetical genomics was first introduced by Jansen and Nap (2001). They proposed that quantitative genomic data could be analyzed using the same quantitative trait loci (QTL) approach used for the analysis of agricultural traits (Sax 1923; Paterson et al. 1988). Specifically, Jansen and Nap suggested that the transcript level data generated from microarray analysis could be used to identify gene expression QTLs, or *e*QTLs. In microarrays, DNA sequences (cDNA or oligonucleotides representing part of the DNA sequence of a gene) are placed individually in known locations on a two dimensional surface (Schena et al. 1995; McGall et al. 1996; Nuwaysir et al. 2002). Because of the high density that can be achieved in this platform, hundreds to thousands of genes can be evaluated in parallel. Labeled cDNA or cRNA, produced from mRNA extracted from a sample is then hybridized to that surface. The signal emitted by the labeled material hybridized to each spot is measured, providing an indirect estimate of the amount of mRNA molecules (i.e. the gene expression), for every given gene, present in the sample.

As in any traditional QTL analysis, an eQTL or genetical genomics study requires: (1) making a cross using genotypes with contrasting transcript levels, to generate a population segregating for alternative alleles associated with the genes' expression; (2) generating a genetic map by genotyping the individuals from the segregating population for genetic markers distributed along the genome; and (3) obtaining gene expression measurements from the segregating population. Ideally, each individual of the segregating population will have inherited half of its genome from each of the parental genotypes, and for any given genetic marker half of the individuals will carry the allele from one parent, and half will carry the allele from the other parent. The identification of the genomic regions, or quantitative trait loci, that control the gene expression of any given gene is then based on identifying genetic markers where the individuals that inherited alternative alleles have different gene expression levels.

Although genetic markers with genome-wide coverage and methods to genotype them in a rapid and cost effective way have become available only in the past two decades, the framework of QTL analysis was established in the beginning of the last century (Altenburg and Muller 1920; Sax 1923). Sophisticated statistical analysis approaches are used today for QTL detection (Mackay 2001; Abiola et al. 2003) and have been applied to a broad spectrum of species and morphological and developmental traits, but they essentially rely on the principle described above. Therefore, the theoretical framework for genetical genomics studies had been well established previously. However, until the mid-1990's the tools to achieve it in a genomic scale were lacking. With the creation and development of microarray methods, the demonstration that this theoretical concept was feasible and applicable to a broad spectrum of eukaryote species was made soon after in the work on yeast, mice, humans and in corn (Brem et al. 2002; Schadt et al. 2003), and in the woody plant *Eucalyptus* (Kirst et al. 2004).

3. eQTL MAPPING IDENTIFIES REGULATORY LOCI OF GENES AND NETWORKS

Mapping of QTLs for gene expression identifies the genomic regions that harbor regulatory sequences that control transcription. Gene expression may be regulated at various levels. The primary mechanism is the interaction of the core promoter region (~50 bases upstream of the transcription initiation start site) with activators and repressors. DNA-binding proteins that recognize specific DNA regulatory elements such as promoters and enhancers (hundreds to thousands of bases upstream or downstream of the transcription start site) recruit and/or stabilize the transcription machinery to the core promoter region. Variation in the DNA sequence of these regulatory elements can lead to modulation of transcription initiation and consequently mRNA amounts, and is referred to as *cis*-acting regulation. In this case, the QTL for the gene's expression will co-localize with the physical position of the gene (Figure 1). Alternatively, gene expression may be modulated by variation in the levels of transcription factors, co-factors and other regulatory proteins. In that case the eQTL will co-localize with the physical position of the gene encoding for the regulatory protein (*trans*-acting regulation, Figure 1). Other mechanisms can play a significant role in gene expression (such as chromatin remodeling and epigenetic factors) but their genetic regulation can only be analyzed by genetical genomics if they display Mendelian inheritance (i.e. they can be studied by QTL analysis).

Schadt et al. (2003) first showed that *cis*-acting eQTLs could be detected for a large number of genes in mice. eQTLs that co-localized with the position of their respective transcriptional units were detected for 34% of the mapped genes, using a log of the odds ratio (LOD) score of 4.3 (p-value of 5×10^{-6}) and standard

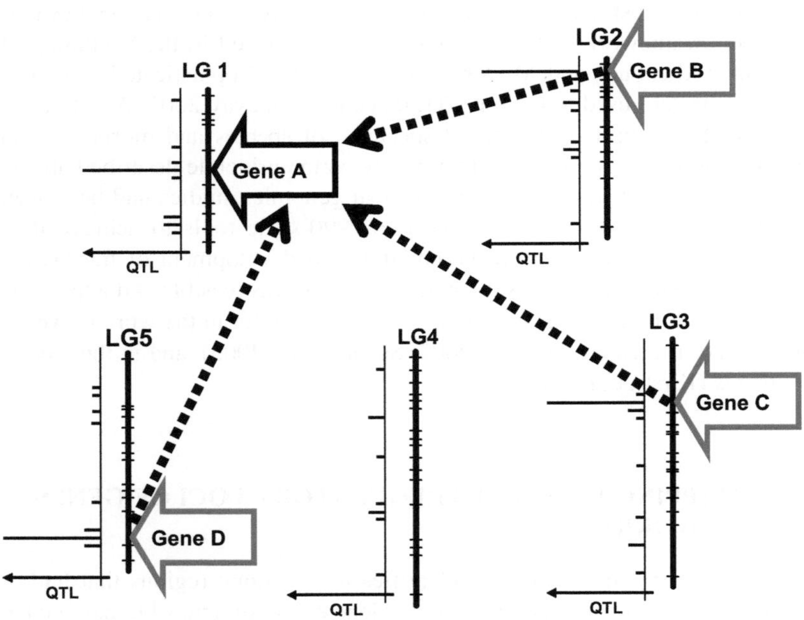

Figure 1. Cis- and *trans*-regulation of gene expression. QTL mapping of transcript levels of a hypothetical Gene A identifies four major QTLs (linkage groups 1, 2, 3 and 5) as evidenced by QTL scans along each chromosome. Gene A and eQTL co-localize in linkage group 1, suggesting *cis*-regulation. The observation of eQTLs for gene A in linkage groups 2, 3 and 5 indicates that expression is also regulated in *trans* by transcription regulators (unknown genes B, C and D)

interval mapping methods. As the stringency used to declare the presence of an eQTL increased to a LOD of 7.0, the frequency of *cis*-regulated genes increased to 71%. The authors associated this difference with the fact that first-order effects (*cis*-acting regulation) should have a more significant effect on gene expression than second-order effects (*trans*-acting regulation). Therefore, defining the proportion of *cis*- and *trans*-regulated genes is directly proportional to the stringency used to declare the presence of an eQTL. The same trend was observed in the data from Hubner et al. (2005), where a ratio of 2:1 in *trans*- to *cis*-regulations ($p < 0.05$) was inverted to a 1:15 ratio at a more stringent threshold ($p < 5\times10^{-6}$). Most studies have identified 20–40% of the genes regulated in *cis* (Table 1). Nonetheless, that number has to be considered with caution as it depends heavily on the threshold used for eQTL detection.

A genetical genomics approach can also identify genetic loci that control transcription regulation of large numbers of genes in *trans*, by finding correlated variation of transcript abundance and co-localized eQTLs (Figure 2). Clusters of co-localized eQTLs have been usually referred to as *eQTL hotspots*. Probabilistically, the likelihood of large numbers of eQTLs co-localizing in a specific genetic interval by chance is relatively small. Co-localized eQTLs indicate that specific genetic loci may contribute to transcript level variation of many genes, suggesting that they are part of genetic network of expression regulation. Such inferences are strengthened when supported by annotation indicating a common molecular function or cellular process. In combination with genomic sequence, eQTL and gene location provide insight into the mechanisms of regulation of gene expression networks, and could lead to the identification of specific genes or loci that control them (Figure 2). Many eQTL hotspots have been shown, in our and other studies (Brem et al. 2002; Schadt et al. 2003; Yvert et al. 2003; Kirst et al. 2004), to include genes associated with specific metabolic and regulatory pathways.

Genomic networks – the lignin biosynthesis pathway as an example of an eQTL hotspot. A microarray study carried out in a mapping population of 91 individuals of the woody plant *Eucalyptus* identified co-localized genetic loci that regulate the expression of genes encoding enzymes of the phenylpropanoid biosynthesis pathway. This pathway is involved in the synthesis of secondary products including the monolignols p-coumaryl, coniferyl and sinapyl, the precursors of the lignin polymer (Higuchi 1990). Analysis of various individuals from the segregating population confirmed that genetic regulation of metabolically related genes correlates with synthesis of the pathway products (Kirst et al. 2004). Genes encoding enzymes of the shikimate and methionine pathway – all of which are involved in providing precursors to the phenylpropanoid pathway – also shared common eQTLs, suggesting a tight regulation of the synthesis of the pathway product. Many of the intermediate products of these pathways are toxic to the cell and need to be processed in a short period of time to avoid deleterious effects. This may justify the mode of regulation seen in this case, but can not be generalized to all the other metabolic or regulatory pathways.

Table 1. Summary of eQTL studies

	Reference	Species (tissue)	n	Platform	Genes on the array	Diff Exp. Genes	Markers[b]	eQTLs (threshold)[c]	cis eQTLs
Animal	Schadt et al. 2003	Mouse (liver)	111	Agilent	23574	7861	13 cM	4339 (gwp 0.05)	34%
	Chesler et al. 2005	Mouse (brain)	35	Affymetrix	12422	608	779	88 (fdr 0.10)	94%
	Bystrykh et al. 2005	Mouse (stem cells)	30	Affymetrix	12422	n.a.	779	1219 (gwp 0.05)	13%
	Hubner et al. 2005	Rat (kidney)	30	Affymetrix	15923	1553	1011	2490 (gwp 0.05)	30%
	Hubner et al. 2005	Rat (fat body)	30	Affymetrix	15923	2046	1011	2118 (gwp 0.05)	35%
Human	Morley et al.	Human (lymphoblastoid cells)	112	Affymetrix	8500	3554	1 cM	142 (gwp 0.001)	19%
	Monks et al, 2004	Human (lymphoblastoid cells)	167	Agilent	23499	2499	4 cM	132 ($p<0.005$)	19%
Yeast	Yvert et al.2003	Yeast	86	Custom ORFs	6200	2294	3114	1001($p<0.0004$)	25%
Plant	Schadt et al. 2003	Maize (earleaf)	76	Agilent	24743	18805	12 cM	7322 (LOD 3.0)	n.a.
	DeCook et al. 2006	Arabidopsis (regenerated shoots)	30	Affymetrix	22787	2637	2 cM	3525 (gwp 0.05)	n.a.
Tree	Kirst et al. 2004, 2005	Eucalyptus (xylem)	91	cDNA	2608	1067	10 cM	1655 (gwp 0.05)	22%

b. number of markers, or approximate marker density in centimorgans.
c. gwp: genome-wide p value; fdr: false discovery rate; LOD: likehood cutoff.

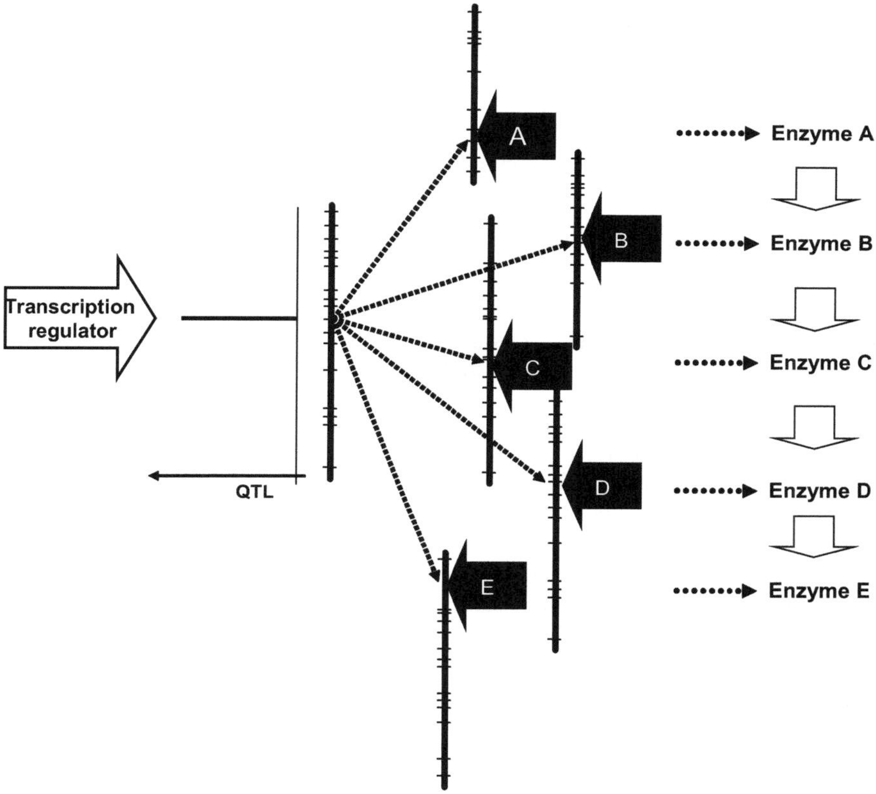

Figure 2. Co-localized eQTLs for several genes define eQTL hotspots. Five hypothetical genes that encode for enzymes A to E share a common eQTL, suggesting that the eQTL hotspot identifies the genetic locus that regulates transcription through the network

4. GENETICAL GENOMICS FOR IDENTIFICATION OF CANDIDATE GENES FOR COMPLEX TRAITS

The identification of genes that regulate complex trait variation using forward genetic approaches has been largely dependent on identifying molecular markers associated with the trait (Figure 3) through QTL analysis (Paterson et al. 1988; Paterson et al. 1991; Stuber et al. 1992). In general, a QTL study identifies genomic regions of 10–20 centimorgans associated with a quantitative phenotype. In *Arabidopsis*, one centimorgan corresponds on average to 200 thousand base pairs (Kbp), or approximately 50 genes. In species with megagenomes such as the gymnosperms, one centimorgan may correspond to more than 15,000 Kbp. Fine-scale mapping can narrow the pool of genes that underlie a QTL by increasing marker density and sampling enough recombination events between markers flanking a broad QTL region. Nonetheless, despite increasing progress

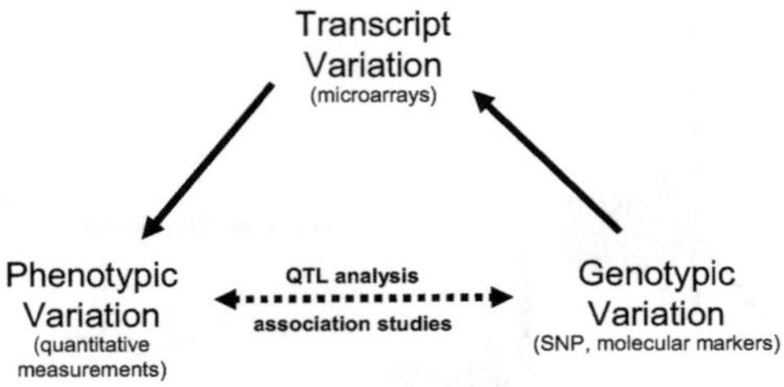

Figure 3. Genetical genomics and the integration of information from genotypic, phenotypic and transcript level variation. Traditional QTL and association genetic studies rely on combining information from phenotypic and genotypic variation. Genetical genomics incorporates transcript level information to the quantitative genetic paradigm to identify candidate genes for complex traits

(Korstanje and Paigen 2002) the path from a broad genomic region to the specific polymorphism underlying a QTL is laborious and costly.

Approaches such as association mapping are creating a new perspective for the identification of specific polymorphisms that regulate complex traits in eukaryotes, by taking advantage of the higher resolution of marker-QTL association generated by historical recombinations (Flint-Garcia et al. 2005). Still, whole-genome scans, which would allow for the evaluation of the entire set of genes of one individual for associations with a quantitative trait, are not yet feasible for most species with high genetic diversity and limited linkage disequilibrium. Until the technical advances become such that all haplotypes can be evaluated for association, these approaches will still rely primarily on the identification of a subset of (candidate) genes to be tested individually for association.

Another limitation of traditional QTL analysis strategies is that they capture mostly only the additive genetic component that contributes to a phenotype. Although the identification of genetic loci that contribute to non-additive components of the genetic variance may be feasible by using specific mating designs and statistical methods, more sophisticated approaches are typically required. For instance, QTL mapping approaches that attempt to assess the epistatic effect of multiple loci, such as Multiple Interval Mapping (Kao et al. 1999) are typically limited in that they can evaluate the combined effect of only few loci. Lower statistical power and the exponential increase in the number of statistical tests as more loci are added to the models limit the capacity of detecting positive interactions and may generate a large number of false-positives. Nonetheless, the relevance of combined effects of multiple loci has been well demonstrated for some traits, such as for accumulation of chlorogenic acid (CGA) in maize silks (Szalma et al. 2005). Genetical genomics may provide not previously foreseen advantages for identifying loci that contribute to additive and non-additive effects, as described below.

Genetical genomics was primarily considered a strategy for describing the genetic control of transcription in individual genes and networks (Jansen and Nap 2001), but its application to understanding the genetic control of complex, quantitative traits was proposed soon after. The integration of genotypic and transcript level data for dissecting the genetic control of allergic asthma and stress responses in mammals was suggested early on by Dumas et al. (2000) and Karp et al. (2000), but the application was limited in scope to a few genes. Shortly after genetical genomics was shown to be a powerful tool for identifying candidate genes for complex diseases and for traits of economic value in forestry (Schadt et al. 2003; Kirst et al. 2004). The strategy proposed relied in identifying *cis*-regulated eQTL that co-localize with a trait QTL. The rationale is that for any trait transcriptionally regulated by a given genomic region (defined by its QTL) there should be a corresponding *cis*-acting *e*QTL for the gene that controls it. Instead of relying on the detection of anonymous markers correlated with a trait, such as in traditional QTL analysis, this approach identifies actual *genes* (Figure 4). This approach was first applied by Schadt et al. (2003) who identified four candidate genes whose eQTLs co-localized with several obesity related QTLs in mice. The homologous region in the human genome had been previously linked to obesity (Lembertas et al. 1997) and two of the four mice candidate genes had homologues in that region, suggesting a possible association with the trait. Following this study, Schadt and colleagues proposed and demonstrated novel strategies for identifying the genes that control complex, quantitative phenotypes, based on gene expression analysis of segregating populations, using specific models (Schadt et al. 2005). The motivation was that an association between transcript levels and trait phenotypic value (demonstrated by the co-localization of eQTL-QTL) may have several origins, some of which are not of interest, and some that may indicate specific target genes. Primarily, the authors defined the different models by which there could be a correlation between gene expression and a trait variation. The two most easily

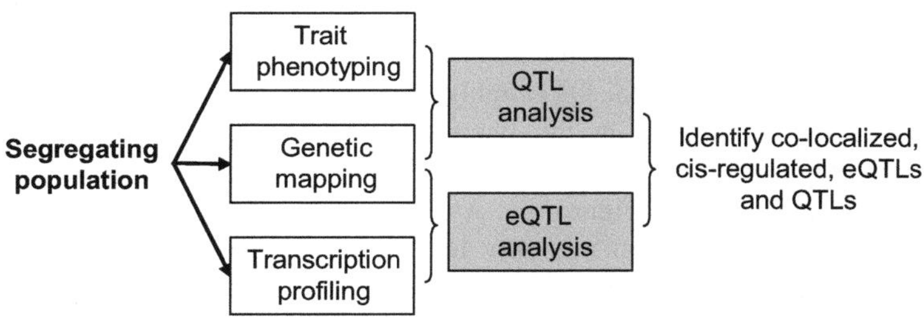

Figure 4. Strategies for identifying candidate genes for quantitative variation combining QTL, eQTL and gene location. Data on genotypic, gene expression and phenotypic variation (white boxes) collected from a segregating population is analyzed by QTL (genotype-phenotype) and eQTL (genotype-gene expression) analysis (grey boxes). Co-localized QTLs and eQTLs suggest a direct (in *cis*) or indirect (in *trans*) association between gene expression and phenotype. The mode of association can be discerned if the genetic or physical position of the gene is known to co-localize with both trait and gene expression QTLs

confoundable models are the (1) causative and (2) reactive models. In the first, a genetic polymorphism (QTL) causes changes in transcription levels at a given gene, which results in phenotypic variation in a trait of interest. In the second, reactive model, the genetic polymorphism causes a change in the phenotype, which has consequences in the transcript level of one or more genes. In both models, genetic polymorphism, eQTL and QTL may be co-localized. In a third, independent model, the genetic polymorphism causes variation in gene expression and trait, but both are unrelated to each other. More complex models were also suggested. The authors developed a likelihood test that uses conditional correlation to evaluate what is the best model to explain the association between gene expression and phenotype data, and demonstrate its use in the identification of genes for total fat pad mass. The function of the genes identified using these procedures was then validated in genetically modified mice.

One observation from most studies (Schadt et al. 2003; Kirst et al. 2004; Schadt et al. 2005) that identified candidate genes that underlie QTLs using genetical genomics is that the observed correlation between candidate gene expression and trait variation is frequently higher than that derived from the QTL analysis. Intuitively, one would expect that the transcript level of a gene that underlies a QTL should explain no more than what is explained by the genotypic value, but that does not always seem to be the case. For instance, we have identified individual QTLs that explained ~20% of the phenotypic variance in tree height (Kirst et al. 2004). However, certain transcript levels associated with the QTL explained more that 30% of the phenotypic variance. Invariably we observed the same trend for other traits such as wood density and wood fiber quality (unpublished data). In the experiments carried out by Schadt et al. (2005) in an F2 mice population, four QTLs that explained together ~39% of the phenotypic variance in adiposity-related traits were detected. However, transcript levels detected in certain genes explained over 60% of the trait phenotypic variance (causative and reactive genes). Several factors may explain this discrepancy (1) correlation coefficients between trait and expression may be overestimated (for some unknown reason), or (2) the QTL analysis of the trait is underestimating the effect of the genetic locus. As pointed out previously, traditional QTL analysis accounts mostly for additive sources of variance that contribute to the phenotype. Alternatively, a genetical genomics approach may add more information to the quantitative genetics paradigm that is the basis for QTL analysis (Figure 3). Analysis of transcript abundance in segregating populations may account not only for additive but also non-additive genetic effects that are not captured in traditional studies. Non-additive variation in transcript levels is significant (Gibson et al. 2004; Auger et al. 2005; Brem et al. 2005; Storey et al. 2005; Vuylsteke et al. 2005). The same non-additive effects that contribute to one gene's expression variation may contribute to phenotypic variation at a trait controlled by that gene. Therefore including transcript level data into the QTL analysis may well explain greater proportions of the phenotypic variance, thereby enhancing statistical power to detect real genotype:phenotype associations.

5. CURRENT STATUS OF GENETICAL GENOMICS STUDIES IN PLANTS AND OTHER SPECIES

Genetical genomics studies have been reported in yeast (Brem et al. 2002; Yvert et al. 2003; Brem and Kruglyak 2005; Brem et al. 2005), mice (Bystrykh et al. 2003; Schadt et al. 2003; Bystrykh et al. 2005; Chesler et al. 2005), rats (Hubner et al. 2005), humans (Schadt et al. 2003; Monks et al. 2004; Morley et al. 2004), maize (Schadt et al. 2003), Eucalyptus (Kirst et al. 2004; Kirst et al. 2005) and Arabidopsis (DeCook et al. 2006). **Table 1** summarizes the results from these studies. Since the pioneering work in the beginning of the decade there has been a steady increase in the number of reports in human and mammal model species, probably reflecting the higher accessibility to the technology and lower cost of microarray platforms. Morley et al. (2004) and Monks et al. (2004) measured baseline levels of gene expression in human lymphoblastoid cell line from 14 and 15 CEPH (Center d'Étude du Polymorphisme Humain) families. Chesler et al. (2005) and Bystrykh et al. (2005) analyzed gene expression in the forebrain and in hematopoietic stem cell from 30 mice recombinant inbreds. Huber et al. (2005) applied the same approach to rat fat and kidney tissues in recombinant inbreds. Yvert et al. (2003) and many others have reported detection of eQTLs in yeast. In contrast, there has been a very limited number of genetical genomics studies in plants – reports have been made for maize, *Arabidopsis* and *Eucalyptus*. A summarized description of these studies is presented below.

5.1. Maize

Schadt et al. (2003) analyzed mRNA transcript abundances in ear leaf tissues collected from 76 F2-derived F3 plants generated originally from two maize inbred lines. A set of 18,805 genes were identified as differentially regulated in the overall population (type I error =0.05), out of 24,743 genes represented in a microarray. The study revealed 6,481 genes with at least one eQTL (LOD $\geq$ 3.0) and 80% of those with a LOD score $\geq$7 were collocated with the gene (when the gene location was known). The authors identified a novel type of epistatic gene-gene interactions in the expression levels, but the methods were not formally described.

5.2. Eucalyptus

In our studies the progeny of a mapping population of 156 individuals from an *Eucalyptus* pseudo-backcross were genotyped for 803 AFLP markers, and genetic maps were created (Myburg et al. 2003). A subset of 91 individuals from the progeny was characterized for gene expression profiles in differentiating xylem using cDNA microarrays comprising 2,608 selected *Eucalyptus* cDNAs related to wood properties and growth (Kirst et al. 2004). Quantitative trait locus analysis using composite and multiple interval mapping (Zeng 1993, 1994; Kao et al. 1999) identified genomic regions that harbor regulatory sequences controlling, in *cis*- or *trans*-, the expression of 1,067 genes. Genetic mapping defined both *cis*- and *trans*-acting mechanisms of regulation. However, the lack of the complete genome sequence reduces the ability to make genomic inferences about mode of transcript regulation for large numbers of genes.

5.3. Arabidopsis

In the latest eQTL study reported in plants, DeCook et al (2006) used 30 recombinant inbred lines derived from a cross of Arabidopsis ecotypes Columbia and Landsberg *erecta*. RNA extracted from root explants in shoot induction medium was hybridized to Affymetrix ATH1 microarrays comprising 22,787 genes. eQTLs were detected by fitting a least-squares linear regression model using the genotype as independent variable and expression as the dependent variables. The hypothesis of slope equal to zero was used to determine the significance of the relationship. A Bonferroni adjustment was used to control the genomewide error rate for each gene expression at the 0.05 level. Then for each threshold a false discovery rate (FDR) was estimated (Storey and Tibshirani 2003). A set of 3,525 eQTLs were detected for 958 genes at an FDR of 2.3%. At a less stringent threshold (FDR of 10.2%) 10,521 eQTLs were detected for 2,637 genes. Five eQTL hot spots were found. The largest hotspot was located on the lower arm of chromosome 5, centered on marker 270, and it involved 34 genes – 23 genes were upregulated and 11 genes were downregulated. The study showed that the upregulated genes are expressed at much higher level when Columbia rather than Landsberg alleles were present at the marker site. Two of the hot spots coincided with previously described QTLs conditioning shoot regeneration (Lall et al. 2004), suggesting that some of the heritable gene expression changes are related to differences in shoot regeneration efficiency between ecotypes. Some of the most significant eQTLs, particularly those at the shoot regeneration QTL sites, tended to show *cis*-chromosomal linkages, whereas many linkages of lesser significance showed *trans*-effects.

Alternative approach derived from the original genetical genomics framework proposed by Jansen and Nap (2001) have been also reported for *Arabidopsis* by Juenger et al. (2006) and Filatov et al (2006). In the first study, instead of relying on the analysis of gene expression in large segregating population, Juenger and colleagues carried out a microarray analysis contrasting the transcript abundance between two near-isogenic lines (NIL) to the recurrent parent (*Ler-2* ecotype). The NILs were selected to contain an allele for increased transpiration and reduced water-use efficiency, identify by a previous QTL study and derived from the Cvi-1 ecotype. Genes differentially regulated between the NIL and the recurrent ecotypes could then be associated with the introgressed segment from Cvi-1, and candidate genes for the target QTL. Using this strategy, the authors identified a pool of 25 differentially regulated genes, many of which were considered strong candidates for the trait regulation. Interestingly the authors found no genes differentially regulated (using a stringent statistical threshold) located outside of the introgression region, suggesting mostly *cis*-regulation of gene expression. Filatov et al. (2006) applied another variation to the original genetical genomics strategy, for identification of genes related to zinc (Zn) hyperaccumulation. From the cross between two Arabidopsis species (*A. halleri* and *A. petreae*) with variable tolerance to high Zn levels, F_2 individuals were grouped in phenotypic classes and allowed to polycross within groups, creating F_3 families. Finally, families were phenotyped for Zn tolerance, and two in each extreme were selected for gene expression analysis in two different levels of Zn. The authors identified approximately 100 genes differentially regulated between hyper- and hypoaccumulator genotypes in roots and leaves, eight of which were common between the two plant organs.

6. PITFALLS AND LIMITATIONS OF GENETICAL GENOMICS

Despite the potential of using genetical genomics to unravel genetic regulatory networks and identify genes that control quantitative traits, QTL analysis of gene expression is plagued by the same problems of traditional QTL studies. In addition, the increase in the scale of analysis – from a few traits in most studies to thousands of genes – exacerbates existing limitations and creates new challenges.

As in traditional QTL analysis, a significant association between genetic marker and trait is only detectable if there is variation in the parents and segregation in the progeny. Therefore, despite the significant effort in carrying out gene expression analysis in large progeny sets, only a limited fraction of the genetic information of interest can be unraveled in one given cross. A population wide survey of the genetic control of gene expression, in an association genetics study, can be achieved but is certainly much more challenging than detecting eQTLs in a single, structured population.

eQTL analysis is also limited in its inferences by the typically low resolution of QTL analysis. An eQTL found co-localized with the corresponding gene location suggests – with a fairly high probability – that polymorphisms at the gene or its immediate regulatory regions are indeed controlling its expression. Nonetheless, when testing thousands of genes a certain proportion will be falsely declared as *cis*-regulated. When the specific gene is regulated in *trans*-, the identification of the specific genetic element that controls it is made difficult by the broad span of QTL regions.

Because genetical genomics involves gene expression profiling of large populations, these studies also carry a significant burden in terms of cost. Large populations are required to generate unbiased estimates of QTL detection (Beavis 1997). To date, genetical genomics studies have relied on relatively small segregating populations in their analysis. To the exception of the work reported by Morley et al. (2004) and Bystrykh et al. (2005), other studies have used segregating populations with less than a hundred individuals. Therefore, it is likely that these genetical genomics studies have suffered from the same limitations of traditional quantitative genetic analysis of small populations, namely a lack of power in detecting small and moderate effect eQTLs, and an overestimation of effects in many cases where eQTLs were detected (Beavis 1997). With a progeny population of fewer than 500 individuals, regardless of marker density, there is little statistical power to identify QTLs of small effect.

Other limiting aspects of a genetical genomics approach are described below.

6.1. Is Gene Expression Heritable?

Heritability refers to the proportion of the phenotypic variance among individuals in a population that can be attributed to the genotypic variance. Most of the genetical genomic studies did not explicitly estimate gene expression heritabilities due to design limitations. Two different groups (Monks et al. 2004; Morley et al. 2004) independently conclude that the large proportion of human lymphoblastoid transcriptome varies in segregating populations. Morley et al. (2004) did not report heritability estimates. Monks et al. (2004) found expression in 762 genes (31%) to be significantly heritable (at a false discovery rate $P < 0.05$), and a median heritability of 34%. However, heritability estimates for all differentially expressed genes, inferred from grandparent to parent, parent to children, or across all three generations, were not correlated (Gibson and Weir 2005). Many reasons could explain this observation. The sample size of the order of a few hundred individuals may be too small to support robust estimates of heritability. There may also be unknown sources of experimental artifact (perhaps relating to sample processing in batches) that produce false estimates of genetic variance components. Finally, the genetic components affecting transcription, and consequently the genetic architecture of transcription, could themselves vary from generation to generation.

6.2. The Confounding Effects of Polymorphisms when Declaring Cis-regulation

Several studies have recently reported the spurious effects of genetic polymorphisms when making inferences about gene transcript levels (Ronald et al. 2005; Rostoks et al. 2005). Those polymorphisms that negatively affect the hybridization kinetics between probe in the microarray and labeled target RNA can lead to the identification of a false *cis*-regulatory loci. In this scenario, one of the two parents used in the cross is polymorphic for a genetic locus in the probe that results in the lack of hybridization of the labeled mRNA – i.e. a weaker signal is detected for the individuals that inherited that allele. As a result, the presence of signal measured in the microarray segregates in the progeny and becomes equivalent to a genetic marker. For instance, some of the most significant *cis*-acting loci detected by Monks et al. (2004) were located in the HLA region, which is highly polymorphic. The authors cautioned that these may not be eQTLs but could be simply associated with sequence differences between the parental lines. This type of marker – referred to as single-feature polymorphism, or SFP – was recently used to generate a genetic map in *Arabidopsis* (West et al. 2006). However, the application of this approach is clearly only conceivable if there is knowledge about the genetic location of each individual gene in the genome. Overall, the issue may not be important for species with low levels of genetic diversity, but could be particularly relevant when carrying out genetical genomics studies in species that are typically genetically diverse, such as maize, poplar and pines.

6.3. Significance Thresholds

The main challenge of any genetical genomics study is to define an appropriate threshold to declare presence of an eQTL. In traditional QTL analysis, the entire genome is scanned for significant marker-trait associations and several hundred statistical tests are carried out (Lander and Botstein 1989). Strategies to address the problems generated by the multiple number of tests have been proposed, the most popular being the use of permutation tests where a null distribution is generated (Churchill and Doerge 1994; Doerge and Churchill 1996). When detecting gene expression QTLs from microarray data, this problem is magnified by 2–3 orders of magnitude, with the analysis of hundreds or thousands of genes, or "traits". To minimize this problem, several studies have taken the approach of evaluating exclusively genes that are identified *a priori* as differentially regulated between the parental genotypes, or in the progeny. This strategy has reduced typically the number of genes tested to less than half the original set, therefore only minimizing the problem but not eliminating it. The drawback of this strategy, as shown by Schadt and colleagues (2003), is that many genes for which eQTLs could have been detected end up not being evaluated. While filtering for genes differentially regulated in the progeny allowed them to detect eQTLs for 2,123 genes, including the entire gene set represented in the microarray increased the number of genes with detectable eQTLs to 3,701. Although many of these may be false-positives, it appears that a significant proportion of the information may be lost by applying a stringent filter prior to the eQTL analysis. In addition to excluding potentially non-differentially regulated genes, other approaches have been proposed to minimize the issue of multiple testing. In our previous studies, we carried out permutation tests for 40 genes randomly selected from the set in the microarray, and used the most conservative threshold as defined by the null distribution, among those genes (Kirst et al. 2005). That threshold was then used to declare presence of an eQTL when analyzing the entire set of genes in the microarray.

One may apply other methods and other authors have proposed alternatives that may be applicable here. Piepho (Piepho 2001) proposed a quick method for estimating approximate genome-wide threshold levels for QTL detection by interval mapping and composite interval mapping. Numerical approaches based on permutation tests (Churchill and Doerge 1994; Doerge and Churchill 1996) that are too slow to be applied to thousands of traits can be complemented by more efficient resampling approaches developed to compute genome-wide significance thresholds (Zou et al. 2004), which are computationally less expensive than permutation tests.

6.4. eQTL Hotspots – Are they Real?

One of the most interesting discoveries from genetical genomics studies has been the observation that eQTLs for large numbers of genes frequently clustered in specific genomic regions or eQTL hotspots. This observation is relevant as it may indicate loci with regulatory effects over a network of genes. These regulatory loci may also represent opportunities for, through genetic manipulation or molecular breeding, alter entire metabolic or regulatory pathways. But are eQTL hotspots real, or a product of biases in the experiments?

Our studies and other reports have shown that mRNA levels are frequently regulated by multiple loci. However, individual *trans*-acting eQTLs exert a relatively weak genetic control over expression levels compared to *cis*-acting eQTLs. Therefore, *trans*-acting eQTLs are more difficulty to detect in small experiments. Among three mouse eQTL studies (Schadt et al. 2003; Bystrykh et al. 2005; Chesler et al. 2005), the location of *trans*-acting eQTLs showed limited overlap. This could be due to a higher proportion of false positive *trans*-acting eQTLs. It has been speculated that the clustering of eQTLs in hotspots, which reflects the highly correlated expression levels of many gene transcripts, may be due to technical or environmental factors that are currently unaccounted for (de Koning and Haley 2005). A simulation study using existing gene expression data from human pedigrees with a randomly generated single-nucleotide polymorphism map showed strong clustering of *trans*-acting eQTLs, and the five most populated bins contained 20% of significant, but spurious eQTLs (Perez-Enciso 2004). Two genetical genomics studies reported the analyses of CEPH cell lines derived from three-generation human pedigrees consisting of four grandparents, two parents and up to ten children (Monks et al. 2004; Morley et al. 2004). Although the studies overlapped for about half (eight) of the CEPH families studied, the results are remarkably different. For example, Morley and colleagues (2004) found significantly more *trans*-acting eQTLs (n =110) than *cis*-acting eQTLs (n = 17), as well as two eQTL hotspots. In contrast, Monks and colleagues (2004) found no evidence of eQTL hotspots. This discrepancy may have many causes (de Koning and Haley 2005; Li and Burmeister 2005). The detection of eQTL hotspot of *trans*-acting eQTLs may be somewhat falsely inflated, because of issues such as multiple testing and the use of a relaxed statistical threshold to define QTL presence.

6.5. Other Limitations

Most genetical genomics studies have presented what seems to be an oversimplified view of transcriptional regulation. Instead, transcriptional variation may be highly polygenic, with many loci controlling each small fractions of the quantitative variation in gene expression. Due to experimental design and/or statistical analysis complexity and limitations, it is likely that most eQTLs remain undetected. For instance, in yeast Brem and Kruglyak (2005) reported that transcription is more likely to be highly polygenic, rather than monogenic. Only 3% of highly heritable transcripts are consistent with single locus inheritance, 18% suggest control by two loci, and > 50% require at least five loci under an additive model (Brem and Kruglyak 2005). Such amazing genetic complexity for a simple eukaryote illustrates the magnitude of challenges lying ahead for higher organisms. In addition, transcript level is obviously not the only linkage between genes and phenotypic traits. A full understanding may require more comprehensive models that include protein and metabolite concentrations and cellular compartmentalization. However, technical and experimental design issues need to be addressed and new statistical tool need to be developed to analyze them.

7. CONCLUSIONS AND PERSPECTIVES

One can argue that one of the primary objectives of genomic sciences is to fully describe the genetic diversity in any given species. As the genetic diversity is defined, the next step would be to describe its implications to at all levels of molecular information – transcript, protein and metabolite levels – and, perhaps most important, higher level phenotypes such as morphological, developmental and adaptive traits. Rapid progress is being made in sequencing the genome of plant species (Arabidopsis Genome 2000; Goff et al. 2002) and novel methods of high throughput genotyping are becoming accessible. Platforms to characterize the genetic variation in several hundred thousand single-nucleotide polymorphisms in a single assay are now already available in humans (Matsuzaki et al. 2004), creating the opportunity for the generation of complete individual haplotype maps. Although characterizing completely the genetic diversity in any plant species will be challenging, particularly for those species with high levels of diversity and low linkage disequilibrium, this may become a reality in the next few years as the density of parallel genotyping platforms increase. While a larger picture of this network of genome, transcriptome, metabolome and proteome has still not emmerged, genetical genomics is one approach that will provide the basis for connecting the information from the transcriptome, with DNA sequence variation.

In the past few years there has been a gradual shift in the paradigm of quantitative genetic analysis of agricultural and forestry traits. The traditional QTL approach that focusses typically in a limited genetic pool for identification of valuable alleles for commercial crops gradualy gives room to other methods that take advantage of extensive natural variation for traits of interest. In this scenario, the natural adaptive process is investigated to identify genes that present signatures of selection that indicate their relevance for the species adaptation (Mitchell-Olds and Schmott, 2006). Although the feasibility of these studies is only now being demonstrated in model plant species (Michell-Olds and Schmitt, 2006), agronomic crops and forestry, the advantages, particularly the improved resolution and allelic diversity sampled may outweight by far the pitfalls and limitations. One can envision future genetical genomics studies in plants where gene expression will be sampled in large association populations to define the specific nucleotides that regulate gene expression. Although it is questionable how well this approach will work for transcript level phenotypes, the evidence from eQTL studies demonstrates that it should not differ significantly from traditional traits. When merged with results from association genetic analysis for phenotypic traits, researchers will be able to unravel the chain of events, from the nucleotide polymorphisms, to gene expression, protein to phenotype, describing a detailed molecular mechanism of phenotypic trait regulation, creating an unifying view of genomics.

REFERENCES

Abiola O, Angel JM, Avner P, Bachmanov AA, Belknap JK, Bennett B, Blankenhorn EP, Blizard DA, Bolivar V, Brockmann GA, Buck KJ, Bureau JF, Casley WL, Chesler EJ, Cheverud JM, Churchill GA, Cook M, Crabbe JC, Crusio WE, Darvasi A, de Haan G, Demant P, Doerge RW, Elliott RW, Farber CR, Flaherty L, Flint J, Gershenfeld H, Gu J, Gu WK, Himmelbauer H, Hitzemann R, Hsu HC, Hunter K, Iraqi FA, Jansen RC, Johnson TE, Jones BC, Kempermann G, Lammert F, Lu L, Manly KF, Matthews DB, Medrano JF, Mehrabian M, Mittleman G, Mock BA, Mogil JS, Montagutelli X, Morahan G, Mountz JD, Nagase H, Nowakowski RS, O'Hara BR, Osadchuk AV, Paigen B, Palmer AA, Peirce JL, Pomp D, Rosemann M, Rosen GD, Schalkwyk LC, Seltzer Z, Settle S, Shimomura K, Shou SM, Sikela JM, Siracusa LD, Spearow JL, Teuscher C, Threadgill DW, Toth LA, Toye AA, Vadasz C, Van Zant G, Wakeland E, Williams RW, Zhang HG, Zou F (2003) The nature and identification of quantitative trait loci: a community's view. Nat Rev Genet 4:911–916

Altenburg E, Muller HJ (1920) The genetic basis of truncate wing – an inconstant and modifiable character in *Drosophila*. Genetics 5:1–59

Arabidopsis Genome Initiative (2000) Analysis of the genome sequence of the flowering plant *Arabidopsis thaliana*. Nature 408:796–815

Auger DL, Gray AD, Ream TS, Kato A, Coe EH, Birchler JA (2005) Nonadditive gene expression in diploid and triploid hybrids of maize. Genetics 169:389–397

Baltimore D (2001) Our genome unveiled. Nature 409:814–816

Beavis WD. (1997). QTL analysis: power, precision and accuracy. In: Paterson AH (ed) Molecular dissection of complex traits. CRC Press, Boca Raton, pp 145–162

Brem RB, Kruglyak L (2005) The landscape of genetic complexity across 5,700 gene expression traits in yeast. Proc Natl Acad Sci USA 102:1572–1577

Brem RB, Storey JD, Whittle J, Kruglyak L (2005) Genetic interactions between polymorphisms that affect gene expression in yeast. Nature 436:701–703

Brem RB, Yvert G, Clinton R, Kruglyak L (2002) Genetic dissection of transcriptional regulation in budding yeast. Science 296:752–755

Bystrykh L, Weersing E, Vellenga E, Manley E, Williams R, Cooke M, De Haan G (2003) A genetical genomics approach to identify transcriptional pathways in hematopoietic stem cells. Exp Hematol 31:137–137

Bystrykh L, Weersing E, Dontje B, Sutton S, Pletcher MT, Wiltshire T, Su AI, Vellenga E, Wang JT, Manly KF, Lu L, Chesler EJ, Alberts R, Jansen RC, Williams RW, Cooke MP, de Haan G (2005) Uncovering regulatory pathways that affect hematopoietic stem cell function using 'genetical genomics'. Nat Genet 37:225–232

Chesler EJ, Lu L, Shou SM, Qu YH, Gu J, Wang JT, Hsu HC, Mountz JD, Baldwin NE, Langston MA, Threadgill DW, Manly KF, Williams RW (2005) Complex trait analysis of gene expression uncovers polygenic and pleiotropic networks that modulate nervous system function. Nat Genet 37:233–242

Churchill GA, Doerge RW (1994) Empirical threshold values for quantitative trait mapping. Genetics 138:963–971

de Koning DJ, Haley CS (2005) Genetical genomics in humans and model organisms. Trends Genet 21:377–381

DeCook R, Lall S, Nettleton D, Howell SH (2006) Genetic regulation of gene expression during shoot development in *Arabidopsis*. Genetics 172:1155–1164

Doebley J, Lukens L (1998) Transcriptional regulators and the evolution of plant form. Plant Cell 10:1075–1082

Doerge RW, Churchill GA (1996) Permutation tests for multiple loci affecting a quantitative character. Genetics 142:285–294

Dumas P, Sun YL, Corbeil G, Tremblay S, Pausova Z, Kren V, Krenova D, Pravenec M, Hamet P, Tremblay J (2000) Mapping of quantitative trait loci (QTL) of differential stress gene expression in rat recombinant inbred strains. J Hypertens 18:545–551

Filatov V, Dowdle J, Smirnoff N, Ford-Lloyd B, Newbury HJ, Macnair MR (2006) Comparison of gene expression in segregating families identifies genes and genomic regions involved in a novel adaptation, zinc hyperaccumulation. Mol Ecol 15:3045–3059

Flint-Garcia SA, Thuillet AC, Yu JM, Pressoir G, Romero SM, Mitchell SE, Doebley J, Kresovich S, Goodman MM, Buckler ES (2005) Maize association population: a high-resolution platform for quantitative trait locus dissection. Plant J 44:1054–1064

Gibson G, Weir B (2005) The quantitative genetics of transcription. Trends Genet 21:616–623

Gibson G, Riley-Berger R, Harshman L, Kopp A, Vacha S, Nuzhdin S, Wayne M (2004) Extensive sex-specific nonadditivity of gene expression in *Drosophila melanogaster*. Genetics 167:1791–1799

Goff SA, Ricke D, Lan TH, Presting G, Wang RL, Dunn M, Glazebrook J, Sessions A, Oeller P, Varma H, Hadley D, Hutchinson D, Martin C, Katagiri F, Lange BM, Moughamer T, Xia Y, Budworth P, Zhong JP, Miguel T, Paszkowski U, Zhang SP, Colbert M, Sun WL, Chen LL, Cooper B, Park S, Wood TC, Mao L, Quail P, Wing R, Dean R, Yu YS, Zharkikh A, Shen R, Sahasrabudhe S, Thomas A, Cannings R, Gutin A, Pruss D, Reid J, Tavtigian S, Mitchell J, Eldredge G, Scholl T, Miller RM, Bhatnagar S, Adey N, Rubano T, Tusneem N, Robinson R, Feldhaus J, Macalma T, Oliphant A, Briggs S (2002) A draft sequence of the rice genome (*Oryza sativa* L. ssp. *japonica*). Science 296:92–100

Higuchi T (1990) Lignin biochemistry: biosynthesis and biodegradation. Wood Sci Technol 24:23–63

Hubner N, Wallace CA, Zimdahl H, Petretto E, Schulz H, Maciver F, Mueller M, Hummel O, Monti J, Zidek V, Musilova A, Kren V, Causton H, Game L, Born G, Schmidt S, Muller A, Cook SA, Kurtz TW, Whittaker J, Pravenec M, Aitman TJ (2005) Integrated transcriptional profiling and linkage analysis for identification of genes underlying disease. Nat Genet 37:243–253

Jansen RC, Nap JP (2001) Genetical genomics: the added value from segregation. Trends Genet 17:388–391

Juenger TE, Wayne T, Boles S, Symonds VV, McKay J, Coughlan SJ (2006) Natural genetic variation in whole-genome expression in *Arabidopsis thaliana*: the impact of physiological QTL introgression. Mol Ecol 15:1351–1365

Kao CH, Zeng ZB, Teasdale RD (1999) Multiple interval mapping for quantitative trait loci. Genetics 152:1203–1216

Karp CL, Grupe A, Schadt E, Ewart SL, Keane-Moore M, Cuomo PJ, Kohl J, Wahl L, Kuperman D, Germer S, Aud D, Peltz G, Wills-Karp M (2000) Identification of complement factor 5 as a susceptibility locus for experimental allergic asthma. Nat Immunol 1:221–226

King MC, Wilson AC (1975) Evolution at 2 levels in humans and chimpanzees. Science 188:107–116

Kirst M, Basten CJ, Myburg AA, Zeng ZB, Sederoff RR (2005) Genetic architecture of transcript-level variation in differentiating xylem of a eucalyptus hybrid. Genetics 169:2295–2303

Kirst M, Myburg AA, De Leon JPG, Kirst ME, Scott J, Sederoff R (2004) Coordinated genetic regulation of growth and lignin revealed by quantitative trait locus analysis of cDNA microarray data in an interspecific backcross of eucalyptus. Plant Physiol 135:2368–2378

Korstanje R, Paigen B (2002) From QTL to gene: the harvest begins. Nat Genet 31:235–236

Lall S, Nettleton D, DeCook R, Che P, Howell SH (2004) Quantitative trait loci associated with adventitious shoot formation in tissue culture and the program of shoot development in *Arabidopsis*. Genetics 167:1883–1892

Lander ES, Botstein D (1989) Mapping mendelian factors underlying quantitative traits using RFLP linkage maps. Genetics 121:185–199

Lembertas AV, Perusse L, Chagnon YC, Fisler JS, Warden CH, PurcellHuynh DA, Dionne FT, Gagnon J, Nadeau A, Lusis AJ, Bouchard C (1997) Identification of an obesity quantitative trait locus on mouse chromosome 2 and evidence of linkage to body fat and insulin on the human homologous region 20q. J Clin Invest 100:1240–1247

Levine M, Tjian R (2003) Transcription regulation and animal diversity. Nature 424:147–151

Li J, Burmeister M (2005) Genetical genomics: combining genetics with gene expression analysis. Hum Mol Genet 14:R163–R169

Mackay TFC (2001) The genetic architecture of quantitative traits. Ann Rev Genet 35:303–339

Matsuzaki H, Dong SL, Loi H, Di XJ, Liu GY, Hubbell E, Law J, Berntsen T, Chadha M, Hui H, Yang GR, Kennedy GC, Webster TA, Cawley S, Walsh PS, Jones KW, Fodor SPA, Mei R (2004) Genotyping over 100,000 SNPs on a pair of oligonucleotide arrays. Nat Meth 1:109–111

McGall G, Labadie J, Brock P, Wallraff G, Nguyen T, Hinsberg W (1996) Light-directed synthesis of high-density oligonucleotide arrays using semiconductor photoresists. Proc Natl Acad Sci USA 93:13555–13560

Mitchell-Olds T, Schmitt J (2006) Genetic mechanisms and evolutionary significance of natural variation in *Arabidopsis*. Nature 441:947–952

Monks SA, Leonardson A, Zhu H, Cundiff P, Pietrusiak P, Edwards S, Phillips JW, Sachs A, Schadt EE (2004) Genetic inheritance of gene expression in human cell lines. Amer J Hum Genet 75:1094–1105

Morley M, Molony CM, Weber TM, Devlin JL, Ewens KG, Spielman RS, Cheung VG (2004) Genetic analysis of genome-wide variation in human gene expression. Nature 430:743–747

Myburg AA, Griffin AR, Sederoff RR, Whetten RW (2003) Comparative genetic linkage maps of *Eucalyptus grandis*, *Eucalyptus globulus* and their F1 hybrid based on a double pseudo-backcross mapping approach. Theor Appl Genet 107:1028–1042

Nuwaysir EF, Huang W, Albert TJ, Singh J, Nuwaysir K, Pitas A, Richmond T, Gorski T, Berg JP, Ballin J, McCormick M, Norton J, Pollock T, Sumwalt T, Butcher L, Porter D, Molla M, Hall C, Blattner F, Sussman MR, Wallace RL, Cerrina F, Green RD (2002) Gene expression analysis using oligonucleotide arrays produced by maskless photolithography. Genome Res 12:1749–1755

Paterson AH, Lander ES, Hewitt JD, Peterson S, Lincoln SE, Tanksley SD (1988) Resolution of quantitative traits into mendelian factors by using a complete linkage map of restriction fragment length polymorphisms. Nature 335:721–726

Paterson AH, Damon S, Hewitt JD, Zamir D, Rabinowitch HD, Lincoln SE, Lander ES, Tanksley SD (1991) Mendelian factors underlying quantitative traits in tomato – comparison across species, generations, and environments. Genetics 127:181–197

Perez-Enciso M (2004) In silico study of transcriptome genetic variation in outbred populations. Genetics 166:547–554

Piepho HP (2001) A quick method for computing approximate thresholds for quantitative trait loci detection. Genetics 157:425–432

Purugganan MD (2000) The molecular population genetics of regulatory genes. Mol Ecol 9:1451–1461

Ronald J, Akey JM, Whittle J, Smith EN, Yvert G, Kruglyak L (2005) Simultaneous genotyping, gene-expression measurement, and detection of allele-specific expression with oligonucleotide arrays. Genome Res 15:284–291

Rostoks N, Borevitz JO, Hedley PE, Russell J, Mudie S, Morris J, Cardle L, Marshall DF, Waugh R (2005) Single-feature polymorphism discovery in the barley transcriptome. Genome Biol 6:R54

Sax K (1923) The association of size differences with seed-coat pattern and pigmentation in *Phaseolus vulgaris*. Genetics 8:552–560

Schadt EE, Monks SA, Drake TA, Lusis AJ, Che N, Colinayo V, Ruff TG, Milligan SB, Lamb JR, Cavet G, Linsley PS, Mao M, Stoughton RB, Friend SH (2003) Genetics of gene expression surveyed in maize, mouse and man. Nature 422:297–302

Schadt EE, Lamb J, Yang X, Zhu J, Edwards S, GuhaThakurta D, Sieberts SK, Monks S, Reitman M, Zhang CS, Lum PY, Leonardson A, Thieringer R, Metzger JM, Yang LM, Castle J, Zhu HY, Kash SF, Drake TA, Sachs A, Lusis AJ (2005) An integrative genomics approach to infer causal associations between gene expression and disease. Nat Genet 37:710–717

Schena M, Shalon D, Davis RW, Brown PO (1995) Quantitative monitoring of gene expression patterns with a complementary-DNA microarray. Science 270:467–470

Storey JD, Tibshirani R (2003) Statistical significance for genomewide studies. Proc Natl Acad Sci USA 100:9440–9445

Storey JD, Akey JM, Kruglyak L (2005) Multiple locus linkage analysis of genomewide expression in yeast. Plos Biol 3:1380–1390

Stuber CW, Lincoln SE, Wolff DW, Helentjaris T, Lander ES (1992) Identification of genetic-factors contributing to heterosis in a hybrid from 2 elite maize inbred lines using molecular markers. Genetics 132:823–839

Szalma SJ, Buckler ES, Snook ME, McMullen MD (2005) Association analysis of candidate genes for maysin and chlorogenic acid accumulation in maize silks. Theor Appl Genet 110:1324–1333

Tautz D (2000) Evolution of transcriptional regulation. Curr Opin Genet Dev 10:575–579

Vuylsteke M, van Eeuwijk F, Van Hummelen P, Kuiper M, Zabeau M (2005) Genetic analysis of variation in gene expression in *Arabidopsis thaliana*. Genetics 171:1267–1275

Wayne ML, McIntyre LM (2002) Combining mapping and arraying: An approach to candidate gene identification. Proc Natl Acad Sci USA 99:14903–14906

West MAL, van Leeuwen H, Kozik A, Kliebenstein DJ, Doerge RW, St Clair DA, Michelmore RW (2006) High-density haplotyping with microarray-based expression and single feature polymorphism markers in *Arabidopsis*. Genome Res 16:787–795

Wray GA, Hahn MW, Abouheif E, Balhoff JP, Pizer M, Rockman MV, Romano LA (2003) The evolution of transcriptional regulation in eukaryotes. Mol Biol Evol 20:1377–1419

Yvert G, Brem RB, Whittle J, Akey JM, Foss E, Smith EN, Mackelprang R, Kruglyak L (2003) Trans-acting regulatory variation in *Saccharomyces cerevisiae* and the role of transcription factors. Nat Genet 35:57–64

Zeng ZB (1993) Theoretical basis for separation of multiple linked gene effects in mapping quantitative trait loci. Proc Natl Acad Sci USA 90:10972–10976

Zeng ZB (1994) Precision mapping of quantitative trait loci. Genetics 136:1457–1468

Zou F, Fine ZP, Hu JH, Lin DY (2004) An efficient resampling method for assessing genome-wide statistical significance in mapping quantitative trait loci. Genetics 168:2307–2316

CHAPTER 12

ANALYSIS OF SALT STRESS-RELATED TRANSCRIPTOME FINGERPRINTS FROM DIVERSE PLANT SPECIES

ASHWANI PAREEK[1], SNEH L. SINGLA-PAREEK[2], SUDHIR K. SOPORY[2] AND ANIL GROVER[3,*]

[1]*Stress Physiology and Molecular Biology Laboratory, School of Life Sciences, Jawaharlal Nehru University, New Delhi 110067, India*
[2]*Plant Molecular Biology, International Centre for Genetic Engineering and Biotechnology, New Delhi 100067, India*
[3]*Department of Plant Molecular Biology, University of Delhi South Campus, Benito Juarez Road, New Delhi 110 021, India*

Abstract: The capacity to sequence genomes and the availability of novel molecular tools such as analysis of transcriptomes has catapulted biological research into eras of genomics and post-genomics. As salinity limits crop production in a major way, large number of studies have been undertaken to analyze how salt stressed plant tissues differ from unstressed tissues in terms of genomic alterations. Bulk of this work has been carried out using the three model plant species namely *Arabidopsis thaliana*, *Oryza sativa* and *Mesembryanthemum crystallinum*. Apart from these plants, others like *Thellungiella halophila*, *Xerophyta humilis*, *Populus euphratica*, *Setaria italica*, *Glycine soja*, *Sorghum bicolor* and *Triticum aestivum* have been exploited to understand the differences in gene expression between saline sensitive and tolerant species. Closer look at the transcriptome based analysis carried out with these systems shows that genes involved in stress reception and stress signaling, regulation of gene expression, protein translation, transport of ions and metabolism (like enzymatic roles, energy production, protective functions and maintenance of osmotic balance) undergo major changes during salt stress conditions. Importantly, there are large numbers of genes that are upregulated under salt stress for which no major function has been ascribed as yet. Ultimately, a system biology based approach needs to be established to understand plant responses towards salinity stress and adaptation in an integrated manner. Future studies must focus on these leads to unravel novel candidate genes for making more-effective salt stress tolerant transgenic crops.

*Corresponding Author: anil.anilgrover@gmail.com

R.K. Varshney and R. Tuberosa (eds.), Genomics-Assisted Crop Improvement: Vol. 1: Genomics Approaches and Platforms, 267–287.

1. INTRODUCTION

Salinity is one of the major stresses that cause significant reductions in plant productivity, and consequent economic losses. Estimations indicate that more than 50% of all the arable land will be affected by serious salinization by the year 2050 (Blumwald and Grover 2006). Owing to the impressive progress made in plant molecular biology and biotechnology research in the past two decades, intensive efforts are in progress to improve plant salt stress tolerance by genetic transformation method (see Singla-Pareek et al. 2001, Grover et al. 2003, Wang et al. 2003, Yamaguchi and Blumwald 2005, Gepstein et al. 2006, Blumwald and Grover 2006). However, in spite of these efforts, it remains true that there is no salt tolerant cultivar made by transgenic research that has been commercially released or being tested on large-scale. Most transgenic salt tolerant plants produced thus far have been tested under laboratory conditions. There appears to be some gap in our capacity to produce "true" salt tolerant transgenic cultivars. What is this gap due to? Is it because our understanding of the biochemistry, physiology and genetics of the salt stress response is far from adequate at this point of time? Is it that single gene being employed for production of salt tolerant plants is too simplistic an approach? Until recently, the prevalent strategy of the molecular genetic methods for raising stress tolerant transgenic plants has been the 'candidate gene' approach which essentially tests the relevance of a single gene only. The bottleneck in producing salt tolerant transgenics with more than one gene does not seem to be the technique part of genetic transformations as there are already reports indicating that this is possible (Halpin 2005). It seems therefore that identification of more, novel candidate genes that should be co-expressed is the limitation; "Gene Discovery" thus appears to be the key event.

Genomics approach places emphasis on integrated analysis of stress-dependent behavior by the entire plant. Genome analysis appears as a possible bridge between molecular biology and whole plant physiology, agronomy and crop breeding (Bohnert et al. 2006, Sahi et al. 2006). High-quality sequences of the whole genomes have been generated for several organisms. Updated and refined tools of bioinformatics have allowed gene predictions with high confidence. At present, high-quality finished genome sequence is available for *Arabidopsis* and rice and efforts are underway for other plants (www.tigr.org). Techniques like microarrays, SAGE, RT-PCR or differential hybridization are the possible ways by which efforts are being made to understand plant responses towards salinity stresses at genome level. In particular, transcript arraying technique has revolutionized the field of transcriptome analysis. Microarrays and macroarrays allow one to have a snap-shot of the transcripts at whole-genome level. Table 1 summarizes some of the representative reports wherein genome-level studies have been carried for salinity stress response using the microarray and macroarray methods. It has been demonstrated that large-scale microarray data can be used to recognize the cross-talk between different signaling pathways providing information that will be useful in elucidating unknown signaling networks (Ma et al. 2006). Though *Arabidopsis* genome sequence was made available to the scientific community in the year 2000, salinity

Table 1. Recent selected reports where salinity stress related transcriptome has been analyzed employing microarrays/macroarrays

Genotype analyzed	Stress treatment	Design of experiment	Comments	Reference
Arabidopsis sp	250 mM NaCl for different time 1, 2, 5, 10 and 24 h	Full length cDNA microarrays	194 genes out of 7000 cDNA (at least five fold change) showed alteration	Seki et al. 2002
	100 mM NaCl, Suspension-cultured cell line	cDNA microarray with 7000 full length cDNAs	57 out of 7000 showed alteration	Takahashi et al. 2004
Glycine sp.	150 mM NaCl	Full length cDNA	2003 ESTs analyzed	Ji et al. 2006
Hordeum sp	Increasing concentration of NaCl starting from 25 mM to 100 mM, 14 d after transplanting until day 17	Microarray containing 22,750 probe sets	Induction of genes involved in jasmonic acid biosynthesis	Walia et al. 2006
Mesembryanthemum sp	500 mM NaCl for 30 and 48 h	EST from cDNAs	3676 out of 9733 EST tags showed alteration	Kore-eda et al. 2004
Oryza sp	150 mM NaCl from 15 minutes to 1 week	cDNA based microarrays	Within 1 h of stress, 10% transcripts upregulated	Kawasaki et al. 2001
	250 mM NaCl, Seedlings	cDNA microarray with 1700 cDNA	57 genes inducible by salinity stress	Rabbani et al. 2003
Populus sp	300 mM NaCl upto 72 h followed by recovery upto48 h	cDNA microarray with preselected by subtractive hybridization	Several transcripts down regulated by 72 h salinity stress but were up regulated after long-term recovery (48h).	Gu et al. 2004
	Leaves of axenically grown plant subjected to dehydration, high salinity, chilling, heat, ABA and H2O2 used for making the cDNA library	4500 ESTs	Thirteen candidates containing ERF/AP2 domain were found. Some showed stress-responsive expression	Nanjo et al. 2004
Triticum sp		60-mer oligo-cDNA microarray	1,811 showed minimum 2-fold change upon salinity treatment	Kawaura et al. 2006
Zea sp	Developing kernels at 15 d after pollination	2500 cDNA macroarray	1.5% up regulated and 0.7% down-regulated under salinity	Andjelkovic and Thompson 2006

related transcriptome analysis was first performed in rice in the year 2001 by the group of Hans Bohnert at University of Arizona (Kawasaki et al. 2001). In this analysis early responses towards salinity stress were analyzed using the cDNA arrays and approximately 10% of the total genes were found to be altered, representing the fact that salinity stress response is controlled by many genes. Nonetheless, it is beyond any doubt that resources and novel tools are rapidly expanding for analysis of the information as available from model plant - *Arabidopsis* (Denby and Gehring 2005). There has been tremendous advancement in this area and curated, publicly available microarray data is serving a large number of researchers in the area. In this regard, AtGenExpress is a multinational effort to profile the transcriptome of Arabidopsis (http://web.uni.frankfurt.de/fb15/botanik/mcb/AFGN/atgenex.htm). Additionally, another database, Genevestigator, is a web-based user-friendly tool that enables researchers to visualize the expression of a relatively small set of genes from a variety of microarray experiments, including AtGenExpress (Zimmerman et al. 2004). The full AtGenExpress data can be accessed via the TAIR (http://www.Arabidopsis.org) and NASC (http://affymatrix.Arabidopsis.info/narrays/experimentbrowse.pl) databases. There are several other monocots and dicots which have been analyzed in the recent past and a host of information has been generated. Besides analyzing the stress related transcriptome in a genome, other meaningful set of data has been generated while comparing the contrasting genotypes. In this chapter, we take a closer look at the transcriptome data emerging from such studies and try to investigate the fine fingerprints related to salinity stress as it emerges from the data from various genomes. Broadly speaking we have selected three genomes – *Arabidopsis*, *Oryza and Mesembryanthemum* - for this detailed analysis. Before we take up the transcriptome-based analysis in detail, we provide some general comments on the utility of these plant species for the genome-related studies.

2. MODEL PLANT SPECIES FOR THE GENOMICS-BASED STUDIES

2.1. Arabidopsis

Nearly twenty years ago, plant biologists were looking for a model organism suitable for detailed analysis using the combined tools of genetics and molecular biology. Plants with effective protocols for regeneration in culture (such as *Petunia* and tomato) were logical candidates, particularly for studies involving *Agrobacterium*- mediated cell transformation. However, attention gradually shifted towards *Arabidopsis*, a small weed in the mustard family that was first chosen as a model genetic organism in Europe and later studied in detail in the United States (Redei, 1975). The shift towards *Arabidopsis* gained momentum in early 1980s with the release of a detailed genetic map and publications stressing upon the value of *Arabidopsis* for research in plant physiology, biochemistry and development (Meinke et al. 1998). Research with *Arabidopsis* has provided valuable insights into all aspects of modern biology (Rhee et al. 2003, Bevan and Walsh 2005). In several cases, long-standing questions in plant physiology and biochemistry have been resolved through genetic and molecular analysis of *Arabidopsis* mutants. In this regard, a special mention needs to be made for the discovery of novel SOS (Salt Overly Sensitive) pathway candidates for signaling under osmotic stress. It is through the combination of genetics, mutations, physiology, biochemistry and molecular biology tools, that we have been been able to get some insight into this signaling cascade (Chinnusamy et al. 2004). With the availability of completely sequenced and annotated genome (*Arabidopsis* Genome Initiative 2000), *Arabidopsis* has turned out to be a wonderful tool for the analysis of salt responsive transcriptomes. Among the various transcript platforms available for this plant, the most complete set is the one which includes the 70mer oligonucleotides representing approximately 26,000 DNA elements, representing the known as well as hypothetical coding regions (http://www.ag.arizona.edu/microarray). Another widely-used platform is the Affymatrix GeneChip with approximately 22,000 genes. Both these platforms have been well exploited for analysis of salt stress responsive transcriptome. Analysis has indicated a highly similar trend in gene regulation patterns, where approximately 80% of the transcripts respond similarly, indicating the comparable usefulness of these platforms (Ma et al. 2006). Recent study suggested that the *Arabidopsis* salt stress response could be categorized into segments distinct from others such as pathogens, cold stress or even ABA application (Ma et al. 2006). Kreps et al. (2002) accomplished analysis of transcriptome changes in response to salinity, osmotic and cold stress using the GeneChip microarray with probe sets for approximately 8,100 genes. This analysis indicated that about 30% of the transcriptome is sensitive to regulation by common stress conditions.

2.2. Rice

Rice has slowly emerged as a model system for monocot studies for reasons such as follows: (1) this species has a compact genome of 430 MB size with ~37 K genes, (2) its complete nuclear genome sequence has been determined (Goff et al. 2002, Yu et al. 2002, IRGSP 2005), (3) its complete chloroplast genome sequence has been determined, (4) ~28,000 rice full-length cDNA have been isolated and sequenced, (5) large number of ESTs have been isolated and sequenced in this species, (6) work on making of knockout mutants for every single gene in this species is in progress, (7) production of transgenics by *Agrobacterium* with high-frequency is demonstrated and (8) detailed molecular marker maps have been constructed. Employing diverse methods, stable genetic transformation of rice is a reality (Bajaj and Mohanty 2005). Availability of complete genome sequence of rice has generated a new spurge of interest. Transcriptome analysis of salt stress response in rice has been carried out using microarray and macroarray based methods by several groups (Table 1). Two types of arrays have been made available for genomic studies in rice, namely oligonucleotides array as well as cDNA array. There are numerous examples wherein detailed studies have been taken in rice using these approaches. Analysis of salt stress-inducible ESTs from salt tolerant rice cultivar Dee-geo-woo-gen revealed several proteins showing homology to proteins functional for detoxification, stress response and signal transduction in plants (Shiozaki et al. 2005). Comparative analysis between different rice genotypes has also been attempted employing salinity tolerant (CSR27 and Pokkali) and sensitive (PB1) cultivars of rice (Sahi et al. 2003). This study highlighted that genes such as SalT, glycine rich RNA binding proteins, ADP ribosylation factor, NADP dependent malic enzyme, Mub ubiquitin fusion protein, tumor suppressor genes, wound inducible genes, ethylene response element binding protein, alanine aminotransferase, copper chaperone, aspartate aminotransferase, ripening regulated protein, metallothionine and Zn finger transcription factor are important constituents of the rice salt stress response. In another study of almost similar nature, comparative analysis between salt-sensitive rice cultivar IR64 and naturally salt tolerant Pokkali revealed several ESTs specifically induced in higher amounts in the stress tolerant Pokkali rice (Pareek et al. unpublished). Importantly, homologous genes from cultivated and wild species of rice have also shown differential response. For example, the PcINOI gene from local wild salinity resistant rice (Porteresia) has been found to possess a short stretch (37 amino acids) which seems to make the protein more tolerant towards salinity stress (Ghosh-Dastidar et al. 2006). This wild homologue has been tested in a range of species and usefulness of the same has been shown (Das-Chatterjee et al. 2006). Overexpression of a serine-rich-protein from Porteresia (PcSrp) in yeast and finger millet improves salinity tolerance (Mahalakshmi et al. 2006).

2.3. Common Ice Plant

Salinity-tolerant model plants have been proposed to be an important resource of salt-stress associated genes with a hope that such investigations will enable future molecular dissections of salt-tolerance mechanisms in important crop plants (Vinocur and Altman 2005). *Mesembryanthemum crystallinum* (the common ice plant – known so because of the icy look due to enlarged bladder cells of leaf epidermis) has been a system of choice with selected laboratories for salt stress-related studies. This facultative halophyte has an inherent unique capability to quickly switch from normal C3 mode to CAM mode of photosynthesis upon exposure to salinity. CAM is a plastic adaptation of photosynthetic carbon fixation found in about 6% of angiosperm species that limits evaporative water loss and photorespiration, and improves water use efficiency under stress conditions (Cushman and Borland 2002, Dodd et al. 2002). Large numbers of reports highlight how this change in photosynthesis behavior is accomplished based on single-gene analysis (Niewiadomska et al. 1999). The ice plant has been used extensively as a system to investigate responses towards salinity stress such as sodium accumulation and partitioning within the cell, potassium uptake, water channels, carbon, nitrogen and amino acid metabolism and transport, and reactive oxygen scavenging mechanisms (Kore-eda et al. 2004 and references therein). In recent years, transcriptome analysis has been carried out in this species for the comprehensive genomic analysis of the salt stress-regulated genes (Kore-eda et al. 2004).

3. SALT STRESS-RELATED TRANSCRIPTOME CHANGES IN *ARABIDOPSIS*, RICE AND COMMON ICE PLANT

A. thaliana and *O. sativa* are believed to have shared a common ancestor ~150 to 200 million years ago. With the availability of (1) whole genome sequence in rice and *Arabidopsis* and (2) wealth of information on ESTs and transcriptome changes in *Arabidopsis*, rice and common ice plant, it becomes a challenge to find out the key components of the stress response pathways operative in these model systems. However, current knowledge is still largely restricted to individual genes and pathways, and the unifying picture remains hidden. The understanding of the salinity stress will be greatly enhanced by identifying the convergent and divergent pathways between salinity and other stress responses and nodes of signaling convergence (Ma et al. 2006). We manually scored the salinity-induced transcriptome changes of *A. thaliana*, *O. sativa* and *M. crystallinum* as revealed through microarray studies available from the published literature, in this study. Our aim was to underscore the commonalities and differences in gene expression profiles in these three genera. Those gene expression changes which were unique to a given species were ignored in the further analysis, considering that such genes might be important for species-specific functions. We emphasized for this study, genes that show alterations in two or all the three species undertaken. This category of genes was thought to be important because such genes might be related to generic functions related to salt stress response and reflect the "commoneome". The above analysis enabled us to broadly put the salt stress related gene expression profile alterations into three categories:

(1) Class I - including genes for which no major change in levels was detected,
(2) Class II - including genes which showed major up-regulation upon stress treatment and
(3) Class III - including genes which showed down-regulation upon stress treatment.

The "Class II" category of genes appear most important for governing salt stress response, considering that such genes might be associated with specific changes in gene expression and thus might represent genes associated with the induced tolerance response. Table 2 presents list of Class II and III genes noted in this study. These genes belong to categories related to diverse functions ranging from stress perception to the effector response. Overall, we find that *A. thaliana* and *O. sativa* shared more overlapping patterns in gene expression as compared to *M. crystallinum*.

Table 2. Up- and down-regulated genes in rice, *Arabidopsis* and ice Plant
Only those candidates have been selected which show response common in at least two of the salinity related transcriptome. This analysis included both salinity up-regulated (+) and down-regulated (-) genes. Artificial grouping of these genes has also been done to reflect their possible physiological role

Gene notation **Up-regulated genes**	Rice	Arabidopsis	Ice Plant
STRESS PERCEPTION AND SIGNALING			
ABA- and stress-induced protein	+	+	
S-adenosylmethionine decarboxylase 2	+	+	
Auxin-regulated protein	+	+	
Calcium-dependent protein kinase	+	+	+
Calcineurin-like phosphatases	+	+	
Calcium-binding EF-hand protein	+	+	
Gibberellic acid–induced gene	+	+	+
Lectin, lectin protein kinase	+	+	
Protein phosphatases 2C	+	+	+
Receptor kinase-like protein	+	+	+
Ser/Thr kinase-like protein	+	+	+
STRESS REGULATION			
AP2 domain-containing trans. factor	+	+	
CC-NBS-LRR resistance protein mla13	+	+	
Myb-like DNA-binding domain	+	+	
NAC-type DNA-binding protein	+	+	
Translation initiation factor	+	+	
Zinc finger protein	+	+	+
bZIP DNA-binding protein	+	+	
GENERAL METABOLISM			
Aldehyde dehydrogenase	+	+	
Anthocyanin biosynthesis	+	+	
Ascorbate peroxidase, cytosolic type	+	+	
Chlorophyll a-b binding protein	+	+	
Cytochrome P450 monooxygenase	+	+	+
Esterase/lipase/thioesterase-like protein	+	+	
Expansin, putative	+	+	
Galactosidase	+	+	
Galactinol-raffinose galactosyltransferase,Galactinol synthase	+	+	+
ß-Glucosidase homolog	+	+	+
Glutamate receptor family protein, glutamate synthase	+	+	+
Glycosyl transferase	+	+	
1,4-hydroxyphenylpyruvate dioxygenase	+	+	
Ribosomal protein	+	+	+
Zeaxanthin epoxidase	+	+	
PROTECTIVE FUNCTION			
Acidic endochitinase	+	+	+
Ankyrin repeat family protein	+	+	
Cold-regulated protein, cor15a	+	+	+
Dehydrin, DREB subfamily	+	+	
ß 1,3-glucanase	+	+	
Glutathione *S*-transferase homolog	+	+	+

Heat shock protein	+	+	+
LEA protein	+	+	+
Lipoxygenase	+	+	
Metallothionein-like protein, OsMT-1	+	+	
Papain cysteine protease	+	+	
Pathogen-responsive -dioxygenase	+	+	
Peroxidase-1	+	+	+
Protease inhibitor	+	+	+
Subtilisin-chymotrypsin inhibitor 2	+	+	
Thioredoxin, Thioredoxin reductase	+	+	
Xyloglucan endotransglycosylase	+	+	
OSMOTIC HOMEOSTASIS			
ABC transporter family protein	+	+	+
F-box family protein	+	+	
Ion transporter,Na/H	+	+	+
MATE efflux family protein	+	+	
Pyrroline 5-carboxylase synthetase	+	+	
Sugar transporter	+	+	+
Trehalose-6-phosphate phosphatase	+	+	
UDP-glucose 4-epimerase	+	+	
V-type ATPase	+	+	+
Water channel protein	+	+	+
FUNCTION YET TO BE DISCOVERED			
Glycine/serine-rich protein	+	+	
O-methyltransferase	+	+	+
Down-regulated genes			
Adenine phosphoribosyltransferase form 3	-	-	
Water channel protein (WCP-III)	-	-	
A-tubulin	-	-	
TMK (gibberellic acid induced)	-	-	-
Peroxidase ATP19a	-	-	-
Putative translation initiation factor eIF-2Ba	-	-	
Receptor-like protein	-	-	

4. LESSONS FROM THE TRANSCRIPTOME DATA: INVESTIGATING THE SALINITY STRESS RELATED "FINGERPRINTS"

From the model species *Arabidopsis*, rice and common ice plant and a host of other species, a large repertoire of genes is noted to be associated with the salt stress response. The bulk of these genes can be grouped into specific categories as follows:

4.1. Genes Involved in Stress Perception and Signaling

The first candidate protein which senses the stress signal is obviously the receptor. Several receptors have been found to be up regulated in salt stress response such as glutamate receptor family protein, receptor kinase like proteins and lectins.

Signaling machinery operative under salt stress conditions plays an important role in determining the survival of the system. Genomic analysis has revealed that a large number of signaling genes are upregulated in response to salt stress. The genes belonging to this category include those induced by ABA application, CDPK, CaM, CaN and Ser/Thr protein kinase and protein phosphatase 2C, indicating the commonality in stress signaling. Some of these genes appeared common while others appeared unique to a specific stress type. Classically, ABA-independent and ABA-dependent pathways have been implicated in the induction of stress genes. Calcium is known to be an important signal transducer for many stress-responsive genes. One important calcium binding protein that modulates the activity of many other proteins in the pathway is calmodulin (CaM). In a recent study, one of the isoform of CaM was found to bind to MYB2 transcription factor, and was reported to enhance its DNA binding activity (Yoo et al. 2005). Over-expression of this isoform of CaM in *Arabidopsis* up-regulated the transcription of MYB2 regulated genes including P5CS1 which is known to confer salt tolerance by over-producing proline (Yoo et al. 2005). The mechanism by which calcium regulates sodium homeostasis via SOS pathway has been well elucidated (Quintero et al. 2002). One of the first components of this pathway is calcineurin like proteins which in turn activate CIP kinases (CIPK). The activated kinase can regulate the expression of a plasma membrane Na^+ antiporter (SOS1) to efflux sodium out of the system (Guo et al. 2004). It has been shown that over-expression of SOS1 can lead to enhanced tolerance of plants to NaCl stress (Shi et al. 2003). Over-expression of yeast calcineurin was also found to confer stress tolerance (Marin-Manzano et al. 2004). Recently, it is reported that over-expression of mouse calcineurin gene in rice results in its higher salt tolerance (Ma et al. 2005). The transgenic plants also showed higher expression of a group 2 LEA protein (Ma et al. 2005). The calcium-dependent protein kinase (CDPK) from rice showed tolerance to salt and drought stress upon over-expression in rice (Saijo et al. 2000). In a recent study, Jin et al. (2005) cloned a MAP kinase gene (*EhHOG*) from a fungus, *Eurotium herbariorum* that grows in Dead Sea in Israel. *EhHOG* was able to complement *hog1* mutant of *S. pombe* for growth under high osmotic stress.

4.2. Gene Regulation

This class includes AP2 domain containing factor, Zn-finger protein, Myb-like DNA-binding proteins, bZIP DNA binding proteins and NAC-type DNA binding proteins. Studies have shown that over-expression of *CBF3* (*DREB1*) confers stress tolerance. It is shown that use of stress-inducible *rd29A* promoter driving *DREB1A* expression conferred both drought and low-temperature tolerance in tobacco (Kasuga et al. 2004). *Arabidopsis CBF3* and *ABF3* increased tolerance to salinity and drought when overexpressed in rice (Oh et al. 2005). ABF2 over-expressing transgenic plants are reported to show tolerance to multiple stresses and altered sensitivity to ABA (Kim et al. 2004). *ABF2* over-expression also promoted glucose-mediated inhibition of seedling development. Gene encoding homeobox- leucine zipper protein *Hahb-4* from sunflower was found to be upregulated in response to drought conditions and to ABA. When overexpressed in *Arabidopsis*, transgenic plants showed shorter stem and internodes and more compact inflorescence. These plants were more tolerant to water stress (Dezar et al. 2005). A novel jasmonate and ethylene responsive factor, JERF3, was found to be induced in response to ethylene, JA, cold, salt and ABA in tomato (Wang et al. 2004). This factor was found to bind to GCC box cis-element that responds to ethylene and JA, and also to DRE element which responds to drought, salinity and cold. Over expression of JERF3 in tobacco resulted in the induction of pathogenesis-related genes and the plants showed enhanced tolerance towards salinity stress. A novel class of transcription factors called NAC (NAM, ATAF1,2, CUC2) are involved in many diverse plant functions (Olsen et al. 2005). It was shown recently that AtNAC2 may be involved in salinity stress tolerance and respond to auxin and ethylene signaling pathway in addition to ABA signaling. Overexpression of AtNAC2 resulted in the promotion of lateral root development and one of the genes that shows upregulation was found to be glyoxalase I (He et al. 2005). Earlier Fujita et al. (2004) had shown that dehydration induced protein RD26 is a NAC transcription factor and it was shown to trans-activate glyoxalase-I promoter. There are reports showing that over-expression of glyoxalase I and II can confer salinity stress tolerance (Veena et al. 1999; Singla-Pareek et al. 2003). Taking a clue from the above findings, we feel that manipulation of NAC transcription may turn out to be another important strategy for developing stress tolerant transgenic plants. In a recent study, the genes downstream to transcription factor – DREB1A/CBF3 have been analyzed employing the microarray technique (Maruyama et al. 2004). From the plants overexpressing DREB1A/CBF3, 38 genes were identified as DREB1A downstream genes in the latter study. Zn-finger protein Zat12 has been isolated and characterized from *Arabidopsis*. This protein has been documented to be responding to a range of biotic and abiotic stresses including salinity (Davletova et al. 2005). Based on microarray analysis, Zat12 has been suggested to play a central role in the metabolism of reactive oxygen species and abiotic stress signaling in *Arabidopsis*.

4.3. Proteins Related to General Metabolism

Several translation initiation factors are found to be up regulated in plants under salt stress. Candidates such as ABC transporter family proteins, sugar transporters, MATE efflux family protein, water channel protein, Na/H ion transporter etc have been found to up regulated upon salinity stress. Several genes coding for enzymes such as ascorbate peroxidase, endochitinase, lipoxygenase, glycosyl transferase, xyloglucan endotransglycosylase, esterase/ lipase thioesterase like protein, UDP glucose 4-epimerase, peroxidase, thioredoxin, zeaxanthin epoxidase, galactosidase and β-1,3 glucanase were found to be stress responsive. Proteins related to energy production such as cytochrome P-450 monoxygenase and V-type ATPase are shown to respond to salt stress.

4.4. Stress Induced Proteins with Some Protective Functions

LEA, HSPs, protease inhibitors, metallothionin like proteins, dehydrins, cold-regulated protein, gibberellic acid induced genes, subtilisin-chymotrypsin inhibitor-2 and papain cysteine protease are other genes which help the plant to cope up with the salt stress. *Nicotiana* HSP-70 (*NtHSP70-1*) was found to be a drought and ABA-inducible gene. Transgenic tobacco plants overexpressing NtHSP70 were found to be tolerant to water stress and with the progression of drought, the retention of optimum water was correlated with the level of the expressed protein (Cho and Hong 2006). Chinese cabbage expressing *B. napus* LEA gene showed enhanced ability to grow under salt and drought stress conditions and also recorded improved recovery upon removal of stress conditions (Park et al. 2005).

4.5. Proteins Related to Maintenance of Osmotic Homeostasis

Genes such as trehalose-6-phosphate phosphatases has been reported to be playing critical role in maintenance of osmotic balance in the cell upon salinity stress and has been found to transcriptionally induced. In yeast, trehalose -6-phosphate synthase, encoded by TPS1 gene, is the key enzyme for trehalose biosynthesis. Cortina and Culianez-Macia (2005) showed that over-expression of yeast TPS1 under the control of CaMV 35S promoter in tomato resulted in enhanced tolerance to drought, salt and oxidative stress. TPS1 gene has also been constitutively expressed in potato and the resulting transgenic plants showed increased drought resistance (Yeo et al. 2000). Introduction of a gene encoding bifunctional fusion (TPSP) of TPS and T-6-P phosphatase (TPP) from *E. coli* was expressed in rice under the control of ubiquitin promoter. The trehalose levels were found to increase and the transgenic plants showed an increased tolerance to drought, salt and cold without having any growth inhibition (Jang et al. 2003). Garg et al. (2002) have shown that overexpression of trehalose biosynthetic genes from *E. coli* into rice resulted in increased amounts of trehalose and sustainable plant growth under salt and drought conditions.

4.6. Proteins with Unknown Function

There are several reports for up-regulation of proteins which either have not been directly correlated with stress protection or whose identity has not yet been established. These include ribosomal proteins, gly-ser rich proteins, pathogen-responsive proteins and several unknown proteins which do not show homology with any entry in the database. The question that how can one get to know functions of these less worked out proteins is being attempted by researchers in several ways. One possible way could be the detailed analysis of coordinately regulated genes as it emerges from transcriptome analysis (Ma et al. 2006). One could also perform a genome wide analysis of the genes having strong homology in domain structures or tissue specific expression patterns employing bioinformatics tools. Comparison of such information between two diverse genera can be of further significance in terms of getting an evolutionary picture as has been done recently for two component signaling candidate genes between *Arabidopsis* and rice (Pareek et al. 2006).

5. ANALYSIS OF STRESS TRANSCRIPTOME FROM OTHER PLANT SPECIES

Transcriptome studies have been extended beyond model crops as well, as new datasets for various plants are being added each day. We now take reports emerging from such examples. Transcript profiling of desiccation tolerance has been attempted in *Xerophyta humilis,* which is a resurrection plant indigenous to Southern Africa (Collett et al. 2004). This plant species is known to tolerate extreme dehydration stress. A total of 55 dehydration-inducible cDNAs were identified from this species, which included candidates such as metallothionins, galactinol synthases, aldose reductase, glyoxalase, LEAs, dehydrins and other desiccation related proteins. This study also reported genes such as putative chloroplast RNA-binding protein and a protein containing SNF2/helicase domains, which have so far not been implicated in dehydration response. *Populus euphratica* is a salt tolerant tree species growing in saline semi-arid areas in Middle East and Asia. It grows at locations with high temperatures and high salt content in the soil salt content. ESTs from normalized and subtracted cDNA libraries prepared from plants exposed to multiple abiotic stress treatments including salt, drought, ozone, cold, freezing and flooding have been analyzed in this study. In striking contrast to other reports where 5 to 30% of the genes have been shown to exhibit altered expression in response to abiotic stresses, the number of *P. euphratica* genes that displayed differential transcript levels was only 1% of the total genes present on the array (Brosche et al. 2005). Some of the genes which showed up-regulation under stress conditions in this species include galactinol synthase, aldehyde dehydrogenase, β-amylase, and ferritin and cysteine protease. These genes have previously been reported to be up-regulated in *Arabidopsis* during combined heat and drought stress (Rizhsky et al. 2004). None of the classical enzymes involved in antioxidant defense, including catalase, ascorbate peroxidase and superoxide dismutase showed altered transcript levels. This was explained on the basis that candidates like aldehyde dehydrogenase and metallothionins might be fulfilling the role as antioxidant defense. Another gene which has been found to be highly up-regulated is cysteine proteases which have also been reported previously to be induced under both salt and drought stress (Koizumi et al. 1993). *Thellungiella halophila* is a closely-related to *A. thaliana.* In sharp contrast with *Arabidopsis*, *Thellungiella* tolerates extreme cold, drought, and salinity (Bressan et al. 2001; Inan et al. 2004; Taji et al. 2004, Amtmann et al. 2005). It has been noted that this naturally-occurring wild plant remains always "ready" to handle stress by keeping, in anticipation, the levels of stress responsive transcripts higher which are otherwise induced by stress signal in *A. thaliana* (Amtmann et al. 2005). In a recent study, the transcriptome of Yukon ecotype of *Thellungiella* has been analyzed employing 6578 ESTs, which represented 3628 unigenes from cDNA libraries of cold-, drought-, and salinity stressed plants (Wong et al. 2005). In-depth analysis indicated that of the 140 common unigenes which are present in all the three libraries, 70% have no known functions demonstrating that *Thellungiella* can be a rich source of genetic information on environmental responses. High DNA sequence homology allowed the use of *Arabidopsis* microarray platforms for the expression profiling of *Thellungiella*. Full-length cDNA arrays (Seki et al. 2002) and arrays based on long oligonucleotides (Maathuis et al. 2003) were therefore used with good results. It was revealed that similar level of salinity induces fewer genes in *Thellungiella* than in *Arabidopsis* (Inan et al. 2004, Taji et al. 2004). Under salt-free conditions, *Thellungiella* orthologs of some stress-related *Arabidopsis* genes showed higher base levels of expression. Transcript intensity analyses and metabolite profiles supported the microarray results, pointing towards a stress-anticipatory preparedness in Thellungiella (Gong et al. 2005). Macroarray containing 620 unigenes from barley have been utilized for analysis of stress response in foxmillet (Sreenivasulu et al. 2004). The differentially-expressed genes under salinity stress in salt tolerant "Prasad" and sensitive "Lepakshi" cultivars of foxtail millet (*Setaria italica* L.) indicated that genes encoding for phospholipids, hydroperoxide, glutathione peroxidase, ascorbate peroxidase, catalase 1 are selectively upregulated in the tolerant cultivar. *Glycine soja* is a salt tolerant variety of soybean found wildly growing in coastal areas. Salinity treated cDNA library constructed from this genotype has been made and compared with the one made from *Glycine max* (salt sensitive) soybean (Ji et al. 2006). It is hoped that ESTs available from the wild soybean type may serve as a useful source for isolation of important salinity responsive genes. Transcriptome analysis has been carried out in *Sorghum bicolor* in response to dehydration, high salinity and ABA (Buchanan et al. 2005). In this analysis, some novel genes were found to be showing alteration in response to stress signal. These genes include β-expansin expressed in shoots, actin depolymerization factor, inositol-3-phosphate synthase and oleosin. These proteins have been reported to be inducible by osmotic stress for the first time. 73, 521 ESTs have been generated from eleven cDNA libraries constructed from wheat plants exposed to various abiotic stresses and at different developmental stages (Houde et al. 2006). Exploiting the naturally occurring halophytes such as *Suaeda salsa*, which can survive the seawater-level salinity, 492 unique clones have been identified from NaCl-treated cDNA library out of which 76 were completely novel (Zhang et al. 2001). Studies on the salt tolerant mangrove species *Avicennia marina* revealed several clones that are implicated earlier in stress responses (Mehta et al. 2005). The expression profiles of some of the antioxidant genes have been monitored and found to be multi stress inducible (Jithesh et al. 2006).

6. CONCLUSIONS AND OUTLOOK

Comparisons of salt stress-related transcriptome alterations in *Arabidopsis*, *Oryza sativa* and *Mesembryanthemum crystallinum* show that there are species-specific alterations and there are generic alterations. Considering that generic alterations may have relation to general salt stress-related protection mechanisms, we have highlighted these in this study. Closer perusal of the transcript profiling data is providing useful clues about the various molecular events responsible for the development of the tolerance phenotype. However, the recent study has shown that such high-throughput monitoring of transcriptional activity of many genes should also take care of small populations of specialized cells within an organ to avoid possible masking of cell specific changes (Poroyko et al. 2007). Based on the analysis carried out in thispaper, it appears to us that processes such as perception and signaling of salt stress may have co-evolved. The candidate genes coding for receptors as well as ABA and/or calcium dependent signaling may be short-listed as potential candidate genes for future functional genomic studies. Another important category of genes appears to be the one involved in maintaining the homeostasis in the cell i.e. the ion transporters and those related to osmotic balance maintenance. An example of this category is the helicase protein: housekeeping gene helicase has been shown to provide salt stress tolerance when over-expressed in tobacco (Sanan-Mishra et al. 2005). Other genes belonging to varied categories have been validated for their contribution towards salinity response in rice as upon over-expression these genes improved tolerance of transgenic plants towards salinity (see Wang et al. 2003; Bajaj and Mohanty 2005; Yamaguchi and Blumwald 2005; Sahi et al. 2006). Detailed analysis taking larger number of species may throw further light on novel candidate genes important for salt tolerance in plants. Emphasis has been

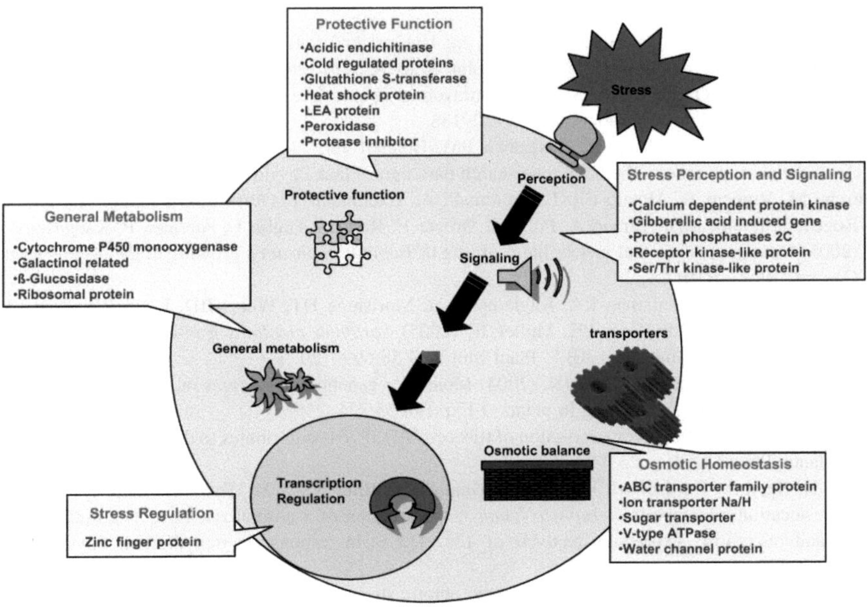

Figure 1. Cartoon depicting the salinity related transcriptome "fingerprints" conserved amongst the three model systems viz. *Arabidopsis*, rice and common ice plant. *(see plate 9)*

laid down on "Allele-mining" which focuses on close relatives of the established models for which sufficient genomic resources are available (Bohnert et al. 2006). There is further possibility that such transgenics might show cross-tolerance to other related and unrelated stresses. Work from our group has shown that overexpression of glyoxalase pathway enzymes confers salinity and heavy metal tolerance in transgenic tobacco (Singla-Pareek et al. 2003, 2006). An example of complex signaling network operative under stress has been indicated by the recent report where a Zn-finger transcription factor has been shown to confer tolerance towards a range of stresses (Mukhopadhyay et al. 2004). Thus it may be concluded that stress responsive transcriptome in plant is quite complex in nature, wherein specific response is an outcome of orchestrated and coordinated fine regulatory networks.

REFERENCES

Amtmann A, Bohnert HJ, Bressan RA (2005) Abiotic stress and plant genome evolution. Search for new models. Plant Physiol 138:127–130

Andjelkovia V, Thompson R (2006) Changes in gene expression in maize kernel in response to water and salt stress. Plant Cell Rep 25:71–79

Arabidopsis Genome Initiative (2000) Analysis of the genome sequence of the flowering plant *Arabidopsis thaliana*. Nature 408:496–815

Bajaj S, Mohanty A (2005) Recent advances in rice biotechnology – towards genetically superior transgenic rice. Plant Biotechnol J 3:275–307

Bevan M, Walsh S (2005) The *Arabidopsis* genome: a foundation for plant research. Genome Res 15:1632–1642

Blumwald E, Grover A (2006) Salt tolerance. In: Halford NG (ed) Plant biotechnology: current and future uses of genetically modified crops. John Wiley & Sons Ltd., UK, pp 206–224

Bohnert HJ, Gong Q, Li P, Ma S (2006) Unraveling abiotic stress tolerance mechanisms—getting genomics going. Curr Opin Plant Biol 9:180–188

Bressan RA, Zhang C, Zhang H, Hasegawa PM, Bohnert HJ, Zhu J-K (2001) Learning from the *Arabidopsis* experience. The next gene search paradigm. Plant Physiol 127:1354–1360

Brosche M, Vinocur B, Alatalo ER, Lamminmaki A, Teichmann T, Ottow EA, Djilianov D, Afif D, Bogeat-Triboulot MB, Altman A, Polle A, Dreyer E, Rudd S, Paulin L, Auvinen P, Kangasjarvi J (2005) Gene expression and metabolite profiling of *Populus euphratica* growing in the Negev desert. Genome Biol 6:R101

Buchanan CD, Lim S, Salzman RA, Kagiampakis I, Morishige DT, Weers BD, Klein RR, Pratt LH, Cordonnier-Pratt MM, Klein PE, Mullet JE (2005) *Sorghum bicolor's* transcriptome response to dehydration, high salinity and ABA. Plant Mol Biol 58:699–720

Chinnusamy V, Schumaker K, Zhu JK. (2004) Molecular genetic perspectives on cross-talk and specificity in abiotic stress signalling in plants. J Exp Bot 55:225–236

Cho EK, Hong CB (2006) Over-expression of tobacco NtHSP70-1 contributes to drought-stress tolerance in plants. Plant Cell Rep 25:349–358

Collett H, Shen A, Gardner M, Farrant JM, Denby KJ, Illing N (2004) Towards transcript profiling of desiccation tolerance in *Xerophyta humilis*: construction of a normalized 11 k X. humilis cDNA set and microarray expression analysis of 424 cDNAs in response to dehydration. Physiol Plant 122:39–53

Cortina C, Culiáñez-Macià FAB (2005) Tomato abiotic stress enhanced tolerance by trehalose biosynthesis. Plant Sci 169:75–82

Cushman JC, Borland AM (2002) Induction of crassulacean acid metabolism by water limitation. Plant Cell Environ 25:297–312

Das-Chatterjee A, Goswami L, Maitra S, Dastidar KG, Ray S, Majumder AL (2006) Introgression of a novel salt-tolerant L-myo-inositol 1-phosphate synthase from *Porteresia coarctata* (Roxb.) Tateoka (PcINO1) confers salt tolerance to evolutionary diverse organisms. FEBS Lett 580:3980–3988

Davletova S, Schlauch K, Coutu J, Mittler R (2005) The zinc-finger protein Zat12 plays a central role in reactive oxygen and abiotic stress signaling in *Arabidopsis*. Plant Physiol 139:847–856

Denby K, Gehring C (2005) Engineering drought and salinity tolerance in plants: lessons from genome-wide expression profiling in *Arabidopsis*. Trends Biotechnol 23:547–552

Dezar CA, Gago GM, Gonzalez DH, Chan RL (2005) Hahb-4, a sunflower homeobox-leucine zipper gene, is a developmental regulator and confers drought tolerance to *Arabidopsis thaliana* plants. Trans Res 14:429–440

Dodd AN, Borland AM, Haslam RP, Griffiths H, Maxwell K (2002) Crassulacean acid metabolism: plastic, fantastic. J Exp Bot 53:569–580

Fujita M, Fujita Y, Maruyama K, Seki M, Hiratsu K, Ohme-Takagi M, Tran LP, Yamaguchi-Shinozaki K, Shinozaki K (2004) A dehydration-induced NAC protein, RD26, is involved in a novel ABA-dependent stress-signaling pathway. Plant J 39:863–873

Garg AK, Kim JK, Owens TG, Ranwala AP, Choi YD, Kochian LV, Wu RJ (2002) Trehalose accumulation in rice plants confers high tolerance levels to different abiotic stresses. Proc Natl Acad Sci USA 99:15898–15903

Gepstein S, Grover A, Blumwald E (2006) Producing biopharmaceuticals in the desert: building an abiotic stress tolerance in plants for salt, heat and drought. In: Knablein J, Muller RH (eds) Modern biopharmaceuticals. Wiley-VCH Verlag GmbH & Co., Weinhaum, pp 967–994

Ghosh-Dastidar K, Maitra S, Goswami L, Roy D, Das KP, Majumder AL (2006) An insight into the molecular basis of salt tolerance of L-myo-inositol 1-P synthase (PcINO1) from *Porteresia coarctata* (Roxb.) Tateoka, a halophytic wild rice. Plant Physiol 40:1279–1296

Goff SA, Ricke D, Lan TH, Presting G, Wang R, Dunn M, Glazebrook J, Sessions A, Oeller P, Varma H, Hadley D, Hutchison D, Martin C, Katagiri F, Lange BM, Moughamer T, Xia Y, Budworth P, Zhong J, Miguel T, Paszkowski U, Zhang S, Colbert M, Sun WL, Chen L, Cooper B, Park S, Wood TC, Mao L, Quail P, Wing R, Dean R, Yu Y, Zharkikh A, Shen R, Sahasrabudhe S, Thomas A, Cannings R, Gutin A, Pruss D, Reid J, Tavtigian S, Mitchell J, Eldredge G, Scholl T, Miller RM, Bhatnagar S, Adey N, Rubano T, Tusneem N, Robinson R, Feldhaus J, Macalma T, Oliphant A, Briggs S (2002) A draft sequence of the rice genome (*Oryza sativa* L. ssp. *japonica*). Science 296:92–100

Gong Q, Li P, Ma S, Rupassara I, Bohnert HJ (2005) Salinity stress adaptation competence in the extremophile *Thellungiella halophila* in comparison with its relative *Arabidopsis thaliana.* Plant J 44:826–839

Grover A, Aggarwal PK, Kapoor A, Katiyar-Agarwal S, Agarwal M (2003) Production of abiotic stress tolerant transgenic crops: present accomplishments and future needs. Curr Sci 84:355–367

Gu R, Fonseca S, Puskas LG, Hackler L Jr., Zvara A, Dudits D, Pais MS (2004) Transcript identification and profiling during salt stress and recovery of *Populus euphratica.* Tree Physiol 24:265–276

Guo Y, Qiu QS, Quintero FJ, Pardo JM, Ohta M, Zhang C, Schumaker KS, Zhu JK (2004) Transgenic evaluation of activated mutant alleles of SOS2 reveals a critical requirement for its kinase activity and C-terminal regulatory domain for salt tolerance in *Arabidopsis thaliana.* Plant Cell 16:435–449

Halpin C (2005) Gene stacking in transgenic plants – the challenge for 21st century plant biotechnology. Plant Biotechnol J 3:141–155

He XJ, Mu RL, Cao WH, Zhang ZG, Zhang JS, Chen SY (2005) AtNAC2, a transcription factor downstream of ethylene and auxin signaling pathways, is involved in salt stress response and lateral root development. Plant J 44:903–916

Houde M, Belcaid M, Ouellet F, Danyluk J, Monroy AF, Dryanova A, Gulick P, Bergeron A, Laroche A, Links M, McCarthy L, Crosby WL, Sarhan F (2006) Wheat EST resources for functional genomics of abiotic stress. BMC Genomics 7:149

Inan, G, Zhang Q, Li P, Wang Z, Cao Z, Zhang H, Zhang C, Quist TM, Goodwin, SM, Zhu J, Shi H, Damsz B, Charbaji T, Gong Q, Ma S, Fredricksen M, Galbraith DW, Jenks MA, Rhodes D, Hasegawa PM, Bohnert HJ, Joly RJ, Bressan RA, Zhu JK (2004) Salt cress: a halophyte and cryophyte

Arabidopsis relative model system and its applicability to molecular genetic analyses of growth and development of extremophiles. Plant Physiol 135:1718–1737

International Rice Genome Sequencing Project (2005) The map-based sequence of the rice genome. Nature 436:793–800

Jang IC, Oh SJ, Seo JS, Choi WB, Song SI, Kim CH, Kim YS, Seo HS, Choi YD, Nahm BH, Kim JK (2003) Expression of a bifunctional fusion of the *Escherichia coli* genes for trehalose-6-phosphate synthase and trehalose-6-phosphate phosphatase in transgenic rice plants increases trehalose accumulation and abiotic stress tolerance without stunting growth. Plant Physiol 131:516–524

Ji W, Li Y, Li J, Dai CH, Wang X, Bai X, Cai H, Yang L, Zhu YM (2006) Generation and analysis of expressed sequence tags from NaCl-treated *Glycine soja*. BMC Plant Biol 6:4

Jin Y, Weining S, Nevo E (2005) A MAPK gene from Dead Sea fungus confers stress tolerance to lithium salt and freezing-thawing: prospects for saline agriculture. Proc Natl Acad Sci USA 102:18992–18997

Jithesh MN, Prashanth SR, Sivaprakash KR, Parida A (2006) Monitoring expression profiles of antioxidant genes to salinity, iron, oxidative, light and hyperosmotic stresses in the highly salt tolerant grey mangrove, *Avicennia marina* (Forsk.) Vierh. by mRNA analysis. Plant Cell Rep 25:865–876

Kasuga M, Miura S, Shinozaki K, Yamaguchi-Shinozaki K (2004) A combination of the *Arabidopsis* DREB1A gene and stress-inducible *rd29*A promoter improved drought- and low-temperature stress tolerance in tobacco by gene transfer. Plant Cell Physiol 45:346–350

Kawasaki S, Borchert C, Deyholos M, Wang H, Brazille S, Kawai K, Galbraith D, Bohnert HJ (2001) Gene expression profiles during the initial phase of salt stress in rice. Plant Cell 13:889–905

Kawaura K, Mochida K, Yamazaki Y, Ogihara Y (2006) Transcriptome analysis of salinity stress responses in common wheat using a 22k oligo-DNA microarray. Funct Integr Genomics 6:132–142

Kim S, Kang JY, Cho DI, Park JH, Kim SY (2004) ABF2, an ABRE-binding bZIP factor, is an essential component of glucose signaling and its overexpression affects multiple stress tolerance. Plant J 40:75–87

Koizumi M, Yamaguchi-Shinozaki K, Tsuji H, Shinozaki K (1993) Structure and expression of two genes that encode distinct drought-inducible cysteine proteinases in *Arabidopsis thaliana*. Gene 129:175–182

Kore-eda S, Noake C, Ohishi M, Ohnishi J, Cushman JC (2004) Transcriptional profiles of organellar metabolite transporters during induction of crassulacean acid metabolism in *Mesembryanthemum crystallinum*. Funct Plant Biol 32:451–466

Kreps JA, Wu Y, Chang HS, Zhu T, Wang X, Harper JF (2002) Transcriptome changes for *Arabidopsis* in response to salt, osmotic, and cold stress. Plant Physiol 130:2129–2141

Ma S, Gong Q, Bohnert HJ (2006) Dissecting salt stress pathways. J Exp Botany 57:1097–1107

Ma X, Qian Q, Zhu D (2005) Expression of a calcineurin gene improves salt stress tolerance in transgenic rice. Plant Mol Biol 58:483–495

Maathuis FJM, Filatov V, Herzyk P, Krijger GC, Axelsen KB, Chen S, Green BJ, Li Y, Madagan KL, Sánchez-Fernández R, Forde BG, Palmgren MG, Rea PA, Williams LE, Sanders D, Amtmann A (2003) Transcriptome analysis of root transporters reveals participation of multiple gene families in the response to cation stress. Plant J 35:675–692

Mahalakshmi S, Christopher GS, Reddy TP, Rao KV, Reddy VD (2006) Isolation of a cDNA clone (PcSrp) encoding serine-rich-protein from *Porteresia coarctata* T. and its expression in yeast and finger millet (*Eleusine coracana* L.) affording salt tolerance. Planta 224:347–359

Marin-Manzano MC, Rodriguez-Rosales MP, Belver A, Donaire JP, Venema K (2004) Heterologously expressed protein phosphatase calcineurin downregulates plant plasma membrane H+-ATPase activity at the post-translational level. FEBS Lett 576:266–270

Maruyama K, Sakuma Y, Kasuga M, Ito Y, Seki M, Goda H, Shimada Y, Yoshida S, Shinozaki K, Yamaguchi-Shinozaki K (2004) Identification of cold-inducible downstream genes of the *Arabidopsis* DREB1A/CBF3 transcriptional factor using two microarray systems. Plant J 38:982–993

Mehta PA, Sivaprakash K, Parani M, Venkataraman G, Parida AK (2005) Generation and analysis of expressed sequence tags from the tolerant mangrove species *Avicennia marina* (Forsk) Vierh. Theor Appl Genet 17:456–464

Meinke DW, Cherry JM, Dean C, Rounsley SD, Koornneef M (1998) *Arabidopsis thaliana*: a model plant for genome analysis. Science 282:662–682

Mukhopadhyay A, Vij S, Tyagi AK (2004) Overexpression of a zinc-finger protein gene from rice confers tolerance to cold, dehydration, and salt stress in transgenic tobacco. Proc Natl Acad Sci USA 101:6309–6314

Nanjo T, Futamura N, Nishiguchi M, Igasaki T, Shinozaki K, Shinohara K (2004) Characterization of full-length enriched expressed sequence tags of stress-treated poplar leaves. Plant Cell Physiol 45:1738–1748

Niewiadomska E, Miszalski Z, Slesak I, Ratajczak R (1999) Catalase activity during C3-CAM transition in *Mesembryanthemum crystallinum* L. leaves. Free Radic Res Suppl:S251–256

Oh SJ, Song SI, Kim YS, Jang HJ, Kim SY, Kim M, Kim YK, Nahm BH, Kim JK (2005) *Arabidopsis* CBF3/DREB1A and ABF3 in transgenic rice increased tolerance to abiotic stress without stunting growth. Plant Physiol 138:341–351

Olsen AN, Ernst HA, Leggio LL, Skriver K (2005) NAC transcription factors: structurally distinct, functionally diverse. Trends Plant Sci 10:79–87

Pareek A, Singh A, Kumar M, Kushwaha HR, Lynn AM, Singla-Pareek SL (2006) Whole-genome analysis of *Oryza sativa* reveals similar architecture of two-component signaling machinery with *Arabidopsis*. Plant Physiol 142:380–97

Park BJ, Liu Z, Kanno A, Kameya T (2005) Transformation of radish (*Raphanus sativus* L.) via sonication and vacuum infiltration of germinated seeds with *Agrobacterium* harboring a group 3 LEA gene from *B. napus*. Plant Cell Rep 24:494–500

Poroyko V, Spollen WG, Hejlek LG, Hernandez AG, Lenoble ME, Davis G, Nguyen HT, Springer GK, Sharp RE, Bohnert HJ (2007) Comparing regional transcript profiles from maize primary roots under well-watered and low water potential conditions. J Exp Bot 58:279–289

Quintero FJ, Ohta M, Shi H, Zhu JK, Pardo JM (2002) Reconstitution in yeast of the *Arabidopsis* SOS signaling pathway for Na^+ homeostasis. Proc Natl Acad Sci USA 99:9061–9066

Rabbani MA, Maruyama K, Abe H, Khan MA, Katsura K, Ito Y, Yoshiwara K, Seki M, Shinozaki K, Yamaguchi-Shinozaki K (2003) Monitoring expression profiles of rice genes under cold, drought, and high-salinity stresses and abscisic acid application using cDNA microarray and RNA gel-blot analyses. Plant Physiol 133:1755–1767

Redei GP (1975) *Arabidopsis* as a genetic tool. Ann Rev Genet 9:111–127

Rhee SY, Beavis W, Berardini TZ, Chen G, Dixon D, Doyle A, Garcia-Hernandez M, Huala E, Lander G, Montoya M, Miller N, Mueller LA, Mundodi S, Reiser L, Tacklind J, Weems DC, Wu Y, Xu I, Yoo D, Yoon J, Zhang P (2003) The *Arabidopsis* information resource (TAIR): a model organism database providing a centralized, curated gateway to *Arabidopsis* biology, research materials and community. Nucleic Acids Res 31:224–228

Rizhsky L, Liang H, Shuman J, Shulaev V, Davletova S, Mittler R (2004) When defense pathways collide: the response of *Arabidopsis* to a combination of drought and heat stress. Plant Physiol 134:1683–1696

Sahi C, Agarwal M, Reddy MK, Sopory SK, Grover A (2003) Isolation and expression analysis of salt stress-associated ESTs from contrasting rice cultivars using a PCR-based subtraction method. Theor Appl Genet 106:620–628

Sahi C, Singh A, Kumar K, Blumwald E, Grover A (2006) Salt stress response in rice: genetics, molecular biology, and comparative genomics. Funct Integr Genom 36:229–239

Saijo Y, Hata S, Kyozuka J, Shimamoto K, Izui K (2000) Over-expression of a single Ca^{2+}-dependent protein kinase confers both cold and salt/drought tolerance on rice plants. Plant J 23:319–327

Sanan-Mishra N, Pham XH, Sopory SK, Tuteja N (2005) Pea DNA helicase 45 overexpression in tobacco confers high salinity tolerance without affecting yield. Proc Natl Acad Sci USA 102:509–514

Seki M, Narusaka M, Ishida J, Nanjo T, Fujita M, Oono Y, Kamiya A, Nakajima M, Enju A, Sakurai T, Satou M, Akiyama K, Taji T, Yamaguchi-Shinozaki K, Carninci P, Kawai J, Hayashizaki Y, Shinozaki K (2002) Monitoring the expression profiles of 7000 *Arabidopsis* genes under drought, cold and high-salinity stresses using a full-length cDNA microarray. Plant J 31:279–292

Shi H, Lee BH, Wu SJ, Zhu JK (2003) Overexpression of a plasma membrane Na^+/H^+ antiporter gene improves salt tolerance in *Arabidopsis thaliana*. Nat Biotechnol 21:81–85

Shiozaki N, Yamada M, Yoshiba Y (2005) Analysis of salt-stress-inducible ESTs isolated by PCR-subtraction in salt-tolerant rice. Theor Appl Genet. 110:1177–1186

Singla-Pareek SL, Reddy MK, Sopory SK (2001) Transgenic approach towards developing abiotic stress tolerance in plants. Proc Ind Natl Sci Acad B67(5):265–284

Singla-Pareek SL, Reddy MK, Sopory SK (2003) Genetic engineering of the glyoxalase pathway in tobacco leads to enhanced salinity tolerance. Proc Natl Acad Sci USA 100:14672–14677

Singla-Pareek SL, Yadav SK, Pareek A, Reddy MK, Sopory SK (2006) Transgenic tobacco overexpressing glyoxalase pathway enzymes grow and set viable seeds in zinc-spiked soils. Plant Physiol 140:613–623

Sreenivasulu N, Altschmied L, Radchuk V, Gubatz S, Wobus U, Weschke W (2004) Transcript profiles and deduced changes of metabolic pathways in maternal and filial tissues of developing barley grains. Plant J 37:539–553

Taji T, Seki M, Satou M, Sakurai T, Kobayashi M, Ishiyama K, Narusaka Y, Narusaka M, Zhu JK, Shinozaki K (2004) Comparative genomics in salt tolerance between *Arabidopsis* and *Arabidopsis*-related halophyte salt stress using *Arabidopsis* microarray. Plant Physiol 135:1697–1709

Takahashi S, Seki M, Ishida J, Satou M, Sakurai T, Narusaka M, Kamiya A, Nakajima M, Enju A, Akiyama K, Yamaguchi-Shinozaki K, Shinozaki K (2004) Monitoring the expression profiles of genes induced by hyperosmotic, high salinity, and oxidative stress and abscisic acid treatment in *Arabidopsis* cell culture using a full-length cDNA microarray. Plant Mol Biol 56:29–55

Veena RVS, Sopory SK (1999) Glyoxalase I from *Brassica juncea*: molecular cloning, regulation and its over-expression confer tolerance in transgenic tobacco under stress. Plant J 17:385–395

Vinocur B, Altman A (2005) Recent advances in engineering plant tolerance to abiotic stress: achievements and limitations. Curr Opin Biotechnol 16:123–32

Walia H, Wilson C, Wahid A, Condamine P, Cui X, Close TJ (2006) Expression analysis of barley (*Hordeum vulgare* L.) during salinity stress. Funct Integr Genom 6:143–156

Wang H, Huang Z, Chen Q, Zhang Z, Zhang H, Wu Y, Huang D, Huang R (2004) Ectopic overexpression of tomato JERF3 in tobacco activates downstream gene expression and enhances salt tolerance. Plant Mol Biol 55:183–192

Wang W, Vinocur B, Altman A (2003) Plant responses to drought, salinity and extreme temperatures: towards genetic engineering for stress tolerance. Planta 218:1–14

Wong CE, Li Y, Whitty BR, Diaz-Camino C, Akhter SR, Brandle JE, Golding GB, Weretilnyk EA, Moffatt BA, Griffith M (2005) Expressed sequence tags from the Yukon ecotype of *Thellungiella* reveal that gene expression in response to cold, drought and salinity shows little overlap. Plant Mol Biol 58:561–574

Yamaguchi T, Blumwald E (2005) Developing salt-tolerant crop plants: challenges and opportunities. Trends Plant Sci 10:615–620

Yeo ET , Kwon HB, Han SE, Lee JT, Ryu JC, Byu MO (2000) Genetic engineering of drought resistant potato plants by introduction of the trehalose-6-phosphate synthase (TPS1) gene from *Saccharomyces cerevisiae*. Mol Cells 10:263–268

Yoo JH, Park CY, Kim JC, Heo WD, Cheong MS, Park HC, Kim MC, Moon BC, Choi MS, Kang YH, Lee JH, Kim HS, Lee SM, Yoon HW, Lim CO, Yun DJ, Lee SY, Chung WS, Cho MJ (2005) Direct interaction of a divergent CaM isoform and the transcription factor, MYB2, enhances salt tolerance in *Arabidopsis*. J Biol Chem 280:3697–3706

Yu J, Hu SN, Wang J, Wong GKS, Li SG, Liu B, Deng YJ, Dai L, Zhou Y, Zhang XQ, Cao ML, Liu J, Sun JD, Tang JB, Chen YJ, Huang XB, Lin W, Ye C, Tong W, Cong LJ, Geng JN, Han YJ, Li L, Li W, Hu GQ, Huang XG, Li WJ, Li J, Liu ZW, Li L, Liu JP, Qi QH, Liu JS, Li L, Li T, Wang XG, Lu H, Wu TT, Zhu M, Ni PX, Han H, Dong W, Ren XY, Feng XL, Cui P, Li XR, Wang H, Xu X, Zhai WX, Xu Z, Zhang JS, He SJ, Zhang JG, Xu JC, Zhang KL, Zheng XW, Dong JH, Zeng WY, Tao L, Ye J, Tan J, Ren XD, Chen XW, He J, Liu DF, Tian W, Tian CG, Xia HG, Bao QY, Li G, Gao H, Cao T, Wang J, Zhao WM, Li P, Chen W, Wang XD, Zhang Y, Hu JF, Wang J, Liu S, Yang J, Zhang GY, Xiong YQ, Li ZJ, Mao L, Zhou CS, Zhu Z, Chen RS, Hao BL, Zheng WM,

Chen SY, Guo W, Li GJ, Liu SQ, Tao M, Wang J, Zhu LH, Yuan LP, Yang HM (2002) A draft sequence of the rice genome (*Oryza sativa* L. ssp. *indica*). Science 296:79–92

Zhang L, Ma, XL, Zhang Q, Ma, CL, Wang PP, Zhao YX, Zhang H (2001) Expressed sequence tags from a NaCl-treated *Suaeda salsa* cDNA library. Gene 267:193–200

Zimmermann P, Hirsch-Hoffmann M, Hennig L, Gruissem W. (2004) Genevestigator. Arabidopsis microarray database and analysis toolbox. Plant Physiol 136:2621–2632 [illegible] Zhang S, Zhu LH, Yuan LP, Yang HM (2002) A draft sequence of the rice genome (*Oryza sativa* L. ssp. *indica*). Science 296:79–92

Zhang L, Ma XL, Zhang Q, Ma CL, Wang PP, Sun YF, Zhao YX, Zhang H (2001) Expressed sequence tags from a NaCl-treated *Suaeda salsa* cDNA library. Gene 267:193–200

Zimmermann P, Hirsch-Hoffmann M, Hennig L, Gruissem W (2004) Genevestigator. Arabidopsis microarray database and analysis toolbox. Plant Physiol 136:2621–2632

CHAPTER 13

AUXIN AND CYTOKININ SIGNALING COMPONENT GENES AND THEIR POTENTIAL FOR CROP IMPROVEMENT

JITENDRA P. KHURANA*, MUKESH JAIN AND AKHILESH K. TYAGI

Interdisciplinary Centre for Plant Genomics and Department of Plant Molecular Biology, University of Delhi South Campus, Benito Juarez Road, New Delhi-110021, India

Abstract: Plant hormones auxin and cytokinin exert pleiotropic effects on growth and development, both individually and in a combinatorial manner. In recent years, our understanding of mechanisms of auxin and cytokinin action has improved considerably. This has largely been due to the molecular genetic analysis of hormone perception and signal transduction mutants of *Arabidopsis*, the model dicot plant, and also the availability of a high quality sequence of its 125 Mb genome. However, little work has been carried out in other plant species. Advances in genomics, particularly on rice, maize, sorghum and tomato, provide new opportunities for investigating these components in crop plants. A comparative analysis of *Arabidopsis* and rice genome sequences is already paying rich dividends and evolutionary relationship among various classes of gene families has been established, including those representing components of auxin and cytokinin signaling. Some of these auxin and cytokinin signaling components are proving to be invaluable genetic tools for manipulation of agronomic traits in crop plants, and it has been illustrated in this article with the help of a few suitable examples.

1. INTRODUCTION

The integration of many environmental and endogenous signals, together with the intrinsic genetic program, regulates plant growth and development. Fundamental to this process are several growth regulators, collectively called the plant hormones or phytohormones. These are the small organic compounds that generally act at very low concentrations to regulate many aspects of plant growth and development. The phytohormones include auxin, cytokinin, gibberellin (GA), abscisic acid (ABA),

*Corresponding Author: khuranaj@genomeindia.org

R.K. Varshney and R. Tuberosa (eds.), Genomics-Assisted Crop Improvement: Vol. 1: Genomics Approaches and Platforms, 289–314.

ethylene, brassinosteroids (BR) and jasmonic acid (JA). Hormones regulate or influence virtually every aspect of plant growth and development. Many of the developmental and physiological processes regulated by plant hormones are of agronomic importance. Some of the examples are the role of ethylene in promotion of fruit ripening, regulation of seed germination and stem elongation by GA, stress responses by ABA, fruit development by auxin and delayed senescence by cytokinin. Because of the central role plant hormones play, they are being used for a long time as growth regulators by exogenous application and, more recently, by targeted modification of hormone signaling pathways in transgenic plants for manipulation of specific agronomic traits to improve crop production.

Among the various hormone signal transduction pathways, auxin and cytokinin signal transduction pathways are best studied. Several biosynthetic and response mutants have been isolated in the model dicot plant *Arabidopsis* to elucidate the molecular mechanisms underlying auxin and cytokinin action (Kakimoto 2003; Grefen and Harter 2004; Woodward and Bartel 2005; Laxmi et al. 2006; Teale et al. 2006). Molecular genetic and biochemical analysis of these mutants have unraveled some of the mysteries underlying phytohormone action by identifying the downstream target genes that mediate hormone-induced changes in growth and development. Although our understanding of auxin and cytokinin signal transduction pathways in crop plants is negligible as compared to *Arabidopsis*, advances in genomics afford new opportunities for investigating these pathways. In addition to the availability of complete rice genome sequence (International Rice Genome Sequencing Project 2005; Vij et al. 2006), a large collection of rice full-length cDNA sequences (Kikuchi et al. 2003), extensive collection of expressed sequence tags (ESTs) from several grasses (http://www.ncbi.nlm.nih.gov/dbEST/) and progress in deciphering gene-rich regions of maize (Messing et al. 2004), gene expression profiling platforms such as microarrays, serial analysis of gene expression (SAGE) and massively parallel signature sequencing (MPSS) (Rensink and Buell 2005), together with advances in proteomics, provide detailed blueprints for crop improvement in future. Furthermore, availability of large collection of transposon and T-DNA insertion mutants, improvements in crop transformation strategies and breeding programs provide additional opportunity to examine and manipulate gene function. This chapter discusses the auxin and cytokinin signaling briefly and the validation of auxin- and cytokinin-responsive genes in crop improvement in terms of physiological processes involved.

2. AUXIN SIGNAL TRANSDUCTION PATHWAY

Auxin plays a pivotal role in many processes throughout the plant life cycle, including embryogenesis, lateral root development, vascular differentiation, apical dominance, tropic responses and flower development. Although auxin was first discovered decades ago (see Pennazio 2002), the information about auxin signaling pathways has accumulated only in the last few years through the combined application of genetic, molecular and biochemical approaches. Recently, the long-sought auxin receptor has also been identified as an F-box protein, TIR1, which is actually a component of SCF^{TIR1} complex involved in ubiquitin-mediated degradation of tagged proteins (Dharmasiri et al. 2005; Kepinski and Leyser 2005).

2.1. Auxin-responsive Genes

Auxin stimulates the transcription of a large number of genes called primary auxin response genes. A large number of auxin-responsive genes have been identified and characterized from different plant species, including pea, soybean, *Arabidopsis*, mung bean, cucumber and rice (Abel and Theologis 1996; Thakur et al. 2001, 2005; Hagen and Guilfoyle 2002; Jain et al. 2006a,b,c). These auxin-responsive genes have been broadly grouped into three gene families: auxin/indoleacetic acid (*Aux/IAA*), *GH3* and small auxin-up RNA (*SAUR*) gene families (Guilfoyle 1999).

2.1.1. Aux/IAA

The *Aux/IAA* genes are present as multigene families in soybean, pea, mung bean, tobacco, tomato, *Arabidopsis* and rice (Liscum and Reed 2002; Jain et al. 2006a). They are rapidly induced by exogenous auxin even in the presence of a translational inhibitor, cycloheximide, indicating that these represent a class of primary auxin-responsive genes (Abel and Theologis 1996; Thakur et al. 2005). These genes encode short-lived nuclear proteins comprising four highly conserved domains, designated as domain I, II, III and IV (Figure 1; Abel et al. 1994). The domain I acts as a strong transcriptional repressor (Tiwari et al. 2004), whereas Domain II is responsible for rapid degradation of Aux/IAA proteins (Worley et al. 2000; Ouellet et al. 2001). Domain III is part of an amphipathic $\beta\alpha\alpha$ -DNA recognition motif found in β-ribbon of DNA binding domain of prokaryotic repressors such as MetJ and Arc (Abel et al. 1994; Phillips 1994). However, its role in DNA binding has not been demonstrated yet. Domains III and IV mediate homo- and hetero-dimerization among the Aux/IAA proteins and auxin response factors (ARFs) (Kim et al. 1997; Ulmasov et al. 1997; Ouellet et al. 2001). Domains II and IV also contain nuclear localization signals (Abel et al. 1994; Abel and Theologis 1996; Jain et al. 2006a). The molecular genetic analyses of several mutants of *Aux/IAA*

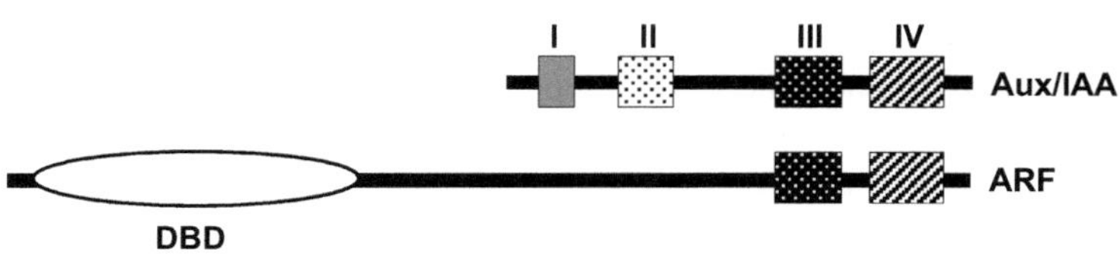

Figure 1. Diagrammatic representation of Aux/IAA and ARF proteins. Aux/IAA proteins contain four conserved domains, I, II, III, and IV. Domain I is a repressor domain, domain II is responsible for protein degradation and domains III and IV are dimerization domains. ARF proteins contain N-terminal conserved DNA-binding domain (DBD) and a middle nonconserved activation or repression domain. ARF proteins share dimerization domains III and IV with Aux/IAA proteins

genes have demonstrated that they play a central role in regulating plant growth and development (Rouse et al. 1998; Tian and Reed 1999; Gray and Estelle 2000; Nagpal et al. 2000; Reed 2001; Rogg et al. 2001; Liscum and Reed 2002). Biochemical studies showed that the conserved region of domain II of Aux/IAA proteins is responsible for rapid degradation. Single amino acid change in domain II resulted in altered auxin response due to increased protein accumulation, suggesting that rapid degradation of Aux/IAA proteins is necessary for a normal auxin response (Worley et al. 2000). The degradation of Aux/IAA proteins was found to be proteasome dependent (Gray et al. 2001; Ramos et al. 2001; Thakur et al. 2005). Auxin treatment in fact promotes degradation of Aux/IAA proteins by enhancing the interaction between SCF^{TIR1} complex and Aux/IAA proteins by affecting the SCF component, TIR1, or its associated proteins (Gray et al. 2001; Kepinski and Leyser 2004). More recently, it has been demonstrated that Aux/IAA protein family has diversified in degradation and auxin responsiveness in *Arabidopsis* (Dreher et al. 2006). Some of Aux/IAA proteins are long-lived and auxin-insensitive, which suggested their novel role in auxin signaling (Dreher et al. 2006).

2.1.2. GH3

The first *GH3* gene was isolated as an early auxin-responsive gene from soybean by differential screening (Hagen et al. 1984). The *GH3* homologues have been found in other plants too (Guilfoyle et al. 1993; Roux and Rechenmann 1997). The *GH3* genes are represented as a small multigene family in the *Arabidopsis* and rice, comprising of twenty and twelve members, respectively (Hagen and Guilfoyle 2002; Jain et al. 2006b). Many of the *Arabidopsis* and rice *GH3* genes are induced by exogenous application of auxin (Hagen and Guilfoyle 2002; Tanaka et al. 2002; Jain et al. 2006b). However, the *GH3* mRNAs, unlike *Aux/IAA* mRNAs, do not accumulate in response to protein synthesis inhibitor cycloheximide (Franco et al. 1990). The GH3 proteins do not possess any known conserved motif or domain. However, they were found to be distantly related to the acyl adenylate-forming firefly luciferase super family (Staswick et al. 2002). Light also regulates the transcript levels of *Arabidopsis GH3* genes through phytochromes A and B (Nakazawa et al. 2001; Tepperman et al. 2001; Tanaka et al. 2002; Takase et al. 2003, 2004). The analysis of an *Arabidopsis* mutant identified FIN219, a GH3 protein, as a phytochrome A signaling component having a crucial role in photomorphogenesis (Hsieh et al. 2000); it negatively regulates COP1, a key repressor of photomorphogenic development. Two other *Arabidopsis* GH3 proteins, DFL1 and YDK1, also identified by mutant analysis, were shown to negatively regulate shoot and hypocotyl cell elongation and lateral root formation (Nakazawa et al. 2001; Takase et al. 2004). *Arabidopsis* GH3 proteins have been reported to adenylate plant hormones such as indole-3-acetic acid, jasmonic acid and salicylic acid (Staswick et al. 2002). Recently, some of the *Arabidopsis GH3* genes have been shown to encode IAA-amido synthetases, which synthesize a diversity of IAA-amino acid conjugates to help maintain auxin homeostasis (Staswick et al. 2005).

2.1.3. *SAUR*

The *SAUR*s have also been isolated and studied in many plants such as soybean, mung bean, pea, *Arabidopsis*, *Zea mays*, and rice (Hagen and Guilfoyle 2002; Jain et al. 2006c). These genes are also induced very rapidly within 2–5 min of application of exogenous auxin. The treatment with the protein synthesis inhibitor, cycloheximide, does not inhibit auxin-induced transcription of *SAUR*s; rather it results in an increase in the abundance of *SAUR* transcripts due to their stabilization (McClure and Guilfoyle 1987; Franco et al. 1990). There are over 70 *SAUR*s in *Arabidopsis* (Hagen and Guilfoyle 2002). In rice also the *SAUR* gene family is comprised of 58 members (Jain et al. 2006c). The *SAUR*s generally do not contain any intron and encode proteins of 9–10 kDa. Many of the *SAUR*s are present in clusters in soybean, *Arabidopsis* and rice (McClure et al. 1989; Hagen and Guilfoyle 2002; Jain et al. 2006c). *SAURs* encode highly unstable mRNAs with a very high turnover rate, which may be due, in part, to the presence of a conserved DST element in the 3'-untranslated region and/or elements within the ORF (McClure et al. 1989; McClure and Guilfoyle 1989; Franco et al. 1990; Newman et al. 1993; Li et al. 1994; Jain et al. 2006c). The function of SAUR proteins is largely unknown. However, there is evidence that they may play some role in auxin signal transduction pathway that involves calcium and calmodulin (Yang and Poovaiah 2000). Two *dst* mutants have been isolated from *Arabidopsis* which show an increased abundance of DST-containing *SAUR-AC1* mRNA (Johnson et al. 2000). The microarray analysis of *dst1* mutant and further studies demonstrated a potential link between DST-mediated decay pathway and circadian rhythm in plants (Perez-Amador et al. 2001; Lidder et al. 2005).

2.1.4. *Other genes*

In addition to *Aux/IAA*, *GH3* and *SAURs*, several other genes are also induced by auxin (Goda et al. 2004). These genes include those encoding cell wall synthesis enzymes, cell wall modifying agents, cell wall component proteins, ethylene biosynthetic enzyme (1-aminocyclopropane-1-carboxylate synthase), cell cycle regulatory proteins and many other genes that still await characterization. Many genes have been shown to be down regulated by auxin application. Some members in this category are annotated as pathogen-related or disease resistance-related genes (Goda et al. 2004).

2.2. Auxin-responsive Promoter Elements and Interacting Factors

The auxin-responsive elements (AuxREs) have been identified in the promoters of several auxin-responsive genes (Guilfoyle 1999; Hagen and Guilfoyle 2002; Jain et al. 2006c). Identification of these AuxREs has been based mainly on conservation of similar sequence elements found in a variety of genes induced by auxin. The smallest element to be identified as auxin-responsive is a six-base pair sequence, TGTCTC (Figure 2; Ulmasov et al. 1997b). This element is conserved and has been shown to function in both simple and composite AuxREs. In composite AuxREs,

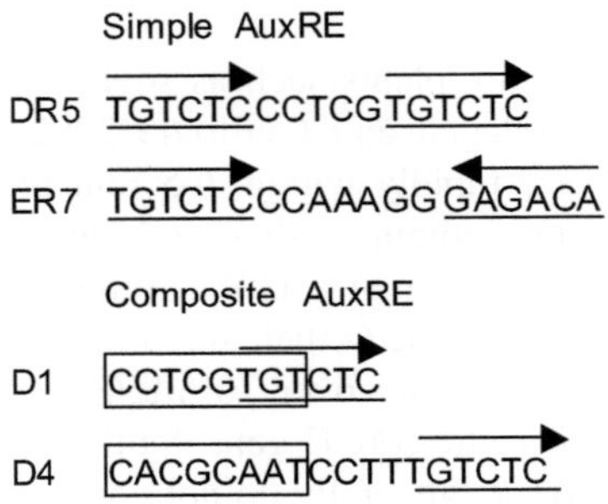

Figure 2. Simple and composite AuxREs. Conserved TGTCTC sequence is underlined. Coupling or constitutive elements in composite AuxREs (D1 and D4) present in *GH3* promoter are indicated in open boxes. ER7 and DR5 represent the simple inverted repeat and simple direct repeat AuxREs, respectively

such as those found in *GH3* promoters D1 and D4, the TGTCTC element is not sufficient by itself to confer auxin-responsive gene expression and requires an adjacent or overlapping coupling element (Figure 2; Guilfoyle 1999). The AuxRE promoter-reporter constructs are used to study organ- and tissue-specific expression patterns of auxin-responsive genes and are valuable tools to study gene expression during growth responses associated with changes in auxin gradients and sensitivities.

Several auxin response factors (ARFs) have been identified that bind with TGTCTC element in a yeast one-hybrid screen of *Arabidopsis* expression library using a synthetic TGTCTC repeat sequence as bait (Ulmasov et al. 1997a). The deduced amino acid sequences of ARFs show the presence of an N-terminal DNA-binding domain followed by a variable region and conserved domains III and IV similar to those present in Aux/IAA proteins, at C-terminus (Figure 1). The DNA-binding domain of ARFs binds to auxin-responsive elements (AuxREs) of auxin-responsive genes and regulates their expression (Kim et al. 1997; Ulmasov et al. 1997a; Tiwari et al. 2003). Domains III and IV mediate interaction with Aux/IAA proteins (Kim et al. 1997; Ulmasov et al. 1997a; Ouellet et al. 2001). There are at least 23 *ARF* genes predicted in each of the genomes of *Arabidopsis* and rice (Liscum and Reed 2002; M Jain, AK Tyagi and JP Khurana, unpublished results). Recently, it has been demonstrated that the interactions of specific pairs of ARF and Aux/IAA proteins generate the specificity of auxin response at different developmental stages and physiological levels in *Arabidopsis* (Weijers et al. 2005).

3. CYTOKININ SIGNAL TRANSDUCTION PATHWAY

Cytokinins regulate various plant growth and developmental processes, including cell division, apical dominance, chloroplast biogenesis, leaf senescence, vascular differentiation, photomorphogenic development, shoot differentiation in tissue cultures and anthocyanin production, primarily by altering the expression of diverse genes (Mok and Mok 2001; Davies 2004). The recent genetic and molecular studies in plants have suggested the involvement of two-component sensor-regulator system in cytokinin signal perception and transduction, comprising sensor histidine kinase (HK) proteins, histidine phosphotransfer (HPt) proteins and effector response regulator (RR) proteins (Hutchison and Kieber 2002; Hwang et al. 2002; Lohrmann and Harter 2002; Oka et al. 2002; Heyl and Schmulling 2003; Kakimoto 2003; Grefen and Harter 2004). Such signal transduction systems, once thought to be restricted to prokaryotes, have also been found in many eukaryotes, including yeast, fungi, slime molds and higher plants (Stock et al. 2000). In *Arabidopsis* and rice, proteins with homology to all the elements of two-component system have been identified (Hwang et al. 2002; Jain et al. 2006d; Pareek et al. 2006).

The current model for cytokinin signaling in plants is similar to the two-component phosphorelay system with which bacteria sense and respond to environmental changes. A simple two-component system involves a His sensor kinase and a response regulator (Stock et al. 2000; West and Stock 2001). The His kinase perceives environmental stimuli via the input domain and autophosphorylates on a conserved His residue within the kinase domain at the C-terminus (Figure 3). The phosphoryl group is subsequently transferred to a conserved Asp residue on the receiver domain of a response regulator, which mediates downstream responses via the output domain (Figure 3). Multicomponent phosphorelay systems occur in most eukaryotic and some prokaryotic systems, which employ His kinase signal transduction in a multistep His-Asp-His-Asp phosphotransfer process (Stock et al. 2000; West and Stock 2001). The *Arabidopsis* cytokinin receptors (CRE1, AHK2 and AHK3) are similar to bacterial His sensor hybrid kinases in two-component signaling, containing a receiver domain fused to the His kinase domain (Inoue et al. 2001; Suzuki et al. 2001; Ueguchi et al. 2001a,b; Yamada et al. 2001). The cytokinin receptors are predicted to signal through His phosphotransfer proteins to ultimately alter the phosphorylation state of the *Arabidopsis* response regulators (ARRs) in a multistep phosphorelay (Hutchison and Kieber 2002). The analysis of *Arabidopsis* genome revealed the existence of 32 putative response regulator genes (Hwang et al. 2002). Based on the predicted protein domain architecture and amino acid composition, the response regulators have been broadly categorized into three distinct families: type-A, type-B and pseudo-response regulators (Hwang et al. 2002; Jain et al. 2006d; Pareek et al. 2006).

3.1. Type-A Response Regulators

The type-A response regulators are relatively small, containing a receiver domain along with small N- and C-terminal extensions (Figure 4; D'Agostino et al. 2000; Jain et al. 2006d). The type-A response regulator genes in *Arabidopsis* (type-A *ARRs*) are rapidly and specifically induced by exogenous cytokinin, although with varying kinetics and have been characterized as primary cytokinin response genes (Taniguchi et al. 1998; Kiba et al. 1999; D'Agostino et al. 2000; Jain et al. 2006d). The transcription of type-A *ARR* genes is regulated in part by type-B ARRs (Hwang and Sheen 2001; Sakai et al. 2001). Some of the type-A ARRs perform partially redundant functions, acting as negative regulators of

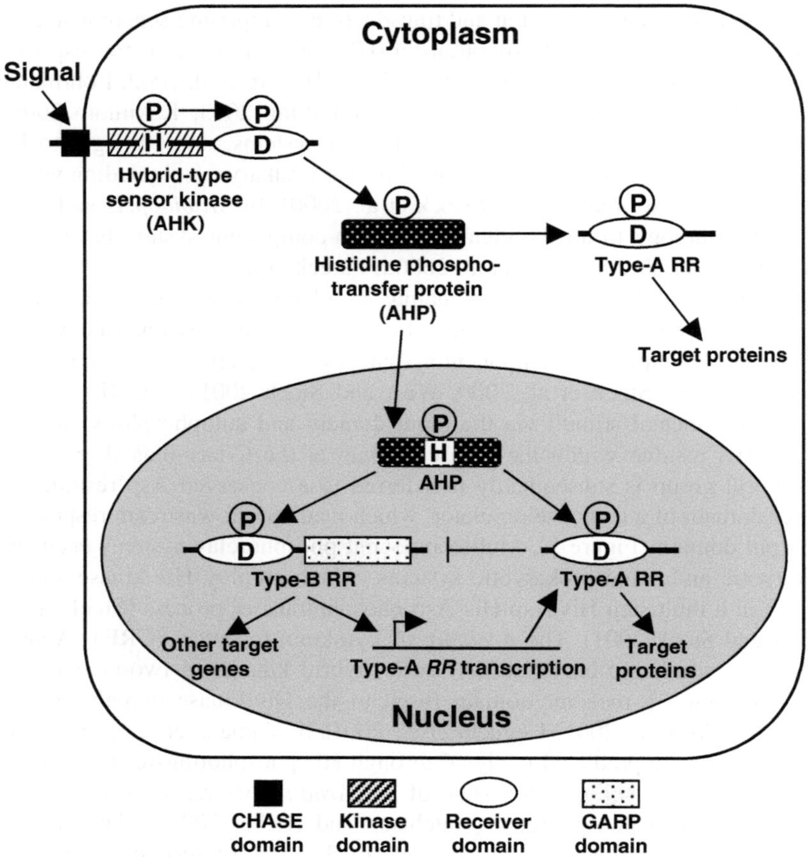

Figure 3. General model of two-component signal transduction in *Arabidopsis* . After signal perception, hybrid histidine kinase undergoes autophosphorylation and the phosphoryl residue is relayed to an AHP. The phosphorylated AHP interacts with the cognate ARR either in the cytoplasm (type-A ARRs) or in the nucleus (type-A and type-B ARRs). By transfer of the phosphoryl group to the receiver domain, the ARRs are activated, which results in regulation of target proteins (type-A ARRs) or target genes (type-B ARRs). H, Histidine residue; P, phosphoryl group; D, aspartate residue

cytokinin responses by a negative feedback mechanism (Hwang and Sheen 2001; Kiba et al. 2003; To et al. 2004). In contrast, ARR4 was claimed to be a positive regulator of cytokinin signaling because its over-expression enhanced the cytokinin responsiveness of transgenic *Arabidopsis* plants (Osakabe et al. 2002). However, the loss-of-function mutant did not reveal a positive role for ARR4 in cytokinin signaling (To et al. 2004) and this discrepancy remains to be resolved. The tissue distribution of ARR4 overlaps to a large extent with that of phytochrome B (PHYB) and it has been found to interact with N-terminus of PHYB to stabilize its active form (Sweere et al. 2001). The transgenic *Arabidopsis* plants overexpressing ARR4 are

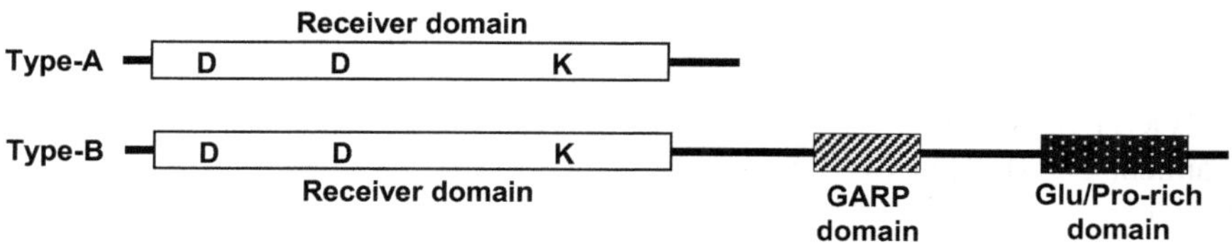

Figure 4. Diagrammatic representation of type-A and type-B response regulators. Type-A response regulators contain a conserved receiver domain and short variable N- and C-terminal extensions. Type-B response regulators contain a conserved N-terminal receiver domain similar to type-A response regulators and middle DNA-binding GARP domain and C-terminal glutamine/proline rich domain

specifically hypersensitive to red light (Sweere et al. 2001), indicating that ARR4 may be involved in integrating red light and cytokinin signaling. Furthermore, some of the rice and *Arabidopsis* type-A response regulators have been implicated in stress signaling (Urao et al. 1998; Jain et al. 2006d). Recently, it has also been demonstrated that WUSCHEL, a homeodomain protein, regulates meristem function by repressing the transcription of several type-A response regulators in *Arabidopsis* (Leibfried et al. 2005).

3.2. Type-B Response Regulators

The type-B response regulators comprise a receiver domain fused to the DNA-binding domain with long C-terminal extensions and are supposed to be transcriptional regulators (Figure 4; Sakai et al. 1998; Sakai et al. 2000; Mason et al. 2004). To date, the strongest evidence for the role of type-B ARRs in mediating cytokinin signal transduction comes from the analysis of ARR1. A null mutation in *ARR1* results in reduced sensitivity to cytokinin in shoot regeneration and root elongation assays (Sakai et al. 2001). Overexpression of either *ARR1* or *ARR2* in *Arabidopsis* plants results in increased sensitivity to cytokinin (Hwang and Sheen 2001; Sakai et al. 2001). ARR2, however, has also been proposed to play a role in mediating ethylene signal transduction based on the analysis of a transposon-induced mutation (Hass et al. 2004). The relatively weak phenotypes revealed by the analysis of individual *ARR* mutations may be the result of functional overlap among the type-B ARRs, a hypothesis consistent with the gene expression patterns (Mason et al. 2004; Tajima et al. 2004). Recently, the analysis of some of mutants of type-B ARRs indicated functional overlap among them and provided evidence that they act as positive regulators of cytokinin signal transduction (Mason et al. 2005).

3.3. Pseudo-response Regulators

The pseudo-response regulators (PRRs) also share significant sequence similarity with the receiver domain of other response regulators but the invariant D-D-K motifs are not present (Hwang et al. 2002). The N-terminal pseudo-receiver domain is followed by a long intervening sequence specific for each PRR, which is followed by another common motif termed as CCT motif (CONSTANS, CONSTANS-like and TOC1) of about 50 amino acids at the very C-terminal end (Mizuno and Nakamichi 2005). This CCT motif is a plant-specific and widespread motif that is found in many apparently unrelated plant proteins, including the CONSTANS family of proteins (Putterill et al. 1995). The molecular functions of PRRs are unknown; however, their biological roles have been well established as the elements of the circadian clock in *Arabidopsis* and rice (Makino et al. 2000; Makino et al. 2002; Matsushika et al. 2002; Murakami et al. 2003; Mizuno and Nakamichi 2005; Nakamichi et al. 2005).

4. DEVELOPMENTAL EFFECTS OF AUXIN

As mentioned earlier, auxin influences nearly every stage of a plant's life cycle from germination to senescence. Historically, the work on phototropism of seedling coleoptiles led to the discovery of auxin (see Khurana 2001; Srivastava 2001). However, only a few of auxin responses that can be potentially exploited for crop improvement are discussed here.

4.1. Apical Dominance

Growth of the shoot apex usually inhibits the development of lateral buds on the stem beneath. The phenomenon is called as apical dominance. If the terminal shoot of a plant is removed, the inhibition is removed, and lateral buds begin to grow. The release of apical dominance enables lateral branches to develop and the plant become bushier. Auxin (IAA) can substitute for the apical meristem in maintaining inhibition of lateral buds. Apical dominance seems to result from the downward (basipetal) transport of auxin produced in apical meristem. Molecular genetic analysis of *Arabidopsis* mutants has helped in the identification of genes encoding regulators of auxin transport. These include permease-like AUX1, plant-specific PIN proteins, and homologs of human multiple drug resistance/P-glycoproteins, PGP1 and PGP19, most-likely involved in efflux of auxin (Bennett et al. 1996; Blakeslee et al. 2005; Paponov et al. 2005). Some of the developmental defects exhibited due to defect in these components can be phenocopied by auxin transport inhibitors, demonstrating that their role is essential as components of auxin transport machinery (Benkova et al. 2003; Paponov et al. 2005).

4.2. Formation of Lateral and Adventitious Roots

Although elongation of the primary root is inhibited by auxin concentrations greater than 10^{-8} M, initiation of lateral and adventitious roots is stimulated by high auxin levels (10^{-6} to 10^{-5} M). Auxin stimulates the cell division in pericycle or deeper lying parenchyma cells from where lateral or adventitious root primordia arise. The dividing cells gradually give rise to root apex and root cap and lateral root grows through the parent root cortex and finally emerges as a rootlet. The stimulatory effect of auxin on the formation of adventitious roots has been very useful in horticulture for the vegetative propagation of plants by cuttings.

Auxin plays an important role in lateral root formation is supported by the fact that mutants defective in polar transport of IAA (for example, *aux1*) or insensitive to auxin (for example, *axr1* and *axr2* of *Arabidopsis*, and *diageotropica* mutant of tomato) are also deficient in production of lateral roots (Marchant et al. 2002). The evidence from these mutants is reinforced by the application of auxin transport inhibitors such as triiodobenzoic acid (TIBA) or naphthyl phthalamic acid (NPA), which inhibit lateral branching.

4.3. Vascular Differentiation

The vascular system constitutes a continuous cellular network consisting of interconnected vascular strands, each composed of two types of conducting tissues, xylem and phloem. The phytohormone, auxin, has been implicated in the formation of vascular strands. The relative amounts of xylem and phloem are regulated by the auxin concentration; high concentration of auxin induces differentiation of xylem and phloem, but only phloem differentiates at low auxin concentration (Aloni 1995). Genetic screens in *Arabidopsis* and other plant species have identified a number of loci with potential functions in vascular differentiation, which suggest a role of auxin in vascular development (Berleth et al. 2000). This role is further supported by the promotion of vascular differentiation by auxin (Aloni 1995), and by the responses of vascular patterns to altered auxin transport during organ development (Mattsson et al. 1999). Many potential vascular patterning genes have been identified by mutant phenotypes, which include three auxin related genes, *MONOPTEROS (MP)/AUXIN RESPONSE FACTOR 5 (ARF5)*, *AUXIN RESISTANT 6 (AXR6)*, and *BODENLOS (BDL)/IAA12*. Mutations in these genes are associated with incomplete vascular systems and defects in the formation of embryo axis (Berleth and Jurgens 1993; Przemeck et al. 1996; Hamann et al. 1999; Hobbie et al. 2000; Mattsson et al. 2003). Interestingly, *MP* and *IAA12/BDL* encode members of two families of transcription factors, ARFs and Aux/IAAs, which are involved in auxin-dependent gene regulation (Hagen and Guilfoyle 2002; Liscum and Reed 2002).

4.4. Flower Development

Auxin is a major controlling signal that synchronizes flower development. Genetic evidences show that the polar transport of auxin controls flower formation and differentiation in *Arabidopsis* (Okada et al. 1991; Bennett et al. 1995; Nemhauser et al. 1998, 2000; Oka et al. 1999; Reinhardt et al. 2003). The treatment of *Arabidopsis* plants with auxin transport inhibitor NPA causes abnormal floral development. Also, the pin-formed mutant *pin1* of *Arabidopsis*, which lacks an auxin efflux carrier in shoot tissues, has abnormal flowers similar to those of NPA-treated plants. Apparently, the developing floral meristem depends on auxin being transported to it from subapical tissues. In the absence of efflux carriers, the meristem is deprived of auxin and normal phyllotaxis and floral development are disrupted (Kuhlemeier and Reinhardt 2001). Recently, it has been demonstrated that anthers are the major sites of high concentrations of free auxin that retard the development of neighboring floral organs to synchronize flower development (Aloni et al. 2006).

4.5. Fruit Development

Fruit development is triggered by pollination and fertilization and auxin signaling is thought to play a dynamic role in the regulation of fruit set and early growth. Fertilization-independent fruit set can also occur either naturally in parthenocarpic fruits or by induction via exogenous application of auxin. Parthenocarpic fruits are generally smaller than seeded fruit, suggesting that auxin is required for full pericarp cell expansion. The role of auxin in fruit development was conclusively demonstrated by Nitsch (1950), who showed that achenes on the surface of strawberry receptacle controlled strawberry fruit development and that auxin could be substituted for the achenes. Later, other studies showed that the accumulation of auxin-regulated polypeptides control strawberry fruit development (Veluthambi and Poovaiah 1984; Reddy and Poovaiah 1987, 1990). During tomato fruit development, two peaks of auxin level occur (Gillaspy et al. 1993). The first auxin peak occurs 10 day after anthesis, coinciding with the beginning of cell expansion. The second auxin peak appears later and coincides with the final phase of embryo development. The development of fruit appears to depend on these sources of auxin. Supporting this hypothesis, the auxin-resistant *diageotropica* (*dgt*) mutant of tomato exhibits delayed onset of fruit development, reduced fruit size and seed production, and the application of auxin or auxin transport inhibitors that cause an increase in auxin in the ovary stimulate fruit set and the development of parthenocarpic fruit (Beyer and Quebedeaux 1974; Balbi and Lomax 2003). It is likely that auxin regulation of fruit development involves gene expression. The expression of some of the members of *Aux/IAA* genes was found to be altered in *dgt* mutants, which demonstrates the significant role of auxin in early fruit development (Balbi and Lomax 2003). Recently, it has been demonstrated that the low expression of another auxin-responsive gene, *IAA9*, results in dramatic alteration of early fruit development in antisense *IAA9* tomato plants (Wang et al. 2005).

5. DEVELOPMENTAL EFFECTS OF CYTOKININ

5.1. Cell Division

Cytokinins are primarily defined by their ability to promote cell division. Actually, the first cytokinin was identified based on its cell division promoting ability only. Several physiological evidences show that the cytokinin levels are higher in plant tissues enriched in mitotically active cells, such as shoot and root meristems, as compared to mature tissues where cell cycle is arrested. Although there may be multiple molecular targets of cytokinins, they regulate cell cycle primarily by inducing the expression of *CYCD3* gene, which encodes a D-type cyclin that regulates G1 to S transition during the cell cycle (Riou-Khamlichi et al. 1999). In fact the constitutive expression of *CYCD3* has been shown to overcome the exogenous cytokinin requirement in callus initiation (Riou-Khamlichi et al. 1999). Other cell cycle genes, including *CDC2* and different B-type cyclins are also regulated by cytokinins (Fuerst et al. 1996). There are evidences that the genes regulating cell cycle are expressed in a tissue-specific manner and that cytokinin effects on the cell cycle vary between different cell types.

The root and shoot apical meristems represent a small group of pluripotent cells that form the lower underground and aerial parts of the plant, respectively. The ability of cytokinins to promote shoot development from undifferentiated cells in culture suggests their role in SAM formation. The ectopic overexpression of cytokinin oxidase, an enzyme involved in cytokinin catabolism, results in reduced shoot development as well as leaf primordia, however, an increase in cytokinin levels leads to proliferation of leaf primordia (Werner et al. 2001, 2003). Surprisingly, the same overexpression of cytokinin oxidase leads to enhanced root growth due to additional rounds of mitosis of root meristem cells. These results indicate that cytokinins have a negative regulatory function in cell proliferation in root meristem contrasting with the promotive role in SAM (Werner et al. 2001, 2003). Cytokinins also elevate the mRNA levels of homeobox genes, *KNOTTED1* (*KNAT1*), *KNOTTED2* (*KNAT2*), and *SHOOT MERISTEMLESS* (*STM*), which are required to maintain meristem cells in indeterminate state (Estruch et al. 1993; Rupp et al. 1999; Shani et al. 2006).

5.2. Stimulation of Axillary Bud Growth

The control of lateral or axillary bud growth has been attributed to the presence of a growing shoot apex. The term apical dominance is used to indicate that the shoot tip exerts an inhibitory control over axillary buds. Through decapitation and/or hormone manipulation experiments, this inhibition has been attributed to the balance of phytohormones auxin and cytokinin. Although apical dominance is determined primarily by auxin, physiological studies indicate that cytokinins play an important role in axillary bud growth. For example, exogenous application of cytokinin to dormant axillary buds of many plant species stimulates cell division and causes them to grow. Also, the phenotypes of cytokinin-overproducing mutants and transgenic plants which tend to be bushy due to reduced apical dominance and growth of lateral buds provide additional evidence for the role of cytokinins in axillary bud induction (Helliwell et al. 2001; Tantikanjana et al. 2001; Khurana et al. 2004).

5.3. Leaf and Cotyledon Expansion and Chloroplast Development

The ability of cytokinins to promote cell enlargement has been demonstrated in the cotyledons of dicots with leafy cotyledons, such as mustard, cucumber and sunflower. Analysis of cytokinin levels in leaf provides a positive correlation between endogenous cytokinins and leaf expansion. A higher concentration of zeatin and ribosylzeatin occurs in basal part of pepper leaves, the region from where leaf expansion takes place, than in distal portion; older, fully expanded leaves had a more uniform distribution of cytokinins. The possible action of cytokinins in leaf expansion is that cytokinins enhance sink strength for the allocation of assimilates. Treatment with kinetin results in the movement of nutrients from the untreated to treated part of tobacco leaf. In addition, the cytokinin-induced cell expansion is associated with an increase in cell wall extensibility (Thomas et al. 1981).

Cytokinins along with other factors, such as light and nutrition, regulate the synthesis of photosynthetic pigments and proteins in the chloroplasts. The evidence comes from the observation that the chloroplasts with more extensive grana, chlorophyll and photosynthetic enzymes are formed when etiolated leaves are treated with cytokinin before being illuminated. Also, the ability of exogenous cytokinins to enhance de-etiolation of dark-grown seedlings is mimicked by mutants that overproduce cytokinins (Chory et al. 1994; Chin-Atkins et al. 1996; Dasgupta 2002).

5.4. Delay in Leaf Senescence

Senescence is a genetically-programmed cell degeneration process that leads to cell death. An array of external (environmental stresses such as extreme temperatures, drought, nutrient deficiency, insufficient light or darkness, and pathogen infection) and internal (age, levels of plant hormones and developmental processes) factors regulate senescence. These factors may act individually or in concert. Plant hormones play important roles in regulating senescence; ethylene, ABA, salicylic acid, jasmonate and brassinosteroids generally promote senescence, whereas cytokinins, auxins and gibberellins retard senescence (Gan 2004). Among these, cytokinins are perhaps the most studied hormone in terms of the regulatory role in leaf senescence. There are three major lines of evidence that support the inhibitory role of cytokinins in leaf senescence. First, the external application of cytokinins to detached or *in planta* leaves can delay their senescence (Gan and Amasino 1996). However, the effect of cytokinins on senescence can vary under different environmental conditions. For example, treatment with 6-benzyladenine retards senescence in *Arabidopsis* leaves in dark but accelerates the loss of chlorophyll in light (Gan and Amasino 1996). Secondly, there is an inverse correlation between the cytokinin levels of leaves before and after the onset of senescence (Gan and Amasino 1996). The cytokinin level falls with the progression of leaf senescence. Third convincing evidence comes from the genetic manipulation of cytokinin production in transgenic plants. For example, the overexpression of *IPT* gene encoding an isopentenyl transferase that catalyzes the first step in cytokinin biosynthesis, leads to overproduction of cytokinins up to 500-fold compared with wild-type plants and display delayed leaf senescence and other physiological and developmental phenotypes characteristic of cytokinin overproduction, including reduced plant and leaf size, less developed vascular and root system and weakened apical dominance (Gan and Amasino 1996). In addition, overexpression of components of the cytokinin signal transduction pathway such as CKI1 (a histidine kinase protein that acts as a cytokinin receptor) and ARR2 (a B-type response regulator) also delays dark-induced leaf senescence in *Arabidopsis*, further confirming the role of cytokinins in delaying leaf senescence (Hwang and Sheen 2001).

5.5. Induction of Bud Formation in Moss

Several studies show that cytokinins stimulate bud initiation in mosses such as *Funaria hygrometrica* and *Physcomitrella patens* (Brandes and Kende 1968). Cytokinin induces nuclear migration followed by an asymmetric division in the target protonema cells (caulonema) leading to the formation of a small number of initial cells which undergo further divisions to form buds. Buds revert to protonemal filaments if cytokinin is removed during the early stages of their development by washing the protonemata. This indicates that the hormone is not acting as a trigger but has to be present until differentiation is stabilized. It has been hypothesized that calcium plays an important role as an intracellular messenger in this developmental event (Saunders and Hepler 1983). Ca^{2+} influx is thought to be an early event in bud formation in moss protonema. One of the cytokinin effects is regulation of the voltage-gated Ca^{2+} channels present on the plasma membrane of moss protonema. A rise in intracellular calcium mediates bud formation in *Funaria*. However, the cytokinin-induced formation of buds can be inhibited by abscisic acid in a concentration-dependent manner (Christianson 2000).

5.6. Morphogenesis in Tissue Cultures

Subsequent to the discovery of kinetin as a potent activator of the proliferation of cultured tobacco pith cells, it became apparent that the ratio rather than the absolute quantity of cytokinin and auxin determines the type of organs regenerated from undifferentiated callus tissue in vitro; high cytokinin to auxin ratio promotes shoot formation while a low cytokinin to auxin ratio promotes root development; a balanced ratio keep the cells in undifferentiated state, i.e. callus (Skoog and Miller 1965). The effect of auxin and cytokinin ratio on morphogenesis has been demonstrated in crown gall (a tumor-like mass of undifferentiated cells that typically occurs near the crown, i.e. junction of stem and root of the plant) and is caused by *Agrobacterium tumefaciens*. Upon infection, the T-DNA of *Agrobacterium* Ti plasmid is incorporated into the plant genome and begins to overproduce auxin and cytokinin, which results in formation of an undifferentiated tumor. Mutation at the *tmr* locus of Ti plasmids blocks zeatin biosynthesis that lowers cytokinin to auxin ratio and the tumor cells show proliferation of roots. In contrast, mutation in *tms* locus required for auxin production increases cytokinin to auxin ratio and the resultant crown gall is shooty in nature (Akiyoshi et al. 1983).

5.7. Other Responses

Cytokinins also stimulate flowering in several plant species, but generally only under favorable (inductive) conditions. However, no general agreement exists between flowering and endogenous cytokinins. Recently, a role for cytokinins has been demonstrated in the formation of nitrogen fixing nodules and nematode induced galls (Lohar et al. 2004). This information is important to researchers seeking to understand the relation of root hormones and nitrogen-fixing bacteria toward improving birdsfoot trefoil cultivars for pasture and livestock. Cytokinins have also been implicated in stomatal responses but supporting evidences are limited.

6. POTENTIAL OF AUXIN AND CYTOKININ SIGNALING COMPONENT GENES FOR CROP IMPROVEMENT

The dwarf stature of plants is an agronomically important trait of great significance as exemplified by the success of green revolution achieved by the advent of dwarf varieties of wheat and rice (Khush 2001). The genes responsible for green revolution dwarf varieties encoded proteins involved in gibberellin biosynthesis or signaling. However, Multani et al. (2003) described a new mechanism that controls plant height in dwarf mutants of maize and sorghum. They showed that genes mutated in *brachytic2* (*br2*) and *dwarf3* (*dw3*) dwarf mutants of maize and sorghum, respectively, encode a P-glycoprotein that modulates polar auxin transport. The basipetal transport of auxin is reduced in *br2* and *dw3* mutants, which results in reduced auxin response, leading to formation of compact lower stalk internodes and dwarfism. The *dw3* mutant has been in use in sorghum plant-breeding programs for several decades and can be further used for effectively engineering sorghum germplasm (Salamini 2003). Likewise, *br2* gene can be manipulated to reduce plant height in maize by employing breeding programs. There are also prospects to use this knowledge for altering plant height in other crop plants.

The role of auxin in fruit development is well established as described earlier. The mutations in auxin-responsive genes, such as *IAA9* and *DIAGEOTROPICA*, alter fruit development dramatically in tomato (Balbi and Lomax 2003; Wang et al. 2005). Parthenocarpy has also been induced in tomato by overproduction of auxin in tomato. The transgenic tomato plants containing *DefH9-iaaM* (*iaaM* gene from *Pseudomonas syringae* under the control of *DefH9* placenta/ovule-specific promoter) produced parthenocarpic fruits that were identical in size and weight to fruit from pollinated flowers (Figure 5; Ficcadenti et al. 1999). The same construct has been used to promote parthenocarpy in egg plant (Rotino et al. 1997).

Recently, a molecular link between auxin signaling and resistance to bacterial pathogens in plants has been demonstrated (Navarro et al. 2006). The exogenous application of auxin enhanced susceptibility to the bacterial pathogens. However, the repression of auxin signaling by miRNA-mediated down regulation of F-box auxin receptors, *TIR1*, *AFB2*, and *AFB3* mRNAs increased antibacterial resistance. The identification of novel miRNAs that downregulate auxin signaling in plants will provide new targets for manipulation of crop plants against pathogens.

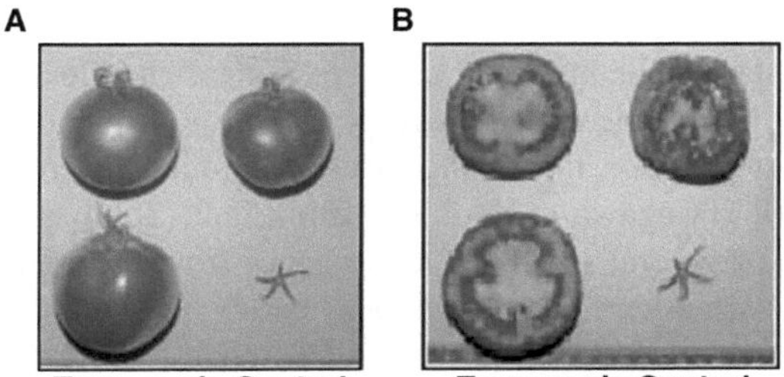

Figure 5. Induction of parthenocarpic tomato fruits by overproduction of auxin. (A) Fruits from pollinated (top) and unpollinated (bottom) flowers from transgenic (transformed with *DefH9::iaaM*) and control plants. (B) Cut fruits from pollinated (top) and unpollinated (bottom) flowers from transgenic (transformed with *DefH9::iaaM*) and control plants. (Adapted from Ficcadenti et al., 1999) *(see plate 10)*

Senescence is an important developmental process in plants meant for mobilization and recycling and to cope with unfavorable environmental conditions. An increased understanding of the genes that control senescence in plants is very important for future agronomic improvements in many crop plants. Delaying senescence, particularly of the flag leaf, in grain crops such as wheat, rice and maize would help to increase grain yield, and such varieties can be used in crop improvement programs. Premature senescence induced by stress also has a detrimental effect on yield, and stay-green plants can exhibit enhanced stress resistance. It has been reported that the autoregulated production of cytokinin in transgenic tobacco plants from senescence-specific *SAG12* promoter inhibited leaf senescence and leaf number and seed yield was also increased (Gan and Amasino 1995). The use of similar system to improve stress tolerance has also been reported. The transgenic *Arabidopsis* plants transformed with *SAG12:IPT* construct exhibited delayed senescence as well as an increased tolerance to flooding (Zhang et al. 2000). In these transgenic plants, at the onset of senescence, *SAG12* promoter is activated, which directs *IPT* expression, resulting in biosynthesis of cytokinins. The increased level of cytokinins inhibits senescence, which in turn, inactivates the *SAG12* promoter, preventing the overproduction of cytokinins. Because the cytokinin production is targeted specifically during senescence, transgenic plants develop normally. The use of such a system provides a very useful example for producing transgenic crop plants with delayed leaf senescence and higher productivity.

The development and survival of plants is constantly challenged by changes in environmental conditions. Abiotic stresses such as drought, high salinity and low temperature are the most common environmental stress factors limiting crop productivity throughout the world. To respond and adapt or tolerate adverse environmental conditions, plants elicit various physiological, biochemical and molecular responses, leading to changes in gene expression. The products of a number of stress-inducible genes such as osmolytes, ion channels, late-embryogenesis-abundant (LEA) proteins, antifreeze proteins, molecular chaperones, transcription factors, protein kinases and detoxification enzymes counteract environmental stresses by regulating gene expression and signal transduction in the stress response. The identification of novel genes involved in environmental stress responses provides us the basis for effective engineering strategies for improving stress tolerance in crop plants (Cushman and Bohnert 2000; Hasegawa et al. 2000; Zhu 2002; Shinozaki et al. 2003). The ectopic expression of several stress-inducible genes from different plant species, including tobacco, *Arabidopsis*, *Brassica*, pea, barley and rice in transgenic plants have been shown to confer multiple stress tolerance (Xu et al. 1996; Holmberg and Bulow, 1998; Kasuga et al. 1999; Veena et al. 1999; Kovtun et al. 2000; Saijo et al. 2000; Mukhopadhyay et al. 2004). Some of the components of cytokinin signal transduction pathway have also been implicated in stress signaling. A transmembrane hybrid-type histidine kinase, AHK1, closely related to cytokinin receptors, acts as a putative osmosensor in *Arabidopsis* (Urao et al. 1999). Moreover, the expression of some of the cytokinin-responsive type-A response regulators is induced under different environmental stress conditions in rice and *Arabidopsis* (Urao et al. 1998; Jain et al. 2006d). The ectopic expression of these genes may be used to engineer stress tolerance in transgenic crop plants.

Many agronomically important traits, including yield, are regulated by a number of genes known as quantitative trait loci (QTLs) derived from natural allelic variations. QTL analysis has been employed as a powerful approach to discover agronomically useful genes. During the past decade, many attempts have been made to characterize QTLs for grain production and plant height; however, the genes involved in these QTLs have not been identified yet. Recently, a QTL, *Gn1a*, that increases grain productivity in rice has been identified as a gene *OsCKX2* encoding a cytokinin oxidase/dehydrogenase enzyme that degrades the phytohormone cytokinin (Ashikari et al. 2005). The reduced expression of *OsCKX2* caused cytokinin accumulation in inflorescence meristems and increased number of reproductive organs, resulting in enhanced grain yield. Furthermore, QTL pyramiding to combine loci for grain number and plant height in the same genetic background generated lines exhibiting both beneficial traits. Identification of agronomically important QTL *Gn1a* as *OsCKX2* and pyramiding of such QTLs presents a useful strategy for efficient crop breeding.

Molecular genetic analyses of *Arabidopsis* have identified many components of auxin and cytokinin signaling. A wealth of information is accumulating on the genomic data of rice, wheat, maize, and tomato, and will help elucidating the significance of auxin and cytokinin signaling components in these plant species of economic importance. Targeting the auxin and cytokinin signaling components may provide an effective strategy for manipulating traits of agronomic importance in crop plants.

REFERENCES

Abel S, Theologis A (1996) Early genes and auxin action. Plant Physiol 111:9–17

Abel S, Oeller PW, Theologis A (1994) Early auxin-induced genes encode short-lived nuclear proteins. Proc Natl Acad Sci USA 91:326–330

Akiyoshi DE, Morris RO, Hinz R, Mischke BS, Kosuge T, Garfinkel DJ, Gordon MP, Nester EW (1983) Cytokinin/auxin balance in crown gall tumors is regulated by specific loci in the T-DNA. Proc Natl Acad Sci USA 80:407–411

Aloni R (1995) The induction of vascular tissues by auxin and cytokinin. In: Davies PJ (ed) Plant hormones: physiology, biochemistry and molecular biology. Kluwer Academic Publishers, The Netherlands, pp 531–546

Aloni R, Aloni E, Langhans M, Ullrich CI (2006) Role of auxin in regulating *Arabidopsis* flower development. Planta 223:315–328

Ashikari M, Sakakibara H, Lin S, Yamamoto T, Takashi T, Nishimura A, Angeles ER, Qian Q, Kitano H, Matsuoka M (2005) Cytokinin oxidase regulates rice grain production. Science 309:741–745

Balbi V, Lomax TL (2003) Regulation of early tomato fruit development by the *diageotropica* gene. Plant Physiol 131:186–197

Benkova E, Michniewicz M, Sauer M, Teichman, T, Seifertova D, Jurgens G, Friml J (2003) Local, efflux-dependent auxin gradients as a common module for plant organ formation. Cell 115:591–602

Bennett SRM, Alvarez J, Bossinger G, Smyth DR (1995) Morphogenesis in *pinoid* mutant of *Arabidopsis thaliana*. Plant J 8:505–520

Bennett MJ, Marchant A, Green HG, May ST, Ward SP, Millner PA, Walker AR, Schulz B, Feldmann KA (1996) *Arabidopsis AUX1* gene: a permease-like regulator of root gravitropism. Science 273:948–950

Berleth T, Jurgens G (1993) The role of the *MONOPTEROS* gene in organizing the basal body region of the *Arabidopsis* embryo. Development 122:575–587

Berleth T, Mattsson J, Hardtke CS (2000) Vascular continuity and auxin signals. Trends Plant Sci 5:387–393

Beyer EM, Quebedeaux B (1974) Parthenocarpy in cucumber: mechanism of action of auxin transport inhibitors. J Am Soc Hortic Sci 99:385–390

Blakeslee JJ, Peer WA, Murphy AS (2005) Auxin transport. Curr Opin Plant Biol 8:494–500

Brandes H, Kende H (1968) Studies on cytokinin-controlled bud formation in moss protonemata. Plant Physiol 43:827–837

Chin-Atkins AN, Craig S, Hocart CH, Dennis ES, Chaudhury AM (1996) Increased endogenous cytokinin in the *Arabidopsis amp1* mutant corresponds with de-etiolation responses. Planta 198:549–556

Chory J, Reinecke D, Sim S, Washburn T, Brenner M (1994) A role for cytokinins in de-Etiolation in *Arabidopsis* (*det* mutants have an altered response to cytokinins). Plant Physiol 104:339–347

Christianson ML (2000) ABA prevents the second cytokinin-mediated event during the induction of shoot buds in the moss *Funaria hygrometrica*. Am J Bot 87:1540–1545

Cushman JC, Bohnert HJ (2000) Genomic approaches to plant stress tolerance. Curr Opin Plant Biol 3:117–124

D'Agostino IB, Deruere J, Kieber JJ (2000) Characterization of the response of the *Arabidopsis* response regulator gene family to cytokinin. Plant Physiol 124:1706–1717

Dasgupta U (2002) Molecular characterization of light signal transduction mutants (*pho*) and analysis of the promoter of a light-regulated gene, $PSBO_A$, from *Arabidopsis*. Ph D Thesis, University of Delhi, India

Davies PJ (2004) Plant hormones: biosynthesis, signal transduction, action. Kluwer Academic Press, The Netherlands.

Dharmasiri N, Dharmasiri S, Estelle M (2005) The F-box protein TIR1 is an auxin receptor. Nature 435:441–445

Dreher KA, Brown J, Saw RE, Callis J (2006) The *Arabidopsis* Aux/IAA protein family has diversified in degradation and auxin responsiveness. Plant Cell 18:699–714

Estruch JJ, Granell A, Hansen G, Prinsen E, Redig P, Van Onckelen H, Schwarz-Sommer Z, Sommer H, Spena A (1993) Floral development and expression of floral homeotic genes are influenced by cytokinins. Plant J 4:379–384

Ficcadenti N, Sestili S, Pandolfini T, Cirillo C, Rotino GL, Spena A (1999) Genetic engineering of parthenocarpic fruit development in tomato. Mol Breed 5:463–470

Franco AR, Gee MA, Guilfoyle TJ (1990) Induction and superinduction of auxin-responsive mRNAs with auxin and protein synthesis inhibitors. J Biol Chem 265:15845–15849

Fuerst RA, Soni R, Murray JA, Lindsey K (1996) Modulation of cyclin transcript levels in cultured cells of *Arabidopsis thaliana*. Plant Physiol 112:1023–1033

Gan S (2004) The hormonal regulation of senescence In: Davies PJ (ed) Plant hormones: biosynthesis, signal transduction and action. Kluwer Academic publishers, The Netherlands, pp 561–581

Gan S, Amasino RM (1995) Inhibition of leaf senescence by autoregulated production of cytokinin. Science 270:1986–1988

Gan S, Amasino RM (1996) Cytokinins in plant senescence: from spray and pray to clone and play. Bioessays 18:557–565

Gillaspy G, Ben-David H, Gruissem W (1993) Fruits: a developmental perspective. Plant Cell 5:1439–1451

Goda H, Sawa S, Asami T, Fujioka S, Shimada Y, Yoshida S (2004) Comprehensive comparison of auxin-regulated and brassinosteroid-regulated genes in *Arabidopsis*. Plant Physiol 134:1555–1573

Gray WM, Estelle I (2000) Function of the ubiquitin-proteasome pathway in auxin response. Trends Biochem Sci 25:133–138

Gray WM, Kepinski S, Rouse D, Leyser O, Estelle M (2001) Auxin regulates SCF(TIR1)-dependent degradation of AUX/IAA proteins. Nature 414:271–276

Grefen C, Harter K (2004) Plant two-component systems: principles, functions, complexity and cross talk. Planta 219:733–742

Guilfoyle TJ (1999) Auxin-regulated genes and promoters. In: Hooykaas PJJ, Hall MA, Libbenga KR (eds) Biochemistry and molecular biology of plant hormones. Elsevier, Amesterdem, The Netherlands, pp 423–459

Guilfoyle TJ, Hagen G, Li Y, Ulmasov T, Liu Z, Strabala T, Gee MA (1993) Auxin-regulated transcription. Aust J Plant Physiol 20:489–502

Hagen G, Guilfoyle T (2002) Auxin-responsive gene expression: genes, promoters and regulatory factors. Plant Mol Biol 49:373–385

Hagen G, Kleinschmidt AJ, Guilfoyle TJ (1984) Auxin-regulated gene expression in intact soybean hypocotyls and excised hypocotyls sections. Planta 16:147–153

Hamann T, Mayer U, Jurgens G (1999) The auxin-insensitive *bodenlos* mutation affects primary root formation and apical-basal patterning in the *Arabidopsis* embryo. Development 126:1387–1395

Hasegawa PM, Bressan RA, Zhu JK, Bohnert HJ (2000) Plant cellular and molecular responses to high salinity. Annu Rev Plant Physiol Plant Mol Biol 51:463–499

Hass C, Lohrmann J, Albrecht V, Sweere U, Hummel F, Yoo SD, Hwang I, Zhu T, Schafer E, Kudla J, Harter K (2004) The response regulator 2 mediates ethylene signalling and hormone signal integration in *Arabidopsis*. EMBO J 23:3290–3302

Helliwell CA, Chin-Atkins AN, Wilson IW, Chapple R, Dennis ES, Chaudhury A (2001) The *Arabidopsis AMP1* gene encodes a putative glutamate carboxypeptidase. Plant Cell 13:2115–2125

Heyl A, Schmulling T (2003) Cytokinin signal perception and transduction. Curr Opin Plant Biol 6:480–488

Hobbie L, McGovern M, Hurwitz LR, Pierro A, Liu NY, Bandyopadhyay A, Estelle M (2000) The *axr6* mutants of *Arabidopsis thaliana* define a gene involved in auxin response and early development. Development 127:23–32

Holmberg N, Bulow L (1998) Improving stress tolerance in plants by gene transfer. Trends Plant Sci 3:61–66

Hsieh HL, Okamoto H, Wang M, Ang LH, Matsui M, Goodman H, Deng XW (2000) *FIN219*, an auxin-regulated gene, defines a link between phytochrome A and the downstream regulator COP1 in light control of *Arabidopsis* development. Genes Dev 14:1958–1970

Hutchison CE, Kieber JJ (2002) Cytokinin signaling in *Arabidopsis*. Plant Cell 14:S47–59

Hwang I, Sheen J (2001) Two-component circuitry in *Arabidopsis* cytokinin signal transduction. Nature 413:383–389

Hwang I, Chen HC, Sheen J (2002) Two-component signal transduction pathways in *Arabidopsis*. Plant Physiol 129:500–515

Inoue T, Higuchi M, Hashimoto Y, Seki M, Kobayashi M, Kato T, Tabata S, Shinozaki K, Kakimoto T (2001) Identification of CRE1 as a cytokinin receptor from *Arabidopsis*. Nature 409:1060–1063

International Rice Genome Sequencing Project (2005) The map-based sequence of the rice genome. Nature 436:793–800

Jain M, Tyagi AK, Khurana, JP (2006c) Genome-wide analysis, evolutionary expansion, and expression of early auxin-responsive *SAUR* gene family in rice (*Oryza sativa*). Genomics 88:360–371

Jain M, Tyagi AK, Khurana JP (2006d) Molecular characterization and differential expression of cytokinin-responsive type-A response regulators in rice (*Oryza sativa*). BMC Plant Biol 6:1

Jain M, Kaur N, Tyagi AK, Khurana JP (2006b) The auxin-responsive *GH3* gene family in rice (*Oryza sativa*). Funct Integr Genomics 6:36–46

Jain M, Kaur N, Garg R, Thakur JK, Tyagi AK, Khurana JP (2006a) Structure and expression analysis of early auxin-responsive *Aux/IAA* gene family in rice (*Oryza sativa*). Funct Integr Genomics 6:47–59

Johnson MA, Perez-Amador MA, Lidder P, Green PJ (2000) Mutants of *Arabidopsis* defective in a sequence-specific mRNA degradation pathway. Proc Natl Acad Sci USA 97:13991–13996

Kakimoto T (2003) Perception and signal transduction of cytokinins. Annu Rev Plant Biol 54:605–627

Kasuga M, Liu,Q, Miura S, Yamaguchi-Shinozaki K, Shinozaki K (1999) Improving plant drought, salt, and freezing tolerance by gene transfer of a single stress-inducible transcription factor. Nat Biotechnol 17:287–291

Kepinski S, Leyser O (2004) Auxin-induced SCF^{TIR1}-Aux/IAA interaction involves stable modification of the SCF^{TIR1} complex. Proc Natl Acad Sci USA 101:12381–12386

Kepinski S, Leyser O (2005) The *Arabidopsis* F-box protein TIR1 is an auxin receptor. Nature 435:446–451

Khurana JP (2001) Cryptic blues: mechanism in sight. Curr Sci 80:189–198

Khurana JP, Dasgupta U, Laxmi A, Kumar D, Paul LK (2004) Light control of plant development by phytochromes: a perspective. Proc Indian Natl Sci Acad B70:379–411

Khush GS (2001) Green revolution: the way forward. Nat Rev Genet 2:815–822

Kiba T, Taniguchi M, Imamura A, Ueguchi C, Mizuno T, Sugiyama T (1999) Differential expression of genes for response regulators in response to cytokinins and nitrate in *Arabidopsis thaliana*. Plant Cell Physiol 40:767–771

Kiba T, Yamada H, Sato S, Kato T, Tabata S, Yamashino T, Mizuno T (2003) The type-A response regulator, ARR15, acts as a negative regulator in the cytokinin-mediated signal transduction in *Arabidopsis thaliana*. Plant Cell Physiol 44:868–874

Kikuchi S, Satoh K, Nagata T, Kawagashira N, Doi K, Kishimoto N, Yazaki J, Ishikawa M, Yamada H, Ooka H, Hotta I, Kojima K, Namiki T, Ohneda E, Yahagi W, Suzuki K, Li CJ, Ohtsuki K, Shishiki T, Otomo Y, Murakami K, Iida Y, Sugano S, Fujimura T, Suzuki Y, Tsunoda Y, Kurosaki T, Kodama T, Masuda H, Kobayashi M, Xie Q, Lu M, Narikawa R, Sugiyama A, Mizuno K, Yokomizo S, Niikura J, Ikeda R, Ishibiki J, Kawamata M, Yoshimura A, Miura J, Kusumegi T, Oka M, Ryu R, Ueda M, Matsubara K, Kawai J, Carninci P, Adachi J, Aizawa K, Arakawa T, Fukuda S, Hara A, Hashizume W, Hayatsu N, Imotani K, Ishii Y, Itoh M, Kagawa I, Kondo S, Konno H, Miyazaki A, Osato N, Ota Y, Saito R, Sasaki D, Sato K, Shibata K, Shinagawa A, Shiraki T, Yoshino M, Hayashizaki Y, Yasunishi A (2003) Collection, mapping, and annotation of over 28,000 cDNA clones from *japonica* rice. Science 301:376–379

Kim J, Harter K, Theologis A (1997) Protein-protein interactions among the Aux/IAA proteins. Proc Natl Acad Sci USA 94:11786–11791

Kovtun Y, Chiu WL, Tena G, Sheen J (2000) Functional analysis of oxidative stress-activated mitogen-activated protein kinase cascade in plants. Proc Natl Acad Sci USA 97:2940–2945

Kuhlemeier C, Reinhardt D (2001) Auxin and phyllotaxis. Trends Plant Sci 6:187–189

Laxmi A, Paul LK, Raychaudhuri A, Peters JL, Khurana JP (2006) *Arabidopsis* cytokinin-resistant mutant, *cnr1*, displays altered auxin responses and sugar sensitivity. Plant Mol Biol 62:409–425

Leibfried A, To JP, Busch W, Stehling S, Kehle A, Demar M, Kieber JJ, Lohmann JU (2005) WUSCHEL controls meristem function by direct regulation of cytokinin-inducible response regulators. Nature 438:1172–1175

Li Y, Strabala TJ, Hagen G, Guilfoyle TJ (1994) The soybean *SAUR* open reading frame contains a cis element responsible for cycloheximide-induced mRNA accumulation. Plant Mol Biol 24:715–723

Lidder P, Gutierrez RA, Salome PA, McClung CR, Green PJ (2005) Circadian control of messenger RNA stability association with a sequence-specific messenger RNA decay pathway. Plant Physiol 138:2374–2385

Liscum E, Reed JW (2002) Genetics of Aux/IAA and ARF action in plant growth and development. Plant Mol Biol 49:387–400

Lohar DP, Schaff JE, Laskey JG, Kieber JJ, Bilyeu KD, Bird DM (2004) Cytokinins play opposite roles in lateral root formation, and nematode and rhizobial symbioses. Plant J 38:203–214

Lohrmann J, Harter K (2002) Plant two-component signaling systems and the role of response regulators. Plant Physiol 128:363–369

Makino S, Matsushika A, Kojima M, Yamashino T, Mizuno T (2002) The APRR1/TOC1 quintet implicated in circadian rhythms of *Arabidopsis thaliana*: i characterization with APRR1-overexpressing plants. Plant Cell Physiol 43:58–69

Makino S, Kiba T, Imamura A, Hanaki N, Nakamura A, Suzuki T, Taniguchi M, Ueguchi C, Sugiyama T, Mizuno T (2000) Genes encoding pseudo-response regulators: insight into His-to-Asp phosphorelay and circadian rhythm in *Arabidopsis thaliana*. Plant Cell Physiol 41:791–803

Marchant A, Bhalerao R, Casimiro I, Eklof J, Casero PJ, Bennett M, Sandberg G (2002) AUX1 promotes lateral root formation by facilitating indole-3-acetic acid distribution between sink and source tissues in the *Arabidopsis* seedling. Plant Cell 14:589–597

Mason MG, Li J, Mathews DE, Kieber JJ, Schaller GE (2004) Type-B response regulators display overlapping expression patterns in *Arabidopsis*. Plant Physiol 135:927–937

Mason MG, Mathews DE, Argyros DA, Maxwell BB, Kieber JJ, Alonso JM, Ecker JR, Schaller GE (2005) Multiple type-B response regulators mediate cytokinin signal transduction in *Arabidopsis*. Plant Cell 17:3007–3018

Matsushika A, Makino S, Kojima M, Yamashino T, Mizuno T (2002) The APRR1/TOC1 quintet implicated in circadian rhythms of *Arabidopsis thaliana*: II characterization with CCA1-overexpressing plants. Plant Cell Physiol 43:118–122

Mattsson J, Ckurshumova W, Berleth T (2003) Auxin signaling in *Arabidopsis* leaf vascular development. Plant Physiol 131:1327–1339

Mattsson J, Sung ZR, Berleth T (1999) Responses of plant vascular systems to auxin transport inhibition. Development 126:2979–2991

McClure BA, Guilfoyle T (1987) Characterization of a class of small auxin-inducible soybean polyadenylated RNAs. Plant Mol Biol 9:611–623

McClure BA, Guilfoyle T (1989) Rapid redistribution of auxin-regulated RNAs during gravitropism. Science 243:91–93

McClure BA, Hagen G, Brown CS, Gee MA, Guilfoyle TJ (1989) Transcription, organization, and sequence of an auxin-regulated gene cluster in soybean. Plant Cell 1:229–239

Messing J, Bharti AK, Karlowski WM, Gundlach H, Kim HR, Yu Y, Wei F, Fuks G, Soderlund CA, Mayer KF, Wing RA (2004) Sequence composition and genome organization of maize. Proc Natl Acad Sci USA 101:14349–14354

Mizuno T, Nakamichi N (2005) Pseudo-response regulators (PRRs) or true oscillator components (TOCs). Plant Cell Physiol 46:677–685

Mok DW, Mok MC (2001) Cytokinin metabolism and action. Annu Rev Plant Physiol Plant Mol Biol 52:89–118

Mukhopadhyay A, Vij S, Tyagi AK (2004) Overexpression of a zinc-finger protein gene from rice confers tolerance to cold, dehydration, and salt stress in transgenic tobacco. Proc Natl Acad Sci USA 101:6309–6314

Multani DS, Briggs SP, Chamberlin MA, Blakeslee JJ, Murphy AS, Johal GS (2003) Loss of an MDR transporter in compact stalks of maize *br2* and sorghum *dw3* mutants. Science 302:81–84

Murakami M, Ashikari M, Miura K, Yamashino T, Mizuno T (2003) The evolutionarily conserved OsPRR quintet: rice pseudo-response regulators implicated in circadian rhythm. Plant Cell Physiol 44:1229–1236

Nagpal P, Walker LM, Young JC, Sonawala A, Timpte C, Estelle M, Reed JW (2000) *AXR2* encodes a member of the Aux/IAA protein family. Plant Physiol 123:563–574

Nakamichi N, Kita M, Ito S, Yamashino T, Mizuno T (2005) Pseudo-response regulators, PRR9, PRR7 and PRR5, together play essential roles close to the circadian clock of *Arabidopsis thaliana*. Plant Cell Physiol 46:686–698

Nakazawa M, Yabe N, Ichikawa T, Yamamoto YY, Yoshizumi T, Hasunuma K, Matsui M (2001) *DFL1*, an auxin-responsive *GH3* gene homologue, negatively regulates shoot cell elongation and lateral root formation, and positively regulates the light response of hypocotyl length. Plant J 25:213–221

Navarro L, Dunoyer P, Jay F, Arnold B, Dharmasiri N, Estelle M, Voinnet O, Jones JD (2006) A plant miRNA contributes to antibacterial resistance by repressing auxin signaling. Science 312:436–439

Nemhauser JL, Feldman LJ, Zambryski PC (2000) Auxin and ETTIN in *Arabidopsis* gynoecium morphogenesis. Development 127:3877–3888

Nemhauser JL, Zambryski PC, Roe JL (1998) Auxin signaling in *Arabidopsis* flower development. Curr Opin Plant Biol 1:531–535

Newman TC, Ohme-Takagi M, Taylor CB, Green PJ (1993) DST sequences, highly conserved among plant *SAUR* genes, target reporter transcripts for rapid decay in tobacco. Plant Cell 5:701–714

Nitsch JP (1950) Growth and morphogenesis of the strawberry as related to auxin. Am J Bot 37:211–215

Oka A, Sakai H, Iwakoshi S (2002) His-Asp phosphorelay signal transduction in higher plants: receptors and response regulators for cytokinin signaling in *Arabidopsis thaliana*. Genes Genet Syst 77:383–391

Oka M, Miyamoto K, Okada K, Ueda J (1999) Auxin polar transport and flower formation in *Arabidopsis thaliana* transformed with indoleacetamide hydrolase (*iaaH*) gene. Plant Cell Physiol 40:231–237

Okada K, Ueda J, Komaki MK, Bell CJ, Shimura Y (1991) Requirement of the auxin polar transport system in early stages of *Arabidopsis* floral bud formation. Plant Cell 3:677–684

Osakabe Y, Miyata S, Urao T, Seki M, Shinozaki K, Yamaguchi-Shinozaki K (2002) Overexpression of *Arabidopsis* response regulators, ARR4/ATRR1/IBC7 and ARR8/ATRR3, alters cytokinin responses differentially in the shoot and in callus formation. Biochem Biophys Res Commun 293:806–815

Ouellet F, Overvoorde PJ, Theologis A (2001) *IAA17/AXR3*: biochemical insight into an auxin mutant phenotype. Plant Cell 13:829–841

Paponov IA, Teale WD, Trebar M, Blilou I, Palme K (2005) The PIN auxin efflux facilitators: evolutionary and functional perspectives. Trends Plant Sci 10:170–177

Pareek A, Singh A, Kumar M, Kushwaha HR, Lynn AM, Singla-Pareek SL (2006) Whole genome analysis of *Oryza sativa* L reveals similar architecture of two-component-signaling-machinery with *Arabidopsis*. Plant Physiol 142:380–397

Pennazio S (2002) The discovery of the chemical nature of the plant hormone auxin. Riv Biol 95:289–308

Perez-Amador MA, Lidder P, Johnson MA, Landgraf J, Wisman E, Green PJ (2001) New molecular phenotypes in the *dst* mutants of *Arabidopsis* revealed by DNA microarray analysis. Plant Cell 13:2703–2717

Phillips SE (1994) The beta-ribbon DNA recognition motif. Annu Rev Biophys Biomol Struct 23:671–701

Przemeck GK, Mattsson J, Hardtke CS, Sung ZR, Berleth T (1996) Studies on the role of the *Arabidopsis* gene *MONOPTEROS* in vascular development and plant cell axialization. Planta 200:229–237

Putterill J, Robson F, Lee K, Simon R, Coupland G (1995) The *CONSTANS* gene of *Arabidopsis* promotes flowering and encodes a protein showing similarities to zinc finger transcription factors. Cell 80:847–857

Ramos JA, Zenser N, Leyser O, Callis J (2001) Rapid degradation of auxin/indoleacetic acid proteins requires conserved amino acids of domain II and is proteasome dependent. Plant Cell 13:2349–2360

Reddy AS, Poovaiah BW (1987) Accumulation of a glycine rich protein in auxin-deprived strawberry fruits. Biochem Biophys Res Commun 147:885–891

Reddy AS, Poovaiah BW (1990) Molecular cloning and sequencing of a cDNA for an auxin-repressed mRNA: correlation between fruit growth and repression of the auxin-regulated gene. Plant Mol Biol 14:127–136

Reed JW (2001) Roles and activities of Aux/IAA proteins in *Arabidopsis*. Trends Plant Sci 6:420–425

Reinhardt D, Pesce ER, Stieger P, Mandel T, Baltensperger K, Bennett M, Traas J, Friml J, Kuhlemeier C (2003) Regulation of phyllotaxis by polar auxin transport. Nature 426:255–260

Rensink WA, Buell CR (2005) Microarray expression profiling resources for plant genomics. Trends Plant Sci 10:603–609

Riou-Khamlichi C, Huntley R, Jacqmar, A, Murray JA (1999) Cytokinin activation of *Arabidopsis* cell division through a D-type cyclin. Science 283:1541–1544

Rogg LE, Lasswell J, Bartel B (2001) A gain-of-function mutation in *IAA28* suppresses lateral root development. Plant Cell 13:465–480

Rotino GL, Perri E, Zottini M, Sommer H, Spena A (1997) Genetic engineering of parthenocarpic plants. Nat Biotechnol 15:1398–1401

Rouse D, Mackay P, Stirnberg P, Estelle M, Leyser O (1998) Changes in auxin response from mutations in an *AUX/IAA* gene. Science 279:1371–1373

Roux C, Perrot-Rechenmann C (1997) Isolation by differential display and characterization of a tobacco auxin-responsive cDNA *Nt-gh3*, related to *GH3*. FEBS Lett 419:131–136

Rupp HM, Frank M, Werner T, Strnad M, Schmulling T (1999) Increased steady state mRNA levels of the *STM* and *KNAT1* homeobox genes in cytokinin overproducing *Arabidopsis thaliana* indicate a role for cytokinins in the shoot apical meristem. Plant J 18:557–563

Saijo Y, Hata S, Kyozuka J, Shimamoto K, Izui K (2000) Over-expression of a single Ca^{2+}-dependent protein kinase confers both cold and salt/drought tolerance on rice plants. Plant J 23:319–327

Sakai H, Aoyama T, Oka A (2000) *Arabidopsis* ARR1 and ARR2 response regulators operate as transcriptional activators. Plant J 24:703–711

Sakai H, Aoyama T, Bono H, Oka A (1998) Two-component response regulators from *Arabidopsis thaliana* contain a putative DNA-binding motif. Plant Cell Physiol 39:1232–1239

Sakai H, Honma T, Aoyama T, Sato S, Kato T, Tabata S, Oka A (2001) ARR1, a transcription factor for genes immediately responsive to cytokinins. Science 294:1519–1521

Salamini F (2003) Hormones and the green revolution. Science 302:71–72

Saunders MJ, Hepler PK (1983) Calcium antagonists and calmodulin inhibitors block cytokinin-induced bud formation in *Funaria*. Dev Biol 99:41–49

Shani E, Yanai O, Ori N (2006) The role of hormones in shoot apical meristem function. Curr Opin Plant Biol 9:484–489

Shinozaki K, Yamaguchi-Shinozaki K, Seki M (2003) Regulatory network of gene expression in the drought and cold stress responses. Curr Opin Plant Biol 6:410–417

Skoog F, Miller CO (1965) Chemical regulation of growth and organ formation in plant tissues cultured in vitro. In: Bell E (ed) Molecular and cellular aspects of development. Harper and Row, New York, pp 481–494

Srivastava LM (2001) Plant growth and development-hormones and environment. Elsevier Science, San Diego, pp 303–339

Staswick PE, Tiryaki I, Rowe ML (2002) Jasmonate response locus *JAR1* and several related *Arabidopsis* genes encode enzymes of the firefly luciferase superfamily that show activity on jasmonic, salicylic, and indole-3-acetic acids in an assay for adenylation. Plant Cell 14:1405–1415

Staswick PE, Serban B, Rowe M, Tiryaki I, Maldonado MT, Maldonado MC, Suza W (2005) Characterization of an *Arabidopsis* enzyme family that conjugates amino acids to indole-3-acetic acid. Plant Cell 17:616–627

Stock AM, Robinson VL, Goudreau PN (2000) Two-component signal transduction. Annu Rev Biochem 69:183–215

Suzuki T, Miwa K, Ishikawa K, Yamada H, Aiba H, Mizuno T (2001) The *Arabidopsis* sensor His-kinase, AHK4, can respond to cytokinins. Plant Cell Physiol 42:107–113

Sweere U, Eichenberg K, Lohrmann J, Mira-Rodado V, Baurle I, Kudla J, Nagy F, Schafer E, Harter K (2001) Interaction of the response regulator ARR4 with phytochrome B in modulating red light signaling. Science 294:1108–1111

Tajima Y, Imamura A, Kiba T, Amano Y, Yamashino T, Mizuno T (2004) Comparative studies on the type-B response regulators revealing their distinctive properties in the His-to-Asp phosphorelay signal transduction of *Arabidopsis thaliana*. Plant Cell Physiol 45:28–39

Takase T, Nakazawa M, Ishikawa A, Manabe K, Matsui M (2003) *DFL2*, a new member of the *Arabidopsis GH3* gene family, is involved in red light-specific hypocotyl elongation. Plant Cell Physiol 44:1071–1080

Takase T, Nakazawa M, Ishikawa A, Kawashima M, Ichikawa T, Takahashi N, Shimada H, Manabe K, Matsui M (2004) *ydk1-D*, an auxin-responsive *GH3* mutant that is involved in hypocotyl and root elongation. Plant J 37:471–483

Tanaka S, Mochizuki N, Nagatani A (2002) Expression of the *AtGH3a* gene, an *Arabidopsis* homologue of the soybean *GH3* gene, is regulated by phytochrome B. Plant Cell Physiol 43:281–289

Taniguchi M, Kiba T, Sakakibara H, Ueguchi C, Mizuno T, Sugiyama T (1998) Expression of *Arabidopsis* response regulator homologs is induced by cytokinins and nitrate. FEBS Lett 429: 259–262

Tantikanjana T, Yong, JW, Letham, DS, Griffith, M, Hussain, M, Ljung, K, Sandberg, G, Sundaresan, V (2001) Control of axillary bud initiation and shoot architecture in *Arabidopsis* through the *SUPER-SHOOT* gene. Genes Dev 15:1577–1588

Teale WD, Paponov IA, Palme K (2006) Auxin in action: signalling, transport and the control of plant growth and development. Nat Rev Mol Cell Biol 7:847–859

Tepperman JM, Zhu T, Chang HS, Wang X, Quail PH (2001) Multiple transcription-factor genes are early targets of phytochrome A signaling. Proc Natl Acad Sci USA 98:9437–9442

Thakur JK, Tyagi AK, Khurana JP (2001) *OsIAA1*, an *Aux/IAA* cDNA from rice, and changes in its expression as influenced by auxin and light. DNA Res 8:193–203

Thakur JK, Jain M, Tyagi AK, Khurana JP (2005) Exogenous auxin enhances the degradation of a light down-regulated and nuclear-localized OsiIAA1, an Aux/IAA protein from rice, via proteasome. Biochim Biophys Acta 1730:196–205

Thomas J, Ross CW, Chastain CJ, Koomanoff N, Hendrix JE (1981) Cytokinin-induced wall extensibility in excised cotyledons of radish and cucumber. Plant Physiol 68:107–110

Tian Q, Reed JW (1999) Control of auxin-regulated root development by the *Arabidopsis* thaliana SHY2/IAA3 gene. Development 126:711–721

Tiwari SB, Hagen G, Guilfoyle T (2003) The roles of auxin response factor domains in auxin-responsive transcription. Plant Cell 15:533–543

Tiwari SB, Hagen G, Guilfoyle TJ (2004) Aux/IAA proteins contain a potent transcriptional repression domain. Plant Cell 16:533–543

To JP, Haberer G, Ferreira FJ, Deruere J, Mason MG, Schaller GE, Alonso JM, Ecker JR, Kieber JJ (2004) Type-A *Arabidopsis* response regulators are partially redundant negative regulators of cytokinin signaling. Plant Cell 16:658–671

Ueguchi C, Koizumi H, Suzuki T, Mizuno T (2001a) Novel family of sensor histidine kinase genes in *Arabidopsis thaliana*. Plant Cell Physiol 42:231–235

Ueguchi C, Sato S, Kato T, Tabata S (2001b) The *AHK4* gene involved in the cytokinin-signaling pathway as a direct receptor molecule in *Arabidopsis thaliana*. Plant Cell Physiol 42:751–755

Ulmasov T, Hagen G, Guilfoyle TJ (1997a) ARF1, a transcription factor that binds to auxin response elements. Science 276:1865–1868

Ulmasov T, Murfett J, Hagen G, Guilfoyle TJ (1997b) Aux/IAA proteins repress expression of reporter genes containing natural and highly active synthetic auxin response elements. Plant Cell 9:1963–1971

Urao T, Yakubov B, Yamaguchi-Shinozaki K, Shinozaki K (1998) Stress-responsive expression of genes for two-component response regulator-like proteins in *Arabidopsis thaliana*. FEBS Lett 427:175–178

Urao T, Yakubov B, Satoh R, Yamaguchi-Shinozaki K, Seki M, Hirayama T, Shinozaki K, (1999) A transmembrane hybrid-type histidine kinase in *Arabidopsis* functions as an osmosensor. Plant Cell 11:1743–1754

Veena Reddy VS, Sopory SK, (1999) Glyoxalase I from *Brassica juncea*: molecular cloning, regulation and its over-expression confer tolerance in transgenic tobacco under stress. Plant J 17:385–395

Veluthambi K, Poovaiah BW (1984) Auxin-regulated polypeptide changes at different stages of strawberry fruit development. Plant Physiol 75:349–353

Vij S, Gupta V, Kumar D, Vydianathan R, Raghuvanshi S, Khurana P, Khurana JP, Tyagi AK (2006) Decoding the rice genome. Bioessays 28:421–432

Wang H, Jones B, Li Z, Frasse P, Delalande C, Regad F, Chaabouni S, Latche A, Pech JC, Bouzayen M (2005) The tomato Aux/IAA transcription factor IAA9 is involved in fruit development and leaf morphogenesis. Plant Cell 17:2676–2692

Weijers D, Benkova E, Jager KE, Schlereth A, Hamann T, Kientz M, Wilmoth JC, Reed JW, Jurgens G (2005) Developmental specificity of auxin response by pairs of ARF and Aux/IAA transcriptional regulators. EMBO J 24:1874–1885

Werner T, Motyka V, Strnad M, Schmulling T (2001) Regulation of plant growth by cytokinin. Proc Natl Acad Sci USA 98:10487–10492

Werner T, Motyka V, Laucou V, Smets R, Van Onckelen H, Schmulling T (2003) Cytokinin-deficient transgenic *Arabidopsis* plants show multiple developmental alterations indicating opposite functions of cytokinins in the regulation of shoot and root meristem activity. Plant Cell 15:2532–2550

West AH, Stock AM (2001) Histidine kinases and response regulator proteins in two-component signaling systems. Trends Biochem Sci 26:369–376

Woodward AW, Bartel B (2005) Auxin: regulation, action, and interaction. Ann Bot (Lond) 95:707–735

Worley CK, Zenser N, Ramos J, Rouse D, Leyser O, Theologis A, Callis J (2000) Degradation of Aux/IAA proteins is essential for normal auxin signaling. Plant J 21:553–562

Xu D, Duan X, Wang B, Hong B, Ho T, Wu R (1996) Expression of a late embryogenesis abundant protein gene, *HVA1*, from barley confers tolerance to water deficit and salt stress in transgenic rice. Plant Physiol 110:249–257

Yamada H, Suzuki T, Terada K, Takei K, Ishikawa K, Miwa K, Yamashino T, Mizuno T (2001) The *Arabidopsis* AHK4 histidine kinase is a cytokinin-binding receptor that transduces cytokinin signals across the membrane. Plant Cell Physiol 42:1017–1023

Yang T, Poovaiah BW (2000) Molecular and biochemical evidence for the involvement of calcium/calmodulin in auxin action. J Biol Chem 275:3137–3143

Zhang J, Toai TC, Huynh L, Priszner J (2000) Development of flooding-tolerant *Arabidopsis thaliana* by autoregulated cytokinin production. Mol Breed 6:135–144

Zhu JK (2002) Salt and drought stress signal transduction in plants. Annu Rev Plant Biol 53:247–273

CHAPTER 14

STATISTICAL ADVANCES IN FUNCTIONAL GENOMICS

REBECCA W. DOERGE*

Departments of Statistics and Agronomy, Purdue University; 150 North University Street, West Lafayette, IN 47907, USA

Abstract: Statistics, agriculture, and genetics share a long successful pre-genomic history that is based on solid principles of experimental design and analysis of variation. In the era of 'omics it is essential that statistical and mathematical standards, as well as guidelines for the experimental design and analysis of biological studies are upheld. The main message of this chapter recalls past statistical issues, discusses current statistical advances that pertain to understanding complex traits, and promotes ideas about both the data and statistical genomic models of the future.

1. INTRODUCTION... THE PAST REVISITED

Whether it is considered as 'statistical advances in function genomics,' or 'statistics advancing functional genomics', most of the progress in functional genomics that has been made through applications of statistics has been the result of modern adaptations and new appreciations of existing statistical methods (namely, experimental design, analysis of variance, and exploratory data analysis) toward the analysis of large-scale data that have resulted from advances in biotechnology (e.g., genetic markers, sequencing, and microarrays). Most of us remember that the marriage of statistics and genetics has a long history of successes in agricultural experiments, most of which were completely void of genomic data for a very long time, but in fact did include applications of proper experimental design and correct statistical analyses of the data. With this history and experience on their side, current day statisticians have adapted many existing statistical methods to meet the modern challenges brought forth by the era of 'omics. In fact, most of the statistical issues that accompany these challenges are not new, they are merely reincarnations of past

*Corresponding Author: doerge@purdue.edu

R.K. Varshney and R. Tuberosa (eds.), Genomics-Assisted Crop Improvement: Vol. 1: Genomics Approaches and Platforms, 315–332.

lessons that need to be remembered. As the phenomenon of collecting more data and building bigger databases happens with record speed, we should all reflect on the past successes that have shaped our foundation and remember the milestones that have been surpassed. For many of us remembering the limitations (e.g., too few polymorphic genetic markers, too small mapping populations, slow and single processor computers) of the past will bring to mind contrasting issues with which we are currently struggling.

My main purpose in supplying this book chapter is to provide a general discussion about the role that Statistics plays in advancing functional genomics. The intention here is not to overwhelm the reader with statistical equations, theories, and complicated graphics that seem to have no connection to reality, but to discuss some of the more current genomic issues that are being addressed by Statistics, and to forecast future grand challenges that will be met by Statistics. These challenges include epistasis, quantitative trait mapping, gene expression trait mapping, epigenomics, and their unification toward molecularly dissecting complex traits. If the reader comes away from this experience with a broader appreciation of Statistics as applied to functional genomics, and a slight confusion due to the nature of these dynamic complex problems, then my vision for this chapter will have been fulfilled.

Many current concerns when dealing with functional genomic experiments and their data share common themes (e.g., accurate measurement, data storage, data annotation, incorrect and missing data, biological replication, design of experiments, sample size, multiple testing issues, etc.) with the past. Even though technologies have improved and expanded to provide a more detailed view of the inner workings of dynamic complex biological systems, the statistical issues and many times the solutions surrounding functional genomics are not novel; they are merely more significant due to the complexity and size of what we are studying. In short, while the modernization of statistical methods has been enabled by high power computing, parallel processing, data mining, machine learning, and massive data application, many of the statistical issues that are thought of as advances are not new, they are simply an evolution of quantitative methods to meet the needs of the broader scientific community. Quantitative biology is at a crossroads of scientific history where the problems are interdisciplinary in nature and the need for responsible statistical analysis essential. Case(s) in point: all levels of 'omics are testing a large number of hypotheses (e.g., gene expression) that routinely fail to acknowledge the multiple testing problem, and as a result undergo too many false positive conclusions; most scientific conclusions are being made in the absence of proper experimental design; and results are being combined and summarized without the benefit of quality control or statistical concern. The level of impact that the discipline of Statistics will have on the quality and quantity of experimental sciences largely depends on the integration of experimental design and statistical methods into current day educational and research programs, as well as federal funding agency assessment criteria and peer reviewed journals.

2. IMPACTING QUANTITATIVE STANDARDS IN LARGE-SCALE BIOLOGY

In a recent editorial by Jorgensen (2006) the dynamic nature of systems biology, which by definition includes functional genomics, was recognized and discussed on a much broader scale. The acknowledgement of Statistics, experimental design, and mathematical modelling is summarized concisely in his following quote, but also outlines the future role of statisticians and mathematicians as both collaborators and contributors in the changing face of science, "True systems biology requires mathematical modeling and simulations of dynamic networks. Although large-scale investigations will be an important contributor to systems biology, they will have to be combined with formal mathematical approaches to produce a true systems perspective. If all goes well, systems biology will lead to the discovery of novel, emergent properties of the molecules and interactions that drive network behavior as well as new higher-order principles of biology." While Jorgensen's vision for the future is an accurate forecast, at this point in time it is not a reality. To date, probably the most significant and influential contribution that is currently being made by Statistics relative to functional genomics, systems biology, and hence large-scale biology is the establishment and maintenance of statistical and mathematical standards and guidelines for the experimental design and analysis of biological studies (Nettleton, 2006). The involvement of formally trained quantitative (e.g., statisticians, mathematicians, and computer scientists) scientists in setting and maintaining obtainable scientific standards are invaluable toward improving peer-review, grant review, and educational programs.

3. FUNCTIONAL GENOMICS... WHO, WHAT, WHEN, WHERE & WHY

If functional genomics is viewed as a study of assigning biological meaning to genomic data (Steinmetz and Davis, 2004), then this biological meaning must include the 'who, what, when, where, and why' list of good investigative (i.e., scientific) reporting. In short, functional genomics aims to identify and locate genes that are expressed at a certain level under certain conditions at a certain time for the purpose of producing some potentially observable phenotypic result. Additional challenges include understanding the interaction (epistasis) of these genes, as well as the regulation and control (epigenetics/epigenomics) of these genes. Genomic data support each level of any functional genomic investigation with the most significant contribution probably being the composition of DNA itself (Miescher, 1871; Watson and Crick, 1953; Trifonov 2000). No one will argue that genetic mapping and sequencing (protein, RNA, DNA) in humans, animal, and plants are obvious successes of the 20th century, or that the magnitude of marker, sequence, genomic, and epigenomics data has already surpassed anything that early pioneers such as Sax (1923), Thoday (1961), and even R.A. Fisher could have imagined.

In crop plants, the identification of genomic regions or quantitative trait loci (QTL) as associated to important economic traits has met success, but has also been criticized as not providing the individual genes that are responsible for important traits (e.g., yield). Initially, there was great expectation placed on knowing genomic sequence information, either from model organisms, crop species, or the relationship between the two (comparative genomics), and that this alone would provide all the necessary information to build and design better crops. However, the "system" itself is not so simple, and determining the nature and function of genes has turned out to be much more complicated than initially thought. The majority of genes do not act alone. They perform much like a well tuned orchestra in concert. Some genes have solos that are supported by other genes; there are duets and interactions of genes; and other complex systematic performances at various stages of the concert (e.g., development, stress, etc.). Furthermore, changing the conductor (or regulator) of the orchestra might in fact change the performance(s) of the gene(s). The components and results of system-wide genome complexity have most certainly created dynamic interactions and associations that are not able to be drawn out by the current level of data, nor the statistical methods (e.g., QTL analysis or association mapping).

As technology pushes forward, we are able to couple genomic information with microarray technology to monitor changes in a gene's transcript (i.e., expression), gene methylation, and/or histone modifications (Lippman et al., 2004). It is exciting to have access to technology that provides data that were unimaginable even twenty years ago. However, now that we do have these technologies, and even more on the horizon, knowing how to both statistically design experiments and analyze these data has become both an interdisciplinary challenge and a bottleneck in science. Some of these data are the result of well designed experiments that are hypothesis driven; meaning there is a question to be tested and answered. Other experiments may not be so well designed, in that they are exploratory missions that are not addressing any questions; they are simply viewing the genome.

3.1. Bayesian or Fisherian Statistics?

The whole idea of applying statistical theory and methods to genetic, genomic, epigenomic data is to either ask questions of your data, or explore your data to look for similarities either with something you already know, or among the data that you have. Designing an experiment to answer a biological question basically provides data that, if analyzed properly, suggests whether the results that you are observing from your experiment are random events or events that are biologically determined, and in fact relevant to the biological phenomena being investigated. Specifically, if the experiment is only performed once (i.e., a single biological sample) it is impossible to separate the biological variation from experimental error/noise. Therefore, there is no way of really assessing whether it is the variation in the biological system (i.e., biological variation) that is creating the observed result, or the variation in the experimental system (i.e., technical variation). Performing the experiment more than once allows the variation in the system to be partitioned according to the sources of variation (i.e., biological and error) that are inherent in any investigation. Understanding the sources of variation well enough so that an experiment can be designed to provide informative data is in fact one of the grand statistical challenges that has been brought forth by functional genomics. In fact, some may feel that the greatest statistical advancement in functional genomics is experimental design.

The most commonly used and most often taught statistical approaches for partitioning variation and testing hypotheses are analysis of variance (ANOVA) and/or linear models (Searle, 1971). These approaches are known as Frequentist approaches, and those that practice these methods are often referred to as Fisherian or Frequentist (Efron, 1986). More recently, Bayesian methods (Bayes, 1763; Berger, 1985) have been employed for the statistical analysis of functional genomic data. Although the majority of the statistics community is quite happy to divide themselves as being either Fisherian or Bayesian, there are many statisticians working in statistical genetics/genomics or functional genomics that have benefited from both sides of this statistical dichotomy. Essentially, Bayesian analyses allow the parameters of the model to have their own distribution, while allowing prior information (as gained from previous experimentation, experience, etc.) to be incorporated into the statistical model. Rather than testing hypotheses, all results are accompanied by levels of certainty, or posterior probabilities that help the user in judging the worth of the analysis and results. Bayesian approaches to functional genomics are becoming more achievable simply because computing has improved so drastically. However, proper application of true Bayesian methods is greatly dependent on the use of informative priors, which in turn means that the user has to think clearly about the process that is defining the system. Designing an experiment so that prior information is incorporated and novel questions addressed remains an important issue in Bayesian experimental design. Although there has been a significant amount of work devoted to developing Bayesian methods, an equal effort in developing experimental designs within a Bayesian framework is lacking (Wu and Lin, 2006). The idea that information from previous studies might aid in supplying updated estimates and prior information to be used for both experimental design and the analysis of experimental data has been slow to catch on, and at times confusing. For example, Bayesian networks do not necessarily require Bayesian approaches. They can often be accomplished within a classical (frequentist) statistics setting (Kaski et al., 2005; Rusakov and Geiger, 2005) and have great potential to incorporate large amounts of data in a meaningful manner to provide statistical models that are capable of establishing both quantitative and qualitative causal relationships between numerous variables. Bayesian approaches, in general are powerful and well suited to take advantage of existing data and existing statistical methods. They exploit prior knowledge through novel applications (e.g., machine learning tools; Segal et al., 2005) that when coupled with optimal experimental design and modern statistical computing lend themselves well to powerful new functional genomic discoveries (e.g., QTL, differential expression, eQTL, differential methylation, etc.).

4. WHERE HAVE ALL THE QTL GONE?

The quantitative trait loci, or QTL, that have been mentioned previously are regions of a genome that are associated with a trait of interest. These traits are often referred to as complex traits in that many QTL and/or genes behave together with the environment within the biological system to influence the measurable/observable phenomena under study. Locating QTL relative to a genetic map of any major crop or experimental population has relied heavily on a variety of statistical approaches and software (Doerge et al., 1997; Doerge, 2002; Liang and Keleman, 2006; Tanksley, 1993) to demonstrate such association(s). However, only a few of these QTL are reproducible across environments, genotypes, or years, and have led to questions about the complexities of the system being studied. The reasons for the lack of consistent major successes in the implementation of applied QTL into major food crops are numerous and range anywhere from epistatic interactions among QTL, to varying genetic backgrounds, to knowledge of QTL being regulatory, to thoughts about epigenomic consequences.

4.1. Epistasis

Epistasis by definition is the interaction of (nonallelic) genes or QTL that result in a phenotypic change. These changes can cause the associated phenotype to no longer be observable, or they can enhance the phenotype well beyond the effects of the genes/QTL involved in the interaction. Epistasis is important in understanding the behavior of traits and diseases under complex control, and if properly incorporated into a statistical model will most likely play a large role in explaining complex regulatory networks.

Standard and accepted approaches for mapping QTL are single marker methods, interval mapping (Lander and Botstein, 1989; 1994), composite interval mapping (Zeng, 1993, 1994; Jansen, 1993), multiple QTL interval mapping (Jansen, 1993; Jansen and Stam, 1994), and multiple trait interval mapping (Kao and Zeng, 1999). Each of these methods uses an algorithmic approach to search for single QTL or sequential multiple QTL, in the absence of epistasis. Once the majority of QTL are identified, epistasis might be dealt with as a second level analysis by including the identified marker/QTL and their interactions into a new model. To date, epistasis is most commonly dealt with this way simply because the number of observations relative to the number of markers under consideration is so disproportioned that the degrees of freedom available become the limiting factor in any further analysis. As one might expect, epistatic QTL play an important role in the genetics and molecular dissection of complex traits. Because of this, there has been a surge of statistical advances and activity in dealing with epistasis. Sophisticated statistical methods have been developed to search the multidimensional (genome) space for associations. However, because of the complexity of the problem, a dimension reduction is typically necessary at the onset (the implications of which are significant on the end result). A common approach is to predefine the number of QTL so that the dimensionality of the problem is quickly maintained and pairwise interactions are easily introduced into the QTL model for the purpose of testing which model (by varying the interactions) best describes the experimental data at hand. With an eye on functional genomics, the idea of parameterizing a problem prior to actually knowing its complexity runs the risk of over simplifying the system and potentially missing important contributors to the process. To address this issue, multidimensional random search QTL (both Bayesian and frequentist) approaches (Broman and Speed, 2002; Carlborg et al., 2000; Carlborg et al., 2001; Carlborg, 2002; Nakamichi et al., 2001;) are showing promise for application to traditional QTL mapping, but more excitingly to functional genomic applications involving gene expression data.

Typically, multidimensional QTL searching methods produce a number of suggested models. To choose the best model, model selection criteria are employed. Many of the model selection criteria are based on choosing the model for which the likelihood of the data given the number of parameters in the model (i.e., dominated by the number of interactions in this case) is maximal. Since the dominating factor, when modelling epistasis, is the number of interactions (i.e., the dimension of the search), typically a penalty is subtracted from the likelihood as an adjustment for dimension (for a further discussion see Bogdan et al., 2004). How to choose this penalty, and thus the model best describing the complexity of the phenotype (i.e., epistasis) under study is an ongoing area of statistical research that will eventually play an important role in the molecular dissection of complex traits using functional genomics.

4.2. Expression Mapping (e-mapping)

Although the impact of functional genomics on biology is still in its infancy, its potential has recently stimulated an area of research that is now referred to as genetical genomics (Jansen and Nap, 2001). Genetical genomics merges the theoretical aspects of quantitative genetics with the power of functional genomics. Specifically, gene expression data measured on a segregating population are understood to be quantitative traits, or expression traits (e-traits), and provide the data for mapping expression QTL (eQTL) onto a genetic map. The implications of using functional genomic technologies, namely microarrays, in a QTL setting are long reaching since microarray technologies are now being used to both genotype (expression markers) and phenotype (expression traits) individuals.

4.3. Expression Markers (e-markers)

It was Nap and Jansen (2001) who first outlined the use of microarray data for both genotyping and phenotyping individuals in a segregating population. They were careful to distinguish between expression profiles of genes by classifying the profile as either quantitative or qualitative. If quantitative (continuous) in its distribution then the gene can be considered an expression trait whose variation is of interest, and can be studied using existing QTL methodology (discussed next). If, in fact the distribution of a gene's profile is qualitative in nature, then the alleles at that gene segregate to provide information for calling the gene's profile a genetic expression marker (see Figure 2 in Nap and Jansen, 2001) and that particular gene's features can be used to genotype individuals in a mapping population. While Nap and Jansen (2001) were careful to distinguish between marker and quantitative trait segregation patterns as supplied from microarray technology, they neglected to discuss the potential pitfalls of gaining multiple correlated data (genotype and phenotype) from the same sample (i.e., array). Specifically, since the same sample/tissue is used for differentiating phenotype and genotype across tens of thousands of features there are a number of underlying technical and statistical issues that may give rise to false (eQTL) associations. The issue of obtaining both genotype and phenotype data from the same sample (array) has yet to be addressed, and remains a concern for the future of eQTL mapping.

The initial uses of microarrray technologies for applications other than expression profiling were based on oligonucleotide arrays to identify (DNA) sequence polymorphisms (Borevitz et al., 2003; Hazen and Kay, 2003; Winzeler et al., 1998). Single feature polymorphisms (SFPs; Borevitz et al., 2003) have quickly become a high-throughput genotyping solution, but are limited in their application in that they rely on short oligo- probes that have potential for reduced hybridization to DNA or cRNA samples when there is in fact a sequence polymorphism. West et al. (2005) circumvented these issues by exploiting the actual transcript level differences in parents of a segregating population to define gene expression markers (GEMs) which are identifiable in the segregating population. While their demonstration relied on Affymetrix technology, GEMs are achievable from any type of DNA microarray. As a further extension of SFP markers (Marilyn West and Dina St. Clair; personal communication) relied on the gain in information as achieved from 148 biologically replicated recombinant inbred lines (RILs) to identify single feature polymorphisms (SFPs) from individual Affymetrix features.

Microarray technologies are providing fast, efficient and relatively cheap means for gaining thousands of genetic markers. It is important to remember that when employing standard QTL mapping methods (that rely on a genetic map) more markers do not necessarily mean more information or more resolution for QTL location. The statistical models that are employed for locating QTL rely on the estimated genetic distance, or recombination, between consecutive (ordered) markers on the genetic map. In the majority of crops, genetic map order is resolved by observed recombination from individuals in a mapping population. Therefore, a bounty of genetic markers is not informative unless enough individuals have been assessed for observable crossover events. Even if a previously known genetic map is available (i.e., map order known), unless there are observable recombinants between genetic markers in the experimental population being studied, no information will be available for estimating genetic distance let alone locating QTL. Equivalently, if two genetic markers are scored identically across all individuals in the mapping population, there is no information available to differentiate these markers. Therefore, although microarray generated e-markers may be plentiful, they may not be polymorphic, or there may be so few observable recombinants that the statistical power to locate QTL is effectively reduced.

4.4. Expression QTL (eQTL)

Quantitative traits as gained from microarray technology, and referred to as e-traits, exhibit variation when measured across a group of individuals or mapping population. Understanding the variation of an individual gene expression observation from microarray technology has potential to allow the molecular dissection of the gene's determinants, or an understanding of the expression level polymorphisms (ELPs; Doerge, 2002) for that gene. How to molecularly dissect the variation that is observed in a gene's expression level has been the intense focus of many interdisciplinary investigations (Borevitz et al., 2003; Brem and Kruglyak, 2005; Doerge, 2002; Jin et al., 2001; Kendziorski and Wang 2006; Kim et al., 2005; Nap and Jansen, 2001; Potokina et al., 2004; Schadt et al., 2003; Steinmetz et al., 2002; Wayne and McIntyre, 2002). Recently in yeast (Yvert et al., 2003), Arabidopsis (Singer et al., 2006), and mouse (Carlborg, 2005), all of which are model organisms, genomic and genetic complexity has been exposed by evaluating e-trait variation. Because of the cost of these experiments (i.e., ideally, one would biologically replicate whole genome microarrays across potentially hundreds of individuals) there have been relative few experiments with adequate sample sizes reported. However for those studies that have been reported, there is a surprising complexity to even the simplest of organisms' genomes.

Thus far, the experimental approaches (that are all supported by statistical analysis) that have been applied to expression or e-trait data can be classified into three experimentally different approaches that result in: non-causal eQTL, quantitative genomics eQTL, and causal eQTL. The first and most straightforward approach (Wayne and McIntyre, 2002) relies on a traditional QTL analysis to first locate QTL. After deletion mapping to resolve QTL location, genes underlying the mapped QTL are surveyed for differential expression between the parental lines using microarray technology. While the Wayne and McIntyre approach is not an application of functional genomics in that biological meaning cannot be assigned to the differentially expressed genes using this approach, it is a fast efficient way to isolate and identify genes that are within QTL regions that are associated with a trait of interest. Whereas Wayne and McIntyre (2002) used the results from QTL mapping as a guide for candidate gene discovery, others have used gene expression variation as an approach for making sense of the complex genetics.

In the absence of QTL results, or a phenotype to guide the investigation, Brem and Kruglyak (2005) employed whole genome yeast (*Saccharomyces cerevisiae*) microarrays to study the genetics and inheritance of e-trait variation. Using a whole genome microarray on each of 112 individuals in a mapping population, e-trait data provided variation for mapping eQTL. Since these associations can be *cis-* acting loci or *trans*-acting loci, results are thought to supply an idea of the genetic complexity across what is commonly referred to as the genetic or genomic "landscape" of the organism. Even though 5700 gene expression traits were mapped, 40% of the highly heritable e-traits showed no QTL association. Of the eQTL that were detected very few (only 3%) of them followed a single locus mode of inheritance. In fact, the majority of e-traits required more than five loci in an additive model to begin to understand the complexity of the variation. While genetic studies of e-traits may reveal the overarching generalities of quantitative genetics, the real question becomes, how much measured transcript variation is actual allelic variation, and is an understanding of the complexity of the genome landscape achieved from knowing this? Does basal genetic variation play an important role in functional consequence (personal communication Lauren McIntyre)? Certainly, millions of years of evolution imply that genetic variation does indeed play a vital role in functional consequences, but can the result of millions of years of evolution be summarized using these approaches? Taking a sample of normal cells and assessing the genetic variation through *cis*- and *trans*- association mapping does not necessarily imply that all genetic variation works through the same (evolutionary) process.

In a study that reaches beyond basic genetic variation of yeast, Steinmetz et al. (2002) employed the complexity of a high-temperature growth (Htg) quantitative trait or phenotype to study the genomic landscape of quantitative variation as associated to the quantitative trait (Htg). Since the trait of interest is known, the idea is to discover genes responsible for its variation in efforts to molecularly dissect a quantitative trait. Using a standard QTL experimental design, strains of yeast both exhibiting and not exhibiting the Htg trait were crossed. The resulting offspring were heterotic, and had alleles that contributed to the presence of the Htg phenotype. By comparing (i.e., those individuals with and without the Htg trait) the progeny using whole genome microarrays, the recombinational breakpoints were thought to have potential to identify genes underlying QTL. The motivation was to identify a locus (or loci) linked to the Htg trait of interest. In fact, three eQTL both in *cis*- and *trans*- were identified as linked, but failed to provide conclusive evidence as to their respective roles in the quantitative variation of the Htg trait.

In an ongoing study that was initially described in Kliebenstein et al. (2006) the benefits of understanding genetic variation relative to functional genomic consequences is enabled by the experimental design of the study, the statistical analysis, and the (Affymetrix) microarray technology. As pointed out by Steinmetz et al. (2002) single marker methods (i.e., "single-gene-per-locus") for statistical analysis most likely are not sufficient to dissect loci that are linked, interacting (i.e., epistatic), or of small effect. These authors hint at the need for narrowing in on a genomic interval while controlling for neighboring linked loci of potentially small and interacting effects. Toward this end, some investigators have turned to standard QTL mapping approaches (Kendziorski and Wang, 2006; Schadt et al., 2003; Schadt et al., 2005) to locate QTL relative to intervals of markers. The two most popular approaches include interval mapping (Lander and Botstein, 1989) and composite interval mapping (Zeng, 1993; Zeng, 1994), with the latter providing the ability to use cofactors beyond a specified window on either side of the testing position to control for additionally linked loci. In the eQTL study that was first reported by Kliebenstein et al. (2006a,b), the experimental design consisted of two inbred parental genotypes of Arabidopsis (Bayreuth-0 and Shahdara) that gave rise to a recombinant inbred population consisting of 211 individuals. The intriguing part about this study relative to understanding whether genetic variation has a functional (genomic) consequence is that by experimental design both a control condition and a treatment condition (salicylic acid; SA) are examined in the parental genotypes, as well as across the 211 biologically replicated RI lines. Essentially, the control condition allows for the same sort of study as the Brem and Kruglyak (2005) study (i.e., natural variation), but the SA treatment is a stress that is related to a well studied defense pathway (Wang et al. 2005) and allows for the examination of e-trait variation in response to a treatment. The obvious comparison between results of two independent eQTL analyses will yield interesting summaries of similarities of genomic variation, and may even address the question whether natural variation is the same as between control and treatment conditions. However, complete statistical analysis of both the control and SA treatment response in a novel statistical application will lend itself well to novel functional genomic discoveries, and may address questions such as how much allelic variation is normal? In this application the RIL experimental (mating) design greatly benefits the functional genomics investigation of complex traits in that individuals of an RIL population have no genetic variation within a line, but are genetically different between lines. The within line replication provides power to estimate non-genetic sources of variation, while the between line replication provides an assessment of allelic variation for estimating genetic sources of variation. For this experiment, the experimental design has greatly advanced and enabled this particular approach by providing the greatest opportunity to measure quantitative variation at the transcript level. In work by Kim et al. (2005) the non-genetic and genetic components of gene expression for this same study are partitioned for the purpose of providing putative regulatory networks that in turn may provide evidence for the genetic basis of complex traits. A multivariate statistical (linear) model (Searle, 1971) was employed. It acknowledged both the genetic (e.g., allelic differences at *cis*- and *trans*- loci) and non-genetic components (i.e., control/treatment), and also included important interactions between the genetic and non-genetic components (e.g., genotype and treatment), as well as technical sources of variation (e.g., array). The model proposed by Kim et al. (2005) acknowledges that the control condition and SA treatment are different "environments" in which the quantitative e-traits are repeatedly (twice) measured. Thus, the multiple observed gene expression data provide evaluations of e-traits across varying environments which in turn lends itself well to a multiple trait analysis where the traits share some correlation structure. Initially, Kim et al. (2005) relied on the multiple trait QTL analysis feature of QTL-Cartographer (Basten et al., 1995), but realized the limitations of such an approach, and thus proposed a novel statistical model that accounts for both the genetic and genomic components, as well as the technical components of the problem. Preliminary results as reported in Kim (2007) indicate greater statistical power when replicate measurements of gene expression are part of the experimental design, and when technical variation is acknowledged from the microarray component of the experiment.

Genetic variation is common, complex, and confusing. As much as we would like to believe that genes act in predictable networks, it is becoming more and more evident that discovery of these network is turning out to be equally complex and confusing. The QTL and/or genes that have been found to be associated with phenotypes may turn out to be associated to genes or e-QTL in a known network or not. What was once thought to be a simple linear process is in fact turning out to be a multidimensional, multilayered, environmentally sensitive system. Understanding the regulation of genes that are cooperating to achieve a biological outcome is quickly becoming the focus of future statistical advances in functional genomics.

5. STATISTICS IN THE FUTURE OF FUNCTIONAL GENOMICS

The future of statistics is bright when considering the impact that it has already had on functional genomics (e.g., gene cluster exploration, testing of differential expression, mapping expression phenotypes or e-traits). Two different, but maybe not independent, avenues will most likely direct future statistical advances that are made in functional genomics. The first area will use existing data (e.g., databases, literature, etc.) and results for the purpose of gaining greater information through statistical data mining and modelling. The second area will depend on new technologies that are giving rise to discoveries in epigenomics (e.g., methylation, histone modification, etc.). The lack of independence between these two areas lies in the fact that it may turn out that epigenomic data, results, and discovery will unify broad concepts for understanding quantitative genetics or quantitative biology in dynamic biological systems.

5.1. Meta-analysis

As a term for analysis, "meta-analysis" has been incorrectly used in the biological literature to mean combining data sets that will be either compared or re-analyzed. Typically, these re-analyses are void of statistical methods, and fail to address many of the important statistical issues that arise when combining data, especially in genomics (e.g., differences in technology, differences between laboratories, differences in replication number). "Meta-analysis" is an area of study in the discipline of Statistics that has existed for years. It was initially made popular in medical and social sciences (Berlin and Colditz, 1999; Glass, 1976; Kassirer, 1992; Segerstrom and Miller, 2004), and refers to combining results (e.g., p-values) from independent analyses of different experimental data sets. Historically, the studies and results that are utilized in a meta-analysis investigate the same basic phenomena, and rely on a huge wealth of, typically published, data results that can be combined using a range of both frequentist and Bayesian approaches. Meta-analysis is independent of the experimental design that generated the data, and furthermore is independent of the statistical approach that provided the results. As outlined by Stevens (2005) the motivation for using a meta-analysis as applied to functional genomic data are three-fold: 1) First, many microarray platforms are now distributed in a standardized manner such that independent laboratories often use the same technology and array design to study the same question. Any differences that do exist between laboratories and environment can be incorporated into the statistical model. As an aside, combining results from different technologies is an open statistical problem that deserves further attention. Second, while obtaining raw data may be easier than in the past, many of the data repositories do not require that raw data be deposited. Therefore, if these data are obtained they may not be the original (not normalized, not transformed) data and thus may not be comparable across studies. Lastly, while a meta-analysis may be equally informative when compared to the combined raw data analysis, often it may be more informative because there is more information to incorporate relative to the specific laboratories contributing the data (for specific statistical details about meta-analysis in functional genomics see Stevens and Doerge, 2005a,b; Stevens, 2005).

5.2. Epigenetics/Epigenomics

Epigenetics was first defined by Waddington (1942) as "the interactions of genes with their environment, which bring the phenotype into being" (Qiu, 2006), and may turn out to be one of the missing links for the explanation of complex traits relative to functional genomics. An updated view of epigenetic research that is currently considered as the study of heritable changes in genome function that occur in the absence of changes to the DNA sequence itself is referred to as epigenomics, the second-code of instruction that affects gene activity in the absence of altered DNA sequence. With regard to functional genomics, epigenomics may indeed supply an explanation for the additional level of variation and regulation that has not been fully appreciated or understood up to this point (Richards, 2006). One of the basic ideas underlying the concept of epigenomic regulation is that chromatin, which is contained in the nuclei of (most) eukaryotic cells, when modified as a response to environment and gene expression might be responsible for heritable changes that are not observable alterations at the DNA sequence level. Since chromatin is composed of DNA that is tightly wound with both histone and non-histone proteins, any changes (e.g., histone modification, DNA methylation, etc.) in chromatin as a result of interacting with the environment or the genome itself (e.g., gene expression) will leave the DNA sequence unchanged, but may have a significant impact on the organism itself.

A great deal of work (Petronis, 2006) has been done in human epigenomics, where one individual of (genetically) identical twins develops a disease (e.g., diabetes, depression, schizophrenia) while the companion twin does not. Clearly, these individuals share the exact same DNA as their genetic make-up, yet the occurrence of the genetically based disease differentiates them. In plants, epigenomic changes have certainly been studied (see Richards, 2006; and references therein for review), but are yet to be connected quantitatively to complex traits. Interestingly, in the study (Kliebenstein et al., 2006) that was previously mentioned, the recombinant inbred population supplies genetically identical individuals within a line, yet genetically different individuals between lines. Since this study was biologically replicated within lines, there is information to further dissect epigenomically observed variation both within and among recombinant inbred lines, if further (epigenomic) data are obtained. This would be especially interesting for this particular study since there are two treatments (control and salicylic acid) that have potential to supply an avenue for testing epigenetic changes that have arisen from environmental changes.

Epigenomic changes have in fact been studied and tested in Arabidopsis using DNA microarray technology (Lippman et al., 2004; Martienssen et al., 2005; Zhang et al., 2006). In the work by Lippman et al. (2004), heterochromatic changes were investigated analytically using statistical methods based on linear models that tested hypotheses of both methylation changes and histone modification changes between an Arabidopsis (WT) genotype that was highly methylated in its heterochromatic structure and a genotype that had mutations in its chromatin remodeling gene (ddm1; Decrease in DNA Methylation 1) that left it having relatively little methylation of its chromatin structure. Although there are similarities (e.g., array, dye, treatment effects) in the statistical models that are used to test differential (gene) expression changes, testing for differential methylation depends on an appreciation of the level of methylation against which one is testing, and is a good example of when testing the incorrect hypothesis will lead to the wrong conclusion.

In the same study by Lippman et al. (2004) the same array design was employed to study not only differential methylation and differential histone modification, but also gene expression. By hybridizing both genomic DNA and cDNA to the same array design (for details see Lippman et al., 2004) methylation, histone modification, and expression data were collected on the same genotypes, thus allowing for a statistical meta-analysis (e.g., vote counting) that can supply global statistically based statements pertaining to epigenetic changes (Yoo and Doerge; unpublished). Data such as these (i.e., same array, different genomics material, highly related questions), as well as other data and experimental design structures (e.g., time series) are motivating new statistical and mathematical models that have potential to combine data, as well as prior knowledge (i.e., Bayesian statistics) while preserving its meaning and extending its interpretation beyond any of the individual, independent analyses. How to perform these analyses so they are statistically responsible and biologically meaningful is a current area of statistical research that is gaining much attention, but has yet to see the type of attention from the statistics community that microarrays did when they first arrived on the scene.

When considered in the content of molecularly dissecting complex traits, epigenomic data combined with quantitative genomic data (e.g., QTL, eQTL) have huge potential to bridge the gap in our current understanding of functional genomics, and in doing so may allow for a more complete understanding of genes, their coordination, response, and their control. With more than 50,000 plant species that are edible, and less than 300 of them actually domesticated, there is an abundance of genetic, epigenetic, and unacknowledged variation that is shared across these species. Dissecting this variation in model organisms such as Arabidopsis will not only enable functional genomics, but also the "system-wide genomics" of the barely fifteen species (e.g., rice, maize, wheat, cotton, soybean, etc.) that make up the majority of the crop plants worldwide. While comparative/translational genomics has not been discussed in detail here, the mechanisms that control and regulate "biological" variation of these systems are most certainly shared. Bringing together information from seemingly remote connections of science is central to advancing our 'omic-assisted understanding of the nature and function of genes, their products, their interactions, and their regulation. Toward this end, Statistics has played a specific and vital role in agriculture (and in all sciences) in the pre-genomic era, has enjoyed a renewed appreciation (and criticism) in the genomic era, and will undoubtedly gain continued recognition and appreciation as a prominent player in this new era of grand-scale data driven biology.

ACKNOWLEDGEMENTS

I thank Drs. Bill Beavis, Lauren McIntyre, Rob Martienssen, and Milos Tandurzic for many conversations about the role of Statistics in genomics. Lastly, I thank my research group who are all much brighter and talented than I can ever imagine being. This work was supported by (NSF: 0501712-DBI and NIH: R01 MHO75041-01A1) grants to RWD.

REFERENCES

Basten C, Weir BS, Zeng Z-B (1995–2004) QTL cartographer. Department of statistics, North carolina State University, Raleigh, NC

Bayes T (1763) An essay towards solving a problem in the doctrine of chances. Philos T Roy Soc 53:370–418

Berger JO (1985) Statistcial decision theory and bayesian analysis. 2nd edn. Springer-Verlag, New York

Berlin JA Colditz GA (1999) The role of meta-analysis in the regulatory process for foods, drugs, and devices. J Am Med Assoc 281:830–834

Bogdan M, Ghosh JK, Doerge RW (2004) Modifying the schwarz bayesian information criterion to locate multiple interacting quantitative trait loci. Genetics 167:989–999.

Borevitz JO, Liang D, Plouffe D, Chang H-S, Zhu T, Weigel D, Berry CC, Winzeler E, Chory J (2003) Large-scale identification of single-feature polymorphisms in complex genomes. Genome Res 13: 513–523

Brem RB, Kruglyak L (2005) The landscape of genetic complexity across 5,700 gene expression traits in yeast. Proc Nat Acad Sci USA 102:1572–1577

Broman KW, Speed TP (2002) A model selection approach for the identification of quantitative trait loci in experimental crosses. JR Stat Soc B 64:641–656

Carlborg Ö (2002) New methods for mapping quantitative trait loci. PhD Thesis, Acta Universitatis Agriculturae Sueciae. Veterinaria 121. Swedish University of Agricultural Sciences, Uppsala, Sweden

Carlborg Ö, Andersson L, Kinghorn B (2000) The use of a genetic algorithm for simultaneous mapping of multiple interacting quantitative trait loci. Genetics 155:2003–2010

Carlborg Ö, Andersson-Eklund L, Andersson L (2001) Parallel computing in interval mapping of quantitative trait loci. J Hered 92:449–451

Carlborg Ö, De Koning DJ, Manly KF, Chesler E, Williams RW, Haley CS (2005) Methodological aspects of the genetic dissection of gene expression. 21: 2383–2393

Doerge RW (2002) Mapping and analysis of quantitative trait loci in experimental populations. Nat Rev Genet 3:43–52

Doerge RW, Zeng Z-B, Weir BS (1997) Statistical issues in the search for genes affecting quantitative traits in experimental populations. Stat Sci 12:195–219

Efron B (1986) What Isn't everyone a bayesian. Am Stat 401:1–5

Glass GV (1976) Primary, secondary, and meta-analysis of research. Educ Res 5(10):3–8

Hazen SP, Kay SA (2003) Gene arrays are not just for measuring gene expression. Trends Plant Sci 8: 413–416

Jansen RC (1993) Interval mapping of multiple quantitative trait loci. Genetics 135:205–211

Jansen RC, Nap JP (2001) Genetical genomics: the added value from segregation. Trends Genet 17:388–391

Jansen RC, Stam P (1994) High resolution of quantitative traits into multiple loci via interval mapping. Genetics 136:1447–1455

Jin W, Riley RM, Wolfinger RD, White KP, Passador-Gurgel G, Gibson G (2001) The contributions of sex, genotype and age to transcriptional variance in drosophila melanogaster. Nat Genet 29:389–395

Jorgensen R (2006) Large-scale biology. Plant Cell 18: 2095–2096

Kao CH, Zeng Z-B, Teasdale RD (1999) Multiple interval mapping for quantitative trait loci. Genetics 152:1203–1216

Kaski S, Nikkila J, Sinkkonen J, Lathi L, Knuttila JEA, Roos C (2005) Associative clustering for exploring dependencies between functional genomic data sets. IEEE/ACM T Comput Bi 2: 203–216

Kassirer JP (1992) Clinical trials and meta-analysis. What do they do for us? New Engl J Med 327–332

Kendziorski C, Wang P (2006) A review of statistical methods for expression quantitative trait loci mapping. Mamm Genome 17:509–517.

Kim, K (2007) Statistical issues in mapping genetic determinants of expression level polymorphisms. PhD Dissertation. Department of Statistics, Purdue University. West Lafayette, IN USA

Kim K, West MAL, Michelmore RW, Clair DAS, Doerge RW (2005) Old methods for new ideas: genetic dissection of the determinants of gene expression levels. In: Gustafson JP, Shoemaker R, Snape JW (eds) Genome exploitation: data mining the genome. The 23rd volume in the stadler symposia. Springer, New York, pp 89–105

Kliebenstein DJ, West MAL, van Leeuwen H, Loudet O, Doerge RW, St. Clair DA (2006) Identification of QTLs controlling gene expression networks defined a priori. BMC Bioinformatics 7:308

Kliebenstein DJ, West MAL, van Leeuwen H, Kim K, Doerge, RW, Michelmore RW, St. Clair DA (2006) Genomic survey of gene expression diversity in *Arabidopsis thaliana*. Genetics 172: 1179–1189

Lander ES Botstein D (1989) Mapping mendelian factors underlying quantitative traits using RFLP linkage maps. Genetics 121:185–199 (1989); erratum (1994) 136:705

Liang Y, Kelemen A (2006) Associating phenotypes with molecular events: recent statistical advances and challenges underpinning microarray experiments. Funct Integr Genomics 6:1–13

Lippman Z, Gendrel A-V, Black M, Vaughn M, Dedhia N, McCombie WR, Lavine K, Mittal V, May B, Kasschau K, Carrington JC, Doerge RW, Colot V, Martienssen R (2004) Transposable elements mediate heterochromatin and epigenetic control. Nature 430:471–476

Martienssen RA, Doerge RW, Colot V (2005) Epigenomic mapping in *Arabidopsis* using tiling microarrays. Chromosome Res 13:299–308

Miescher F (1871) Ueber die chemische Zusammensetzung der Eiterzellen. Med Chem Unt 4:441–460

Nakamighi R, Ukai Y, Kishino H (2001) Detection of closely linked multiple quantitative trait loci using genetic algorithm. Genetics 158:465–475

Nettleton D (2006) A discussion of statistical methods for design and analysis of microarray experiments for plant scientists. Plant Cell 18:2112–2121

Petronis A (2006) Epigenetics and twins: three variations on the theme. Trends Genet 22:347–350

Potokina E, Caspers M, Pradad M, Kota R, Zhang H, Sreenivasulu N, Wang M, Graner A (2004) Functional association between malting quality trait components and cDNA array based expression patterns in barley (*Hordeum vulgare* L.). Mol Breeding 14:153–170

Qiu J (2006) Unfinished symphony. Nature 441:143–145

Richards, EJ (2006) Inherited epigenetic variation – revisiting soft inheritance. Nat Genet Rev 7:395–401

Rusakov D, Geiger D (2005) Asymptotic model selection for naive bayesian networks. J Mach Learn Res 6:1–35

Sax K (1923) The association of size differences with seed-coat pattern and pigmentation in *Phaseolus vulgaris*. Genetics 8:552–560

Schadt EE, Monks SA, Drake TA, Lusis AJ, Che N, Collnayo V, Ruff TG, Milligan SB, Lamb JR, Cavet G, Linsley PS, Mao M, Stoughton RB, Friend SH (2003) Genetics of gene expression surveyed in maize, mouse, and man. Nature 422:297–302

Schadt EE, Lamb J, Yang X, Zhu J, Edwards S, GuhaThakurta D, Sieberts SK, Monks S, Reitman M, Zhang C, Lum PY, Leonardson A, Thieringer R, Metzger JM, Yang L, Castle J, Zhu H, Kash SF, Drake TA, Sachs A, Lusis AJ (2005) An integrative genomics approach to infer causal associations between gene expression and disease. Nat Genet 37:710–717

Searle S (1971) Linear models. John Wiley & Sons, New York

Segal E, Pe'er D, Regev A, Koler D, Friedman N (2005) Learning module networks. J Mach Learn Res 6:557–588

Segerstrom SC, Miller GE (2004) Psychological stress and the human immune system: a meta-analytic study of 30 years of inquiry. Psychol Bull 130:601–630

Singer T, Fan Y., Chang H-SC, Zhu T, Hazen SP, Briggs SP (2006) A high-resolution map of Arabidopsis recombinant inbred lines by whole-genome exon array hybridization. PLoS Genet 2:1–10

Steinmetz LM, Davis RW (2004) Maximizing the potential of functional genomics. Nat Genet Rev 5:190–201

Steinmetz LM, Sinha H, Richards DR, Spiegelman JI, Oefner PJ, Mc-Cusker JH, Davis RW (2002) Dissecting the architecture of a quantitative trait locus in yeast. Nature 416:326–330

Stevens JR (2005) Meta-analytic approaches for microarray data. PhD dissertation. Department of Statistics, Purdue University, West Lafayette, IN USA

Stevens JR, Doerge RW (2005a) Combining affymetrix microarray results. BMC Bioinformat 6:57

Stevens J, Doerge RW (2005b) Meta-analysis combines affymetrix microarray results across laboratories. Compar Funct Genom 6:116–122

Tanksley SD (1993) Mapping polygenes. Annu Rev Genet 27:205–233

Thoday JM (1961) Location of polygenes. Nature191:368–370

Trifonov EN (2000) Earliest pages of bioinformatics. Bioinformatics 16:5–9

Waddington C (1942) The epigenotype. Endeavor 1: 18–20

Wang D, Weaver ND, Kesarwani M, Dong X (2005) Induction of protein secretory pathway is required for systemic acquired resistance. Science 308:1036–1040

Watson JD, Crick FHC (1953) Molecular structure of nucleic acids. A structure for deoxyribose nucleic acid. Nature 171:737–738

Wayne ML, McIntyre LM (2002) Combining mapping and arraying: an approach to candidate gene identi_cation. Proc Nat Acad Sci USA 99:14903–14906

West MAL, van Leeuwen H, Kozik A, Kliebenstein DJ, Doerge RW, St. Clair DA, Michelmore RW (2006) High-density haplotyping with microarray-based expression and single feature polymorphism markers in *Arabidopsis*. Genome Res 16:787–795

Winzeler EA, Richards DR, Conway AR, Goldstein AL, Kalman S, McCullough MJ, McCusker JH, Stevens DA, Wodicka L, Lockhart DJ et al (1998) Direct allelic variation scanning of the yeast genome. Science 281:1194–1197

Wu R, Lin M (2006) Functional mapping – how to map and study the genetic architecture of dynamic complex traits. Nat Genet Rev 7:229–237

Yvert G, Brem RB, Whittle J, Akey JM, Foss E, Smith EN, Mackelprang R, Kruglyak L (2003) Trans-acting regulatory variation in Saccharomyces cerevisiae and the role of transcription factors. Nat Genet 35:57–64

Zeng Z-B (1993) Theoretical basis of precision mapping of quantitative trait loci. Proc Natl Acad Sci USA 90:10972–10976

Zeng Z-B (1994) Precision mapping of quantitative trait loci. Genetics 136:1457–1468

Zhang H, Yazaki J, Sundaresan A, Cokus S, Chan S, Chen H, Henderson IR, Shinn P, Pellegrini M, Jacobsen SE, Ecker JR (2006) Genome-wide high-resolution mapping and functional analysis of DNA methylation in *Arabidopsis*. Cell (Resource) 126:1–13

CHAPTER 15

TILLING AND ECOTILLING FOR CROP IMPROVEMENT

BRADLEY J. TILL[1], LUCA COMAI[2,*] AND STEVEN HENIKOFF[1,3]

[1]*Basic Sciences Division, Fred Hutchinson Cancer Research Center, Seattle, Washington 98109, USA*
[2]*University of California Davis Genome Center, Davis, California 95616, USA*
[3]*Howard Hughes Medical Institute*

Abstract: The modern crop scientist has a large amount of available nucleotide sequence information to identify genes of potential agronomic importance. Using reverse genetic approaches, specific genes can be disrupted, and hypotheses regarding gene function directly tested *in vivo*. Although a number of reverse genetic methods have been introduced, many are limited in application because they are organism-specific, expensive, transgenic or only transiently disrupt gene function. However, traditional mutagenesis using chemical mutagens has been widely used as a forward genetics strategy to create many new crop plant varieties at relatively low cost. Mutagens such as ethyl methanesulphonate (EMS), cause stable point mutations and thus produce an allelic series of truncation and missense changes that can provide a range of phenotypes. TILLING (Targeting Induced Local Lesions IN Genomes) uses traditional mutagenesis and SNP discovery methods for a reverse genetic strategy that is high in throughput, low in cost, and applicable to most organisms. Over the past six years, TILLING has moved from proof-of-concept to production with the establishment of publicly available services for *Arabidopsis*, maize, lotus, and barley. Pilot-scale projects have been completed on several other plant species, including wheat. The protocols developed for TILLING have been adapted for the discovery of natural nucleotide diversity, a method termed EcoTILLING. Like TILLING, EcoTILLING is general and has been applied to plants as diverse as *Arabidopsis* and poplar. We review here current TILLING and EcoTILLING technologies and discuss the progress that has been made in applying these methods to many different plant species.

1. INTRODUCTION

Nucleotide sequence variation is a major determinant of heritable phenotypic difference and has been exploited by humans for crop improvement since the dawn of domestication. Variation can either be natural, from divergent populations, or

*Corresponding Author: comai@u.washington.edu

R.K. Varshney and R. Tuberosa (eds.), Genomics-Assisted Crop Improvement: Vol. 1: Genomics Approaches and Platforms, 333–349.

induced through treatment with mutagens. An important goal of modern crop science is to use nucleotide sequence variation to improve crops. This goal is furthered by the accumulation of large-scale sequence data. EST sequencing projects have been completed or are underway for many crop plants, and large-scale genome sequencing projects are now complete for a few plants, with more in progress or planned (http://www.ncbi.nlm.nih.gov/entrez/query.fcgi?db=genomeprj). For the crop scientist, a major challenge will be to use these sequence resources to create improved varieties that meet the demands of a growing population and a changing climate. The availability of extensive sequence resources fuels the demand for technologies that can use sequence to probe gene function. Tools that are applicable to many species can have an especially broad impact on plant science.

The generation of mutations in specific genes that can then be assayed for phenotypes (reverse genetics) is a powerful strategy for elucidating gene function and for creating new varieties. Many reverse genetic approaches have been developed for plants (An et al. 2005, Burch-Smith et al. 2004, Kusaba 2004, Henikoff and Comai 2003). Some involve targeting genes one at a time, such as transformation with hairpin constructs for RNAi-mediated knock-down, which can be used to dominantly knock down multiple homologous genes with a single construct. Others involve creating a population of mutagenized individuals that can be screened for generally recessive DNA lesions, including insertions, deletions and point mutations.

T-DNAs have been a very successful tool for insertional mutagenesis in *Arabidopsis*, and more recently for rice (http://signal.salk.edu/cgi-bin/tdnaexpress and http://signal.salk.edu/cgi-bin/RiceGE). However, the low copy number of T-DNAs means that large populations are required to provide a high probability of finding an insertion in a specific gene. Probabilities reduce as gene sizes decrease, and even with a population of 360,000, there are many *Arabidopsis* genes smaller than 1 kb for which no T-DNA insert has been isolated (http://signal.salk.edu/database/T-DNA/). When moving from basic research to crop breeding, transgenic approaches may prove to be undesirable because of regulatory costs and the continuing GMO debate. Non-transgenic reverse genetic approaches include the use of endogenous transposons as a method of insertional mutagenesis (McCarty et al. 2005, May et al. 2003, Hirochika et al. 2004). As with T-DNAs, insertions are identified via PCR screening using one element-specific primer and one gene-specific primer. Transposons have been utilized in organisms such as maize and rice, but for many plants, the development of a large library of plants with transposon insertions may be impractical due to the lack of suitable transposons, or methods to activate transposon movement. For example, the Tos17 transposon of rice can be used for insertional mutagenesis (Hirochika et al. 2004), but activating the transposon requires passage through tissue culture, making the process of generating a large library challenging.

Genomic deletions have also been exploited for reverse genetics. This method has the advantage that it can be used to target blocks of tandemly repeated genes for excision. To discover a deletion, PCR is performed with primers that hybridize to the genome. Amplification of DNA from plants with deletions between the primer binding sites will produce an amplicon of lower molecular weight than a plant with no deletion. Fast neutron mutagenesis has been used to generate deletions in both *Arabidopsis* and rice that could be recovered with this PCR screening method (Li and Zhang, 2002). A strength of the method is that the increased efficiency of amplifying a smaller fragment allows for detection in pools of up to ~1000 plants. Furthermore, deletions can be larger than a single gene, and the method is suitable to knock out tandemly repeated genes. However, the spectrum of deletion sizes caused by fast neutron mutagenesis is not well characterized, and it is likely that deletions by fast neutron mutagenesis will not be tolerated at a high density due to the deleterious nature of large deletions. Therefore, like T-DNAs, the population size required to ensure a deletion in a specific target gene is likely to be quite large.

TILLING (Targeting Induced Local Lesions IN Genomes) is a non-transgenic reverse genetic technique that is suitable for most plants (McCallum et al. 2000a). For TILLING, mutations are created by treatment with the same chemical mutagens that have been successfully employed in mutation breeding programs for decades. By using chemical mutagens that induce primarily random point mutations at high density, allelic series of missense and truncation mutations can be discovered with TILLING (Greene et al. 2003). Thus with only a small population, multiple alleles may be obtained regardless of the size of the gene. Gene regions are targeted for mutation discovery using PCR and standard SNP discovery methods. The use of general techniques for the generation and discovery of mutations means that the method should be applicable to a wide variety of organisms. TILLING methodology can also be used to uncover natural nucleotide variation linked to important phenotypic traits, a process termed EcoTILLING (Comai et al. 2004). The current status of various plant TILLING and EcoTILLING projects discussed in this review show that the methods are generally applicable across the plant kingdom.

2. TILLING FOR MUTATIONS

TILLING consists of three main steps: 1) Development of a mutagenized population, 2) DNA preparation and pooling, and 3) mutation discovery (Figure 1).

2.1. Developing a Mutagenized Population

Plants are ideally suited for TILLING. The ability to store the organism in a dormant state as seed allows for continual mutational screening without the need for continual plant propagation. Additionally, there is a rich heritage of mutagenesis in a variety of plant species, and traditional mutagenesis techniques have been used to create many new crop varieties [for example, Stadler 1928, Maluszynski et al. 2000]. Protocols for mutagenesis already exist for many plants, and only details of importance to TILLING are reviewed here.

The ideal mutagen for TILLING is one that randomly induces single nucleotide substitutions, or small insertions/deletions (~<30 nucleotides) at a high frequency

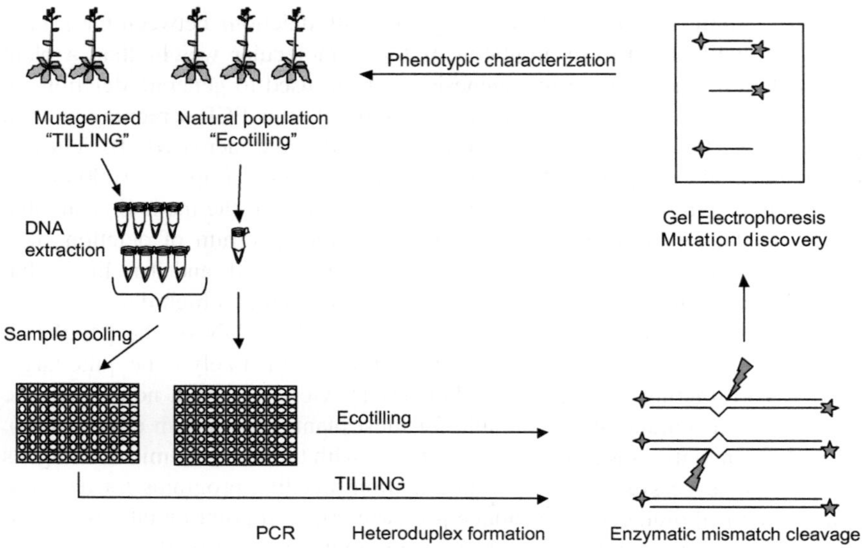

Figure 1. Outline of the basic steps for typical TILLING and EcoTILLING assays. DNA is collected from a mutagenized population (TILLING), or a natural population (EcoTILLING). For TILLING, DNAs from up to eight individuals are pooled. Typical EcoTILLING assays do not use sample pooling, but pooling has been used to discover rare natural single-nucleotide changes (Till et al., 2006). After extraction and pooling, samples are typically arrayed into a 96-well format. The target region is amplified by PCR with gene-specific primers that are end-labeled with fluorescent dyes. Following PCR, samples are denatured and annealed to form heteroduplexes that become the substrate for enzymatic mismatch cleavage. Cleaved bands representing mutations or polymorphisms are visualized using denaturing polyacrylamide gel electrophoresis. Plants with mutations predicted to affect protein function can be carefully analyzed for phenotypic abnormalities

in the genome. The chemical mutagen ethyl methanesulfonate (EMS) generates mostly SNPs, and can be controlled to produce a high density of point mutations, causing a variety of lesions including nonsense and missense mutations (Greene et al. 2003, Koornneef et al. 1982). Indeed, EMS has been the mutagen of choice for most plant TILLING projects (section 3). The effect of treatment with EMS is highly predictable; G:C->A:T transition changes represent the majority of induced mutations in most organisms. This is especially striking in *Arabidopsis* and wheat where > 99% of mutations identified by TILLING are G:C->A:T transitions (Greene et al. 2003, Slade et al. 2005).

For many plants, seed mutagenesis is most practical (Figure 2a). Seed are soaked in a dilute solution of chemical mutagen for approximately 10–24 hrs. Due to the multicellular nature of the embryo that is mutagenized, different tissues in the resulting adult plant (termed the "M1" generation), will contain different genotypes (Henikoff and Comai 2003). Thus, mutations present in the somatic tissue will not match those in the germinal tissues, and this generation is not suitable for TILLING screens. Mutations in the M1 germline will be heterozygous, and therefore M2

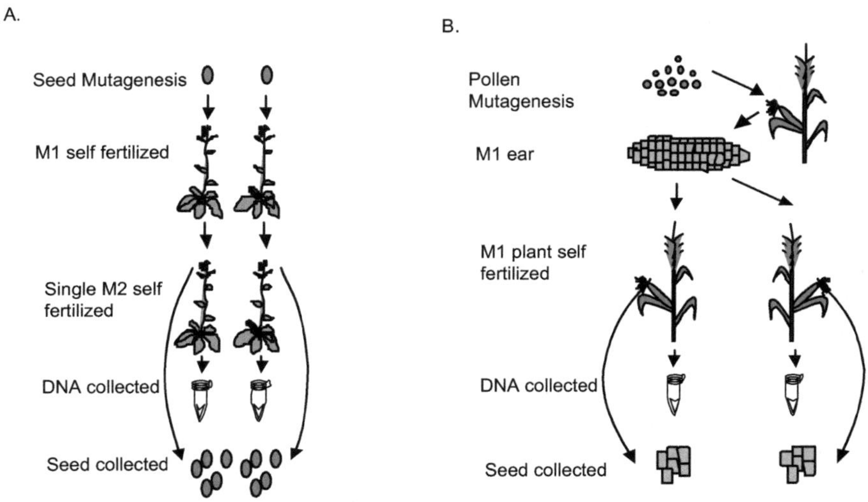

Figure 2. Seed (A) and pollen (B) mutagenesis strategies used for TILLING. A) For seed mutagenesis, a single M2 plant per line is typically included in the TILLING population. The M1 generation is chimeric for mutations and is unsuitable for TILLING. DNA and seed are collected from the M2 generation. When mutations are identified, the M3 seed can be germinated for phenotypic analysis. B) When mutagenizing pollen, the M1 generation is not chimeric and so can be screened for TILLING. All M1 progeny from a single cross should carry distinct heterozygous mutations as each pollen grain will have accumulated mutations randomly. Seed from a self-cross of the M1 is collected for subsequent phenotypic analysis

progeny from a self cross of the M1 should segregate mutations in a typical 1:2:1 Mendelian ratio. For *Arabidopsis*, only one M2 sibling was chosen at random from the progeny of a single M1 self cross (Colbert et al. 2001). The single seed descent approach provides a predictable ratio of mutant to wild-type alleles in a single individual (either 1:1 for heterozygotes or 2:0 for homozygous alleles) and allows for straightforward segregation analysis of the M3 generation. A similar approach has been used for other plant TILLING projects (section 3). For some species, pollen mutagenesis can be considered (Figure 2b). This approach was used in generating the populations for the Maize TILLING Project service (Till et al. 2004). After crossing the mutagenized pollen onto a non-mutagenized ear, each developed kernel contains unique heterozygous mutations. The M1 plants are therefore suitable to use in a TILLING screen and hundreds of unique lines are possible from a single ear. Pollen mutagenesis therefore requires fewer field resources than the seed mutagenesis approach. A potential additional advantage of pollen mutagenesis is that the nearly quiescent pollen might be less sensitive to cytotoxic effects from chemical mutagens.

The density of induced mutations is a major factor in the efficiency and cost of mutation discovery. For example, to discover 10 mutations in Maize TILLING populations, with a density of 1 mutation per 500 kb, will take approximately twice as long and cost twice as much as to discover the same number of mutations as in *Arabidopsis* TILLING populations with a density of ~1 mutation per 250 kb. Differences in mutation density between organisms may result from differences in the uptake of, cytotoxic response to, and repair of lesions induced by treatment with the mutagen. Following established mutagenesis protocols would therefore seem a good route to success, however, an approximately 1.5-fold range in mutation density was observed when following a standard protocol for *Arabidopsis* seed mutagenesis [Till et al. 2003, Till, Reynolds, and Young, unpublished]. To control for variability in mutagenesis, phenotypic markers can be used to predict the level of induced mutations before investing in the resources required for a large population. For *Arabidopsis*, a correlation between embryo lethality in the M2 seed and density of mutations was found [Till et al. 2003 and Till, Reynolds, Young, Comai and Henikoff unpublished].

For many organisms, the appropriate measure of mutagenesis will not be known in advance and can only be determined through careful observation of phenotypes, followed by TILLING screens to determine the density of induced mutations. This strategy may not be practical in some circumstances. To balance the risks of scale, propagation costs, and reproducibility, Muehlbauer and colleagues working with *Cicer anteritum* (chickpea) chose to mutagenized approx. 9000 seed but only propagate ~800 to test for mutation density (F. Muehlbauer and P.N. Rajesh, personal communication). The population is currently being tested at our TILLING facility in Seattle (Seattle TILLING Project, STP). If the test set proves suitable for a large-scale TILLING project, the entire mutagenized population has already been prepared, thus avoiding the risk of batch-to-batch variability in mutation frequency. If not, the cost of DNA extraction and propagation for 8200 lines has been avoided.

2.2. DNA Pooling

In addition to the density of mutations, sample pooling will directly affect the efficiency and cost of mutation discovery. With similar false positive and false negative discovery rates, screening four samples pooled together will take approximately twice as long and cost twice as much as screening a pool of eight samples. Factors that affect the ability to pool include the quality of genomic DNA, the accuracy of sample quantification, and the method used for SNP discovery. Blindly screening samples in various sized pools will allow an unbiased determination of the optimal level of pooling. Various TILLING groups have performed screens utilizing two-, three-, four-, six-, and eight-fold pooling (section 3). At the STP, all samples are currently pooled eight-fold.

Two basic pooling strategies have been most often used by the STP. For large scale services, we typically use a one-dimensional pooling strategy where each individual sample is represented in only one pool. When a mutation is identified in a pool of eight individuals, each member of the pool is then screened independently to identify the individual harboring the mutation (Colbert et al. 2001). The other approach is to pool samples two-dimensionally such that each sample is present in two unique pools. The STP has used a two-dimensional pooling strategy for smaller scale projects and for the larger scale Maize TILLING service. Although duplicating each sample reduces the throughput of detection in pools by half, the sample harboring the mutation is unambiguously determined in the pool screening step, so that there is no need to screen individual samples as is done in the one-dimensional strategy. Also, because two-dimensional pooling involves screening each sample with two-fold coverage, potential false positive and false negative errors are minimized at the initial screening step, rather than when individuals are screened in the second step with one-dimensional pooling. The current approach for STP is to perform small scale pilots using two-dimensional pooling, where error rates are unknown. Before moving to a large-scale operation such as a public TILLING service, the advantage of higher throughput using one-dimensional pooling is weighed against the advantage of one-step determination and a decision is made on a case-by-case basis.

2.3. Mutation Discovery

SNP discovery technologies include array-based methods, denaturing HPLC, mass spectroscopy, denaturing gradient capillary electrophoresis and enzymatic mismatch cleavage (Comai and Henikoff 2006). In theory, any accurate SNP discovery method can be used for TILLING. In practice, the method must be both robust and cost effective. Given that the highest density of induced point mutations yet reported for TILLING a diploid species is ~1 mutation/250 kb (Greene et al. 2003), screening several thousand mutant individuals will likely be required to ensure a high probability of identifying at least one deleterious mutation (for example see http://tilling.fhcrc.org:9366/files/user_fees.html). Because of this limitation, whole genome scanning methods such as SNP-chips and polony based sequencing are too error-prone to be cost-competitive at the present time.

Targeted scanning methods allow screening resources to be spent only on SNP discovery in candidate genes and thus provide a large cost savings over whole genome methods. Sanger sequencing, denaturing HPLC, denaturing gradient capillary electrophoresis (DGCE), and enzymatic mismatch cleavage have all been used as SNP discovery methods for reverse-genetic screens (McCallum et al. 2000a, Colbert et al. 2001, Wienholds et al. 2002, Slade and Knauf 2005). The most common method used for TILLING has been enzymatic mismatch cleavage and resolution on polyacrylamide gels to detect the cleaved fragments (Figure 3). In a typical reaction, a ~1.5-kb gene target is amplified by PCR with gene specific primers. Primers are end-labeled with fluorescent dyes for downstream visualization. After PCR, products are denatured and annealed to create heteroduplexes between wild-type and mutant DNA strands. Mismatches are cleaved by incubation with a nuclease and products are visualized using denaturing polyacrylamide gel electrophoresis and a gel readout platform such as the Li-Cor DNA analyzer. Although a number of nucleases have been described for mismatch cleavage (Fuhrmann et al. 2005, Youil et al. 1995, Mashal et al. 1995, Till et al. 2004). The CEL I nuclease extracted from celery is most commonly used (Oleykowski

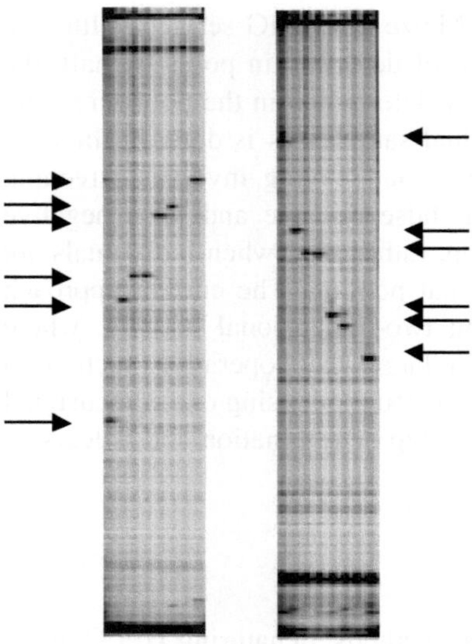

Figure 3. Discovery of induced *Arabidopsis thaliana* mutations by TILLING. The target region was amplified with a forward primer that was end-labeled with the fluorescent IRDye 700 dye, and the reverse was labeled with IRDye 800. After heteroduplex formation, samples were incubated with a celery juice extract to cleave mismatches. Gel electrophoresis was performed using a Li-Cor DNA analyzer. For each gel run, two images are produced, one for DNA labeled with IRDye 700 (A), and one for IRDye 800 (B). For any one mutation, the molecular weight of the cleaved fragment in the IRDye 700 image plus the molecular weight of the IRDye 800 fragment should add up to the molecular weight of the full length PCR product. Seven *Arabidopsis* mutations are shown, each with a corresponding fragment in the complementary fluorescent channel (marked by arrows)

et al. 1998). CEL I is a single-strand specific nuclease related to S1 nuclease, and CEL I, S1 and mung bean nucleases have all been shown to be usable for mutation discovery using standardized TILLING methods (Till et al. 2004). It is noteworthy that the choice of enzyme and readout platform can potentially affect the optimal level of sample pooling. Analysis of mutations identified by STP indicates that using crude CEL I extract and the Li-Cor DNA analyzer allows efficient discovery of heterozygous mutations in samples pooled eight-fold [a dilution of 1 in 16, Greene et al 2003].

Once a mutations are discovered, they are sequenced to determine the precise base change. An important advantage of the mismatch cleavage system is that the location of each mutation is determined within a few nucleotides, unlike methods such as denaturing HPLC, which can detect a mismatch but does not identify where it lies in the sequence. By pinpointing the location of the putative mutation, the mismatch cleavage method allows for confident identification of each mutation, whether heterozygous or homozygous, with a single sequencing run, priming with the nearer of the amplifying primers.

3. TILLING THE PLANT KINGDOM

3.1. *Arabidopsis thaliana*

TILLING was first applied to *Arabidopsis thaliana* (McCallum et al. 2000a, McCallum et al. 2000b). A mutagenized population was created by treating seed with EMS, using the single seed descent strategy described in section 2.1. Proof of concept was shown by the discovery of novel alleles in two cytosine methyltransferase genes. This initial work was done using a denaturing HPLC readout platform and five-fold sample pooling. To facilitate gene modeling and primer design, a computational tool termed CODDLe (Codons Optimized to Deliver Deleterious Lesions, http://www.proweb.org/coddle/) was developed. CODDLe obtains genomic and protein-coding information from public databases or from the user, constructs gene models, and analyzes them to determine the region that has the highest density of predicted deleterious nucleotide changes (McCallum et al. 2000b). With the success of the basic TILLING system, the goal became to develop a large population and offer TILLING to the *Arabidopsis* community as a public service. To meet the expected demand, the STP explored alternative SNP discovery methods and decided on the use of the single-strand specific nuclease CEL I and the Li-Cor readout platform (Colbert et al. 2001). Throughput was increased by lengthening the PCR amplicon size (currently at ~1.5 kb), and by increasing the sample pooling from five- to eight-fold. Throughput was also increased as machine run time per sample was decreased approximately four-fold compared to denaturing HPLC. These improvements allowed the creation of the first public TILLING service known as the Arabidopsis TILLING Project (see section 4.1).

3.2. *Lotus japonicus*

Perry and colleagues adapted the TILLING method for the model legume *Lotus japonicus* (Perry et al. 2003). Seeds were treated with EMS similar to what was done for *Arabidopsis*. Samples were pooled three-fold and CEL I was used to digest SNPs followed by readout using the ABI377 denaturing polyacrylamide slab gel system. Their work showed that a different readout platform can be used for mismatch cleavage-based TILLING. They also introduced a phenotypic enrichment strategy to reduce the amount of screening to find mutations of interest. A database was created containing the phenotypes of M2 plants. For the pilot screen, a target gene was chosen that was known to give non-nodulating phenotypes (*SYMRK*). A population of 288 plants with nodule and root-specific phenotypes was selected for screening, and 15 mutants were identified with homozygous missense changes plus one mutant with a homozygous splice site acceptor mutation. Some M2 individuals included in the screen were siblings and a total of 6 novel alleles were identified. While the density of induced mutations is not easily inferred, it is clear that this approach will be more efficient for finding functional alleles than blindly screening the entire population, provided of course that one assumes the correct phenotype. A phenotypic database is also advantageous in that alleles found when screening the population can be immediately associated with a previously described homozygous phenotype. This allows for TILLING to become a largely *in silico* process.

3.3. *Zea mays*

As part of a NSF-funded research project to ascertain the suitability of plant populations for TILLING, the STP screened maize populations donated by Clifford Weil and Nathan Springer (Till et al. 2004). Samples were screened in four-fold and eight-fold pools using CEL I and the Li-Cor platform. Seventeen EMS-induced mutations were identified in six gene target regions of approximately 1 kb. From this pilot project we estimated a density of ~1 mutation per 500 kb. Although the density was approximately two-fold less than that observed in *Arabidopsis*, it was considered suitable for a high-throughput service, and we subsequently developed an NSF-funded public service in collaboration with Clifford Weil (see section 4.2). Importantly, we were able to discover mutations in multiple gene targets without the availability of complete genome sequence, suggesting that a lack of sequence information is not an insurmountable impediment to a TILLING project. In addition, our work with maize showed that genome size was not an important factor for TILLING, insofar as the maize genome is ~20-fold larger than the *Arabidopsis* genome. Maize mutations could be discovered as easily as *Arabidopsis* mutations by simply increasing the amount of genomic DNA in PCR reactions to maintain the proper ratio of primer to target molecules.

3.4. Wheat

The feasibility of TILLING in a polyploid species was shown for wheat by Slade and colleagues (Slade et al. 2005). Starting with seed mutagenized with EMS, they developed TILLING populations in tetraploid and hexaploid wheat. To target genes, the group designed homeolog-specific primers. Samples were pooled 2-, 4- or 6-fold, mismatches were cleaved using CEL I, and fragments were visualized using a Li-Cor DNA analyzer. Over 200 mutations were discovered in the pilot screen and the estimated mutation densities were exceptionally high: 1 mutation / 40 kb in tetraploid and 1/24 kb in hexaploid wheat. The ~10-fold increase in density compared to other TILLING populations is likely attributable to the protective effects against mutation by increased ploidy (Stadler 1929). As with maize, the large genome size of polyploid wheat did not have an effect on the ability to TILL it. As with *Arabidopsis*, >99% of EMS induced mutations were G:C->A:T transitions. Importantly, Slade and colleagues were able to use TILLING to generate a wheat variety with reduced amylose production, which demonstrates the utility of the method for breeding programs, especially those where polyploids are used.

3.5. Other Plant Species

The number of plant species in which TILLING has been successfully applied continues to grow. Caldwell and colleagues used a combination of denaturing HPLC and CEL I to identify mutations in barley (Caldwell et al. 2004). In collaboration with Tom Tai, the STP has developed TILLING for rice (Till et al. 2007). In collaboration with Fred Muehlbauer at Washington State University, we have identified EMS induced point mutations in chickpea (Muehlbauer, Rajesh, Till, Cooper, Comai and Henikoff, unpublished). For soybean, we have screened three independent populations produced by Khalid Meksem, Niels Nielsen, and Kristin Bilyeu and found each population to have a high density of chemically induced mutations (Cooper, Till, Laport, Darlow, Kleffner, Jamai, El-Mellouki, Liu, Ritchie, Nielsen, Bilyeu, Meksem, Comai, and Henikoff, Unpublished). Other groups continue to develop TILLING for even more plant species, including the group of Doug Cook (UC Davis) which has identified a large number of EMS induced mutations in the model legume Medicago truncatula (D. Cook, personal communication). Similarly, a TILLING population is being developed for pearl millet at ICRISAT (R. K. Varshney, personal communication). Several European laboratories recently reported their progress on TILLING a variety of different plant species at the 1st International GABI-TILL Workshop held at the IPK in Gatersleben Germany (http://meetings.ipk-gatersleben.de/gabi2006/schedule.php). Slade and Knauf have reported success with a variety of other species including soybean, tomato, peanut, and castor (Slade and Knauf 2005). There are additional groups working toward establishing TILLING projects in important crops, and in the near future more successful applications of TILLING will undoubtedly be reported.

4. PUBLIC PLANT TILLING SERVICES

4.1. The *Arabidopsis* TILLING Project

The *Arabidopsis* TILLING project (ATP) was established in August 2001, allowing the international research community access to induced point mutations in *Arabidopsis thaliana* (Till et al. 2003). A series of user-friendly web-based tools were developed that have allowed for an automated system for placing TILLING orders and for the analysis of mutations found in TILLING screens (Figure 4). To place an order, users create gene models and search for protein homology models using CODDLe. Genes are then scored to find the best region to TILL using CODDLe, primers are designed using Primer 3 (Rozen and Skaletsky 2000), and orders are placed. All steps occur within the browser window. Order confirmations are sent automatically via email along with payment forms. Once mutations have been identified, their sequence identities are automatically sent to the user along with a summary of the mutations identified. The program SIFT is used to predict if missense changes are likely to have a deleterious affect on protein function (Ng and Henikoff, 2003). Users are also provided a link to the program PARSESNP which graphically displays mutations, provides additional missense scoring, and lists restriction endonuclease sites either created or destroyed by the point mutation (Taylor and Greene 2003). Such restriction site differences can be exploited for subsequent genotyping. By mid-2006, ATP had delivered > 6700 mutations in > 480

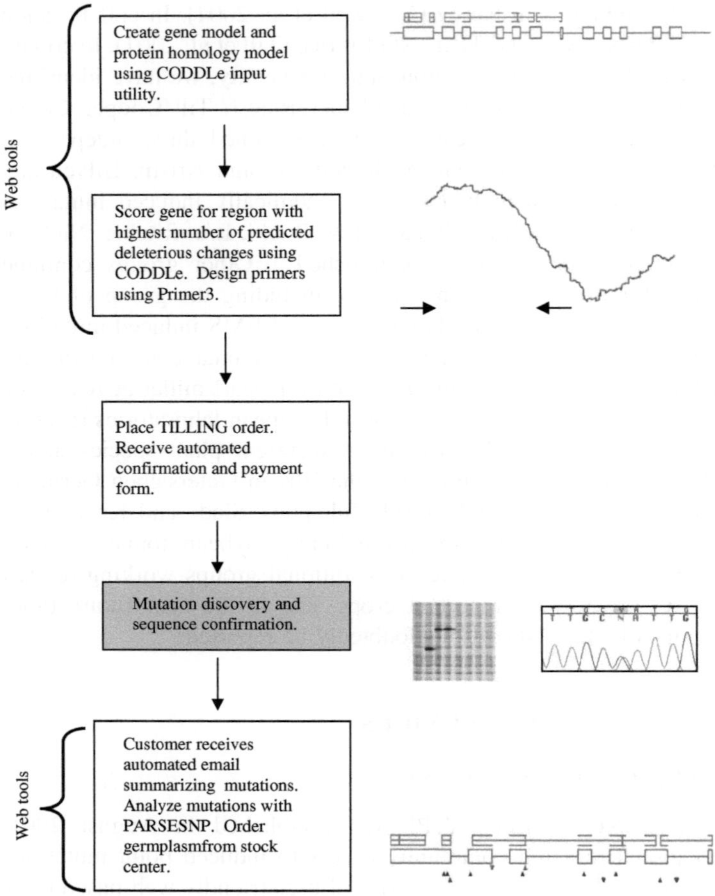

Figure 4. Outline of the steps involved in a public TILLING service. A series of web-based tools have been developed or adapted for the system. The process starts when a user creates a gene model and obtains and aligns homologous protein sequences by using the CODDLe input utility. CODDLe then identifies the region of the gene containing the highest density of potential nucleotide changes that could damage the protein when mutated. Primers design is accomplished with the program Primer3, and the researcher enters the selected primers. All of these steps are performed within the web browser window. The researcher receives and automated email confirmation of the submitted order, and a payment form. The primer order is automatically sent to the oligonucleotide supplier, and primers are shipped to the TILLING facility. Screening commences, and mutations identified by TILLING are sequence-verified. The results are automatically emailed to the customer who placed the order. A link to PARSESNP output is provided in the report. PARSESNP graphically displays the location and type of mutations, predicts the severity of missense mutations, and provides restriction sites that are either gained or lost by the induced mutation (Taylor and Greene 2003) *(see plate 11)*

gene targets. Approximately 90% of primers designed by users produce sufficient amplification product for TILLING, and users with failing primers are allowed a free second trial. The current time to delivery is ~10.5 weeks after the order is placed (updated information is available at http://tilling.fhcrc.org:9366/arab/status.html). ATP was established with generous support from NSF, however, since October 2005, ATP costs associated with a TILLING screen have been recovered from user fees. From October 2005 through May 2006, a fee of $1500 applied for screening 3000 lines. As a result of improvements in throughput gained from moving to a 384-well liquid handling platform, ATP lowered its costs and now offers screening of its entire population of 6144 mutagenized individuals for a fee of $2000.

4.2. The Maize TILLING Project

In collaboration with Clifford Weil of Purdue University, the STP created a maize TILLING service modeled after what was created for the *Arabidopsis* community. To meet the expected demand, an independent facility was developed at Purdue. Testing began in Seattle and work was gradually transitioned to the new facility, which is managed by Rita Monde, called the Maize TILLING Project (MTP, http://genome.purdue.edu/maizetilling/). At the start of service in January 2005, bench work was split 50% between the two facilities. By September 2005, 100% of the bench work was being performed by MTP. The major technical challenge with the maize service has been the lack of available genomic sequence. This has driven primer failure above the 10% range observed for *Arabidopsis* and necessitated a primer pre-screening step performed by MTP (Rita Monde, personal communication). Pre-screening is performed using unlabeled primers to avoid unnecessary expenditures on more expensive fluorescently labeled primers that fail in PCR. MTP uses the same computational and informatics tools developed for the ATP service, an illustration of the general applicability of TILLING tools.

4.3. Other TILLING Services

In addition to the services mentioned above, there are several TILLING services currently being offered by groups not affiliated with the STP. The Lotus TILLING facility (Perry et al 2003) began accepting screening orders in June of 2003 (http://www.lotusjaponicus.org/tillingpages/Homepage.htm), with a population of 5000 M2 plants along with smaller populations enriched for defects in symbiosis and in starch accumulation. The Lotus facility has switched from the ABI377 to the Li-Cor DNA Analyzer and has reported the discovery of over 100 EMS induced mutations for the Lotus community (http://www.lotusjaponicus.org/ tilling-pages/Developments.htm). The work reported for barley (Caldwell et al. 2004), has been expanded and the Barley TILLING facility at the Scottish Crops Research Institute now offers a screening service for 8600 cv. Optic lines mutagenized with EMS (http://www.scri.sari.ac.uk/programme1/BarleyTILLING.htm). The group is currently using the Li-Cor system as a readout platform and offers their service to the international barley community. The Meksem lab at Southern Illinois University Carbondale offers a TILLING service for both the Forrest and Williams82 genetic backgrounds of soybean (http://www.soybeantilling.org/). Access to screening services by the international plant biology community provides an efficient use of resources, and thus we envision the establishment of more TILLING services in the near future.

4.4. Centralized Core TILLING Facilities

In March of 2005 the STP began operating a TILLING service for mutations on the third chromosome of *Drosophila melanogaster*, and in May of 2006 opened a service for the second chromosome (Fly-TILL, http://tilling.fhcrc.org:9366/fly/). The same methods, protocols and machine settings are used for all species screened by STP, making the implementation of a multi-organism TILLING facility a straightforward task.

Arabidopsis, *Drosophila* and maize are supported by large communities of investigators, and demand is sufficient to justify establishing and maintaining a TILLING service. For many organisms that can be effectively TILLed, the size of the community and/or the total number of target genes may be too small to justify an independent facility. Fortunately, the generality of the TILLING methods and services encourage the establishment of core facilities that TILL multiple organisms. Because new organisms can be easily added to the production pipeline, the total number of organisms screened can be determined by the throughput of the facility and number of candidate targets per organism. Thus, a facility with the capacity of 100 targets/year can choose to screen 100 targets in one organism or 20 targets in 5 different organisms. We believe that this strategy will allow TILLING services in agronomically important crops that are studied by relatively few investigators.

5. ECOTILLING

5.1. EcoTILLING of *Arabidopsis thaliana*

The enzymatic mismatch cleavage method used for TILLING should be applicable to any heteroduplexed DNA target regardless of the source of the nucleotide polymorphism. Therefore, the same methods should be applicable to the discovery of natural nucleotide variation in populations. However, if cleavage of mismatches were complete, then only the closest polymorphism to the labelled end would be detectable, and this method would be unsuitable for polymorphism discovery. Fortunately, cleavage is only partial at any site, and TILLING methodology has been used to score multiple mismatches within single end-labelled heteroduplexes. To determine if the method could be used to accurately catalogue natural diversity, 196 *Arabidopsis* ecotypes were screened for nucleotide variation in 5 gene target regions (Comai et al. 2004). To uncover homozygous polymorphisms that are thought to predominate in self-fertile species, an equal amount of DNA from the sequenced Columbia accession was added to each test sample so that heteroduplexes could be created. To unambiguously assign haplotypes, individual accessions were not pooled together. Fifty-five distinct haplotypes were discovered. In addition to SNPs, small insertions/deletions and a satellite repeat number polymorphism were identified. The study indicated a low false positive and false negative rate of discovery when compared to sequencing. Based on this work, the modification of TILLING for the discovery and genotyping of natural nucleotide polymorphisms was termed EcoTILLING.

5.2. Automation of TILLING Gel Analysis

The major difference between TILLING and EcoTILLING is the amount of information present in gel data from a single run. For TILLING, there are typically 3–5 mutations per 1.5 kb *Arabidopsis* gene per 768 pooled individuals screened on a gel. With this amount of information, manual gel analysis and database entry is practical. For EcoTILLING, over 100 polymorphic bands could be identified in a single gel run, when screening 96 samples with a 1-kb target fragment. This large increase in data points created a serious bottleneck and increased the possibility of human error during manual analysis. To facilitate the gel reading and data entry processes, Zerr and Henikoff developed the PC/Mac program, GelBuddy (http://www.gelbuddy.org) (Zerr and Henikoff 2005). GelBuddy provides automated lane identification and molecular weight calibration. Bands are scored with a click of the mouse, and reports containing lane and molecular weight information for each band are automatically generated. A fully automated version that includes band scoring has recently been introduced as part of an effort to apply EcoTILLING to the high-throughput discovery of rare human SNPs and cancer mutations (Till et al., 2006).

5.3. EcoTILLING for *Populus trichocarpa*

As with TILLING, EcoTILLING is general, and should be applicable to most species. Gilchrist and colleagues recently reported a genotyping analysis of western black cottonwood populations (*Populus trichocarpa*), using EcoTILLING for SNP identification (Gilchrist et al. 2006). Sixty-three novel SNPs were identified in 9 target genes, for 41 tree accessions. From these data, the group estimated the degree of linkage disequilibrium, heterozygosity, and nucleotide diversity. Much can be learned from studying natural nucleotide diversity; new markers will be generated from EcoTILLING projects, and non-synonymous SNPs may be identified that provide a beneficial phenotype. For species in which mutagenesis is impractical, exploiting natural nucleotide diversity will be invaluable for crop improvement.

6. CONCLUDING REMARKS

TILLING and EcoTILLING are high-throughput and low-cost methods for the discovery of induced mutations and natural polymorphisms. The methods are general and have successfully been applied to many plants, including crops. With sequence data and general tools such as TILLING, reverse genetics can be applied to lesser-studied species. Now that successes have been reported in a variety of important plant species, the next challenge will be to use the technology to develop improved crop varieties. The utility of induced mutations and natural polymorphism has already been established for crop breeding and so the task is mostly one of implementation.

ACKNOWLEDGEMENTS

Work performed by the Seattle TILLING Project was supported by grants from the Plant Genome Research Program and *Arabidopsis* 2010 initiative of the National Science Foundation, the Genome Program of the US Department of Agriculture-National Research Initiative, and from the Rockefeller Foundation

REFERENCES

An G, Jeong DH, Jung KH, Lee S, (2005) Reverse genetic approaches for functional genomics of rice. Plant Mol Biol 59:111–123

Bradley JT, Reynolds SH, Weil C, Springer N, Burthner C, Young K, Bowers E, Coclomo, CA, Enns LC, Odden AR, Greena EA, Comai L, Henikoff S (2004) Discovery of induced point mutations in maize genes by TILLING. BMC Plant Biol 4:12

Burch-Smith TM, Anderson JC, Martin GB, Dinesh-Kumar SP (2004) Applications and advantages of virus-induced gene silencing for gene function studies in plants. Plant J 39:734–746

Caldwell DG, McCallum N, Shaw P, Muehlbauer GJ, Marshall DF, Waugh R (2004) A structured mutant population for forward and reverse genetics in Barley (*Hordeum vulgare* L.). Plant J 40:143–50

Colbert T, Till BJ, Tompa R, Reynolds S, Steine MN, Yeung AT, McCallum CM, Comai L, Henikoff S, (2001) High-throughput screening for induced point mutations. Plant Physiol 126:480–484

Comai L, Young K, Till BJ, Reynolds SH, Greene EA, Codomo CA, Enns LC, Johnson JE, Burtner C, Odden AR, Henikoff s (2004) Efficient discovery of DNA polymorphisms in natural populations by EcoTILLING. Plant J 37:778–786

Comai L, Young K, Till BJ, Reynolds SH, Greene EA, Codomo CA, Enns LC, Johnson JE, Burtner C, Odden AR, Henikoff S (2006) TILLING: practical single-nucleotide mutation discovery. Plant J 45 684–694.

Fuhrmann M, Oertel W, Berthold P, Hegemann P (2005) Removal of mismatched bases from synthetic genes by enzymatic mismatch cleavage. Nucleic Acids Res 33:e58

Gilchrist EJ, Haughn GW, Ying CC, Otto SP, Zhuang J, Cheung D, Hamberger B, Aboutorabi F, Kalynyak T, Johnson L , Bohlmann J, Ellis BE, Douglas CJ, Cronk QC (2006) Use of EcoTILLING as an efficient SNP discovery tool to survey genetic variation in wild populations of populus trichocarpa. Mol Ecol 15:1367–1378

Greene EA, Codomo CA, Taylor NE, Henikoff JG, Till BJ, Reynolds SH, Enns LC, Burtner C, Johnson JE, Odden AR, Comai L, Henikoff S (2003) Spectrum of chemically induced mutations from a large-scale reverse-genetic screen in *Arabidopsis*. Genetics 164:731–740

Henikoff S, Comai L (2003) Single-nucleotide mutations for plant functional genomics. Annu Rev Plant Biol 54:375–401

Hirochika H, Guiderdoni E, Gynheung AN, Hsing Y-I, Moo Young E, Han C-D, Upadhyaya N, Ramachandran S, Qifa Z, Pereira A, Sundaresan V, Hei L, (2004) Rice mutant resources for gene discovery. Plant Mol Biol 54:325–334

Koornneef M, Dellaert LW, van der Veen JH (1982) EMS- and radiation-induced mutation frequencies at individual loci in *Arabidopsis thaliana* (L.) Heynh. Mutat Res 93:109–123

Kusaba M (2004) RNA interference in crop plants. Curr Opin Biotechnol 15:139–143

Li X, Zhang Y (2002) Reverse genetics by fast neutron mutagenesis in higher plants. Funct Integr Genomics 2:254–258

Maluszynski M, Nichterlein K, Van Zanten L, Ahloowalia BS (2000) Officially released mutant varieties – the FAO/IAEA database. Mutat Breeding 20:1–88

Mashal RD, Koontz J, Sklar J (1995) Detection of mutations by cleavage of DNA heteroduplexes with bacteriophage resolvases. Nat Genet 9:177–183

May BP, Liu H, Vollbrecht E, Senior L, Rabinowicz PD, Roh D, Pan X, Stein L, Freeling M, Alexander D, Martienssen R (2003) Maize-targeted mutagenesis: a knockout resource for maize. Proc Natl Acad Sci USA 100:11541–11546

McCallum CM, Comai L, Greene EA, Henikoff S (2000a) Targeted screening for induced mutations. Nat Biotechnol 18:455–457

McCallum CM, Comai L, Greene EA, Henikoff S (2000b) Targeting induced local lesions IN genomes (TILLING) for plant functional genomics. Plant Physiol 123:439–442

McCarty DR, Settles MA, Suzuki M, Tan B.C., Latshaw S, Porch TG, Robin K, Baier J, Avigne W, Lai J, Messing J, Koch KE, Hannah, LC (2005) Steady-state transposon mutagenesis in inbred maize. Plant J 44:52–61

Ng PC, Henikoff S (2003) SIFT: predicting amino acid changes that affect protein function. Nucleic Acids Res 31:3812–3814

Oleykowski CA, Bronson Mullins CR, Godwin AK, Yeung AT (1998) Mutation detection using a novel plant endonuclease. Nucleic Acids Res 26:4597–4602

Perry JA, Wang TL, Welham TJ, Gardner S, Pike JM, Yoshida S, Parniske M (2003) A TILLING reverse genetics tool and a web-accessible collection of mutants of the legume Lotus japonicus. Plant Physiol 131:866–871

Rozen S, Skaletsky H (2000) Primer3 on the WWW for general users and for biologist programmers. Methods Mol Biol 132:365–386

Slade AJ, Knauf VC (2005) TILLING moves beyond functional genomics into crop improvement. Transgenic Res 14:109–115

Slade AJ, Fuerstenberg SI, Loeffler D, Steine MN, Facciotti DA (2005) Reverse genetic, nontransgenic approach to wheat crop improvement by TILLING. Nat Biotechnol 23:75–81

Stadler LJ, (1928) Genetic effects of X-rays in maize. Proc Natl Acad Sci USA 14:69–75

Stadler LJ (1929) Chromosome number and the mutation rate in avena and triticum. Proc Natl Acad Sci USA 15:876–881

Taylor NE, Greene EA (2003) PARSESNP: a tool for the analysis of nucleotide polymorphisms. Nucleic Acids Res 31:3808–3811

Till BJ, Burtner C, Comai L, Henikoff S (2004) Mismatch cleavage by single-strand specific nucleases. Nucleic Acids Res 32:2632–2641

Till BJ, Zerr, T, Bowers E, Greene EA, Comai L, Henikoff S (2006) High-throughput discovery of rare human nucleotide polymorphisms by Ecotilling. Nucleic Acids Res 34:e99

Till BJ, Cooper J, Tai TH, Colowit, P, Greene EA, Henikoff S, Comai L (2007) Discovery of chemically induced mutations in rice by TILLING. BMC Plant Biol 7:19

Till BJ, Reynolds SH, Greene EA, Codomo CA, Enns LC, Johnson JE, Burtner C, Odden AR, Young K, Taylor NE, Henikoff JG, Comai L, Henikoff S (2003) Large-scale discovery of induced point mutations with high-throughput TILLING. Genome Res 13:524–530

Wienholds E, Schulte-Merker S, Walderich B, Plasterk RH (2002) Target-selected inactivation of the zebrafish rag1 gene. Science 297:99–102

Youil R, Kemper BW, Cotton RG (1995) Screening for mutations by enzyme mismatch cleavage with T4 endonuclease VII. Proc Natl Acad Sci USA 92:87–91

Zerr T, Henikoff S (2005) Automated band mapping in electrophoretic gel images using background information. Nucleic Acids Res 33:2806–2812

CHAPTER 16

CHARACTERIZATION OF EPIGENETIC BIOMARKERS USING NEW MOLECULAR APPROACHES

MARIE-VÉRONIQUE GENTIL AND STÉPHANE MAURY*

Laboratoire de Biologie des Ligneux et des Grandes Cultures, UPRES EA 1207, rue de Chartres. BP 6759. Faculté des sciences, Université d'Orléans. 45067 Orléans Cedex 2, France

Abstract: Plants exhibit a polymorphism of DNA methylation status in their genomes in relation to various breeding traits and phenotypes. Evidence for relationships between DNA methylation and given phenotypes can be shown through the variations of phenotypes after treatments that alter DNA methylation percentages or through the variations of methylation percentages in different phenotypes. The corresponding "epialleles" are potential biomarkers for plant breeding selection. The target genes of these epigenetic modifications could be identified with a genome scanning approach using methyl-sensitive enzymes or methyl-binding affinity columns. Correlations between DNA methylation polymorphism and phenotypes could be tested using various methods such as bisulfite sequencing, physiological and genetic analyses. Identification of methylation biomarkers by these new molecular approaches have been successfully applied to human cancer detection and should be now envisaged for plant breeding selection.

Keywords: Bisulfite sequencing; DNA methylation; Epiallele; Plant breeding; Restriction Landmark Genome Scanning

1. INTRODUCTION: THE FUNDAMENTAL ROLE OF EPIGENETIC EVENTS IN PLANT DEVELOPMENT

Contrary to animals, plants are static organisms and consequently exhibit high developmental plasticity in response to environmental variations. Plasticity is a term used to describe the ability of organisms to change form or shape and growth in response to environmental changes (Pigliucci, 2001). Variations of phenotypes without modifications of DNA sequences correspond to epigenetic

*Corresponding Author: stephane.maury@univ-orleans.fr

R.K. Varshney and R. Tuberosa (eds.), Genomics-Assisted Crop Improvement: Vol. 1: Genomics Approaches and Platforms, 351–370.

phenomena. Epigenetic defines all mitotically and meiotically heritable changes in gene expression that are not coded in the DNA sequence itself (Holliday, 1990). In plants, epigenetic phenomena controlled many biological processes such as development, morphogenesis, genomic imprinting, somaclonal variations, heterosis, transgene silencing and stress responses. Epigenetic regulation is mediated by DNA methylation and histones modifications that regulate chromatin condensation and consequently gene expression.

In plants, addition of a methyl group from S-adenosyl-L-methionine to cytosine residues occurs at CpG or CpNpG sequences (where N could be any nucleotide). This reaction is catalysed by DNA methyltransferases. *Arabidopsis* MET1 (homologue of the DNA methyltransferase 1 in mammals) and plant-specific chromomethylases are responsible for the methylation of hemi-methylated sites during DNA replication and are referred to maintenance methylation (Finnegan and Kovac, 2000). Maintenance of CpG methylation and CpNpG are respectively catalyzed by MET1 and chromomethylases. The domain rearranged methyltransferase family (similar to the mammalian DNA methyltransferase 3 family) is involved in the methylation of unmethylated DNA, a process called *de novo* methylation (Cao et al., 2000; Tariq and Paszkowski, 2004). In plants, the demethylation of cytosine residues seems to occur both through the action of a DNA glycosylase and through cell division without maintenance methylation (Choi et al., 2002).

Another mechanism controlling gene expression is the covalent modifications of histones. Thus, free N-terminal tails of histones protrude from the octameric protein's core and are subjected to various posttranslational modifications, including acetylation, phosphorylation, methylation, ribosylation and ubiquitinylation (Meyer, 2001; Loidl, 2004). All these modifications constitute the "histone code" that presents more combinatorial possibilities in plants than in animals. DNA methylation and histone methylation on lysine 9 of histone H3 (H3-K9) are generally associated to the condensed chromatin status that prevents the binding of transcription factors and induces transcriptional gene silencing. On the contrary, histone acetylation, phosphorylation and methylation on lysine 4 of histone H3 (H3-K4) are present in the decondensed chromatin status that allows gene transcription (Meyer, 2001) (Figure 1).

Polymorphism in DNA methylation status leads to differences in genes expression and confers phenotypic effects (Ronemus et al., 1996; Kakutani et al., 1999). For instance, *SUPERMAN* gene plays a role in floral morphology in *Arabidopsis*. *SUPERMAN* gene presents several *clark kent* alleles which correspond to different phenotypes. However, if all these alleles exhibit the same DNA sequence they differ in their methylation status and correspond to epiallele (Kalisz and Purugganan, 2004). The polymorphism associated to these epialleles constitutes biomarkers for various applications. Indeed, a biomarker is a substance or a process that is indicative of a phenotype or a biological event (Laird, 2003). For example, in human cancers, epialleles are used as biomarkers for early detection or characterization of cancer types. In plants, epigenetic inheritance is a source of polymorphism that could be exploited for selection and plant breeding (Tsaftaris et al., 2005). Nevertheless, only

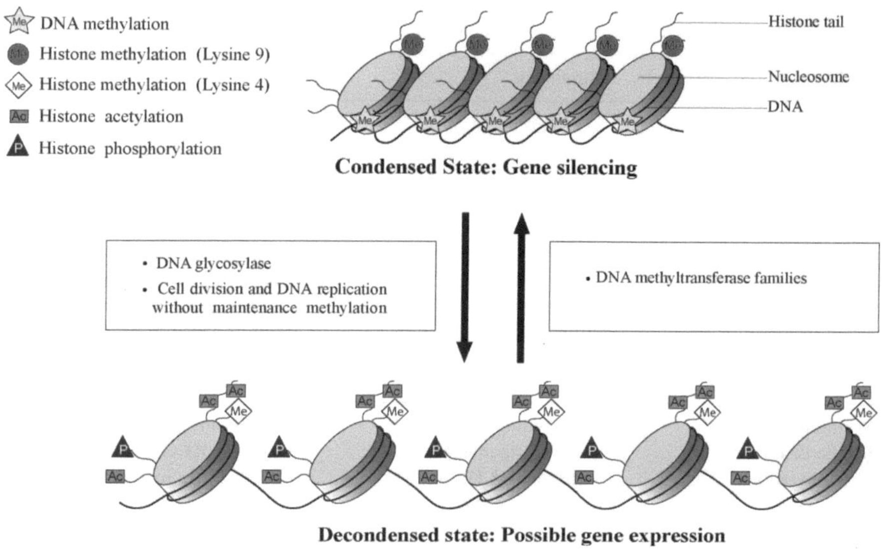

Figure 1. Model for the regulation of chromatin structure in plants. Only the processes controlling DNA methylation status are indicated *(see plate 12)*

few data are available concerning epialleles in plants (Finnegan, 2001). Indeed, current methods of molecular biology such as DNA sequencing do not distinct methylation status. In this chapter, we present new molecular approaches to establish relationships between phenotype or breeding trait and DNA methylation status. The discovery of methylation biomarkers or epialleles by scanning approaches is also presented. Finally, the validation of methylation biomarkers for a given phenotype or breeding trait by an original gene candidate approach will be discussed.

2. DNA METHYLATION AND PLANT BREEDING: EVIDENCE FOR AN EPIGENETIC REGULATION

In order to establish relationships between a breeding trait or a phenotype and DNA methylation polymorphism, two strategies have been used (Table I). Firstly, the modification of genomic DNA methylation levels by physical, chemical treatments or genetic manipulations and the analysis of the consequences on the phenotypes. Secondly, the determination of the global genomic DNA methylation percentage in distinct phenotypes.

2.1. Modification of Genomic DNA Methylation Levels

2.1.1. DNA hypo- or hyper-methylation treatments

DNA hypomethylation is obtained using analogues of cytosine such as 5-azacytidine, 5-azacytosine or 5-azadeoxycytosine (Causevic et al., 2005). Nitrogen on the position 5 in the pyrimidine ring forbidden the methylation of these

Table I. A synopsis of the study of methylation biomarkers

Steps	Strategies	Methods
1. Evidence for an epigenetic regulation by DNA methylation	– Modification of DNA methylation levels	– **DNA hypo- or hyper-methylating treatments** (Causevic et al., 2005) – **Genetic transformation** (Vongs et al., 1993; Ronemus et al., 1996; Finnegan et al., 1996)
	– Determination of the global DNA methylation levels	– **Chromatographic analysis: HPLC or HPCE** (Fraga et al., 2000; Causevic et al., 2005; Johnston et al., 2005)
2. Scanning approach for the discovery of methylation biomarkers	– Screening for genomic sequences with distinct DNA methylation status	– **Methylation-sensitive enzymes: RLGS, SPM or MSAP, tiling microarrays** (Xiong et al., 1999; Costello et al., 2002; Shiraishi et al., 2004a; Lippman et al., 2005; Martienssen et al., 2005; Causevic et al., 2006; Takamiya et al., 2006) – **Affinity chromatography**: MBD or 5mC (Cross, 2002; Shiraishi et al., 2004b; Salzberg et al., 2004)
	– Cloning of the sequences	– **Adapters ligation and PCR amplification** (Causevic et al., 2006) – **Biotinylated linkers and PCR amplification** (Takamiya et al., 2006) – **Promoter library and PCR amplification** (Yu et al., 2004)
3. Gene candidate (GC) approach for the validation of methylation biomarkers	– Choosing the GC(s)	– **Physiological studies** (Causevic et al., 2006) – **Scanning approach**(Causevic et al. , 2006)
	– Determination of the methylation status for GC(s)	– **Methylation-sensitive enzymes**: Southern blotting (Moore, 2001; Causevic et al., 2006) – **Bisulfite-PCR: COBRA, MS-PCR**...cloning and sequencing or direct pyrosequencing. (Herman et al., 1996; Xiong and Laird, 1997; Shiraishi et al., 2002; Laird, 2003; Dupont et al., 2004; Causevic et al., 2006; Ogino et al., 2006)

	– Validation of the methylation biomarkers (should be define in each case)	– **Physiological analyses** of GC(s) expression at mRNA (northern blotting, microarrays, qRT-PCR) or protein (activity, western blotting, 2D electrophoresis-mass spectrometry) levels – **Genetic transformation** (mutagenesis, sense or antisense transgene, RNAi, homologous recombination) and functional complementation – **Analysis of the epigenetic inheritance**: estimate the extent to which methylation status of the markers are linked to the phenotypic variation among individuals within a selection population

molecules. The incorporation of these analogues in DNA during cell replication induces a progressive DNA hypomethylation status in the daughter cells. Treatments with cold temperatures were also shown to induce DNA hypomethylation (Finnegan et al., 1998). DNA hypermethylation is achieved using hydroxyurea which inhibits ribonuclease and replication. Modifications of the methyl donor (S-adenosyl-L-methionine) contents also lead to efficient variations of global DNA methylation.

A recent report has evaluated the potential of all these treatments to modify DNA methylation level in one plant system. Thus, eight distinct treatments such as three analogues of cytosine, cold treatment, ethionine, diaminobutanone, 2,4-dichlorophenoxyacetic acid and hydroxyurea have induced DNA hypo- or hyper-methylation on three sugarbeet cell lines displaying distinct morphogenetic status (Causevic et al., 2005). In this collection of treated lines with ± 10% methylcytosine percentages, variations of morphogenetic status were observed: loss of organogenic potential and dedifferentiation. Altogether, these results give evidence for a relationship between DNA methylation levels and morphogenesis status in sugarbeet cell lines. Such relationship has also been reported in many other plant systems (Burn et al., 1993; Lambé et al., 1997; Kaeppler et al., 2000).

2.1.2. Genetic transformation

The analysis of mutants and transgenic plants has allowed establishment of correlations between DNA methylation levels and plant development. In antisense DNA methyltransferase 1 lines, a progressive loss of genomic DNA methylation from generation to generation induces deleterious phenotypes, such as reduction in fertility and altered apical dominance (Finnegan et al., 1996). In *ddm* (decrease in

DNA methylation) *Arabidopsis* mutant encoding a chromatin remodelling protein, many morphological abnormalities were reported (Vongs et al., 1993). All these data demonstrate the importance of DNA methylation control for plant development.

2.2. Determination of Global DNA Methylation Levels

Two techniques have been improved these last years to determine the global DNA methylation levels after genomic DNA hydrolysis: High-performance liquid chromatography (HPLC) and High-performance capillary electrophoresis (HPCE). They allow the quantification of methylcytosine and the calculation of a percentage of DNA methylation in the genome.

HPLC is considered as the most reliable and sensitive technique to determine DNA methylation. DNA is digested by enzymatic, thermic or acid treatments. Nucleotides, nucleosides or bases are then separated in by HPLC (Figure 2A). Quantification is achieved by using UV detection (Causevic et al., 2005; Johnston et al., 2005) or laser induced fluorescence system (Wirtz et al., 2005), where lower sample amounts are required for analysis. Nucleoside analysis is recommended because nucleotides and bases are more polar and consequently more difficult to separate by HPLC. Identification of cytosine (C) and methylcytosine (mC) is assessed by co-migration with commercial standards under the same HPLC conditions. The methylcytosine percentages are calculated using the following formula: $\%mC = (mC/(C+mC)) \times 100$ (Figure 2B). HPLC analyses have been used to characterize DNA methylation levels in many phenotypes (Causevic et al., 2005; Johnston et al., 2005).

HPCE using a sodium dodecyl sulfate micelle system allows separation of bases from acid hydrolyzed genomic DNA (Fraga et al., 2000). This method gives a faster and a better separation than HPLC for the quantification of cytosine and methylcytosine in genomic DNA of plants.

Overall, HPLC and HPCE allow rapid, accurate and readily automatable quantitative results by measuring overall methylcytosine contents from appropriately hydrolyzed DNA samples. However, they require relatively large amounts of genomic DNA, limiting the applicability of the methods. Furthermore, they could only provide information on methylation levels of genome and gene-specific information would be masked.

Several other methods have also been described such as liquid chromatography-mass spectrometry separation, anti-methylcytosine immunological techniques and bisulfite conversion. Nevertheless, their use was restricted to punctual applications (Laird, 2003).

2.3. Perspectives: Research of Epigenetic Biomarkers for Plant Breeding

In plants, epigenetic mechanisms control gene expression during various biological processes and can be transmitted over many generations. Polymorphism of the

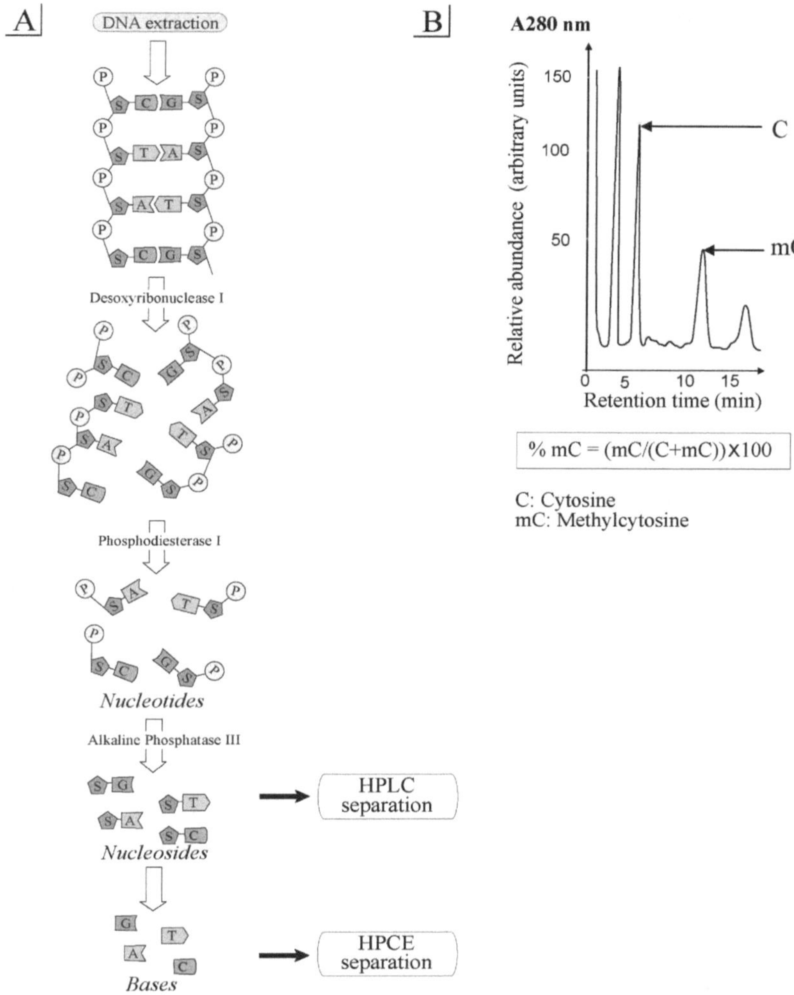

Figure 2. Determination of global genomic DNA methylation levels: A, Enzymatic DNA hydrolysis. B, HPLC chromatogram for the determination of methylcytosine percentage. P: Phosphate group. S: Sugar. A, T, C and G: Bases *(see plate 13)*

epialleles could be used to distinguish individuals or genotypes within population. Detection of global methylation levels allows calculation of methylcytosine percentages that could be correlated to the variations of a quantitative trait. Nevertheless, this result gives no indication on the target genes of the DNA polymorphism. The research of methylation biomarkers and their validation for various applications is well developed in the field of human cancer and could now be envisaged for plant breeding.

3. THE GENOME SCANNING APPROACH FOR DISCOVERY OF METHYLATION BIOMARKERS

3.1. Scanning Approach

DNA methylation biomarkers will be of great interest for plant breeding. How to identify these modifications among millions of CpG dinucleotides and tens of thousands of gene-associated CpG islands? that are 500-base-pair windows with a G:C content of at least 55% and an observed over expected CpG frequency of at least 0.65 (Laird, 2003).

The objective of this part is to present an approach allowing the discovery of methylation biomarkers in plants. Therefore, the methodology must scan the CpG islands within the genome in order to select loci with distinct methylation profile between two or more biological plant samples. Several techniques have been developed on animal systems and only few of them have actually been applied on plant systems (Costello et al., 2002; Frühwald and Plass, 2002; Mills and Ramsahoye, 2002; Shiraishi et al., 2002; Laird, 2003; Shiraishi et al., 2004b). This part presents a subset of these techniques with potential applications for the research of plant breeding markers (Table I). Among them, the Restriction Landmark Genome Scanning (RLGS) method has retained specific attention. Thus, RLGS method provides a quantitative epigenetic assessment of several gene-associated CpG islands in a single gel without prior knowledge of gene sequence and has been successively applied on animal and plant systems (Hatada et al., 1991; Matsuyama et al., 2000; Costello et al. 2002; Rush and Plass, 2002; Matsuyama et al. 2003; Causevic et al., 2006; Takamiya et al., 2006).

3.2. Identification of Methylation Biomarkers by Restriction Landmark Genome Scanning (RLGS)

After isolation of concentrated solution of pure genomic DNA without mechanical breakings, an enzymatic processing is performed with an infrequently cutting restriction enzyme that can not cleave methylated CpG sites (Figure 3). *Not* I was selected as the landmark enzyme since most of it sites (GCGGCCGC) is within CpG islands (Costello et al., 2002). Recently, an alternative approach was published using two isoschizomers *Hpa* II and *Msp* I that recognize the same sequence (CCGG), but have different methylation sensitivity. *Msp* I cleaves this site if the second C of CCGG is methylated or not. *Hpa* II could only cut if the site is not methylated. This method directly discriminates methylation polymorphism from sequences (Takamiya et al., 2006). Cohesive extremities of these restriction fragments are filled with radionucleotides. A second restriction enzyme can be used as *Eco*RV (blunt end) to increase the number of fragments before the first electrophoretic separation on 30 cm thin 2 nm diameter agarose gel tube. Then, DNA is *in gel* digested by *Hinf* I, a more frequently cutting enzyme, and electrophoresed in a second perpendicular direction

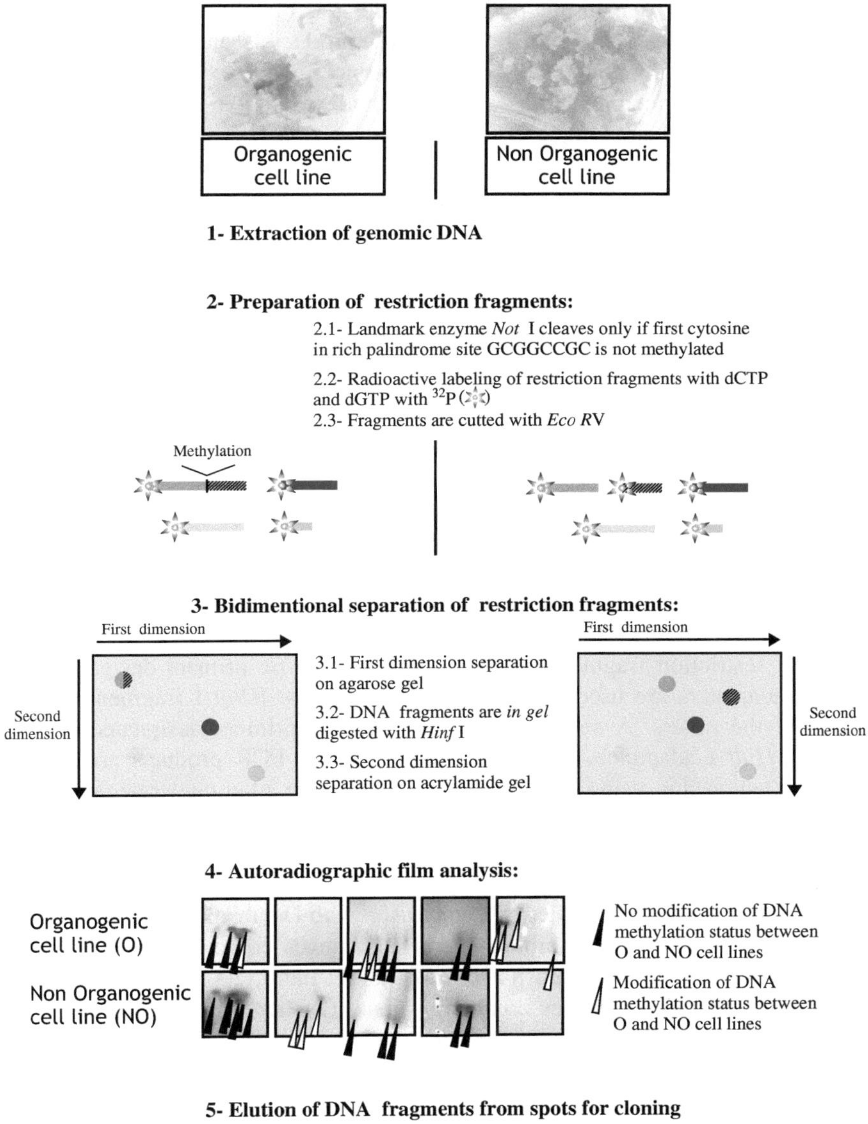

Figure 3. Principle of Restriction Landmark Genome Scanning (RLGS) method for the discovery of methylation biomarkers. RLGS sections were obtained with DNA extracted from organogenic or non-organogenic sugarbeet lines. Spots indicated by arrows correspond to fragments that can be superposed (black) or not (white) on the RLGS sections obtained with both lines. (Adapted from Causevic et al., 2006) *(see plate 14)*

on 30 × 40 cm non denaturing polyacrylamide gel. Finally, the gel is autoradiographed (Figure 3). The profiles are highly reproducible and display both the copy number and methylation status of the sequences. The comparison of autoradiogrammes of distinct phenotypes allows the identification of two groups of spots: (1) spots that are unique in one of the two RLGS profiles, suggesting remodelling of the methylation status of some CpG rich loci and (2) spots that can be superposed in both RLGS profiles and that present no polymorphism of DNA methylation on *Not* I sites. In this last case, the intensity of the spots reflects the copy numbers that can differ between the biological samples.

3.3. Cloning and Sequencing of the Methylation Biomarkers

The DNA fragments screened by RLGS have a high probability of containing a gene and are of ideal length for cloning and sequencing (Costello et al, 2002). Three restriction enzymes have successively cut genomic DNA and the labelling was performed on *Not* I extremities. Therefore, the cloning strategy of these fragments, that are present in the gel in low amount, should specifically amplify *Not* I/*Not* I and *Not* I/*Hinf* I fragments among all the generated fragments (*Eco*RV/*Eco*RV; *Eco*RV/*Hinf* I; *Eco*RV/*Not* I; *Hinf* I/*Hinf* I) (Figure 4) (Causevic et al., 2006). The first step consists to the ligation of *Not* I and *Hinf* I adaptaters at the extremities of the restriction fragments (Table I). Then, specific primers designated on the *Not* I adaptaters are used to enrich by PCR the *Not* I/*Not* I fragments at the expense of the others. A second PCR with specific primers designated on the *Not* I and *Hinf* I adaptaters is then performed. These PCR products are cloned in an adapted vector using available kits. Alternative cloning strategies using biotinylated linkers or promoter library have also been reported for the cloning of RLGS fragments obtained from tumorous cells (Yu et al., 2004; Takamiya et al., 2006).

The results of homology studies by comparison with Databanks allow annotation for these sequences. RLGS exhibits several advantages compared to PCR-based method (Costello et al., 2002; Rush and Plass, 2002; Laird, 2003): (1) high probability of CpG islands sequences containing gene, (2) large number of sequence could be analyzed simultaneously, (3) quantitative and qualitative high reproducible information, (4) no biased or difficulty of PCR amplification. Nevertheless, this technique that depends on methyl-sensitive restriction enzyme is poorly suited for routine analysis and is more designated for methylation biomarkers discovery.

3.4. Applications and Perspectives of Methylation Biomarkers Isolated by RLGS

Several publications demonstrate the successful use of RLGS to identify modifications of DNA methylation, particularly in tumorous cells (for a review see

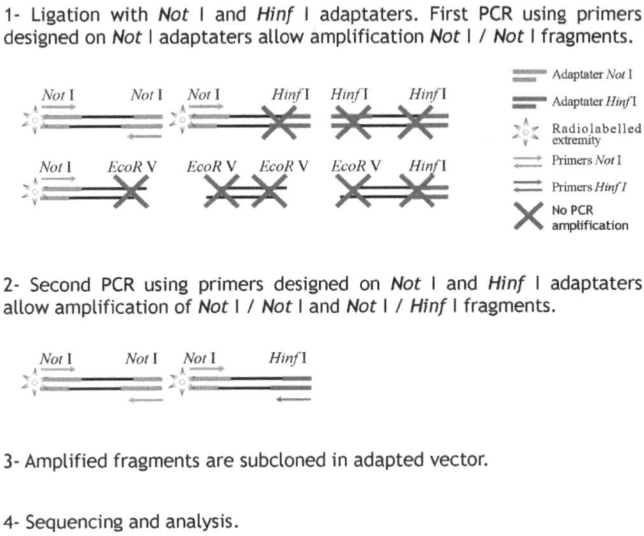

Figure 4. Cloning strategy for epigenetic biomarkers screened by RLGS using adaptaters and PCR amplifications *(see plate 15)*

Rush and Plass 2002). Thus, DNA methylation changes occurred in carcinogenesis and are potentially good early indicators of disease, especially those where symptoms appeared lately as ovarian, pancreatic and lung cancer (Laird, 2003). In plants, RLGS has already been used to screen the global methylation status in the *Arabidopsis* genome (Matsuyama et al., 2003), to analyze DNA deletions in an albino mutant genome (Abe et al., 2002), to discriminate for methylation polymorphism in the *Arabidopsis* genome (Takamiya et al., 2006) and to screen for methylation biomarkers of *in vitro* morphogenesis in sugarbeet cell lines (Causevic et al., 2006). In this last situation, an experimental system composed of two *in vitro* sugarbeet callus lines originating from the same mother plant and exhibiting different status of differentiation was used. One line is organogenic (O) with continuous production of leafy shoot and active photosynthesis. Another line is non-organogenic (NO) and photosynthetically active (Hagège et al., 1991; Causevic et al., 2005; Causevic et al., 2006). A relationship between the differentiation status of these lines and their genomic DNA methylation levels was first demonstrated (Causevic et al., 2005). These cell lines represent an interesting model for the research of methylation biomarkers of plant *in vitro* morphogenesis. Among potential sugarbeet morphogenesis DNA markers that were cloned and sequenced, five are homologous to genes involved in cell cycle and embryogenesis and two in metabolic functions. These results are in good agreement with the potential of these sequences as biomarkers of morphogenesis in sugarbeet cell lines. The two last biomarkers presents homologies with protein of unknown function and display a great interest since it is the first report of a functional annotation for these sequences.

3.5. Other Techniques for the Discovery of Methylation Biomarkers

Several methods have been developed to screen the genome for modifications of CpG islands methylation (Table I; Frühwald and Plass, 2002; Shiraishi et al., 2002; Laird, 2003). Some of them used also methyl-sensitive enzymes as Methylation-Sensitive Amplified Polymorphism (MSAP) (Xiong et al., 1999) and Segregation of Partly Melted molecules (SPM) (Shiraishi et al., 2004a). This last one is a convenient and efficient method to isolate CpG islands methylated sequences on denaturing gradient gel electrophoresis. Recently, an epigenomic mapping including a DNA methylation profiling in the *Arabidospsis* genome was performed using tiling microarrays (Lippman et al., 2005; Martienssen et al., 2005). This powerful tool is composed of genomic tiling microarrays which represent contiguous stretches of chromosomes without bias toward coding sequencing. After shearing of genomic DNA by nebulization, DNA is digested with McrBC that allows DNA to be depleted of methylated sequences. The digested DNA and an untreated sample are size-fractionated, differentially labelled and hybridized to genomic tiling microarrays. This method allows DNA methylation pattern of all sequence types to be assayed simultaneously at high resolution. Methods have also been developed to map epigenetic quantitative trait loci (QTL) defines as QTL activated by an epigenetic event and that exhibit the potential to alter the developmental trajectory of a growth trait (Pigliucci, 1998; Wu et al., 2002).

Another group of methods using methyl binding affinity chromatography has recently retained specific attention. These methods are insensitive to the methylation status of specific internal recognition sites and provide useful information on CpG islands methylation (Shiraishi et al., 2004b). Methyl-CpG Binding Domain (MBD) column consists of an affinity matrix containing a polypeptide derived from the methyl-CpG binding domain of MeCP2 protein. MBD synthesized *in vitro* contains an additional six consecutive histidine residues attached to the amino terminus bound to nitrilotriacetic acid-agarose by chelation with nickel ion (Cross et al., 1994; John and Cross, 1997; Cross, 2002; Shiraishi et al., 2004b). DNA fragments are loaded onto this affinity column and are eluted by a linear or stepwise gradient of sodium chloride. The eluted DNA fragments are subjected to PCR amplification or Southern experiments allowing the identification of methyl-CpG sequences in a given genome providing interesting information on the functional organization of genomes. As one MBD protein will bind to a single CpG sequence, number and density of methylated CpG sites determine the separation. The sequence preference of MBD column is not clear since contradictory results have been reported *in vitro* and *in vivo* (Shiraishi et al., 2004b). Furthermore, hemimethylated or methylated CpNpG sequences in plants could not bind to such column. Therefore, another methyl binding affinity column was developed: an anti-5-methylcytosine affinity column. This system that binds all methylated sequences (hemimethylated or not, CpG and CpNpG) has been successfully applied on several systems (Salzberg et al., 2004). Application of this method on plants for crop improvement and comparison to the other scanning methods should be performed (Salzberg et al., 2004; Shiraishi et al., 2004b).

4. THE GENE CANDIDATE APPROACH FOR VALIDATION OF EPIGENETIC BIOMARKERS

4.1. Gene Candidate Approach

Many agronomic traits, resulting from the interactions of multiple genes under environnemental influence, show quantitative inheritance. Nevertheless, the partial effect of each gene on the phenotypic variation and their imprecise localization on genomic maps lead researchers to use a gene candidate approach to characterize QTL instead of positional cloning or insertional mutagenesis (Pfiegler et al., 2001). The gene candidate approach corresponds to the use of sequenced genes of known function (structural genes or regulatory genes) that could correspond to major loci. The working hypothesis is that a molecular polymorphism within the candidate genes (CGs) is linked to the major loci or QTLs, or is statistically associated with the variations of the trait. The gene candidate approach is composed of three steps: (1) choosing the CG(s) using physiological or linkage data, (2) screening the CG(s) by revealing a polymorphism and (3) validating the CG(s) using various physiological or genetic analyses.

In the context of the study of epigenetic biomarkers, the working hypothesis of the CG approach becomes the research of a polymorphism of methylation status within the CG(s) that is linked to the major loci or QTLs, or that is statistically associated with the phenotypic variations. The three classical steps are adapted as follow (Table I): (1) choosing the CG(s) according to physiological data or genomic screening of sequences done by scanning approach, (2) Determining of polymorphism that corresponds to variation of the status of methylation within CG(s) performed by Southern blot or bisulfite-PCR, (3) validating of the CG using physiological, genetic and inheritance analyses in a selection population.

4.2. Analysis of Methylation Status

4.2.1. Southern blot

This technique is still widely used due to its robustness, reproducibility and relative simplicity (Moore, 2001). Genomic DNA samples digested with methylation-sensitive enzymes (that do not cut if their recognition sequence is methylated) such as *Pvu* II, *Taq* I, *Not* I and *Hpa* II are electrophoretically separated, blotted on membrane and hybridized with probes corresponding to a part of the CG sequences. If distinct patterns of bands are observed between samples, it reveals a methylation polymorphism and suggests an epigenetic control of the expression of this CG. This basic method has been used, for example, to show that several CGs related to cell wall differentiation or cell redox status have distinct methylation status depending on the morphogenetic status of sugarbeet cell lines (Causevic et al., 2005; Causevic et al., 2006). Nevertheless, the weak number of methylation-sensitive restriction sites in CGs, the variable efficiency of the cutting, the availability of specific probes in many crop plants, the high amounts of DNA needed, the sensitivity and scalability of this method are clear limitations.

4.2.2. *Bisulfite sequencing*

A revolutionary method employing bisulfite treatment of genomic DNA and subsequent PCR amplification has been introduced in the nineties (Frommer et al., 1992; Clark et al., 1994). This method, contrary to Southern blot, is insensitive to the methylation status of specific internal recognition sites and as a consequence provides a robust qualitative and quantitative method for detailed methylation profiling (Hajkova et al., 2002; Shiraishi et al., 2002; Laird, 2003; Shiraishi et al., 2004b; Causevic et al., 2005). The bisulfite reaction leads to the conversion of cytosines into uracil residues after hydrolytic deamination (Figure 5). However, methylated cytosine remains largely intact. Subsequent PCR amplification, cloning and sequencing convert nonmethylated cytosine to thymine, while methylated cytosine are detected as cytosine. This enables to determine the methylation status at any CpG or CpNpG sites on both DNA strands in a CG. Quantitative information could be obtained by sequencing several individual clones of bisulfite-PCR products or by using the pyrosequencing technology (Dupont et al., 2004; Causevic et al., 2006). An example is shown in Figure 5, where a specific primer from the 5' region of one CG for the *in vitro* morphogenesis, catalase sequence, was designed. This figure shows distinct patterns of methylation in the three sugarbeet cell lines at distinct morphogenetic status (Causevic et al., 2006). Furthermore, the frequencies of methylated CpG dinucleotides were variable between cell lines. Thus, the proportion of methylated CpG sites scaled negatively with the levels catalase activities measured in the three sugarbeet cell lines (Figure 5; Causevic et al., 2006). Bisulfite sequencing confirmed the distinct methylation profiles of catalase, which was previously proposed by Southern blot analysis. Altogether, these results demonstrated a direct connection between epigenetic regulation and the expression of these biomarkers during morphogenesis.

In spite of richness of informations obtained from bisulfite-mediated sequencing, the method is relatively expensive, time-consuming and difficult to apply to a several-kb chromosomal region. In that sense, several modifications of the original protocol improving the sensitivity and quality of the results have been published (Mills and Ramsahoye, 2002; Shiraishi et al., 2002; Laird, 2003). Several alternative methods, based on bisulfite treatment and subsequent PCR amplification, have been developed to overcome sequencing high cost. Combined bisulfite restriction analysis (COBRA; Xiong and Laird, 1997) and methylation-specific PCR (MS-PCR; Herman et al., 1996) methods are fast and cost-effective, which would be suitable for screening or discrimination of a large set of samples (Mills and Ramsahoye, 2002). Employment of real-time PCR improves also quantitativeness of these bisulfite based methods (Eads et al., 2000; Ogino et al., 2006).

4.3. Validation of Epigenetic Biomarkers

Few reports are now available to compare the results between scanning and/or candidate approaches (Shiraishi et al., 2002; Salzberg et al., 2004; Yu et al., 2004; Mils and Ramsahoye, 2002; Laird, 2003; Causevic et al., 2006). Nevertheless,

A

Bisulfite sequencing

- Extraction of genomic DNA.

-Treatment by hydroxyquinone/bisulfite in order to deaminate unmethylated cytosine into uracile.

- PCR amplification with specific primers on genomic DNA treated or not.

- Subcloning of PCR products in a vector.

- Sequencing of about 10 clones by sequence.

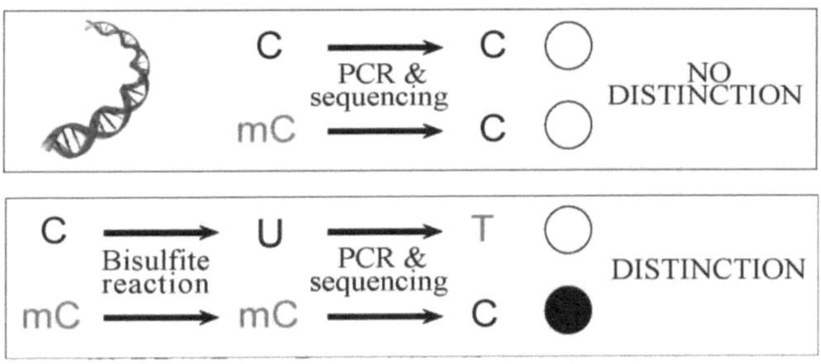

B

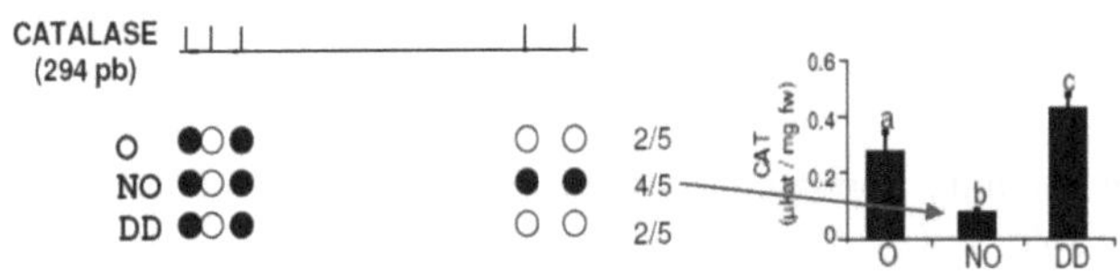

Figure 5. A, Principle of bisulfite-PCR sequencing method for the determination of the methylation status of gene candidates. B, Results of the methylation analysis of 5' regions of sugarbeet catalase gene by bisulfite sequencing. The potential methylated CpG sites in the sequence are indicated by perpendicular lines. For the three cell lines organogenic (O), non organogenic (NO) and dedifferentiated (DD), 6 to 10 PCR products were subcloned and sequenced. Five CpG sites were considered to be methylated when more than half the clones retained an unmodified cytosine at that position. Methylated CpG sites (Filled circles) and unmethylated CpG sites (open circles) are shown. The proportions of methylated CpG sites are indicated on the right for catalase activity was measured in the O, NO and DD sugarbeet cell lines. Data are means ± SE from three independent replicates. Values marked with different letters are significantly different between cell lines ($P \leq 0.05$) as determined by oneway ANOVA. *fw* fresh weight. (Adapted from Causevic et al., 2006)

it should be mentioned that in many cases, methylation status varies depending on target sequences and the approach used. All the authors usually agree that combination of methods compensates for the deficiencies associated with each method and enables a more accurate characterization of the methylation status of CpG islands.

Validation is more or less complex according to the nature of the breeding trait. Physiological studies will determine the expression of the epigenetic markers at the mRNA level (quantitative RT-PCR, northern blotting or microarray analyses) and/or protein level (enzymatic activity, western blotting or two-dimensional electrophoresis coupled to mass spectrometry) in various genotypes, in individuals within a selection population or under various environmental conditions. Genetic transformation (sense or antisense strategies, RNAi or homologous recombination) is the ultimate way to validate these markers. All these analyses display limitations and will only provide arguments but not undisputed evidence for or against the role of the epigenetic markers (Pfiegler et al., 2001). Finally, the analysis of the inheritance of epigenetic polymorphism and its statistical relationship with variations of the breeding traits in individuals of a selected population should be done.

5. CONCLUSION: THE POWER AND PROMISE OF DNA METHYLATION BIOMARKERS FOR PLANT BREEDING

DNA methylation is a key event for the regulation of gene expression. Quantitative differences in gene expression control different physiological processes and consequently phenotypic diversity (Tsaftaris et al., 2005). Heritable phenotypic variation in a population is the basis for selection and breeding. DNA methylation is affected by the developmental stage, growth conditions and genotype, and consequently is a major source of variation for selection. DNA methylation biomarkers (epialleles) reflect genetic variation and environmental effects but also contribute actively to the phenotype.

Three successive complementary approaches should be followed to characterize methylation biomarkers (Table I). (1) A relationship must be established between the variations of a breeding trait and the global level of DNA methylation and/or the status of methylation of some genes, (2) few genes must be selected (CG approach or discovery approach by genome scanning), (3) the methylation status of these genes and / or their expression profile must be followed in many genotypes under specific experimental conditions in order to validate these biomarkers.

As opposed to plants, in mammals most of the epigenetic alterations are associated to disease and are rarely inherited (Jones and Baylin, 2002; Chong and Whitelaw, 2004; Robertson, 2005; Zilberman and Henikoff, 2005). Thus, many methods for the discovery and validation of methylation biomarkers were developed on tumorous cells allowing sensitive detection of disease or markers associated with disease progression (Laird, 2003). All these methods display specific advantages and limitations. Therefore, a combination of methods should be used and few additional reports are also needed to compare methods among them (Frühwald and Plass, 2002; Shiraishi et al., 2002; Laird, 2003).

Many efforts have already been made to use methylation biomarkers in human cancer diagnostic. However, only a small amount of preliminary data is actually available regarding plants, for identification and varieties or hybrids creation and marker-assisted selection. Furthermore, the identification of epigenetic markers may lead to the cloning of new genes of agronomic importance and will allow a better understanding of the biological mechanisms controlling the development and the growth of plants of economic interest. Moreover, analysis will help to elucidate mechanisms controlled by epigenetic phenomena such as somaclonal variation, heterosis, parental imprinting, transgene silencing and environmental responses that constitute tremendous commercial interest for breeders and industrials (Tsaftaris et al., 2005; Varshney et al., 2005; Grant-Downton and Dickinson, 2006). The constant increase of interest for epigenetic phenomena in plants and the application of analytical methods on plant systems are very promising in the next year for a great development of such approach.

ACKNOWLEDGEMENTS

A Ph.D. grant (M.-V. Gentil) was supported by the Conseil Régional de la Région Centre (France) and SES-Europe (Tienen, Belgium). The authors thank S. Barnes, F. Brignolas, A. Delaunay, F. Delmotte, D. Hagège, C. Joseph, M. Lefebvre and G. Moreau for their participation and continuous interest.

REFERENCES

Abe T, Matsuyama T, Sekido S, Yamaguchi I, Yoshida S, Kameya T (2002) Chlorophyll-deficient mutants of rice demonstrated the deletion of a DNA fragment by heavy-ion irradiation. J Radiat Res 43:157–161

Burn JE, Bagnall DJ, Metzger JD, Dennis ES, Peacock WJ (1993) DNA methylation, vernalization and initiation of flowering. Proc Natl Acad Sci USA 90:287–291

Cao X, Springer NM, Muszinsky MG, Phillips RL, Kaeppler S, Jacobsen SE (2000) Conserved plant genes with similarity to mammalian de novo methyltransferases. Proc Natl Acad Sci USA 97:4979–4984

Causevic A, Delaunay A, Ounnar S, Righezza M, Delmotte F, Brignolas F, Hagège D, Maury S (2005) DNA methylating and demethylating treatments modify phenotype and cell wall differentiation status in Sugarbeet cell lines. Plant Physiol Biochem 43:681–691

Causevic A, Gentil M-V, Delaunay A, El-Soud WA, Garcia Z, Pannetier C, Brignolas F, Hagège D, Maury S (2006) Relationship between DNA Methylation and histone acetylation levels, cell redox and cell differentiation status in Sugarbeet lines. Planta 224:812–827

Choi Y, Gerhing M, Johnston L, Hannon M, Harada JJ, Goldberg RD, Jacobsen SE, Fischer RL (2002) DEMETER, a DNA glycosylase domain protein, is required for endosperm gene imprinting and seed viability in *Arabidopsis*. Cell 110:33–42

Chong S, Whitelaw E (2004) Epigenetic germline inheritance. Curr Opin Genet Dev 14:692–696

Clark SJ, Harrison J, Paul CL, Frommer M (1994) High sensitivity mapping of methylated cytosines. Nucl Ac Res 22:2990–2997

Costello J, Smiraglia D, Plass C (2002) Restriction landmark genome scanning. Methods 27:144–149

Cross SH (2002) Isolation of CpG islands using a methyl-CpG binding column. Methods Mol Biol 200:111–130

Cross SH, Charlton JA, Nan X, Bird AP (1994) Purification of CpG islands using a methylated DNA binding column. Nat Genet 6:236–244

Dupont J-M, Tost J, Jammes H, Glynne Gut I (2004) De novo quantitative bisulfite sequencing using the pyrosequencing technology. Anal Biochem 333:119–127

Eads CA, Danenberg KD, Kawakami K, Saltz LB, Blake C, Shibata D, Danenberg PV, Laird PW, (2000) MethyLight: a high-throughput assay to measure DNA methylation. Nucl Ac Res 28:E32

Finnegan EJ (2001) Epialleles: a source of of random variation in time of stress. Curr Opin Plant Biol 5:1001–1106

Finnegan EJ, Kovac KA (2000) Plant DNA methyltransferases. Plant Mol Biol 43:189–201

Finnegan EJ, Peacock W, Dennis E, (1996), Reduced DNA methylation in *Arabidopsis thaliana* results in abnormal plant development, Proc. Natl Acad Sci USA 93:8449–8454

Finnegan EJ, Genger R, Kovac K, Peacock W, Dennis E (1998) DNA methylation and the promotion of flowering by vernalization. Proc Natl Acad Sci USA 95:5824–5829

Fraga MF, Rodriguez R, Canal MJ (2000) Rapid quantification of DNA methylation by high performance capillary electrophoresis. Electrophoresis 21:2990–2994

Frommer M, McDonald LE, Millar DS, Collis CM, Watt F, Grigg GW, Molloy PL, Paul PL (1992) A genomic sequencing protocol that yields a positive display of 5-methylcytosine residues in individual DNA strands. Proc Natl Acad Sci USA 89:1827–1831

Frühwald MC, Plass C (2002) Global and gene-specific methylation patterns in cancer: aspects of tumor biology and clinical potential. Mol Gen Met 75:1–6

Grant-Downton RT, Dickinson HG (2006) Epigenetics and its implications for plant biology 2. The "epigenetic epiphany": epigenetics, evolution and beyond. Ann Bot 97:11–27

Hagège D, Kevers C, Crevecoeur M, Tollier M, Monties B, Gaspar T (1991) Peroxidases, growth and differentiation of habituated sugarbeet cells. In: Lobarzewski HG. J, Penel C, Gaspar T (eds) Biochemical molecular and physiological aspects of plant peroxidase. University of Geneva, Geneva, Switzerland, pp 281–290

Hajkova P, El-Maari O, Engemann S, Oswald J, Olek A, Walter J (2002) DNA-methylation analysis by the bisulfite-assisted genomic sequencing method. Methods Mol Biol 200:143–154

Hatada I, Hayashizaki Y, Hirotsune S, Komatsubara H, Mukai T (1991) A genomic scanning method for higher organisms using restriction sites as landmark. Proc Natl Acad Sci USA 88:9523–9527

Hermann JG, Graff JR, Myohanen S, Nelkin BD, Baylin SB (1996) Methylation specific PCR: a novel PCR assay for methylation status of CpG islands. Proc Natl Acad Sci USA 93:9821–9826

Holliday R (1990) DNA methylation and epigenetic inheritance. Philos Trans R Soc Lond B Biol Sci 326:329–338

John RM, Cross SH (1997) Gene detection by the identification of CpG islands, in genome analysis: a laboratory manual. In: Birren B, Green ED, Klapholz S, Myers RM, Roskams J, (eds) Detecting genes Vol. 2. Cold Spring Harbor Laboratory Press, Cold Spring Harbor, NY, pp 217–285

Johnston JW, Harding K, Bremner DH, Souch G, Green J, Lynch PT, Grout B, Benson EE (2005) HPLC analysis of plant DNA methylation: a study of critical methodological factors. Plant Physiol Biochem 43:844–853.

Jones PA, Baylin SB (2002) The fundamental role of epigenetic events in cancer. Genetics 3:415–428

Kaeppler S, Kaeppler H, Rhee Y (2000) Epigenetic aspects of somaclonal variation in plants. Plant Mol Biol 43:179–188

Kakutani T, Munakata K, Richards EJ, Hiroshika H (1999) Meiotically and mitotically stable inheritance of DNA hypomethylation induced by ddm1 mutation of *Arabidopsis thaliana*. Genetics 151: 831–838

Kalisz S, Purugganan D (2004) Epialleles via DNA methylation: consequences for plant evolution. Trends Ecol Evol 19:309–314

Laird PW (2003) The power and the promise of DNA methylation markers. Nature Reviews Cancer 3:253–264

Lambé P, Mutambel H, Fouche J, Deltour R, Foidart J, Gaspar T (1997) DNA methylation as a key process in regulation of organogenic totipotency and plant neoplastic progression?, In Vitro Cell Dev Biol Plant 33:155–162

Lippman Z, Gendrel A-V, Colot V, Martienssen R (2005) Profiling DNA methylation patterns using genomic tiling microarrays. Nature Meth 2:219–224

Loidl P (2004) A plant dialect of the histone language. Trends Plant Sci 9:84–90

Martienssen RA, Doerge RW, Colot V (2005) Epigenomic mapping in *Arabidopsis* using tiling microarrays. Chrom Res 13:299–308

Matsuyama T, Abe T, Bae C-H, Takahashi Y, Kiucchi R, Nakano T, Asami T, Yoshida S (2000) Adaptation of restriction landmark genomic scanning (RLGS) to plant genome analysis. Plant Mol Biol Report 18:331–338

Matsuyama T, Kimura M, Koike K, Abe T, Nakano T, Asami T, Ebisuzaki T, Held W, Yoshida S, Nagase H (2003) Global methylation screening in the *Arabidopsis thaliana* and *Mus musculus* genome: Applications of virtual image restriction landmark genomic scanning (Vi–RLGS). Nucl Ac Res 31:4490-4496

Meyer P (2001) Chromatin remodelling. Curr Opin Plant Biol 4:457–462

Mills K, Ramsahoye B (2002) DNA methylation protocols: methods in molecular biology. Humana Press, New Jersey, pp 1–189

Moore T (2001) Southern analysis using methyl-sensitive restriction enzymes. Methods Mol Biol 181:193–203

Ogino S, Kawasaki T, Brahmandan M, Cantor M, Kirkner GJ, Spielgelman D, Makrigiorgos GM, Weisenberger DJ, Laird PW, Loda M, Fuchs CS (2006) Precision and performance characteristics of bisulfite conversion and real-time PCR (MethylLight) for quantitative DNA methylation analysis. J Mol Diagn 8:209–217

Pflieger S, Lefebvre V, Causse M (2001) The candidate gene approach in plant genetics: a review, Mol Breed 7:275–291

Pigliucci M (2001) Phenotypic plasticity: beyond nature and nurture. The Johns Hopkins University Press. Baltimore, Maryland, USA

Pligliucci M (1998) Developmental phenotypic plasticity: where internal programming meets the external environment. Curr Opin Plant Biol 1:87–91

Robertson KD (2005) DNA methylation and human disease. Genetics 6:597–610

Ronemus M, Galbiati M, Ticknor C, Chen J, Dellaporta S (1996) Demethylation-induced developmental pleiotropy in *Arabidopsis*. Science 273:654–657

Rush L, Plass C (2002) Restriction landmark genomic scanning for DNA methylation in cancer: past, present and futures applications. Anal Biochem 307:191–201

Salszberg A, Fisher O, Siman-Tov R, Ankri S (2004) Identification of methylated sequences in genomic DNA of adult *Drosophila melanogaster*. Biochem Biophys Res Comm 322:465–469

Shiraishi M, Oates AJ, Li X, Chuu YH, Sekiya T (2004a) Segregation of partly melted molecules: isolation of CpG islands by polyacrylamide gel electrophoresis. Biol Chem 385:967–973

Shiraishi M, Sekiguchi A, Oates AJ, Terry MJ, Miyamoto Y, Sekiya T (2004b) Methyl-CpG binding domain column chromatography as a tool for the analysis of genomic DNA methylation, Anal Biochem 329:1–10

Shiraishi M, Sekiguchi A, Oates AJ, Terry MJ, Miyamoto Y, Tanaka K, Sekiya T (2002) Variable estimation of genomic DNA methylation: a comparison of methyl-CpG binding domain column chromatography and bisulfite genomic sequencing. Anal Biochem 380:182–185

Takamiya T, Hosobuchi S, Asai K, Nakamura E, Tomioka K, Kawase M, Kakutani T, Paterson AH, Murakami Y, Okuizumi H (2006) Restriction landmark genome scanning method using isoschizomers (*Msp*I/*Hpa*II) for DNA methylation analysis. Electrophoresis 27:2846–2856

Tariq M, Paszkowski J (2004) DNA and histone methylation in plants. Trends Genet 6:244–251

Tsaftaris AS, Polidoros AN, Koumproglou R, Tani A, Kovacevic N, Abatzizou E (2005) Epigenetic mechanisms in plant and their implications in plant breeding. In: Tuberosa R, Phillips RL, Gale MA (eds) In the wake of the double helix: from the green revolution to the gene revolution. Avenue Media, Bologna, Italy, pp 157–172

Varshney RK, Graner A, Sorrells M (2005) Genomic-Assisted breeding for crop improvement, Trends Plant Sci 10:621–630

Vongs A, Kakutani T, Martienssen R, Richards E (1993) *Arabidopsis thaliana* DNA methylation mutants. Science 260:1926–1928

Wirtz M, Schumann CA, Schellentrager M, Gab S, Vom Brocke J, Podeschwa MA, Altenbach HJ, Oscier D, Schmitz OJ (2005) Capillary electrophoresis-laser induced fluorescence analysis of endogenous damage in mitochondrial and genomic DNA. Electrophoresis 26:2599–2607

Wu R, Ma C-Z, Zhu J, Casella G (2002) Mapping epigenetic quantitative trait loci (QTL) altering a developmental trajectory. Genome 45:28–33

Xiong Z, Laird PW (1997) COBRA: a sensitive and quantitative DNA methylation assay. Nucl Ac Res 25:2532–2534

Xiong LZ, Xu CG, Saghai Maroof MA, Zhang Q (1999) Patterns of cytosine methylation in an elite rice hybrid and its parental lines, detected by a methylation-sensitive amplification polymorphism technique. Mol Gen Genet 261:439–446

Yu L, Liu C, Bennett K, Wu YZ, Dai Z, Vandeusen J, Opavsky R, Raval A, Trikha P, Rodriguez B, Becknell B, Mao C, Lee S, Davuluri RV, Leone G, Van den Veyer IB, Caligiuri MA, Plass C (2004) A *Not* I-*Eco*RV promoter library for studies of genetic and epigenetic alterations in mouse models of human malignancies. Genomics 84:647–660

Zilberman D, Henikoff S (2005) Epigenetic inheritance in *Arabidopsis*: selective silence. Curr Opin Genet Dev 15:557–562

APPENDIX I

LIST OF CONTRIBUTORS

Ramesh K. Aggarwal Centre for Cellular and Molecular Biology (CCMB), Uppal Road, Hyderabad– 500 007, India

Paulo Arruda Centro de Biologia Molecular e Engenharia Genética (CBMEG)/ Departamento de Genética e Evolução, Universidade Estadual de Campinas (UNI CAMP), 13083-970, Campinas, SP, Brazil

Rupakula Aruna International Crops Research Institute for the Semi-Arid Tropics (ICRISAT), Patancheru-502 324, India

Pere Arús IRTA (Institut de Recerca i Tecnologia Agroalimentàries), Carretera de Cabrils Km.2, E-08348 Cabrils, Spain

Michael Baum Integrated Gene Management Mega-Project, International Center for Agricultural Research in the Dry Areas, Aleppo, Syria

Sabhyata Bhatia National Institute for Plant Genome Research (NIPGR), Aruna Asaf Ali Marg, New Delhi-110 067, India

Stéphane Bieri Institute of Plant Biology, University of Zürich, Zollikerstrasse 107, 8008 Zürich, Switzerland

Andreas Börner Leibniz Institute of Plant Genetics and Crop Plant Research (IPK), Corrensstrasse 3, D-06466 Gatersleben, Germany

Eligio Bossolini Institute of Plant Biology, University of Zürich, Zollikerstrasse 107, 8008 Zürich, Switzerland

Edward S. Buckler Institute for Genomic Diversity, Cornell University, Ithaca, NY 14853, USA USDA-ARS, US Plant, Soil and Nutrition Laboratory, Ithaca, NY 14853–2901

Salvatore Ceccarelli Integrated Gene Management Mega-Project, International Center for Agricultural Research in the Dry Areas, Aleppo, Syria

Wafa Choumane Integrated Gene Management Mega-Project, International Center for Agricultural Research in the Dry Areas, Aleppo, Syria

Luca Comai University of California Davis Genome Center, Davis, CA 95616, USA

Mark Cooper Pioneer Hi-Bred International. 7250 NW 62nd Ave, Johnston, IA 50131-0552, USA

Enzo DeAmbrogio Società Produttori Sementi Bologna, Divisione Ricerca, Argelato, Italy

Rebecca. W. Doerge Departments of Statistics and Agronomy, Purdue University; 150 North University Street, West Lafayette, IN 47907, USA

Gebisa Ejeta Department of Agronomy, Purdue University, 915 W. State Street, West Lafayette, IN 47907, USA

Elhan S. Ersoz Institute for Genomic Diversity, Cornell University, Ithaca, NY 14853, USA

Catherine Feuillet UMR INRA-UBP 1095, Amélioration et Santé des Plantes, Domaine de Crouelle 234, Avenue du Brézet, 63100 Clermont-Ferrand, France

Vidal Fey Centre for Biotechnology, Tykistökatu 6, FIN-20521 Turku, Finland

Majid R. Foolad Department of Horticulture, The Pennsylvania State University, University Park, PA 16802, USA

Wolfgang Friedt Institute of Crop Science and Plant Breeding I, Justus-Liebig-University, Heinrich-Buff-Ring 26-32, 35392 Giessen, Germany

Martin W. Ganal TraitGenetics GmbH, Am Schwabeplan 1b, D-06466 Gatersleben, Germany

Susan Gardiner HortResearch, Private Bag 11030, Palmerston North, New Zealand

Pooran M. Gaur International Crops Research Institute for the Semi-Arid Tropics (ICRISAT), Patancheru-502 324, India

Marie-Véronique Gentil Laboratoire de Biologie des Ligneux et des Grandes Cultures, UPRES EA 1207, rue de Chartres. BP 6759. Faculté des sciences, Université d'Orléans. 45067 Orléans Cedex 2 France

M. V. Channabyre Gowda University of Agricultural Sciences (UAS), Dharwad, 500 006, India

Silvana Grandillo CNR-IGV, Institute of Plant Genetics, Portici Via Università 133, 80055 - Portici (NA), Italy

Stefania Grando Integrated Gene Management Mega-Project, International Center for Agricultural Research in the Dry Areas, Aleppo, Syria

Anil Grover Department of Plant Molecular Biology, University of Delhi South Campus, Benito Juarez Road, New Delhi 110 021, India

Peiguo Guo Integrated Gene Management Mega-Project, International Center for Agricultural Research in the Dry Areas, Aleppo, Syria College of Life Science, Guangzhou University, Guangzhou 510006, China

Prasad S. Hendre Centre for Cellular and Molecular Biology, Uppal Road, Hyderabad-500 007, India

Steven Henikoff Basic Sciences Division, Fred Hutchinson Cancer Research Center, Seattle, Washington 98109, USA, Howard Hughes Medical Institute

David A. Hoisington International Crops Research Institute for the Semi-Arid Tropics (ICRISAT), Patancheru-502 324, India

Mukesh Jain Interdisciplinary Centre for Plant Genomics and Department of Plant Molecular Biology, University of Delhi South Campus, Benito Juarez road, New Delhi-110021, India

Sini Junttila Centre for Biotechnology, Tykistökatu 6, FIN-20521 Turku, Finland

Beat Keller Institute of Plant Biology, University of Zürich, Zollikerstrasse 107, 8008 Zürich, Switzerland

Jitendra P. Khurana Interdisciplinary Centre for Plant Genomics and Department of Plant Molecular Biology, University of Delhi South Campus, Benito Juarez road, New Delhi-110021, India

Moon Young Kim Department of Plant Science/Research Institute for Agriculture and Life Sciences, Seoul National University, Seoul 151-921, The Republic of Korea

Matias Kirst School of Forest Resources and Conservation, University of Florida, PO Box 110410, Gainesville, FL 32611, USA

Joseph E. Knoll Department of Agronomy, Purdue University, 915 W. State Street, West Lafayette, IN 47907, USA

Maria van Korff Integrated Gene Management Mega-Project, International Center for Agricultural Research in the Dry Areas, Aleppo, Syria

Domenico Lafiandra Department of Agrobiology and Agrochemistry, University of Tuscia, Viterbo, Italy

Berhane Lakew Integrated Gene Management Mega-Project, International Center for Agricultural Research in the Dry Areas, Aleppo, Syria

Thomas Lübberstedt University of Århus, Faculty of Agricultural Sciences, Research Centre Flakkebjerg, 4200 Slagelse, Denmark

Michael Lee Iowa State University, Ames, IA 50011, USA

Suk-Ha Lee Department of Plant Science/Research Institute for Agriculture and Life Sciences, Seoul National University, Seoul 151-921, The Republic of Korea

Kean Jin Lim Centre for Biotechnology, Tykistökatu 6, FIN-20521 Turku, Finland

Lang Luo Pioneer Hi-Bred International. 7250 NW 62nd Ave, Johnston, IA 50131-0552, USA

Marco Maccaferri Department of Agroenvironmental Sciences and Technology, University of Bologna, Viale Fanin 44, 40127 Bologna, Italy

David J. Mackill International Rice Research Institute, DAPO Box 7777, Metro Manila, Philippines

Thudi Mahendar International Crops Research Institute for the Semi-Arid Tropics (ICRISAT), Patancheru- 502324, India

Hideo Matsumura Iwate Biotechnology Research Center, 22-174-4, Narita, Kitakami, Iwate 024-0003, Japan

Stéphane Maury Laboratoire de Biologie des Ligneux et des Grandes Cultures, UPRES EA 1207, rue de Chartres. BP 6759. Faculté des sciences, Université d'Orléans. 45067 Orléans Cedex 2, France

Trilochan Mohapatra National Research Centre on Plant Biotechnology (NRCPB), Indian Agricultural Research Institute (IARI), New Delhi-110 012, India

Henry T. Nguyen National Center for Soybean Biotechnology and Division of Plant Sciences, University of Missouri, Columbia, MO 65211, USA

Shyam N. Nigam International Crops Research Institute for the Semi-Arid Tropics (ICRISAT), Patancheru-502 324, India

Frank Ordon Institute of Epidemiology and Resistance Resources, Federal Centre for Breeding Research on Cultivated Plants, Theodor-Roemer-Weg 4, 06449 Aschersleben, Germany

Ashwani Pareek Stress Physiology and Molecular Biology Laboratory, School of Life Sciences, Jawaharlal Nehru University, New Delhi 110067, India

Md S. Pathan National Center for Soybean Biotechnology and Division of Plant Sciences, University of Missouri, Columbia, MO 65211, USA

Johan Peleman Nunhems B.V., P.O. Box 4005, 6080 AA Haelen, The Netherlands

Dean W. Podlich Pioneer Hi-Bred International. 7250 NW 62nd Ave, Johnston, IA 50131-0552, USA

Carlo Pozzi Fondazione Parco Tecnologico Padano, via Einstein Loc. C.na Codazza, 26900 Lodi, Italy

Thaís Rezende Silva Centro de Biologia Molecular e Engenharia Genética (CBMEG)/Departamento de Genética e Evolução, Universidade Estadual de Campinas (UNICAMP), 13083-970, Campinas, SP, Brazil

Marion S. Röder Leibniz Institute of Plant Genetics and Crop Plant Research (IPK), Corrensstr. 3, D-06466 Gatersleben, Germany

Stephen Rudd Centre for Biotechnology, Tykistökatu 6, FIN-20521 Turku, Finland

Francesco Salamini Fondazione Parco Tecnologico Padano, via Einstein Loc. C.na Codazza, 26900 Lodi, Italy Università degli Studi di Milano, Dip. Produzione Vegetale, via Celoria 2, 20133 Milano, Italy

Jérôme Salse UMR INRA-UBP 1095, Amélioration et Santé des Plantes, Domaine de Crouelle 234, Avenue du Brézet, 63100 Clermont-Ferrand, France

Silvio Salvi Department of Agroenvironmental Sciences and Technology, University of Bologna, Viale Fanin 44, 40127 Bologna, Italy

Maria Corinna Sanguineti Department of Agroenvironmental Sciences and Technology, University of Bologna, Viale Fanin 44, 40127 Bologna, Italy

Kulbhushan Saxena International Crops Research Institute for the Semi-Arid Tropics (ICRISAT), Patancheru-502 324, India

Haitham Sayed Integrated Gene Management Mega-Project, International Center for Agricultural Research in the Dry Areas, Aleppo, Syria

Niroj K. Sethy National Institute for Plant Genome Research (NI), Asaf Ali Marg, New Delhi-110 067, India Present address: Defence Institute of Physiology and Allied Sciences (DIPAS), Defence Research and Development Organization (DRDO), Timarpur, Delhi-110 054, India

Prakash C. Sharma University School of Biotechnology, Guru Gobind Singh Indraprastha University, Delhi 110006, India

Nagendra K. Singh National Research Centre on Plant Biotechnology (NRCPB), Indian Agricultural Research Institute (IARI), New Delhi-110 012, India

Sneh L. Singla-Pareek Plant Molecular Biology, International Centre for Genetic Engineering and Biotechnology, New Delhi 100067, India

Sudhir K. Sopory Plant Molecular Biology, International Centre for Genetic Engineering and Biotechnology, New Delhi 100067, India

Anker P. Sørensen Keygene N.V., P.O. Box 216, 6700 AE, Wageningen, The Netherlands

Jeroen Stuurman Keygene N.V., P.O. Box 216, 6700 AE, Wageningen, The Netherlands

Steve D. Tanksley Department of Plant Breeding and Department of Plant Biology, Cornell University, Ithaca, NY 14853, USA

Ryohei Terauchi Iwate Biotechnology Research Center, 22-174-4, Narita, Kitakami, Iwate 024-0003, Japan

Bradley J. Till Basic Sciences Division, Fred Hutchinson Cancer Research Center, Seattle, WA 98109, USA

Roberto Tuberosa Department of Agroenvironmental Sciences and Technology, University of Bologna, Viale Fanin 44, 40127 Bologna, Italy

Akhilesh K. Tyagi Interdisciplinary Centre for Plant Genomics and Department of Plant Molecular Biology, University of Delhi South Campus, Benito Juarez Road, New Delhi-110021, India

Sripada M. Udupa Integrated Gene Management Mega-Project, International Center for Agricultural Research in the Dry Areas, Aleppo, Syria

Hari D. Upadhyaya International Crops Research Institute for the Semi-Arid Tropics (ICRISAT), Patancheru-502 324, India

Vincent Vadez International Crops Research Institute for the Semi-Arid Tropics (ICRISAT), Patancheru-502 324, India

Babu Valliyodan National Center for Soybean Biotechnology and Division of Plant Sciences, University of Missouri, Columbia, MO 65211, USA

Kyujung Van Department of Plant Science, Seoul National University, Seoul 151-921, The Republic of Korea

Jeroen Rouppe van der Voort Enza Zaden B.V, P.O. Box 7, 1600 AA Enkhuizen, The Netherlands

Rajeev K. Varshney International Crops Research Institute for the Semi-Arid Tropics (ICRISAT), Patancheru-502 324, India

Tri D. Vuong National Center for Soybean Biotechnology and Division of Plant Sciences, University of Missouri, Columbia, MO 65211, USA

Xiaolei Wu National Center for Soybean Biotechnology and Division of Plant Sciences, University of Missouri, Columbia, MO 65211, USA

Nabila Yahiaoui Institute of Plant Biology, University of Zürich, Zollikerstrasse 107, 8008 Zürich, Switzerland

Jianming Yu Institute for Genomic Diversity, Cornell University, Ithaca, NY 14853, USA

Qibin Yu School of Forest Resources and Conservation, University of Florida, PO Box 110410, Gainesville, FL 32611, USA

Dani Zamir Faculty of Agriculture, The Hebrew University of Jerusalem, PO Box 12, Rehovot 76100, Israel

APPENDIX II

LIST OF REVIEWERS

Amalia Barone, University of Naples "Federico II", Portici, Italy

Alain Charcosset, UMR, INRA UPS-XI INAPG CNRS Génétique Végétale, Gif-sur-Yvette, France

Ayman A. Diab, Agricultural Genetic Engineering Research Institute (AGERI), Modern Science and Art University (MSA), Egypt

Ivo Grosse, Leibniz Institute of Plant Genetics and Crop Plant Research (IPK), Gatersleben, Germany

B Jayashree, International Crops Research Institute for the Semi-Arid Tropics (ICRISAT), Patancheru, India

Guenter Kahl, University of Frankfurt, Frankfurt am Main, Germany

Beat Keller, University of Zurich, Zurich, Switzerland

Takao Komatsuda, National Institute of Agrobiological Sciences, Tsukuba, Japan

Thomas Lübberstedt, Árhus University, Research Centre Flakkebjerg, Slagelse, Denmark

Jianxin Ma, Purdue University, West Lafayette, IN 47907, USA

Tudi Mahender, International Crops Research Institute for the Semi-Arid Tropics (ICRISAT), Patancheru, India

Henry Nguyen, National Centre for Soybean Research, University of Missouri, Columbia, MO 65211, USA

Antoni Rafalski, DuPont Co. Crop Genetics and University of Delaware, Newark, DE 19716, USA

Saurabh Raghuvanshi, University of Delhi South Campus, New Delhi, India

Ajoy K Roy, Indian Grassland and Fodder Research Institute (IGFRI), Jhansi, India

Joy K Roy, University of Minnesota, St. Paul, MN 55108, USA

Nese Sreenivasulu, Leibniz Institute of Plant Genetics and Crop Plant Research (IPK) Gatersleben, Germany

Brian Steffenson, University of Minnesota, St. Paul, MN 55108, USA

WT Bill Thomas, SCRI, Dundee, Scotland

Roberto Tuberosa, University of Bologna, Italy

Vincent Vadez, International Crops Research Institute for the Semi-Arid Tropics (ICRISAT), Patancheru, India

Rajeev K Varshney, International Crops Research Institute for the Semi-Arid Tropics (ICRISAT), Patancheru, India

Desh Pal S. Verma, The Ohio State University, Columbus, OH 43210, USA

San Ming Wang, Center for Functional Genomics, Northwestern University, Evanston, IL 60208, USA

Clifford Weil, Purdue University, West Lafayette, IN 47907, USA

Ramakrishna Wusirika, Michigan Technological University, Houghton, MI 49931, USA

INDEX

MIX
Papier aus verantwortungsvollen Quellen
Paper from responsible sources
FSC® C105338

If you have any concerns about our products,
you can contact us on
ProductSafety@springernature.com

In case Publisher is established outside the EU,
the EU authorized representative is:
Springer Nature Customer Service Center GmbH
Europaplatz 3, 69115 Heidelberg, Germany

Printed by Libri Plureos GmbH
in Hamburg, Germany